The Ecology and Evolution of *Heliconius* Butterflies

The Ecology and Evolution of *Heliconius* Butterflies

Chris D. Jiggins
Department of Zoology, University of Cambridge, UK

OXFORD
UNIVERSITY PRESS

Great Clarendon Street, Oxford, OX2 6DP,
United Kingdom

Oxford University Press is a department of the University of Oxford.
It furthers the University's objective of excellence in research, scholarship,
and education by publishing worldwide. Oxford is a registered trade mark of
Oxford University Press in the UK and in certain other countries

First Edition published in 2017

Impression: 1

Published in the United States of America by Oxford University Press
198 Madison Avenue, New York, NY 10016, United States of America

British Library Cataloguing in Publication Data
Data available

Library of Congress Control Number: 2016946838

ISBN 978–0–19–956657–0

DOI 10.1093/acprof:oso/9780199566570.001.0001

Printed and bound by
CPI Litho (UK) Ltd, Croydon, CR0 4YY

Preface

I had been working in the western part of Ecuador for about three months before I got around to heading east, taking the road from Baños, high in the Andes and overshadowed by volcanoes, down through the narrow valley towards Puyo. At the time, this road was famous with backpackers because of a waterfall that fell right onto the middle of the road. Travellers who opted for the open-air option and sat on the roof of the bus would get soaked on the way down. My first stop was in a town called Shell—in the 1930s, it was a frontier oil town, named after the oil company responsible for its establishment. I walked through the dusty streets towards a stream just at the edge of the town, which I had heard was a good place to collect *Heliconius*.

It was getting towards dusk, and sure enough, as I waded along the shallow stream, there was suddenly a gentle fluttering of butterflies all around me. I had come across a roost of *Heliconius erato*, just starting to gather before settling on a twig to sleep together for the night. Even by this stage, early in my PhD, I felt that I was starting to get to know this species fairly well. I had already collected several hundred of them across south-western Ecuador, as part of my PhD project.

But these butterflies were wildly different. For a few moments I was completely confused—I knew the shape, the flight pattern, even the strange habit of gathering in streams at dusk. But the butterfly I knew from western Ecuador, *Heliconius erato cyrbia*, was iridescent blue with red patches. These butterflies had pink and white spots. I had seen lots of pictures, so logically I knew that these were *Heliconius erato notabilis*, a geographic race of the same species. But nothing quite prepared me for the dramatic difference in appearance between what I was familiar with and what was fluttering around my head. It was the moment in which the geographic variation in form that *Heliconius* are famous for came vividly to life. I hope that in this book I can convey some sense of the awe that I felt that day about the unusual biology of these butterflies.

Along the way, I hope to address some current questions in evolutionary biology, and highlight areas in which we remain ignorant. In the first chapter, I outline some of the questions that work on *Heliconius* has helped to address, or in some cases might contribute to in the future. These topics are largely focused on the genetic basis for adaptation and speciation that have been the motivation for my own work on *Heliconius*. However, this brief overview of topics is not meant to be comprehensive, and throughout the book I will also describe many other areas of ongoing and past *Heliconius* research.

The second chapter introduces the butterflies and their relationships, as well as briefly summarizes the history of *Heliconius* research (with not a little trepidation, as I have certainly missed some moments that others would consider critical). The next four chapters review many aspects of the basic biology of *Heliconius*, before moving on to wing pattern evolution and finally speciation.

This book is based on the premise that a complete understanding of any topic in biology—but evolution particularly so—requires an intimate understanding of the natural biology of the organism studied. I therefore review the different aspects of *Heliconius* natural history that have been studied over the years, before coming back to the topics of speciation and the genetic basis of adaptation, which have come to the fore in *Heliconius* research. I hope that readers primarily interested in the latter will gain a richer understanding of *Heliconius* speciation biology from reading about the ecology and natural history of these butterflies. Conversely, readers interested in the natural history of these brightly coloured butterflies—so common in tropical butterfly displays around the world—will hopefully be pleasantly surprised to learn that they are also the subject of evolutionary studies in genomics and developmental biology.

It has taken me a long time to write this book. Looking back through emails, I found that my first correspondence with Ian Sherman at OUP dates back to November 2005. In the meantime, *Heliconius* research has surged ahead. I had originally planned that the book would summarize the older literature on the natural history of *Heliconius*, and only briefly touch upon later developments. However, the genomic and developmental studies have progressed so much that I now feel compelled to include more of them here than was originally planned. Undoubtedly, some of these sections will become dated fairly quickly, but I hope that our understanding of wing pattern development and speciation genomics is sufficiently advanced now that the basics will remain relevant for some time to come.

One of the sources of inspiration as I was thinking of writing this book was Peter Grant's *Ecology and Evolution of Darwin's Finches*. In his Preface, Grant mentions a number of other systems that might also be excellent case studies for understanding biological diversification, one of which is *Heliconius* butterflies. Over 30 years after the first publication of the *Ecology and Evolution of Darwin's Finches*, I am hoping that the butterflies can finally catch up with the finches.

Acknowledgements

This book has been several years in gestation, and over that period very many people have contributed. I am grateful to Margarita Beltran, Grace Wu, Luana Maroja, John Davey, Markus Möst, Adriana Briscoe, Richard Merrill, and Claire Mérot for reading chapters. Special thanks to Arnaud Martin and Joe Hanly, who provided very constructive comments on the development chapter, and Richard Merrill and Simon Martin, for reading and commenting on the final two chapters. Patricio Salazar wrote an introduction to his PhD thesis that was very helpful in thinking about the history of *Heliconius* wing pattern genetic studies. Other members of my group and the wider *Heliconius* community have contributed; I am grateful to Camilo Salazar, Carolina Pardo-Diaz, Krzystof Kozak, Laura Ferguson, Simon Baxter, Jamie Walters, Ana Pinharanda, Kathy Darragh, Emma Curran, Liz Evans, Nicola Nadeau, Denise Dellaglio, Arnaud Martin, Marcus Kronforst, Ricardo Papa, Robert Reed, Susan Finkbeiner, Kanchon Dasmahapatra, Violaine Llaurens, Catalina Estrada, and Tim Thurman for various discussions.

I am especially grateful to Keith Willmott, Andrei Sourakov, and Tom Emmel at the MacGuire Center, University of Florida, for providing access to the Neukirchen collection in order to take photos of specimens. I am also indebted to Walter Neukirchen for amassing such a wonderful collection of perfect specimens. Thanks to Derya Akkaynak for help with image manipulation. Gerardo Lamas generously shared his most recent taxonomy for the group, reproduced here in updated form as Chapter 12.

In Panama, I am grateful to Liz Evans and Oscar Paneso for raising caterpillars and keeping an eye out for unusual species that helped with obtaining images of early stages. I thank Ana Carolina Aymone, Arnaud Martin, Harald Krenn, Neil Rosser, Keith Willmott, Adriana Briscoe, Richard Wallbank, Steven Montgomery, and Andrei Sourakov for sharing figures and images, and Patricio Salazar, Krzystof Kozak, Joe Hanly, and Claire Mérot for the use of unpublished figures and data from their respective PhD theses. I also thank the Tupper library and staff, notably Angel Aguirre and Elizabeth Sanchez, for support in obtaining obscure literature while I was in Panama.

I thank the institutions that have supported me during the writing, especially St John's College and the Department of Zoology in Cambridge, but also the Smithsonian Tropical Research Institute in Panama, where the bulk of the writing was completed. I also thank Ian Sherman, Helen Eaton and Lucy Nash at OUP for their continued patience and support and Indumadhi Srinivasan for support in the final stages.

I thank my colleagues in the *Heliconius* community who carried out the research described here and have provided many stimulating conversations over the years; in particular, Mathieu Joron, Owen McMillan, Richard Merrill, Mauricio Linares, John Turner, Larry Gilbert, and Jamie Walters deserve a special mention. Most importantly, I thank Jim Mallet, who got me hooked on *Heliconius* in the first place and who certainly should have written this book himself.

Finally, I thank my family, Margarita, Catalina, and Manuela, and my parents, Roger and Sylvia, who have supported my time spent on this project. And finally, my brother Frank, whose scepticism spurred me on to finish the project.

Contents

CHAPTER 1

Some evolutionary problems

The *Heliconius* butterflies are also known as longwing, postman, or passion vine butterflies. All these names are fitting—longwing describes their characteristic shape and evokes their slow determined flight. Postman is derived from the black and red uniforms worn by the Trinidad postal service, similar in colour to the local forms of *Heliconius erato* and *Heliconius melpomene*, and describes their bright showy wing patterns. Passion vines are the sole food for the caterpillars. These butterflies have been the focus of evolutionary studies since the nineteenth century, when they provided some early support for Darwin's theory of evolution by natural selection.

Evolutionary biologists seek to explain the diversity of life that makes our planet so unique, from viruses containing just a few molecules, through to conscious beings that can study images of those viruses, from the deepest oceans to the tropical rainforests. This diversity of life has arisen over the course of about 3 billion years, since the first replicating molecules arose somewhere in the ancient oceans. The challenge of evolutionary biology is to make sense of a process responsible for a vast diversity of different forms, and which occurred over such a vast span of time.

Underlying all of this diversity are two related processes, adaptation and speciation. Adaptation describes the 'fit' of species to their environment—the wide variety of ways in which organisms manage to survive and reproduce—and the only process known to generate adaptation is natural selection. Inspired by their travels in the tropics, Charles Darwin and Alfred Russel Wallace both recognized the inevitability of natural selection: if organisms vary in characteristics that influence their survival or reproduction, and those characteristics are at least in part heritable, then evolution by natural selection is a virtually inevitable outcome.

Henry Walter Bates, with whom Wallace first travelled to the tropics, used the striking wing patterns and geographic variability of *Heliconius* as an example both of natural selection in action and of the process of formation of new species (Bates, 1862). This prompted Darwin to write, in a review of Bates's paper, 'It is hardly an exaggeration to say, that whilst reading and reflecting on the various facts given in this Memoir, we feel to be as near witnesses, as we can ever hope to be, of the creation of a new species on this earth' (Darwin, 1863, p. 223). I hope to show that *Heliconius* have continued to be at the forefront of evolutionary biology, and remain a remarkable case study of evolution in action.

Influenced by the gradualist views of the geologist Charles Lyell, Darwin emphasized the gradual, almost imperceptible nature of evolution by natural selection. He believed that great changes were possible only over the span of geological time. An important development during the first half of the twentieth century was the realization that natural selection can be strong enough to measure in contemporary populations (Haldane, 1924). Studies of evolution in action—such as the rapid changes in appearance of the peppered moth in response to industrial pollution in England—contributed to the development of our modern understanding of evolution, often termed the modern synthesis.

As long as there is genetic variation between individuals that influences survival or reproduction, then selection will be acting. One arena in which it has been possible to directly study the action of natural selection is in natural hybrid zones, where populations with divergent characteristics come into contact, mate together, and produce hybrids.

The Ecology and Evolution of Heliconius Butterflies. Chris D. Jiggins, Oxford University Press (2017).
 DOI 10.1093/acprof:oso/9780199566570.001.0001

One of my favourite examples is the carrion crow, one of the commonest birds in the English countryside. The carrion crow is generally all black, but as you travel northwards through Scotland (preferably by train), the crows suddenly become strikingly different—pale grey on their body, with black wings and head. These are hooded crows, a distinct form (species, if you like), which 'takes over' the role of being a crow in eastern Europe, northern Scotland, and Ireland. In between, there is a narrow region where the two forms come into contact and birds with mixed characteristics can be seen (Meise, 1928; Poelstra et al., 2014). These hybrids are often highly variable, commonly intermediate between their parents, and provide an opportunity to study how genetic differences between individuals influence survival and reproduction (Barton & Hewitt, 1989). *Heliconius* butterflies, with their bright, conspicuous wing patterns, and geographic colour pattern 'races' separated by hybrid zones, have played a key role in the study of selection in hybrid zones.

Another opportunity to measure natural selection is provided by polymorphism, where individuals within a population are genetically variable in some characteristic. Polymorphism has long fascinated evolutionary biologists, and much effort has been expended in attempting to explain its persistence. For example, the banded land snail, *Cepaea nemoralis*, is common in fields and gardens in England, where it has widely variable forms. Some are pale yellow; others pale pink; many have variable numbers of brown stripes. On the face of it, such polymorphism is paradoxical—if natural selection is so powerful at selecting for the 'optimum', then how can stable polymorphism be maintained?

For a long time such variation was largely dismissed as neutral, or irrelevant to the survival of the organism. However, in *Cepaea* it was found that different colour forms were more common in different habitats. Banded shells do better in hedgerows and grassland, while plain shells tend to do better in woodland. This implies that visual predation was at least one of the factors maintaining variation, by favouring different forms in different locations (Cain & Sheppard, 1954). This was one of the examples that led to a shift towards the acceptance of adaptive explanations for the maintenance of polymorphism.

Commonly, adaptive, polymorphisms result from variable selection acting either through space or time, or even multiple solutions to living in the same place at the same time. Such polymorphisms have therefore provided opportunities to document the action of natural selection in the wild. In Chapter 6, I will explain why stable polymorphisms are unusual in *Heliconius*, and how the few examples that do exist are maintained. This is one of the many areas in the book where modern genomic approaches are providing new insights into long-established evolutionary questions.

Another major advance in evolutionary biology has been to recognize the speed with which adaptation can occur. Unsurprisingly, some of the most striking examples are in microbes. For example, experimental evolution in bacteria and yeast can lead to rapid and dramatic changes as populations adapt to their environment (Elena & Lenski, 2003). The terrifying real-world parallel to this is the evolution of antibiotic resistance in human pathogens. For example, in 1941, 10,000 units of penicillin four times a day for four days would cure pneumococcal pneumonia. By the early 1990s, a patient could receive 24 million units of penicillin and die of pneumococcal meningitis (Neu, 1992).

Rapid evolution can also occur in more complex organisms. In parallel with the rise in antibiotic resistance, agricultural pests have rapidly evolved resistance to the pesticides that we use to try to control them. For example, the diamondback moth, *Plutella xylostella*, is a major pest of brassica crops such as canola and cabbages, and has evolved resistance to almost every class of insecticide that has been thrown at it (Talekar & Shelton, 1993). The most bizarre resistance trait is something known as 'leg drop' where the moths drop their legs on contact with the insecticide (Moore & Tabashnik, 1989). Wherever we try to challenge populations of pests or pathogens, biology finds a way to fight back in unexpected ways.

Strong selection pressures and rapid evolution can also be seen in wild populations. Perhaps the most famous example are Darwin's finches. In one case, a rapid evolutionary change in beak size was driven by changes in food availability over the course of a single season, due to climate events caused by an El Niño event in the Pacific ocean

(Grant & Grant, 1995; Grant, 1999). Similarly rapid changes have been seen in the plants that butterflies prefer to lay their eggs on (Singer, Thomas, & Parmesan, 1993), and in dispersal ability as insects colonize new habitats in response to changing climate (Thomas et al., 2001). Experimental introductions of guppies into streams with different predation regimes also show rapid changes in life history over just a few years (Reznick et al., 1997). In *Heliconius* we have observed changes in wing patterns happening among contemporary populations that offer insights into how novel wing patterns arise and diverge (Blum, 2002).

1.1 Coevolution

Some of the strongest selection pressures come from interactions with other organisms, whether they are parasites, predators, or mutualists (Whitney & Glover, 2013). When these interactions are sustained over a significant period of time, it can lead to coevolution, where two species mutually influence one another's evolution. *Heliconius* offer a number of excellent examples of coevolution, both with their *Passiflora* host plants—which have evolved a wide variety of chemical, physical, and psychological defences against herbivory—and with *Psiguria* vines, which provide pollen in return for pollination by *Heliconius*. Many of the classic examples of coevolution involve tight mutualisms in which two species interact with a specific partner. For example, fig trees and fig wasps could not continue without one another. The wasps are the sole pollinators of fig trees, and in return figs provide food and shelter for the wasp grubs that are adapted to eat developing fig seeds. This leads to matching patterns of diversification in figs and their wasps.

However, such tight one-to-one coevolution may actually be rather rare. In pollinator interactions, for example, most flowers are pollinated by a variety of different insects, and most pollinators visit several different species of flower. Coevolutionary networks are therefore often rather diffuse, with interacting partners visiting one another and then moving on to a different partner. Nonetheless, these diffuse interactions can still lead to coevolution. For example, the Japanese bee *Andrena lonicerae* has an extraordinarily complex proboscis, finely tuned for extraction of pollen from *Lonicera gracilipes* flowers, despite the fact that the bee also commonly visits other species (Shimizu et al., 2014). The interactions of *Heliconius* with their caterpillar foodplants and adult pollen sources are similar, in that they do not involve one-to-one interactions, but have nonetheless led to the evolution of highly specialized adaptations in both parties. This kind of diffuse coevolution may be common, and is especially likely to be important in tropical forests where species diversity means that there are so many potentially interacting species.

1.2 The genetic basis of adaptation

Of course, we now know that variation in genes, encoded in DNA molecules, provides Darwin's missing mechanism for the transmission of information between generations. However, even after the discovery of DNA, evolutionary biologists were for a long time largely constrained to speculation regarding the link between genetic variation at the molecular level and the kind of phenotypic variation that leads to evolutionary change. In the past, students of evolution were asked to 'imagine an allele that causes the giraffe to have a longer neck'. Imagination is no longer required. We now have a wealth of examples of DNA sequence variants that have dramatic effects on the phenotype of the organism. Perhaps the simplest examples are changes in the DNA that influence the amino acid sequence of the protein. Somewhat unhelpfully, DNA that codes for proteins is called coding sequence, although this definitely does not mean that the rest of the DNA does not code for anything.

For example, amino acid changes in the haemoglobin protein of deer mice in mountain populations lead to demonstrable effects on the haemoglobin-oxygen affinity of the molecule, adapting them to life at high altitudes (Storz et al., 2009). Similarly, milkweed is a highly toxic plant that only a few unrelated insect species can feed on, including the monarch butterfly, but also the milkweed tussock moth, the milkweed stem weevil, various milkweed bugs, and the oleander aphid. Many of these species have evolved exactly the same amino acid changes in the same enzyme, which allow them to detoxify cardenolide poisons found in milkweed (Zhen et al., 2012).

Evolution can influence not just the amino acid sequence of proteins, but also the genetic switches that turn the production of these proteins on and off. For example, all human babies can digest lactose, the sugar molecule found in milk. However, in many human populations, adults are unable to digest lactose, leading to unpleasant consequences when they drink milk. In the African and Middle Eastern populations that domesticated cattle, mutations were favoured that 'switch on' the lactase enzyme even in adulthood (Tishkoff et al., 2007). These populations, and their modern-day descendants, are therefore able to drink milk throughout life. In this case, the enzyme itself does not change, but rather the switch (in the gene promotor) that determines when it is expressed.

Changes in the deployment of enzymes are not the only thing that can be influenced by altering gene switches. Such switches are also the key to understanding the development of an organism, and the mind-boggling process of turning information in a four-letter DNA code into a complex living organism. As evolutionary biologists, we would like to understand not merely how this happens in one organism, but how the process varies between individuals and species. We would like to be able to explain the huge variety of form and shape in the natural world. For example, how do genes control variation in facial features? As every mother-in-law knows, babies look exactly like their fathers, so there can be no denying that genes do somehow influence our facial characteristics but the mechanisms that cause this variation remain mysterious (Daly & Wilson, 1982). Evolutionary biologists might be more concerned about the spots on a peacock's tail, or the shape of a finch's beak, but we remain surprisingly ignorant of how differences between individuals in their shape and form are regulated by changes in DNA sequence.

In order to understand how DNA changes the form of organisms through evolutionary time, we need to focus on characteristics with significance for the fitness of the organism that are also genetically tractable—amenable to genetic and developmental analyses. Good examples that have been studied to date include the loss of eyesight in Mexican cave fish, the loss of armour plates and spines in freshwater threespine stickleback, and the gain of horns in fighting beetles (Colosimo et al., 2005; Protas et al., 2007; Jeffery, 2009; Moczek & Rose, 2009). We are also now beginning to understand the way in which different *Heliconius* wing patterns are controlled during development, and alongside these other organisms, we can begin to understand the process of evolution all the way from a specific mutation in DNA, to changes in appearance, right through to how those changes influence fitness in the wild.

So what can we learn from identifying genetic variants underlying a particular trait? If the answer is just a list of genes X that affect trait Y, then the enterprise becomes little more than a very expensive exercise in stamp collecting. So it is important to think carefully about the evolutionary questions that can be addressed (Rausher & Delph, 2015). One such long-standing question concerns the effect size of the mutations causing evolutionary change. This problem echoes early debates between gradualists and mutationists in the first part of the twentieth century (Provine, 2001).

Following Darwin, gradualists emphasized the slow, cumulative nature of evolutionary change. In contrast, mutationists studied major effect mutations in laboratory populations, and argued that important evolutionary changes primarily occurred in large jumps driven by mutation. This view was taken to its extreme by Richard Goldschmidt, who used major effect genes influencing butterfly mimicry to argue that evolution was primarily driven by large mutations, or 'hopeful monsters' (Goldschmidt, 1945). It is clear that Goldschmidt was mistaken in downplaying the importance of natural selection. On the other hand, in some cases single mutations can have major evolutionary effects, and the reality is likely somewhere in between the extreme positions taken by gradualists and mutationists. Indeed, recent theory has suggested that a distribution of effect sizes might be expected as populations adapt to a new challenge—with some large effect mutations and many of smaller effect (Orr, 1998, 2005). Although we should be wary of the assumptions underlying any single general theory of adaptation (Rockman, 2012), this approach provides a way of thinking about the problem of adaptation.

In reality, there is now good evidence for major effect mutations in at least some cases. For example, in freshwater stickleback fish populations, deletion of a small section of DNA that normally acts to switch on the *Pitx1* gene removes pelvic spines and is favoured by selection wherever populations adapt to freshwater (Chan et al., 2010). Similar, but non-identical deletions have occurred repeatedly in different populations. Furthermore, this locus explains around 70% of the variation in pelvic spine length in stickleback families, so this truly seems to be a single mutation with a massive and evolutionarily important phenotypic effect (Shapiro et al., 2004).

At the other extreme, some traits are under highly polygenic control, with any one locus making only a slight statistical contribution to the overall differences between individuals. Height in humans is perhaps the archetypal example of this kind of trait, where there is a very strong genetic contribution, but individual genes each have very small effects (Yang et al., 2010).

Resolving the relative importance of major versus minor effect mutations in adaptation will require detailed dissection of loci controlling adaptive phenotypes. For example, analysis of the *agouti* locus, which regulates coat colour in mice, has shown that individual mutations at this same gene have subtly different effects on the phenotype (Linnen et al., 2013). Therefore, another likely outcome is that although individual mutations can have relatively minor effects, if they are located at a single genetic locus, it can act as a single gene of major effect. *Heliconius* wing patterns have also become a widely cited example where major effect loci control adaptive traits—and although we have yet to dissect the relative effects of individual mutations, they have made a major contribution to this debate.

Another open question relates to the source of genetic variation for evolution. Variation is essential for evolution to occur, and a central plank of Darwin's argument for evolution by natural selection was the existence of heritable variation in wild populations. But where does that variation come from? Essentially there are three possibilities.

The first is that adaptation needs to wait for new mutations. A population experiences a new environmental challenge, and then has to wait for the relevant mutation(s) to occur. A likely example of this is insecticide resistance, where resistance alleles are commonly highly deleterious in the absence of insecticide (Gazave et al., 2001). In many cases, therefore, the actual mutation occurs while the insecticide treatment is underway and immediately confers an advantage to insects that would otherwise be killed (Daborn et al., 2002).

The second major source of genetic variation is known as 'standing variation'. Under this model, there are already genetic variants present in the population at low frequency when the new environmental challenge arises, but are either mildly deleterious or perhaps have no influence on phenotype (neutral) (Barrett & Schluter, 2008). These variants then rise in frequency as they are favoured by selection. A likely example are rapid changes in beak shape in populations of Darwin's finches. They happened over such a short timescale that the contribution of novel mutation is minimal, and beak size variation was known to be already present in the original population (Grant & Grant, 1995). Artificial selection experiments also demonstrate the surprisingly large amount of hidden genetic variation that can be present in a small population. For example, artificial selection on the size of eyespots on the wing of one butterfly species, *Bicyclus anynana*, can produce as much variation as that seen naturally in the eyespots of all the species in the genus *Bicyclus* (Beldade, Koops, & Brakefield, 2002).

Finally, novel variants can arise through gene flow from other populations or even different species (Welch & Jiggins, 2014). Darwin recognized the importance of hybridization in facilitating adaptation from his conversations with pigeon breeders, who would typically start the process of developing a new breed by crossing two or more existing breeds in order to introduce variation. Some would consider such gene flow as a form of standing variation, but I prefer to treat it as a distinct process. In effect, hybridization forms a continuum. At one extreme, hybridization can look similar to novel mutation. If there is extremely rare hybridization with a different species, then the new variant will be introduced at very low frequency and with low probability—so it will look more like a novel mutation (if the idea of hybridization between species makes you uncomfortable, there is more on this to come). If hybridization occurs with a nearby population of the same species and is ongoing at a fairly high rate, then the new variants might be commonly present at low frequency in the target population, and this will look more like standing variation. *Heliconius* butterflies provide some of the best examples of both of these processes, and have led the way in a general shift in our appreciation of hybridization as a source of variation for adaptation.

Why should we care about all of this? One reason is that these different processes make very different predictions about how populations will respond to a changing world. If a population has to wait for a rare novel mutation, then an immediate response may be unlikely, while a population with standing variation may respond much more rapidly to selection. This could influence our strategies to aid populations to survive changes such as climate warming, or, on the other hand, to prevent adaptation that is bad for us, such as antibiotic resistance. Recommendations for drug treatment programmes of infections such as malaria might differ dramatically depending on whether resistance arises from de novo mutations or from spread of existing resistance variants (Read & Huijben, 2009; Norris et al., 2015). Identifying the genetic variants that contribute to adaptation in wild populations will help us to distinguish between these different processes in natural populations and determine their relative importance.

1.3 The species problem

In addition to adaptation, the second plank of Darwinian evolution is speciation. This is the process whereby one species separates into two, resulting in an overall increase in biodiversity. Estimates vary wildly, but there are 10–30 million species on this planet (Caley, Fisher, & Mengersen, 2014), most of them yet to be described—speciation has clearly been rampant. And yet speciation remains a controversial area of evolutionary theory.

Despite the centrality of species to evolutionary biology, there is no clear consensus on how to define a species. To some extent, that's okay. If evolution is a gradual process, then we may disagree on exactly where to draw the line in a few marginal cases, but nonetheless agree on recognizing the majority of species. However, if different criteria are used to recognize species, then this can influence the approach used to study speciation, and also have implications for conservation policy where species are typically the units recognized for protection (but see Garnett & Christidis, 2007).

Whatever the definition used, it is important to be clear about how species are recognized in any particular group of organisms. The most commonly used approach is the 'biological species concept' popularized by Ernst Mayr (1963), who developed earlier ideas from Poulton (1904) and Dobzhansky (1937) (Mallet, 1995). This defines species as 'groups of actually or potentially interbreeding individuals'. In principle, this means that biologists need to find coexisting populations in the wild and determine whether they are interbreeding with one another.

If the populations do not overlap in the wild—for example, if they are found on different islands—the only option is to bring species together in the laboratory and assess their potential to interbreed (hence the word 'potentially' in the definition). However, this is also problematic, as species such as ducks often interbreed when kept in captivity, even though they rarely do so in the wild. In addition, although most first-year undergraduates know that species can never interbreed, this is in fact not the definition used by biologists, who are happy to accept some degree of interbreeding between species. How much interbreeding is permitted is never entirely clear. Studying closely related groups of species such as *Heliconius* can inform our understanding of what we mean by the term species, and has led to new ideas about species that are widely applicable. I will return to this problem of species and speciation in the final chapters of this book.

1.4 Hybridization and gene exchange

As is clear from the preceding discussion, interbreeding does occur between species even when defined under the 'biological species concept' (Abbott et al., 2013). This raises the question of the importance of hybridization in evolution. How common is hybridization between species? And does it result in any significant degree of genetic exchange between species?

The answer to the first question is that hybridization is undoubtedly relatively frequent on a per species level—large surveys of well-studied taxa, such as all European butterflies, birds of the world, and vascular plants of the UK, indicate that around 10% of animal species and 25% of plant species hybridize with at least one other species (Grant & Grant, 1992; Mallet, 2005). In some animal groups the rates are much higher—at the extreme, 16 of the 21 duck species in the UK hybridize with at least one other species, and rates are similarly high in birds of paradise (Mallet, 2005). *Heliconius* are also more like plants in this respect, with some 26% of species known to hybridize (Mallet et al., 2007). Although in some sense the exact percentages are not especially meaningful, since more recent radiations of closely related species are expected to show higher rates of hybridization, these surveys do indicate that hybridization can be common and widespread, even, contrary to popular belief, in animals.

In answer to the second question—whether genetic exchange occurs between species that hybridize—there is also abundant evidence that this is the case. It has been commonly observed that mitochondrial DNA can move between species (Toews & Brelsford, 2012)—although it is sometimes perceived that this molecule, which is passed through the female line, is unusual. However, analyses using much larger DNA data sets (in some cases, covering the whole genome) also indicate that gene flow between species is very widespread. Thus, for example, parts of our own genome have been inherited from Neanderthals (Green et al., 2010; Sankararaman et al., 2014). If the mixing of lineages is limited to a few per cent of the genome, then it can perhaps be dismissed as froth on the surface that is unlikely to perturb the deeper signal of relationships. But as we collect genomic data for recent radiations, I think that we will uncover increasing evidence of pervasive mixing between species due to hybridization.

For example, recent analysis of the malarial mosquito *Anopheles gambiae* indicates that less than 10% of the genome follows the relationships of species known from patterns of interbreeding and genetic compatibility. The sex chromosomes, which are less influenced by mixing, show a much larger proportion of sequence that follows the 'species tree' pattern of relationships (Fontaine et al., 2015). Extensive gene flow between species across most of the genome is the most likely explanation for these patterns, and the 'species tree' therefore represents just a small proportion of the genome.

Our work on *Heliconius* similarly indicates that very large proportion of the genome can show mixed relationships, often very different to patterns inferred from ecology, appearance, and mating preference (Martin et al., 2013). In summary, there is an evolving perspective on the question of 'What is a species?' We are moving towards a much more fluid and exciting view of species identity, and evidence from *Heliconius* butterflies has played a key role in this shifting perspective.

1.5 Mechanisms of speciation

Populations can become isolated in many different ways. These range from not recognizing one another as potential mates, living in different habitats, flying at different times of the year, through to producing offspring that fail to survive or reproduce. One or more of these forms of isolation may result as populations adapt to different conditions. As these differences accumulate, speciation will result.

One of the major areas of debate in speciation biology has been surrounding the importance of geographic isolation. One of the most influential figures in this area is Ernst Mayr, an ornithologist who spent much of his early research career in remote regions of Papua New Guinea. Influenced by his studies of endemic bird species found on Pacific islands, Mayr believed that geographic isolation played a key role in driving speciation (a process known as allopatric speciation) (Mayr, 1963). In other words, populations need to be separated—by a mountain, a river, or some other barrier—in order for them to diverge into new species. This led to an intense debate over whether speciation without geographic isolation (sympatric speciation) was possible at all.

Nonetheless, it has been clear for a long time that sympatric speciation is theoretically plausible (Maynard-Smith, 1966; Felsenstein, 1981; Kirkpatrick & Ravigné, 2002), and that all the processes necessary were indeed occurring in natural populations. So the debate really revolved around trying to prove that particular empirical examples had been entirely sympatric or otherwise. This turns out to be rather difficult, as in most cases it is very hard to know the geographic ranges of species throughout their history. Only in a few cases, such as isolated cichlid fish living in crater lakes (Schliewen, Tautz, & Pääbo, 1994; Barluenga et al., 2006), did it seem certain that diverging populations have indeed been sympatric throughout their history. And it turns out that even these iconic examples may be more complex than has been thought (Martin et al., 2015). This 'sympatric versus allopatric' speciation debate has largely fizzled out, not so much because it has been resolved, but rather because speciation biologists have moved on to more tractable and interesting questions.

The more interesting questions entail dissecting the processes that contribute to speciation. One of the dominant themes has been the role of ecological divergence and the multitudinous ways in which changes in the ecology of an organism can lead to reproductive isolation as a side effect. *Heliconius*, with their bright wing patterns that are used both in signalling to predators and in recognizing mates, have become an excellent example of this (Jiggins, Naisbit, et al., 2001a; Jiggins, Emelianov, & Mallet, 2006; Kronforst et al., 2006a; Merrill et al., 2012). Similar processes are occurring in species such as plant-feeding insects that mate on their host plants (having switched to a new host plant, individuals will automatically mate with others who have made a similar choice), or Darwin's finches, whose beaks are shaped by their need to manipulate food items but which also influence the mating calls they produce (Podos, 2010).

However, ecological adaptation and divergence between populations often do *not* lead to speciation (Nosil, Harmon, & Seehausen, 2009). Why is this, and can we start to predict where ecological speciation is most probable? There are therefore a variety of open questions surrounding the ecological context of speciation that remain unanswered, including the relative importance of selection vs drift, and the kinds of ecological or genetic scenarios that promote speciation (The Marie Curie Speciation Network, 2012).

Another set of questions in speciation biology covers the genetic control of reproductive isolation. Species differences are controlled primarily by genes (notwithstanding epigenetics and plasticity), and it turns out that the arrangement of genes in the genome, the number of genes and their effect size, and patterns of recombination can all influence the likelihood of speciation (Kirkpatrick & Ravigné, 2002; Kirkpatrick & Barton, 2006; Smadja & Butlin, 2011a).

Theoretical models have consistently shown that speciation becomes more probable when species differences are controlled by a few genes of major effect, or linked genes with low recombination. For example, in *Laupaula* crickets from Hawaii, several very closely related species differ primarily in their songs. Genetic mapping has shown that the genes for the songs sung by males and the genes controlling female preferences are located in the same region of the genome (Shaw & Lesnick, 2009). This means that if the two species make hybrids, those hybrids tend to inherit compatible combinations (either fast songs and a preference for fast songs, or slow songs and a preference for slow songs). This tends to make speciation easier, as hybridization does not mix up species differences so quickly. In *Heliconius*, there is also evidence that genes for wing pattern and those for mate preference are linked, which is also expected to promote speciation (Kronforst et al., 2006a; Merrill et al., 2011b). However, there are relatively few such examples and a general lack of information on the genomic organization of differences between species with which to empirically test how common such patterns are.

1.6 Tree-thinking and tracing the history of adaptations

The idea that speciation can occur through strong selection on just a few linked genes—but in the meantime, gene flow can continue across much of the genome—also has implications for our understanding of the relationships between species. DNA sequences have provided an excellent tool for tracing the history of populations and species. For the most part, sequences from homologous regions derived from different populations can be readily aligned with one another, and a wealth of methods for reconstructing relationships based on these sequence alignments are available. Biologists can then use these patterns of relationships to make inferences about the history of the populations concerned, and the characteristics that differentiate them.

For example, some stick insects are wingless, while others have wings. Since virtually all other insects have wings, we can infer that the ancestral stick insect was winged. A phylogeny of the species involved can then be used to address how many times wings have been lost (or, more controversially in this case, perhaps also regained) (Stone & French, 2003; Whiting, Bradler, & Maxwell, 2003). If all the wingless stick insects were one another's closest relatives, then we could reach the straightforward conclusion that wings had been lost only once in the phylogeny. A more complex pattern might imply a greater number of losses or gains. This kind of phylogenetic approach to studying evolution has become so mainstream that it can be hard to publish results on the adaptation of a trait without incorporating some comparative phylogenetic analysis (Losos, 2011).

Nonetheless, there are problems with a 'tree-thinking' approach. First is the assumption of a single 'true' tree of life. In some organisms, this is clearly false. Although bacteria reproduce asexually, they can also exchange genetic material between one another in a process known as horizontal gene transfer. This exchange mixes the evolutionary histories of genes in different lineages, and can be sufficiently pervasive to virtually erase the signature of underlying branching (Ochman, Lawrence, & Groisman, 2000; He et al., 2010). This shuffling of genes can have important consequences for the ecology of bacteria, and it is one factor that makes them such a moving target for antibiotics (He et al., 2010). Similarly, plant biologists have also long seen the world as something of a network rather than a branching tree (Arnold, 1997; Rieseberg & Carney, 1998). Hybridization between plant species is common, and plant biologists have often been sceptical of the biological species concept.

In animals, faith in a single bifurcating tree of life has been more persistent. However, at the level of species relationships, it is increasingly realized that different parts of the genome may have very different evolutionary histories (see earlier for evidence for gene exchange between species). This is true even at a deeper scale, such that, for example, in an analysis of bird genomes, none of the trees derived from individual genes were identical to the overall consensus tree derived from the data as a whole (Jarvis et al., 2014; Zhang et al., 2014). It remains an open question the extent to which a single bifurcating tree is actually a reasonable representation of animal life, but irrespective of the answer to that question, and whatever the exact reason for the discrepancy between different parts of the genome, it remains fundamentally true that all organisms are a mosaic of segments of DNA with different histories. The so-called species tree can therefore only ever be an approximation to reality—a simplification of a complex pattern of relationships. There is no such thing as a single 'true' species tree.

The other side of the coin is that even when most of the genome does follow a consistent, bifurcating species tree, the traits that we are commonly interested in studying are those under strong selection, and hence might show a very different history. In humans, it has been speculated that gene regions that help high-altitude populations live in Tibet may have arisen through genetic mixing between different human populations and introgression from archaic humans (Jeong et al., 2014; Racimo et al., 2015). Similarly, genes involved in transmission of malaria or insecticide resistance can be transferred between species of *Anopheles* mosquitoes (Norris et al., 2015).

At a much deeper level, genes can even be transferred between kingdoms, such as fungal genes that allow some aphids to generate carotenoid pigments (Moran & Jarvik, 2010). Mapping these traits onto a bifurcating species tree would indicate convergent evolution, whereas in fact the trait has a common genetic origin. Indeed, these discoveries raise questions about the definition of the word 'convergence', which turns out to cover quite a variety of distinct processes (Stern, 2013). The wing patterns of *Heliconius* have become an exemplar case of adaptive traits being transferred between species by introgression, a theme that will reappear later in this book.

1.7 The model organism

So, returning to the question at the beginning of this chapter: How do we go about studying such a diverse and multitudinous process as evolution, across such a diversity of organisms and such a vast span of time? One approach is to focus on a single organism or group of organisms, and try to understand different facets of its biology, to provide an exemplar case of processes that must be more widespread. Such a focus is absolutely necessary—we do not have the resources or time available to study all processes in all species, and a holistic understanding of ecology, behaviour, and genetics is needed to study processes such as speciation.

Needless to say, I will argue that the *Heliconius* butterflies are just such a useful system in which to study both adaptation and speciation. However, I will not be using the term 'model species'. I don't think that this term is helpful in evolutionary biology. The term comes from the medical literature, where, for example, a 'mouse model' of diabetes would be an experimental procedure or strain of mice that has been developed to mimic the human disease. The 'model' is just that—a tractable, toy version of the real thing. However, the term has somehow taken on a new meaning in biology, to refer to a particular species in which biological processes are studied. The intended implication seems to be that whatever is discovered can be inferred to be more generally true across the tree of life. This approach has, in fact, been rather extraordinarily successful in cellular biology, where, for example, basic cellular processes uncovered in the fruit fly have proven crucial in understanding cancer in humans.

However, some of the processes that we uncover in one group of organisms will turn out to be rather unusual, whereas others will be more widely shared across many organisms. There are many examples that illustrate this, but one of my favourites is the circadian clock. This is a molecular mechanism that regulates the daily rhythm of animals. Studies in mice and fruit flies seemed to suggest that there were major differences in the way that the clock functioned between insects and vertebrates. However, when the circadian clock was studied in a butterfly, it was realized that the fruit fly is unusual. The fruit fly has lost components of the clock that are otherwise shared between mice and butterflies (Zhu et al., 2008). This suggests that most insects are probably much more like vertebrates in this particular aspect of their biology than was thought from studying the fruit fly.

So even for basic cellular processes that are shared across animals, such as the circadian clock, we can be seriously misled by studying one or two 'model' species. This must be inherently more true of evolutionary processes such as speciation: some of what we learn from one particular group of organisms will turn out to be generally true, while other processes will turn out to be unique. As biologists, we certainly hope that we can draw general conclusions from our work, but the reality is that some aspects will turn out to be rather arcane and specific. Only once we have studied many different kinds of organisms will the general patterns become clear. Of course it is the diversity of life that is the primary fascination of the evolutionary biologist, so it is best not to get too downhearted about this. Let's celebrate the diversity of organisms being studied, rather than try to compete to be the best 'model' system. I will try to highlight areas where I think we can draw general conclusions across multiple species, but also acknowledge that some aspects of *Heliconius* biology are quite specific. So this is a book that celebrates a single group of species, but also acknowledges that this group is a tiny slice of the diversity of life—just 0.02% of described Lepidoptera species are in the genus *Heliconius*.

1.8 Why study in the tropics?

Cellular biologists focus on just a few species, but evolutionary biologists and ecologists suffer from a similar bias in terms of the ecosystems they study. Most evolutionary biologists live in the wealthy nations of the temperate world, in Western Europe, Eastern Asia, and North America, whereas the vast majority of the world's species are located in the wet tropics. This has led to an inevitable bias in ecological and evolutionary studies towards the temperate regions. Of course this is understandable, but to an overwhelming extent, terrestrial biodiversity is located in the wet tropics. For example, the UK has approximately 60 resident butterfly species, whereas Ecuador, a comparable size, has some 2,700. To understand speciation, we need to study tropical biodiversity.

Another potential bias is that several of the classic tropical systems for the study of speciation are located on islands (e.g. Darwin's finches, *Anolis* lizards). There are good reasons for this also—island ecosystems are depauperate and therefore simple, so ecological interactions can be more readily understood. However, again, most diversity is in the mainland tropics. These ecosystems are far more complex and difficult to work in. Species can be hard to identify and there are more complex ecological interactions. Individuals of any particular species tend to be rare and difficult to study in large numbers. Nonetheless, despite the challenges, studies of the diverse mainland tropics must play a crucial role in any complete understanding of evolution.

CHAPTER 2

Meet the butterflies and the biologists

2.1 Characteristics

Heliconius are generally straightforward to recognize in the field. They have a highly distinctive long wing shape and a gentle, sailing flight pattern, often gliding for periods with wings held outstretched, and, unlike other butterflies, often seem to have a clear idea of where they are going. They can also fly extremely fast when disturbed, or flutter persistently when searching for egg-laying sites. They have relatively long antennae and large eyes compared to other butterflies, and although their wing patterns are very diverse, they are mostly distinctive (although identification of species within *Heliconius* and *Eueides* can be challenging due to mimicry). The most difficult group to distinguish are the species that mimic ithomiine butterflies, where the mimicry can be extremely precise. Species such as *H. numata* that are mimetic with members of the genera *Melinaea* and *Mechanitis* can be most readily distinguished from the ithomiines by their head size, which is generally wider than the body. Species in the related genera *Dryas, Dione, Philaethria,* and *Agraulis* are fast-flying, ubiquitous members of neotropical butterfly fauna, and are found widely across different habitats.

Identification of early stages is greatly helped by their specific association with *Passiflora* host plants, such that learning to find the plants is the best way to start finding eggs and larvae. The eggs are yellow or white, an elongated oval shape with vertical ridges, and range in size from 0.5 mm for some of the smallest *Eueides* species up to nearly 2 mm in some of the larger *Heliconius* species. The larvae are all spiny, with two spines on the head capsule, and in *Heliconius* are most commonly white or yellow in background colour, although there is considerable variation (Figure 2.1a). They often rest on exposed parts of the *Passiflora* vine, although some species may rest at the base of the host plant during the day and feed at night. The pupae are generally brown and cryptic, with considerable variation between species in form (Figure 2.1b). Some hang vertically while others are held horizontally. Pupae of the smaller genera are not spiny, while *Heliconius* and *Eueides* pupae are spiny, often with gold or silver metallic spots on the ventral side. Unlike larvae and adults, which are mostly warningly coloured, the pupae are cryptic and mimic either dead leaves or in some cases bird droppings (e.g. *Dryas iulia*). Pupae are typically hard to find in the field, as caterpillars commonly move off the host plant in order to pupate, so they are not necessarily associated with host plants.

2.2 Suitability as research organisms

Heliconius have been studied primarily by evolutionarily biologists in a wide variety of fields, ranging from ecology, behaviour, genetics, genomics, and neurobiology. A search on Google Scholar in June 2016 generated more than 10,000 hits for *Heliconius*. As is the case for all well-studied groups, this has come about in part due to historical accident, but also due to their suitability as research organisms. It is therefore worth considering briefly what has made them suited to research and, conversely, what is difficult to study.

The Ecology and Evolution of Heliconius Butterflies. Chris D. Jiggins, Oxford University Press (2017).
 DOI 10.1093/acprof:oso/9780199566570.001.0001

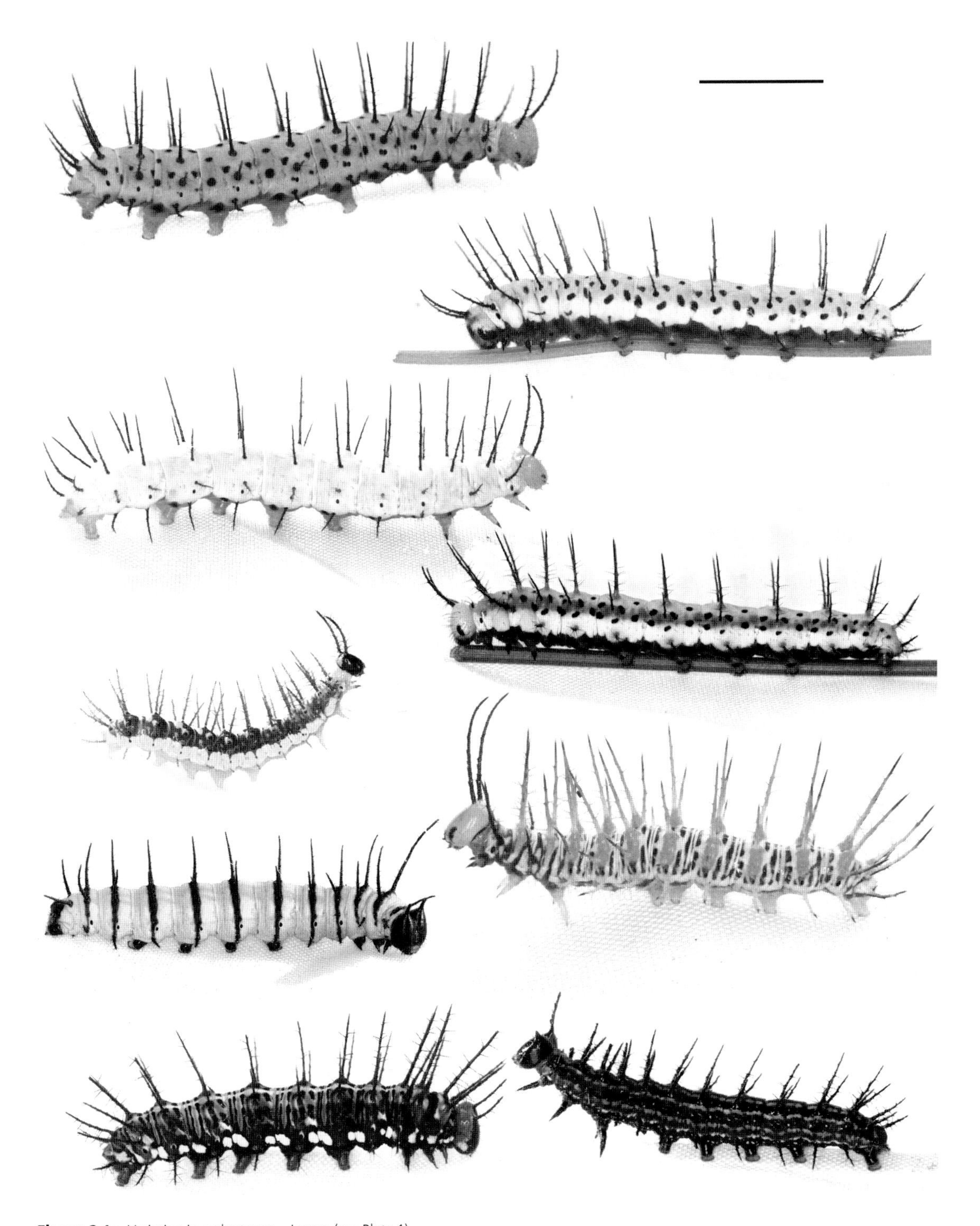

Figure 2.1a Variation in early stages—Larvae (see Plate 1)

From top: *H. cydno; H. charithonia; H. hecale; H. erato; Eueides aliphera; Philaethria dido; Heliconius doris; Agraulis vanillae; Dryas iulia* (lower left). All specimens were raised and photographed in Gamboa, Panama. Images are not precisely scaled but scale bar indicates approximately 1 cm.

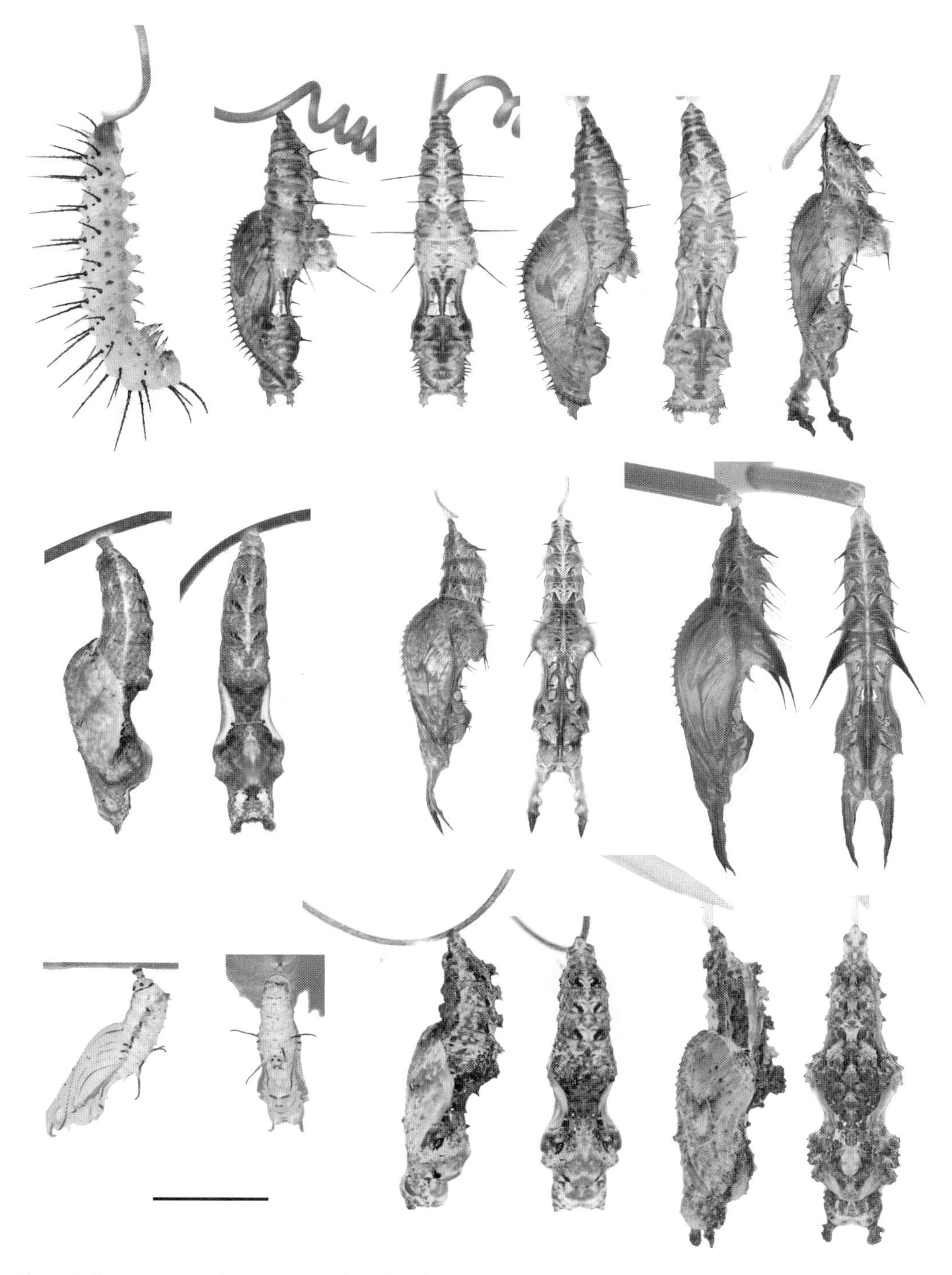

Figure 2.1b Variation in early stages—Pupae (see Plate 2)

First row: *Heliconius cydno* prepupa; *H. cydno* pupa (side, front); *H. hecale* pupa (side, front); *H. charithonia*. Second row: *Agraulis vanillae* (side, front); *H. erato* (side, front); *H. hecalesia* (side, front). Third row: *Eueides aliphera* (side, front); *Dryas iulia* (side, front); *Philaethria dido* (side, front). All specimens were raised and photographed in Gamboa, Panama. Images are not precisely scaled but scale bar indicates approximately 1 cm.

Initially there were two reasons for study: first, their relative abundance in a wide variety of neotropical habitats, which meant they were represented in good numbers in early collections; second, their bright, visually attractive and highly diverse colour patterns, which also attracted collectors. Subsequently, it was recognized that many similarly looking forms were actually mimics—unrelated forms that had evolved a superficial wing pattern similarity—and this phenomenon became a prime example of evolution by natural selection.

Heliconius are also amenable to ecological field studies. They can be readily marked on the wings using permanent marker pens, and are sufficiently long-lived that it is possible to track individuals in natural populations through mark-recapture studies. Their larval host plants, the passion vines, are relatively abundant and readily identifiable in the field from leaf shape, without the need for flowers or fruits. Many species grow along roadsides, with eggs and larvae being fairly easy to find in the field. Nonetheless, *Heliconius* are large, mobile tropical species, and individuals can range over several kilometres. Large-scale studies of natural selection in the wild are very challenging and have been attempted only a few times. Only once has a single site been sufficiently well documented and accessible for a long-term ecological study—the Sirena site in the Osa Peninsula in Costa Rica, intensively studied by the Gilbert lab in the 1980s.

Heliconius can also be raised in insectaries. This is easier in the tropics, but is also possible in temperate climates with temperature-controlled greenhouses (Crane & Fleming, 1953; Turner, 1974). The main requirement is an abundant supply of fresh *Passiflora* leaves to feed the caterpillars. The adults also require relatively large enclosures, especially when recently captured from the wild. Both of these factors can be a major limitation on the scale of the operation. Hence, genetic studies of wing pattern variation involving mapping families of several hundred individuals (up to a few thousand when families are combined) have been possible. However, genetic studies of behavioural traits, when pheotyping of offspring requires large amounts of space, have to date been limited to far smaller sample sizes.

More recently, it turns out that *Heliconius* also have relatively small genomes. *H. melpomene* has 270 million base pairs in its haploid genome, which is bigger than *Drosophila* (120 million base pairs) (Adams et al., 2000), but smaller than many insects, which can range up to several gigabases (Gregory et al., 2007). This is important as it makes genome sequencing studies on a large scale highly feasible. A number of other quirks also help with genetic studies—notably, the lack of recombination in females, common to all Lepidoptera, which means that chromosomes are inherited intact in the maternal line. In addition, chromosome number is stable across most species within the genus *Heliconius* ($n = 21$), which can facilitate comparative genomics. The large size of individual butterflies is also an advantage. Relatively large amounts of DNA can be extracted from a single individual, and visualization of gene expression patterns, especially in pupal wings, is readily achievable. It seems likely that in the future, studies of the neurobiology of vision and learning may also be made easier by the relatively large size of the *Heliconius* eye and head (Montgomery, Merrill, & Ott, 2016).

2.3 Geographic range

Heliconius are found across the New World tropics, from northern Argentina, through South and Central America, up to the southern USA, with *Heliconius charithonia* just reaching into Florida and Texas. Some of their relatives get a little further north, such as *Agraulis vanillae*. Across this range they occur in most habitats, from lowland wet forest to dry llanos or savannahs, being absent only from the very driest areas. Many species do well in the face of moderate disturbance as a result of farming or timber extraction, so often the best habitats are in areas where there is a mosaic of undisturbed forest in an anthropogenic landscape. Nonetheless, species vary considerably in their tolerance of disturbance, with some requiring undisturbed forest, whereas others, such as *Heliconius erato*, can be found even in semi-urban landscapes. Only one species, *Heliconius nattereri*, is considered endangered, and is on the IUCN Red List as critically endangered (as of January 2015). This species is restricted to the highly threatened Atlantic forest habitat of south-eastern Brazil (Brown, 1972a). *Heliconius* occur from sea level up to about 2,000 m, and each species has a characteristic altitudinal range.

Species diversity peaks in the upper Amazon basin, in common with diversity patterns for most plants and animals in the neotropics (Figure 2.2). Alpha-diversity describes the number of coexisting taxa at a single site, and is highest in the foothills of the Andes and the upper Amazon lowland forest (Rosser et al., 2012). In contrast, beta-diversity describes geographically structured diversity, and is higher in the Andes and to a lesser extent Central America. Moving up altitudinal gradients, crossing the Andes from east to west, or even in some cases moving between adjacent valleys can make apparent dramatic changes in the heliconiine species and subspecies composition. In contrast, the Amazonian fauna is broadly similar across thousands of kilometres.

Heliconiines also occur across the Caribbean islands, although species diversity is greatly reduced, and there are no strongly differentiated island endemic forms (Davies & Bermingham, 2002). In effect, the island faunas represent a subset of mainland species, typically those species that are tolerant of disturbed habitats and are more highly dispersive (sometimes endearingly called trashy species). Similarly, many Andean species extend into Central America, but there is only one species found solely in Central America, *Heliconius hortense*.

The biogeographic distribution of the butterflies therefore provides no evidence for the evolution of new species on islands or diversification at the periphery of the range. Instead, since current species

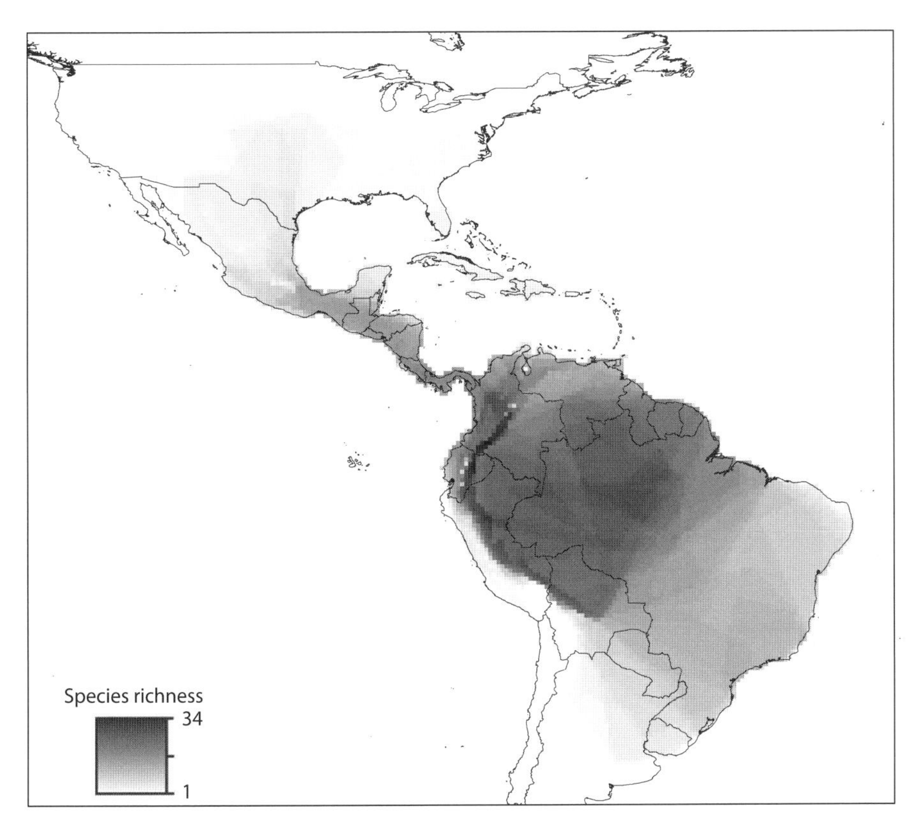

Figure 2.2 Patterns of species richness (see Plate 3)

The diversity of heliconiine species reaches a peak in the upper Amazon, where lowland Amazonian taxa overlap with Andean species. This region corresponds to peak diversity in many taxa. Image courtesy of Neil Rosser.

diversity is primarily found in the Andes and upper and central Amazon basin, it seems most likely that diversification has mainly occurred in these same areas of high biological diversity, rather than areas at the periphery with high levels of geographic isolation.

2.4 Taxonomy

The genus *Heliconius* forms part of the family Nymphalidae, also known as the brush-footed butterflies, which are one of the most diverse butterfly families, containing around 6,000 species. The Nymphalidae are characterized by reduced forelegs, such that they appear at first glance to have only four legs. The reduced forelegs are often used by the females for tasting host plants before laying eggs. Within the Nymphalidae, the subfamily Heliconiinae contains four tribes: the Heliconiini, Argynnini, Vagrantini, and Acraeini.

The Heliconiini tribe is the main focus of this book; it consists of eight genera: *Agraulis, Dryas, Dryadula, Podotricha, Dione, Philaethria, Eueides*, and *Heliconius* that together contain 77 species (Figure 2.3). Their closest relatives within the Heliconiinae are the Indopacific genus *Cethosia* and the Acraeini, found throughout the tropics (Wahlberg et al., 2009). Of the genera, by far the most diverse are the *Heliconius*, with 48 species, followed by *Eueides*, with 12 species. In this book, I will use the term 'heliconiine' to describe those butterflies in the tribe Heliconiini, rather than the broader subfamily Heliconiinae. The definitive taxonomic checklist for the group has been compiled by Gerardo Lamas, working in Lima. I use this with some slight modifications as a reference throughout (Chapter 12).

Although I have challenged the assumption of a single true bifurcating tree of life, our system of naming species is dependent on the underlying assumption of a bifurcating tree, which gives biological meaning to the nested Linnean system of zoological names. A taxonomic name should correspond to a group of organisms with a single common ancestor, known as a monophyletic clade. Despite the complexity, I think the evidence suggests, in animals at least, that it is generally possible to reconstruct a consistent phylogeny for most of the genome, most of the time. A consensus phylogeny therefore remains a reasonable approximation to the evolutionary history of most species and provides a guide for taxonomic decisions.

The most recent Heliconiini species phylogeny is derived from analysis of DNA sequence data from 22 gene markers (Kozak et al., 2015), although similar results are obtained from whole-genome sequences (Kozak, 2015) (Figures 2.3 and 2.4). The tree is broadly similar to earlier results derived from smaller datasets (Beltran et al., 2007; Brower, 1994a; Brower & Egan, 1997). The main discrepancy from Andrew Brower's first molecular phylogeny based on solely mitochondrial DNA sequences was the position of the genus *Eueides*, which was placed within *Heliconius*. However, Brower's placement is not supported by the nuclear genome or ecological and morphological characters (Penz, 1999). This example highlights the potential for discrepancy between gene markers within the tree, and the need to base phylogenetic analysis on multiple genomic regions, although in this case the difference may also be partly due to differences in analysis methods (Beltran et al., 2007).

The consensus molecular phylogeny is largely concordant with an extensive analysis of morphological characters (Penz, 1999). The main exceptions are the formerly recognized genera *Laparus* and *Neruda* (Penz, 1999). They clearly group within *Heliconius* in the molecular analyses, but were placed in a clade with *Eueides* in the morphological study. Morphological characters that link these taxa include the horizontal position of the pupae, white egg colour, and lack of pollen-feeding behaviour, which are shared between *Neruda* and *Eueides*. These characters might be due to convergence, shared ancestry, or genomic incongruence. Nonetheless, the lack of morphological characters strongly supporting monophyly of *Heliconius* (to the exclusion of *Laparus* and *Neruda*), suggests a general lack of strong morphological support for relationships between these taxa (Penz, 1999). Here I will consider Laparus and Neruda as informal groups within the genus *Heliconius* (see later). It is worth noting that this is consistent with John Turner's original description of Neruda as a subgenus of *Heliconius* (Turner, 1976).

I will use the following informal species groupings throughout the book to describe clades of related species within *Heliconius* (Figure 2.4). Thus,

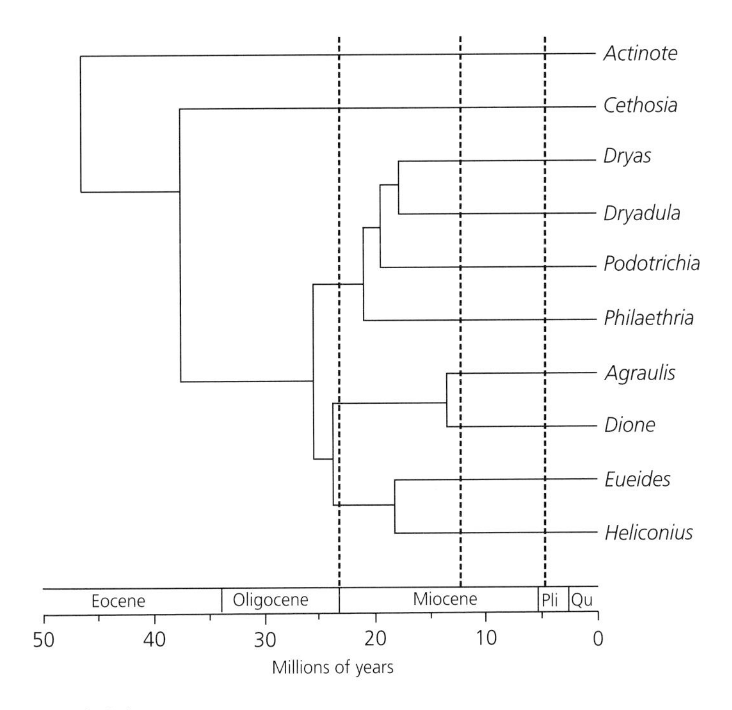

Figure 2.3 Relationships among the heliconiine genera

These relationships are inferred from partial sequences for 22 genes. Pli = Pliocene and Qu = Quaternary. Redrawn from Kozak et al. (2015).

the melpomene clade includes the widespread species *H. melpomene* and its Andean and Central American relatives including *H. cydno* and *H. timareta*. Sister to the melpomene clade is the silvaniform clade, a diverse group of 11 species that mostly have characteristic orange, yellow, and black patterns. These patterns form part of mimicry rings dominated by ithomiine butterflies. These species are found throughout the Amazon, Andes, and Central America.

The outgroups to the melpomene/silvaniform clades are a group of species that have often been called 'primitive' species in the literature, reflecting their phylogenetic position (Brown, 1981). These include species previously considered as two distinct genera, *Laparus* and *Neruda*, that are now included within *Heliconius* based on molecular data (Beltran et al., 2007; Kozak et al., 2015). These species fall into three clades I will call the burneyi clade, primarily Amazonian forest species; the Laparus clade, which is primarily Andean, apart from the widespread *Heliconius doris*; and the Neruda clade, which includes the three species previously considered as a separate genus *Neruda*.

The remaining two lineages in *Heliconius* are the erato clade, consisting of the widespread species *Heliconius erato* and its relatives. These are either primarily Andean (*H. hecalesia, H. clysonymus, H. telesiphe*), dry forest endemics (*H. himera, H. hermathena*), or, in one unusual case, restricted to northern Central America (*H. hortense*). Sister to the erato clade is the diverse sapho clade, which includes widespread species such as *H. sara* and *H. charithonia*, Andean species such as *H. sapho* and *H. eleuchia*, and Amazonian species such as *H. demeter* and *H. ricini*.

2.4.1 A note on Linnean terminology at the subspecies-species level

Heliconius are also well known for considerable within-species diversity in wing pattern. I will commonly refer to these forms using the term 'races', which is largely synonymous with the

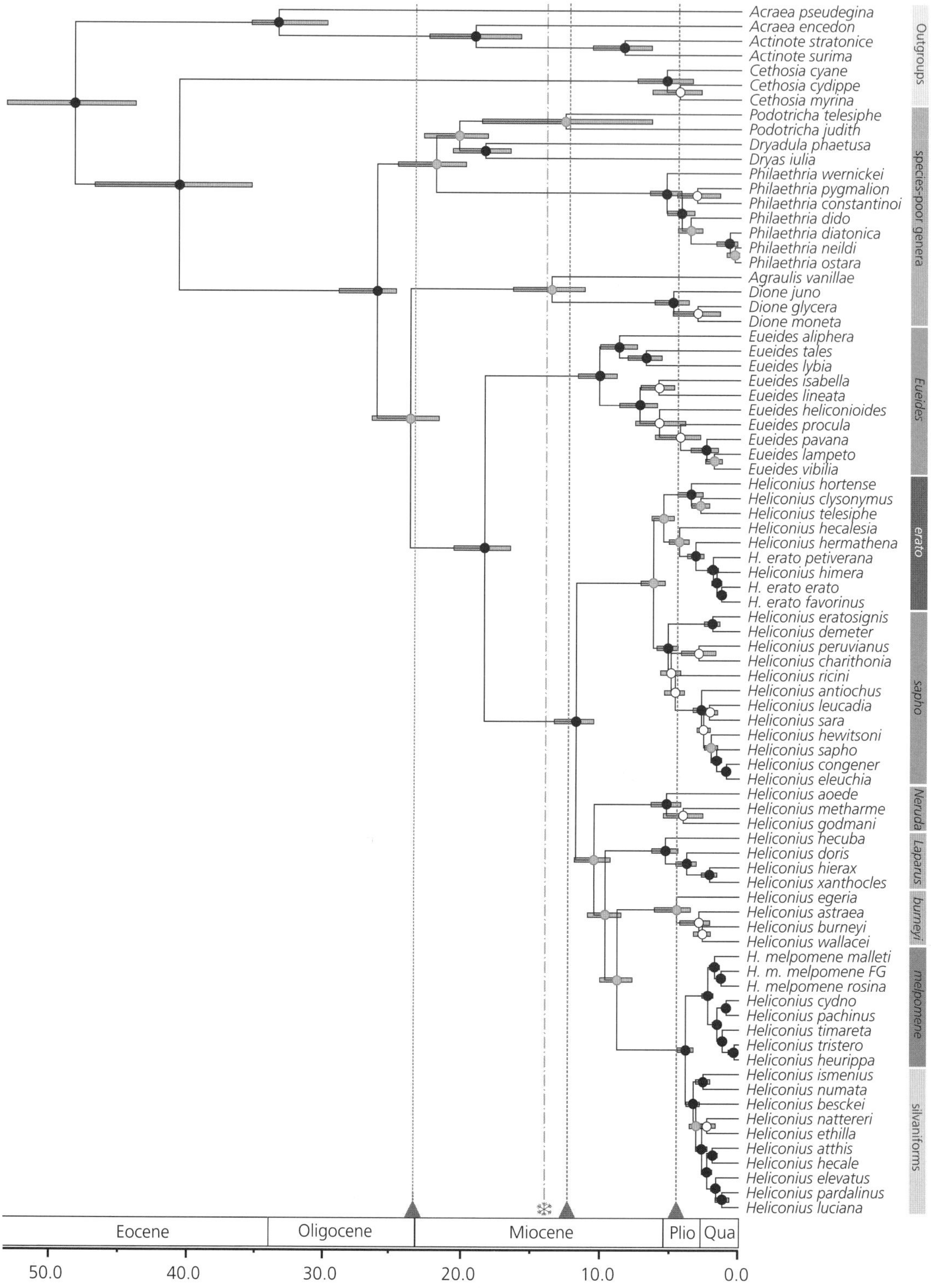

Acraea pseudegina
Acraea encedon
Actinote stratonice
Actinote surima
Cethosia cyane
Cethosia cydippe
Cethosia myrina
Podotricha telesiphe
Podotricha judith
Dryadula phaetusa
Dryas iulia
Philaethria wernickei
Philaethria pygmalion
Philaethria constantinoi
Philaethria dido
Philaethria diatonica
Philaethria neildi
Philaethria ostara
Agraulis vanillae
Dione juno
Dione glycera
Dione moneta
Eueides aliphera
Eueides tales
Eueides lybia
Eueides isabella
Eueides lineata
Eueides heliconioides
Eueides procula
Eueides pavana
Eueides lampeto
Eueides vibilia
Heliconius hortense
Heliconius clysonymus
Heliconius telesiphe
Heliconius hecalesia
Heliconius hermathena
H. erato petiverana
Heliconius himera
H. erato erato
H. erato favorinus
Heliconius eratosignis
Heliconius demeter
Heliconius peruvianus
Heliconius charithonia
Heliconius ricini
Heliconius antiochus
Heliconius leucadia
Heliconius sara
Heliconius hewitsoni
Heliconius sapho
Heliconius congener
Heliconius eleuchia
Heliconius aoede
Heliconius metharme
Heliconius godmani
Heliconius hecuba
Heliconius doris
Heliconius hierax
Heliconius xanthocles
Heliconius egeria
Heliconius astraea
Heliconius burneyi
Heliconius wallacei
H. melpomene malleti
H. m. melpomene FG
H. melpomene rosina
Heliconius cydno
Heliconius pachinus
Heliconius timareta
Heliconius tristero
Heliconius heurippa
Heliconius ismenius
Heliconius numata
Heliconius besckei
Heliconius nattereri
Heliconius ethilla
Heliconius atthis
Heliconius hecale
Heliconius elevatus
Heliconius pardalinus
Heliconius luciana

Outgroups
species-poor genera
Eueides
erato
sapho
Neruda
Laparus
burneyi
melpomene
silvaniforms

Eocene | Oligocene | Miocene | Plio | Qua

50.0 40.0 30.0 20.0 10.0 0.0

Figure 2.4 Dated phylogeny of the heliconiines

Full species phylogeny for the heliconiines. Dates shown below in Mya and error bars for divergence times are shown on each node. The timing of rapid climate cooling in the mid-Miocene is indicated. Species groups are named as used throughout this book. These relationships are inferred from partial sequences for 22 genes. Redrawn from Kozak et al. (2015).

formal zoological classification 'subspecies'. Thus, for example, *H. melpomene rosina* is the geographic wing pattern race of *H. melpomene* found in Central America from the Canal zone in Panama westwards. The term race is commonly used to describe these forms in *Heliconius* because their differences are superficial and largely limited to wing pattern. This interpretation is supported by recent genomic analysis showing little differentiation between adjacent wing pattern races apart from at a few wing patterning loci (Martin et al., 2013).

In addition to extensive geographic variation, some species are also genetically variable in wing pattern (polymorphic) within populations. These forms or morphs are not given subspecies status. Thus, for example, the east Ecuadorean subspecies *Heliconius timareta timareta* has three morphs that result from segregation of alleles at a single genetic locus. These morphs are generally classified as 'forms' rather than subspecies, despite the fact that they differ in a similar manner, and at the same genetic loci, as geographic subspecies. The difference is solely in their patterns of geographic distribution. Hence, the name becomes *Heliconius timareta timareta* f. *contigua*, which is a bit of a mouthful at the best of times. The one exception is *H. numata*, where extensive, geographically structured polymorphism blurs the boundaries between races and forms, and I have chosen to give distinct morphs subspecies names.

2.5 Taxonomic uncertainty

Twenty years ago, it seemed that the taxonomy of *Heliconius* was well resolved, but extensive fieldwork and molecular analysis has continued to uncover new taxa and change our understanding over recent years. In some cases this has involved recognition of cryptic species. *Heliconius demeter* is a widespread Amazonian species in which a deep DNA sequence split was discovered between northern and southern populations (Dasmahapatra et al., 2010). This was confirmed by relatively slight morphological differences; the southern form is now considered a distinct species, *H. eratosignis*.

Perhaps most dramatic has been the recognition of the wide range of the species *H. timareta*. This was formerly known only from a polymorphic population restricted to a narrow area in eastern Ecuador in the Andes, above the town of Puyo. In 1996, Andrew Brower described a new species, *H. tristero*, collected near to the town of Mocoa in Colombia (Brower, 1996b). This was extremely similar to the local population of *H. melpomene bellula*, but had a mitochondrial DNA sequence more similar to *H. timareta*. Subsequently, similar populations were recognized in Colombia, Peru, and Ecuador that were mimetic with local *H. melpomene* but genetically more similar to *H. timareta* (Lamas, 1997; Giraldo et al., 2008; Mérot et al., 2013). These are all now considered as races of *H. timareta*, which is therefore recognized as a far more widespread east Andean species with greater wing pattern diversity than was appreciated previously. These changes also call into question the species status of two well-studied taxa, *H. pachinus* and *H. heurippa*. It is now clear that these are geographic replacements of the more widespread species *H. cydno* and *H. timareta*, respectively.

In order to investigate their species status, samples are needed from zones of parapatry to determine the extent of reproductive isolation, but unfortunately both are difficult to access (due to widespread deforestation, and the location of the city of San José in the region of putative contact, in the case of *H. pachinus*, and due to political instability in the case of *H. heurippa* in Colombia). I here retain the traditional separation of these taxa into distinct species, which is supported by the observation of significant reproductive isolation in insectary experiments (Kronforst et al., 2006a; Mavárez et al., 2006).

Other areas of uncertainty include the genus *Philaethria*, in which several new species have been described recently (Barão et al., 2014), and the recently described but poorly known species *Heliconius (Neruda) metis* and *Heliconius lalitae* (Brévignon, 1996; Moreira & Mielke, 2010). In summary, heliconiine taxonomy is by no means set in stone, and undoubtedly will continue to evolve with further discoveries.

Heliconius are famous for their diversity of wing patterns and for convergence between populations due to mimicry. Despite the huge diversity, the wing patterns show a number of recurrent themes and can be classified into a few broad classes (Figure 2.5).

Figure 2.5 Main wing pattern forms (see Plate 4)

Representative taxa are shown for major wing patterning forms with upper (ventral; above) and underside (dorsal; below) wing patterns. Note that this is not a definitive classification of pattern diversity, but rather is intended as a rough guide to the terms that will be used throughout the book to refer to different types of patterns. Row 1, Postman: *H. ricini insulanus, H. erato hydara*; Dennis-ray: *H. xanthocles napoensis*; Sara: *H. leucadia pseudorhea*; Row 2, Cydno: *H. cydno cydnides, H. eleuchia eleusinus*, Metharme: *H. metharme metharme*, Nattereri: *H. nattereri*; Row 3, Tiger: *H. numata silvana, H. pardalinus tithoreides, H. numata aristiona*, Hecuba: *H. hecuba hecuba*; Row 4, Dryas: *Dryas iulia alcionea*, Agraulis: *Agraulis vanillae maculosa*, Philaethria: *Philaethria dido dido*, Eueides: *Eueides lybia olympia.* Note that the genus *Eueides* contains a large diversity of wing patterns including forms with postman, dennis-ray, tiger, and dryas patterns. Scale bar = 1 cm.

The degree of convergence due to mimicry is such that these wing pattern themes are widely distributed across the tree. The dennis-ray pattern, which dominates the Amazonian mimicry rings, is found in at least one species in every single one of the major clades mentioned earlier, representing a remarkable frequency of convergent evolution. Similarly, the postman patterns, with red, yellow, and black bands, are also found in virtually every clade.

Nonetheless, there is some phylogenetic conservatism in wing pattern theme. Most notably, in the silvaniform clade most species have 'tiger' patterns with orange, black, and yellow patches (Figure 2.5) and the sapho clade includes species with primarily yellow and/or white patterns on black backgrounds ('sara' and 'cydno' patterns in Figure 2.5). However, even within these clades there is considerable diversity. For example, the silvaniform *Heliconius hecale* has 27 named wing pattern subspecies.

2.6 A brief history of *Heliconius* biology

The first known illustration of a *Heliconius* butterfly dates back to 1705, published by the English collector Petiver (Turner, 1967a). This shows the Central American form now named after Petiver, *Heliconius erato petiverana*. Petiver's collection is still held at the Natural History Museum in London, and includes several other heliconiines. The first depiction of the life history of a heliconiine was by Maria Sybilla Merian (Merian, 1705) in her book on the insects of Surinam. Her drawings showed the life cycles of *Agraulis vanillae* and *Heliconius ricini*. The former was shown on both a *Passiflora* and a vanilla vine, while the larvae and pupae for the latter were shown on a castor oil plant (*Ricinus*). It seems that the latter was intended solely as an attractive backdrop for the painting, but Linnaeus later incorrectly interpreted these paintings as host-plant associations, which gave the two species their Latin names, *vanillae* and *ricini* (Turner, 1967a). Much later, Turner named a form of *Heliconius melpomene* found in the Guiana shield in honour of Merian, recognizing her early contribution to *Heliconius* biology (Turner, 1967a).

The first taxonomic descriptions of *Heliconius* species were by Linnaeus. The name *Heliconius* means 'dweller of Mount Helicon', which is where the nymphs and muses of ancient Greek mythology resided. The tradition of naming the species after Greek nymphs was begun by Linnaeus and continued by subsequent authors (the exceptions to this rule in Linnaeus's work are the two species mentioned earlier, named after plant associations depicted in illustrations).

However, as Turner has pointed out, the early taxonomy was very confused—perhaps unsurprisingly for such a difficult group, and at a time when descriptions were based on just a few specimens sent to Europe by collectors (Turner, 1967a). For example, *Heliconius erato* and *Heliconius doris* were confused in the early literature because Linnaeus attributed the name *Papilio (Heliconius) erato* to an illustration by Clerck (which we now recognize as the red form of the polymorphic species, *H. doris*), but also, in later editions of *Systema Naturae*, to an illustration by Cramer (Cramer, 1775) that Cramer had labelled *Papilio vesta*. Only later, in 1882, did Aurivillius fix the Cramer figure as the type specimen for *Heliconius erato* (Turner, 1967a). All of this led to considerable confusion in the literature, with many eighteenth-century studies tending to use the name *vesta* for the species we now call *Heliconius erato*, and *erato* for the red form of *Heliconius doris*.

The classically derived names include *melpomene*, the muse of tragedy, *erato*, the muse of lyric poetry, and *charithonia*, named after Charites (the graces). Zeus is a mere subspecies, *Heliconius hecale zeus*. One of my favourite names is the subgenus Neruda, named after the Chilean poet Pablo Neruda by John Turner, and continuing in the long tradition of naming species after poets, but with a modern Latin American twist.

Undoubtedly the most famous nineteenth-century naturalist who studied *Heliconius* was Henry Walter Bates (Figure 2.6). In 1848 Bates sailed for the Amazon with Alfred Russel Wallace, who later went on to co-discover the idea of natural selection at the same time as Darwin. Neither Bates nor Wallace were from wealthy families, and their trip was to be financed by the sale of their natural history collections to wealthy collectors in Europe. The two spent about a year collecting together around Belem, before agreeing to split up

Figure 2.6 Henry Walter Bates

Plate from *The Naturalist on the River Amazons* (Bates, 1864).

and go their separate ways. Wallace returned to Europe in 1852, but lost virtually his entire collection in a shipwreck on the way home. Meanwhile, Bates stayed in the Amazon for 11 years, eventually returning in 1859 (and wisely sending his collections in three separate ships). These included over 14,000 specimens, a large proportion of which were new to science. Bates was especially inspired by his observations on heliconiine and ithomiine butterflies, and wrote about them soon after his return from the Amazon (Bates, 1862). Bates recognized that these two groups were distinct, but followed convention at the time and treated them both as Heliconidae, which he termed Acraeoid Heliconidae and Danaiod Heliconidae, respectively.

Bates's most famous contribution to biology was the recognition of the protective value of mimicry. He had noticed striking resemblances between many groups of unrelated butterflies, but his argument focused on similarity between members of the Heliconidae and a group he termed the *Leptalides* (*now known as the genus Dismorphia*). The latter group are pierid butterflies, which includes the whites and yellows. Bates argued that *Leptalides* had evolved to mimic members of the Heliconidae in order to benefit from a resemblance to unpalatable prey and therefore be avoided by predators.

He presented a series of observations that supported this argument: (1) *Leptalides* were 'exceedingly rare; they cannot be more than as 1 to 1,000 with regard to the Ithomiae'. (2) Despite considerable geographic variability, the mimic forms of *Leptalides* occurred only in the same geographic region as the species that they mimicked. (3) Most pierid butterflies are white or yellow, in marked contrast to the few forms that showed mimetic patterns, suggesting the latter had evolved recently in the pierids. In contrast, the heliconids are all brightly coloured, suggesting these patterns are more ancient in this group. (4) Finally, the Heliconidae appeared to be unpalatable, unlike *Leptalides*.

Bates's heliconid specimens were rarely eaten by 'vermin' in his camp, and he observed predation by birds much more commonly on the pierid butterflies as compared to the Heliconidae. Bates argued therefore that rare and edible species would evolve to mimic common protected species found in their local environment. He argued that natural selection provided the only plausible mechanism to explain such mimicry. Although Bates did not carry out any experimental studies to support his hypotheses, he used detailed natural history and taxonomic knowledge to build sophisticated arguments in favour of evolution by natural selection.

Another extraordinary character who made an important contribution at this time was Johann Friedrich Theodor ('Fritz') Müller. He studied medicine in Germany, but became an atheist and refused to swear the religious oath of graduation at the end of his studies. This left him without a profession, so he emigrated to southern Brazil in 1852 where he lived as a smallholder for some time. After travelling and botanizing around southern Brazil, he was eventually appointed as travelling naturalist at the National Museum in Rio de Janeiro. This gave him a salary and allowed him to study natural history and publish while travelling around southern Brazil. Müller was a frequent correspondent with Charles Darwin and strong supporter of evolution by natural selection. He published on a huge variety of topics, including mutualism of ants and *Cecropia* trees and morphological dimorphism in midges, and contributed to Darwin's botanical publications.

However, Müller is most famous for his contribution to our understanding of mimicry. Bates had always seen mimicry as a one-sided affair, with one species being more abundant or more toxic, and acting as a model for the mimic. Using a mathematical model, Müller showed that there was a benefit to mimicry between two equally abundant and toxic species. Essentially, his insight was that as more individuals share the same warning colour pattern their fitness increases. By assuming that predators need to sample a fixed number of chemically defended prey in order to learn to associate a bright warning colour pattern with a bad experience, he showed that toxic prey species all benefit from mimicry. As more species join in the mimicry ring, each individual has a lower chance of being sampled during learning. This form of mimicry is mutualistic, as all the species involved benefit from the similarity (Müller, 1879). The two forms of mimicry are now termed Batesian and Müllerian mimicry after Henry Walter Bates and Fritz Müller, respectively.

By the early twentieth century, there was a much greater availability of specimens in European collections from all across the New World, which began to give a more comprehensive view of the diversity of *Heliconius*. When Dr Eltringham studied the collections of Lord Rothschild and the British Museum, he was able to reduce the number of species from over 70 to about 30, by describing transitional forms that linked conspecific populations, and by studying the structure of the male genitalia (Eltringham, 1916). These are hard structures that can be readily dissected from museum specimens, and show enough variation at the level of species to assist taxonomic identification, but are not so variable as the wing patterns within species. This improved understanding of the species relationships and led both Eltringham and Kaye to recognize the general phylogenetic pattern of mimicry, whereby members of different clades within *Heliconius* converge on each other in mimetic pattern (Kaye, 1907; Eltringham, 1916).

Nonetheless, there remained considerable confusion, with several forms that we now recognize as intraspecific hybrids being figured by Eltringham as distinct species, and modern species entities such as *H. melpomene* being split into several distinct species. At about the same time, important contributions were also made by Seitz, and by Stichel and Riffarth, although these studies primarily relied on wing pattern characters and described many more species (Stichel & Riffarth, 1905; Seitz, 1913; Emsley, 1965).

William Beebe was an American naturalist who had become famous in the 1930s for his descent into the deep ocean near the Bahamas in a bathysphere, becoming one the first biologists to observe deep-sea animals in their natural habitat. A new era of *Heliconius* studies was begun in 1949, when he established a field station of the New York Zoological Society at Simla in the Arima Valley, Trinidad. This led to the first systematic attempts to study *Heliconius* in captivity, and offered the first comprehensive insights into their life history and behaviour (Crane & Fleming, 1953; Crane, 1957a; Beebe, Crane, & Fleming, 1960). Most notable in the 1950s was the work of Jocelyn Crane, who carried out pioneering work on the courtship behaviour of *Heliconius erato* and on the biological importance of colour in mating (Crane, 1954, 1955). William Beebe also carried out the first studies showing the inheritance of wing pattern variants, by raising the offspring of *Heliconius melpomene* and *Heliconius erato* females that had been caught in a wild hybrid zone in Surinam (Beebe, 1955). This work helped clarify the taxonomy of these geographically variable species, but also began a tradition of studying the genetic basis of wing pattern variation that continues to this day.

The field station in Arima set the stage for subsequent work in *Heliconius*. In terms of taxonomy, the next major advance was by Michael Emsley, who was encouraged to work on *Heliconius* by Jocelyn Crane (Emsley, 1964, 1965). He studied a wider variety of characters than had been attempted before in order to work out the relationships between populations and species. In particular, the arrangement of androconial patches on the wings that are involved in scent production proved to be informative, as well as a more detailed study of the genitalia than had been attempted by Eltringham 60 years earlier. This led to the first clear resolution of the pattern of geographic variation across the range of the two co-mimic species *Heliconius melpomene* and *Heliconius erato*.

For the first time, the considerable taxonomic confusion that had reigned through most of the twentieth century began to clear, and it was recognised that many of the names had been assigned to occasional hybrid phenotypes (Emsley referred to variation in hybrid zones as 'polychromatism'). Our current understanding of these species as a geographic mosaic of wing pattern races separated by narrow zones of hybridization started to come into focus at this time. Emsley also went some way towards resolving the species boundaries across *Heliconius* and *Eueides* (which he considered as a single genus) (Emsley, 1965). Also worth noting is the earlier work by Michener which clarified most of the currently recognized genera of heliconiines (Michener, 1942).

Although resolving the taxonomy of *Heliconius melpomene* and *H. erato* was a challenge, the problem pales into insignificance compared to that posed by the silvaniform species. It is a real testament to the natural history skills of two American biologists working mainly in Brazil, Keith Brown and Woody Benson, that the relationships among these species was resolved before the advent of DNA taxonomy (Brown & Benson, 1974; Brown, 1976). Unlike *Heliconius melpomene* and *H. erato*, which are monomorphic outside of narrow hybrid zones, many of the silvaniform species are highly variable across large parts of their range. Some species, such as *H. numata*, are characterized by stable polymorphisms found within populations, in which large numbers of wing pattern morphs are controlled by genetic polymorphism. These forms were often previously described as distinct species, and the only way to resolve their relationships was by raising the offspring from wild females, thereby demonstrating that different forms were actually members of the same species.

The opposite was also true. In the words of Keith Brown:

> Extensive biosystematic work in the field and insectary . . . has shown that near-identical adults can have very easily distinguishable eggs or larvae, and that the insects themselves perceive behavioural barriers between species which are invisible to the taxonomist who works only on museum specimens.

Keith Brown was a remarkably hard-working field biologist, who even wrote a paper on 'Maximising Daily Butterfly Counts' which includes a number of practical tips on how to see as many butterfly species in a day as possible (Brown, 1972b).

Brown travelled extensively in Brazil and elsewhere with another pioneering *Heliconius* field biologist, Larry Gilbert. A brash Texan, Gilbert began his field studies in the Arima Valley in Trinidad around William Beebe's field station, with his PhD adviser Paul Ehrlich (who later became a highly influential ecological campaigner) (Ehrlich & Gilbert, 1973). Gilbert has spent his career at Austin, Texas, where he has influenced most of the current generation of *Heliconius* biologists in some way or another. He bred *Heliconius* extensively in Texas, and also established a field study site at Sirena, on the Osa Peninsula in Costa Rica. These parallel research endeavours in the laboratory and field led to a remarkable succession of insights into *Heliconius* biology, with many of the characteristic aspects of their ecology and behaviour uncovered during this period—notably, pollen feeding, pupal mating, traplining, and use of anti-aphrodisiac pheromones (Ehrlich & Gilbert, 1973; Gilbert, 1972, 1976; Boggs & Gilbert, 1979). As well as all that, Gilbert was the PhD adviser to a remarkably successful group of PhD students, including James Mallet, Mauricio Linares, Chris Thomas, Carla Penz, Jack Longino, and, more recently, Marcus Kronforst and Catalina Estrada.

As has been the case in many strands of evolutionary biology, there have been parallel English and American cultures of *Heliconius* research. John Turner, an eccentric English man, could not be more different from Larry Gilbert, but has been similarly influential. Like Emsley, he was also spurred to study *Heliconius* by Jocelyn Crane, who sent a box of specimens to Philip Sheppard and Sir Cyril Clarke in Liverpool. They passed the box on to Turner, then a young student in the lab who was inspired to try to breed them in the UK. Turner carried out early studies of the genetic basis of wing patterns and other aspects of leidopteran genetics (Turner & Sheppard, 1975; Johnson & Turner, 1979; Sheppard et al., 1985), made significant contributions to *Heliconius* taxonomy (Turner, 1966a, 1968a), and carried out early field studies of dispersal and home-range behaviour (Turner, 1971a). However, following in the tradition of English ecological genetics, perhaps his major contribution was reaffirming mimicry in *Heliconius* as a prime example of evolution by natural selection within the modern view of Darwinian evolution that had been established in the mid-twentieth century (Turner, 1973, 1977, 1981).

Modern *Heliconius* research combines molecular genetics, developmental biology, population genetics, neurobiology, and ecology. Over the last 20 years, there has been a proliferation of *Heliconius* research and researchers, which I will not attempt to document here, but will hopefully become evident in the rest of the book. However, I will mention two individuals who have contributed in vital ways to *Heliconius* biology and to this book, but who might otherwise be overlooked.

The first is Gerardo Lamas, a Peruvian taxonomist based in Lima who has kept up with the remarkable task of revising the taxonomy of the Heliconiini and tracking the nomenclature (and indeed of all neotropical butterflies). The most recent taxonomy of the group in Chapter 12 is largely his work. The second is Walter Neukirchen, a German amateur collector who amassed the most impressive and complete *Heliconius* collection in the world. This collection is now housed at the MacGuire Center in Gainesville, Florida, and is a wonderful resource for *Heliconius* research, as well as providing most of the butterfly photographs in this book. It is always worth remembering the crucial role that taxonomy and museum collections play in underpinning modern evolutionary biology.

CHAPTER 3

The passion: niche differentiation, coexistence, and coevolution

Chapter summary

The *Heliconius* butterflies show complex coevolutionary relationships with their *Passiflora* host plants. As is the case for most plant-feeding insects, *Heliconius* butterflies specialize on a specific group of host plants, in this case the family Passifloraceae or passion vines. *Heliconius* species share out the available host plants, each having a distinct strategy of host-plant use. The passion vines are characterized by an unusually high diversity of different cyanogenic compounds, which have evolved in response to herbivores such as *Heliconius*. In turn, *Heliconius* butterflies detoxify the cyanogens, sequester them for their own defence, and disable the plant β-glucosidase enzymes to prevent release of cyanide.

Other plant defences include hooked trichomes that kill unwary caterpillars, structures that mimic butterfly eggs to deter egg-laying by female *Heliconius*, extra-floral nectaries that attract aggressive ants, diverse leaf shapes that confuse the visual searching of butterflies, and unpredictable growth patterns. These provide some of the best examples of structures that have arisen through coevolution. There is a general lack of specificity in coevolutionary interactions between *Passiflora* and their *Heliconius* herbivores, which may be typical of interactions in most ecosystems, but can nonetheless have important consequences for both parties to the interaction. Phylogenetic analysis of *Passiflora* and *Heliconius* indicates little evidence for coevolution in terms of matching phylogenetic diversification.

Perhaps a quarter of all known species on this planet are herbivorous insects (Janz, Nylin, & Wahlberg, 2006), and a large proportion of those live in the tropics. Evolutionary interactions between tropical plant-feeding insects and their host plants therefore play a crucial role in the evolution and ecology of biodiversity on our planet. In a classic paper, Ehrlich and Raven (1964) used the relationships between butterflies and their host plants to illustrate how coevolutionary dynamics could promote adaptive radiation. John Thompson (2001) has characterized the Ehrlich and Raven model as 'escape and radiate'. The hypothesis is that a group of herbivores colonizes a new group of host plants, perhaps by evolving to overcome specific chemical defences employed by those plants. In doing so, the herbivores are freed from competition and can radiate into multiple species as a result. During this period of radiation, shifts between hosts may promote reproductive isolation and lead to the evolution of new species (Drès & Mallet, 2002). As well as promoting diversification, patterns of host-plant use also play a critical role in permitting coexistence and therefore promoting ecological stability. In this chapter I will describe how heliconiine butterflies interact with their host plants and explore the importance of this interaction for their diversification.

3.1 Caterpillar host plants

The Heliconiini virtually all feed on the plant family Passifloraceae, or passion vines, and primarily

The Ecology and Evolution of Heliconius Butterflies. Chris D. Jiggins, Oxford University Press (2017).

DOI 10.1093/acprof:oso/9780199566570.001.0001

the genus *Passiflora*. These are heavily defended plants, containing an array of toxic chemicals including cyanogenic glycosides. The heliconiines are one of only a few groups of herbivores that have cracked this chemical defence—the others include a few flea-beatles (Chrysomelidae), a family of moths (*fff*: Dioptinae), some butterflies in the family Riodinidae, and some coreid bugs. However, these other groups are fairly rare, so the heliconiines pretty much have this food resource to themselves, presenting them with a great ecological opportunity (Turner, 1981). The only exception is *Eueides procula*, which is known to feed on a host plant in the related family Turneraceae (Janzen, 1983), the only Heliconiini host-plant record outside the Passifloraceae.

3.2 A brief introduction to Passiflora

The best way to find heliconiine larvae in the field is to learn to recognize the local passion vine species. Many have very distinctive leaf shapes and habits, so this can be fairly easy for the commoner species. Conversely, rare *Passiflora* species can sometimes be discovered by following female butterflies. Unlike many butterflies, which can often seem to visit plants rather indiscriminately in their search for oviposition sites, a female *Heliconius* that is seen fluttering persistently around some vegetation undoubtedly knows that there is (or was) a host-plant in the vicinity.

The common name for *Passiflora* is passion vine, which refers to the passion of Christ. The flowers are supposed to represent the crown of thorns (corona), the three nails (stigma), and the five wounds (stamens). This symbolism was apparently used by the conquistadors as a sign that the New World should be converted to Christianity, although one presumes that they would have gone ahead anyway even in the absence of support from the passion flowers. There are more than 540 *Passiflora* species (and 750 spp. of Passifloraceae), with only 20 species in the Old World. Most are found in the neotropics, where they range from the southern United States down to southern Argentina. Although most are herbaceous or woody vines with tendrils, some of those in the subgenus Astrophea are woody shrubs or small trees. The leaves are opposite and tendrils are axillary (i.e. located in the axil between the leaf and stem, rather than opposite the leaf). These two characters generally serve to distinguish *Passiflora* from other common vines found in the forest, such as those in the Cucurbitaceae.

3.3 Passiflora taxonomy and phylogeny

The only comprehensive taxonomy of *Passiflora* was by Killip (1938), who divided the genus into 22 subgenera. Although this classification has been used very widely, it is outdated and does not include a large number of species described in the last 80 years. Recent molecular phylogenetics has shown that traditional groups need updating; there is a clear need for a comprehensive revision of the taxonomy (Yockteng & Nadot, 2004; Hansen et al., 2006; Feuillet & MacDougal, 2007; Krosnick et al., 2013).

Nonetheless, some broad patterns are clear. The *Passiflora* genus itself is monophyletic. Within the genus there are two major subgenera, *Passiflora* (250 spp.) and *Decaloba* (referred to previously as *Plectostemma*, with 230 spp.), which contain most of the species diversity (Krosnick et al., 2013). The *Passiflora* subgenus includes economically important and tasty passion fruits including maracuyá (*P. edulis*) and granadilla (*P. ligularis*). The *Decaloba* subgenus includes widespread and abundant species such as *P. biflora* and *P. capsularis*, which have characteristic two-winged leaves and commonly have glands on the leaf blade. Two smaller clades are most deeply branching in the genus, the subgenera *Astrophea* and *Deidamioides*. The latter includes the Tryphostemmatoides group (Krosnick et al., 2013). *Astrophea* includes unusual woody shrub species such as *P. arborea* and *P. macrophylla*, which are host plants for several *Heliconius* species in Colombia and Ecuador. *P. tryphostemmatoides* is a delicate species with branched tendrils that is also a *Heliconius* host plant. Outside *Passiflora*, the sister genus *Dilkea* contains just 12 species, which are the host plants for species in the Neruda group. As is clear from this brief summary, the Heliconiini feed on a wide variety of species across the Passifloraceae. I will return to a comparison of the phylogenies of *Heliconius* and *Passiflora* at the end of this chapter, but first I will describe the natural history of the relationships between these two ecological enemies.

3.4 Strategies of host-plant exploitation

Female *Heliconius* butterflies seem to spend most of their time searching for somewhere to lay their eggs (Benson, 1978) (Figure 3.1)—so much so, in fact, that it is often possible to tell the sex of a butterfly in the wild within seconds, just from observing its behaviour. Unlike many butterflies, in which females seem to go around almost arbitrarily tapping leaves until they come across the right host, *Heliconius* quite commonly fly directly to a host plant before searching avidly for the growing shoots (Benson, Brown, & Gilbert, 1975; Gilbert, 1975). This suggests that they remember the location of host plants and 'stake them out' for future visits. Choosing a site for oviposition involves careful visual inspection of the plant, repeatedly fluttering around the shoot and landing, touching the leaves with antennae, and 'drumming' with the forelegs (Benson et al., 1975; Benson, 1978). The female may find the new shoot but continue to inspect the whole plant before coming to a decision. Eventually, she will either abandon the plant or land, often hanging under the shoot to lay an egg (or many, depending on the species—see later). The inordinate effort invested in choosing a site for oviposition strongly suggests that this is a critical decision that is under strong natural selection.

Figure 3.1 Oviposition behaviour (see Plate 5)

Heliconius charithonia laying on *Passiflora biflora* and *Heliconius melpomene melpomene* laying on *Passiflora menispermifolia.* Both photographs were taken in the insectary in Gamboa, Panama.

Across different communities, there is a close relationship between the diversity of *Heliconius* and that of *Passiflora* (Thomas, 1990). The more *Passiflora* species present, the more *Heliconius* species that can coexist. This implies that niche segregation in host-plant use is a primary determinant of heliconiine diversity. However, although some species are complete host-plant specialists, there is not a one-to-one relationship between *Passiflora* and *Heliconius*. Instead, every coexisting heliconiine species in any community has a distinct strategy of host-plant exploitation.

For example, in central Panama, the species found in open, sunny areas are mostly specialists, feeding on only a single host species in some cases (*Heliconius hecale, Philaethria dido, H. melpomene, H. sara*), or a handful of related *Passiflora* species (*Dryas iulia* and *H. erato* on Decaloba group hosts). Some of the species that share hosts feed on different parts of the plant, such as *Philaethria dido* and *H. hecale*, which feed on old and young leaves, respectively (Table 3.1).

In contrast, the abundance of suitable shoots in neighbouring forested areas is much lower. Those species that fly in the forest are either host-plant generalists feeding on shoots of small *Passiflora* plants that colonize light gaps (e.g. *Heliconius cydno*), or specialists on large canopy vines (e.g. *H. doris* and *H. sapho*). The latter species generally have gregarious larvae, allowing them to take advantage of the occasional availability of abundant food resource in the form of large and rapidly growing shoots. Similar patterns have been described from other communities that have been studied (Benson, 1978). This partitioning of the host-plant niche strongly suggests that interspecific competition is playing a major role in organizing local patterns of host use (Benson, 1978) (Table 3.1 and 3.2).

Some of these patterns can be understood by looking at host-plant use across the butterfly phylogeny. The strategy that may be ancestral to the

Table 3.1 Host use by the *Heliconius* community along Pipeline Road in Panama

	menispermifolia	vtifolia	auriculata	coriacea	biflora	ambigua	nitida	tryphostemmatoides	foetida
H. cydno	14	15	19	–	–	1	9	–	–
H. melpomene	146	1	1	–	–	–	–	–	–
H. hecale	4	126	1	–	–	–	2	–	–
H. erato		–	3	13	12	–	–	–	–
H. ismenius	–	–	–	–	–	11	–	–	–
H. hecalesia	–	–	–	–	2	–	–	1	–
H. sara	–	–	46	–	–	–	–	–	–
Laparus doris	–	–	–	–	–	205	–	–	–
Eueides aliphera	1	11	–	–	–	–	1	–	–
Dryas iulia	–	1	33	8	6	–	–	–	–
Philaethria dido	1	19	–	–	–	–	–	–	–
Dione juno	–	70	–	–	–	–	–	–	–
Agraulis vanillae	–	–	–	–	–	–	–	–	1
Riodinidae	–	3	38	5	7	–	–	–	–
Josia	3	–	–	–	–	–	–	–	–

Host-plant records from the Pipeline Road area of Panama showing that coexisting species all employ distinct strategies of host use. Numbers indicate individual larvae and eggs collected from each species on each host plant. Data from Naisbit (2001) and Merrill et al. (2013).

Table 3.2 *Heliconius* host records

Genus	Species	Host plants
Agraulis	***vanillae***	*Passiflora affinis, alata, auriculata, bahiensis, bicornis, biflora, caerulea, candida, capsularis, cincinnata, coccinea, coriacea, costaricensis, cubensis, cyanea, deidamioides, edulis, foetida, ichthyura, incarnata, jilekiii, kermesina, laurifolia, lonchophora, ligularis, lutea, maliformis, manicata, mansoi, menispermifolia, microcarpamiersii, misera, mixta, tripartita, mucronata, nitida, ovalis, palmeri, pectinata, platyloba, quadrangularis, rhamnifolia, rubra, sidaefolia, speciosa, suberosa, tenuiloba, urbaniana, vellozii,* and *vitifolia.* (x, b)
Dione	***glycera***	*Passiflora alnifolia, caerulea, cyanea, edulis, tripartita, ligularis,* and *mixta.* (b, v14)
	juno	*Passiflora alata, alnifolia, ambigua, bahiensis, bicornis, caerulea, capsularis, cincinnata, coccinea, coriacea, cornuta, deficiencis, edulis, gracilenshelleri, lancetillensis, laurifolia, ligularis, microcarpa, misera, mixta, tripartita=tripartita mollissima mucronata, nitida, oerstedii, pittieri, platyloba, quadrangularis, quadriglandulosa, sanguinolenta, seemannii, serratifolia, serratodigitata, sidaefolia, speciosa, suberosa, veraguasensis, vitifolia,* and *Erblichia odorata.* (x, b, pb89)
	moneta	*Passiflora adenopoda, caerulea, capsularis, edulis, lobata, misera, tripartita=tripartita mollissima, morifolia, oerstedii, serratodigitata, sexflora,* and *warmingii.* (b, x, b77)
Dryadula	***phaetusa***	*Passiflora bicornis, biflora, caerulea, capsularis, costaricensis, foetida, jilekiii, mansoi, misera, morifolia, mucronata, organensis, punctate, quadrangularis, rubra, sidaefolia, suberosa, talamancensis, tuberosa, vespertilio,* and *vitifolia.* (b, x, s08)
Dryas	***iulia***	*Passiflora actinia, adenopoda, alata, ambigua, auriculata, bicornis, biflora, caerulea, capsularis, cerradensis, cobanensis, coriacea, costaricensis, costata, cubensis, edulis, elegans, foetida, glandulosa, helleri, ichthyura, lancifolia, lutea, miersii, misera, tripartita, morifolia, organensis, pittieri, platyloba, pohlii, punctata, quadrangularis, rhamnifolia, rubra, sexflora, sidaefolia, suberosa, talamancensis, trifasciata truncate, tryphostemmatoides, tuberosa, vespertilio,* and *vitifolia.* (x, b)
Philaethria	***diatonica***	*Passiflora ambigua.* (b)

continued

Table 3.2 *Continued*

Genus	Species	Host plants
	dido	*Passiflora acuminata, ambigua, auriculata, bahiensis, coccinea, costata, cyanea, edulis, guazumaelifolia, laurifolia, mucronata, platyloba, skiantha,* and *vitifolia.* (b, x)
	wernickei	*Passiflora actinia, caerulea, elegans, jilekiii, mansoi, mucronata, quadrangularis, rhamnifolia,* and *sidaefolia.* (b)
Eueides	***aliphera***	*Passiflora amethystine, auriculata, bahiensis, bicornis, caerulea, capsularis, coccinea, coriacea, costaricensis, cyanea, foetida, kermesina, laurifolia, lonchophora, manicata, misera, morifolia, oerstedii, quadrangularis, quadriglandulosa, rovirosae, rubra, setacea, sexflora, sidaefolia, tuberosa,* and *vitifolia.* (x, b)
	emsleyi	*Passiflora tryphostemmatoides.* (b)
	isabella	*Passiflora acuminata, adenopoda, alata, ambigua, amethystine, arborea, bahiensis, bicornis, biflora, caerulea, capsularis, coriacea, edulis, foetida, kermesina, laurifolia, ligularis, lonchophora, maliformis, mayarum, tripartita=tripartita mollissima, mucronata, multiformis, oerstedii, pedata, platyloba, quadriglandulosa, rubra, seemannii, serratifolia, serratodigitata, subpeltata, triloba, tryphostemmatoides,* and *veraguasensis.* (b, x)
	lampeto	Granadilla
	lineata	*Passiflora eueidipabulum, lancetillensis, microstipula, pedicellaris,* and *sp. nov.* Gilbert 1978 (Corcovado). (b, md03, m82)
	lybia	Distephana and Granadilla *Passiflora acuminata, coccinea, glandulosa, guazumaefolia, nitida, palenquensis* and *vitifolia.* (b, x)
	pavana	Subgenus *Astrophea* and *polyanthea* *Passiflora rhamnifolia,* and *sidaefolia.* (x, b)
	procula	Granadilla section Quadrangulares, Digitales, Laurifoliae *Passiflora arborea, edulis, helleri, ligularis, maliformis, tripartita=tripartita mollissima, ocanensis, oerstedii, tryphostemmatoides,* and *Erblichia odorata.* (b)
	tales	*Passiflora acuminata,* and *laurifolia.* (x)
	vibilia	*Dilkea, Passiflora arborea, auriculata, costata, glaziovii, mansoi,* and *pittieri.* (b, m82, x)
Heliconius	***antiochus***	*Passiflora araguensis, costata, leptopoda, pymantha,* and *spinosa.* (x, b)
	astraea	*Passiflora coccinea,* and *glandulosa.* (x)
	atthis	*Passiflora resticulata.* (x)
	besckei	*Passiflora actinia, caerulea, menispermifolia, organensis, sidaefolia, suberosa,* and *vellozii.* (b, x)
	burneyi	*Passiflora coccinea, quadriglandulosa,* and *vitifolia.* (x, b)
	charithonia	*Passiflora adenopoda, apetala, bicornis, biflora, capsularis, citrina, coriacea, cubensis, edulis, foetida, hahnii, laurifolia, lobata, lutea, menispermifolia, ovalis, perfoliata, pulchella, punctata, quadrangularis, rubra, suberosa,* and *tacsonioides.* (x, b, s82, j97)
	clysonymus	*Passiflora antioquensis, apetala, biflora, capsularis, cuneata, filipes, helleri, kalbreyeri, lyra, membranacaea, nubicola, standleyi,* and *tryphostemmatoides.* (b, x)
	congener	Astrophea
	cydno	*Passiflora antioquensis, apetala, arborea, auriculata, biflora, caerulea, capsularis, coriacea, edulis, guazumaefolia, helleri, lancearia, laurifolia, ligularis, lobata, maliformis, menispermifolia, nitida, oerstedii, palenquensis, pittieri, quadrangularis, tryphostemmatoides,* and *vitifolia.* (b, x, s82, st78, s82).
	demeter	*Dilkea* and *Passiflora auriculata.*
	doris	*Passiflora acuminata, ambigua, biflora, laurifolia, ligularis, maliformes, oerstedii, praeacuta, riparia,* and *serrato-digitata.* (x, st78, b, r05)
	egeria	*Passiflora glandulosa.* (x)
	eleuchia	*Passiflora macrophylla* and *tica.* (um04)

Genus	Species	Host plants
	elevatus	*Passiflora laurifolia.* (x)
	erato	Amazon and Eastern slope: *Dilkea, Passiflora actinia, alata, arborea, biflora, caerulea, candollei, capsularis, chelidonea, cuspidifolia, edulis, eichleriana, elegans, jilekiii, miersii, misera, mucronata, organensis, ovalis, pohlii, resticulata, rhamnifolia, rubra, sidaefolia, suberosa, tricuspis, trifasciata, truncata,* and *tryphostemmatoides.* (0, b, romanovsky85, barreto85, b77, k05, m96) Panama to Guyanas: *auriculata, biflora, capsularis, coriacea, edulis, foetida, hahnii, laurifolia, leptopoda, suberosa, tryphostemmatoides, tuberosa,* and *vespertilio.* (b, x) Central America and West: *Dilkea, Passiflora auriculata, bicornis, biflora, caerulea, capsularis, coriacea, costaricensis, foetida, helleri, panamensis, pittieri, punctata, rubra, standleyi, talamancensis, vitifolia,* and *xiikzodz.* (b, x)
	eratosignis	*Dilkea* and *Passiflora citrifolia.* (b, dm10)
	ethilla	*Passiflora actinia, alata, amethystine, bahiensis, cerradensis, cornuta, cyanea edulis, eichleriana, garckei, glandulosa, jilekiii, kermesina, lonchophora, miersii, nitida, oerstedii ovalis, picturata, racemosa, recurva, rhamnifolia, setacea, subpeltata, tarapotina (sp. Nov. Mallet),* and *vellozii.* (x, b)
	hecale	*Passiflora acuminata, alata, auriculata, bicornis, biflora, capsularis, cirrhiflora, coriacea, edulis, filipes, foetida, guazumaefolia, laurifolia, menispermifolia, oerstedii, pedata, platyloba, quadrangularis, quadriglandulosa, spinosa, seemannii, tryphostemmatoides,* and *vitifolia.* (x, b)
	hecalesia	*Passiflora arbelaezii, biflora, gracillima, lancearia, obovata,* and *tryphostemmatoides.* (b, x, bb75, sg78)
	hecuba	Simplicifoliae, Lobatae, and Kermesinae.
	hermathena	*Passiflora faroana* and *hexagonocarpa.* (x)
	heurippa	*Passiflora laurifolia,* and *oerstedii.* (x)
	hewitsoni	*Passiflora pittieri.* (x)
	himera	*Passiflora auriculata* and *sanguinolenta.* (b)
	hortense	*Passiflora bicornis, capsularis, edulis,* and *foetida.* (b)
	ismenius	*Passiflora alata, ambigua, apetala, mayarum, pedata, platyloba, quadrangularis, salvadorensis, serratifolia,* and *vitifolia.* (b, j97, x)
	leucadia	*Passiflora auriculata.* (b)
	melpomene	*Dilkea, Passiflora acuminata, alata, bahiensis, cerradensis, coccinea,* cyanea, *edulis, eichleriana, garckei, glandulosa, laurifolia, ligularis, lonchophora, maliformis, mansoi, menispermifolia, misera, nitida, oerstedii, ovalis, pedata, quadriglandulosa, seemannii, serratodigitata, spinosa, triloba tuberosa,* and *vitifolia.* (x, b)
	nattereri	*Dilkea* and *Passiflora ovalis.* (x, tol)
	neruda	*Dilkea.* (x)
	numata	*Dilkea, Passiflora actinia, acuminata, alata, ambigua, auriculata, candida, cerradensis, cirrhiflora, citrifolia, coccinea, costata, edulis, eichleriana, faroana, glandulosa, jilekiii, laurifolia, lonchophora, nitida, orstedii, ovalis, quadriglandulo, rhamnifolia, setacea, sidaefolia, tricuspis, vellozzi,* and *vitifolia.* (x, b, bbg75)
	pachinus	*Passiflora alata, ambigua, auriculata, coriacea, costaricensis, menispermifolia, oerstedii, pittieri, quadrangularis, tryphostemmatoides,* and *vitifolia.* (b, x)
	pardalinus	*Passiflora spinosa.* (x)
	peruvianus	*Passiflora adenopoda, manicata, rubra,* and *tuberosa.* (j98)
	ricini	*Passiflora capparidifolia* and *laurifolia.* (x)
	sapho	*Passiflora gigantofolia, macrophylla,* and *pittieri.* (bbg75, x, sg78)
	sara	*Passiflora auriculata, biflora, candida, cerradensis, cirrhiflora, citrifolia, costata, edulis, faroana, mansoi, mucronata, ovalis, pentagona, rhamnifolia, sidaefolia,* and *spinosa.* (x, s82, b)
	telesiphe	*Passiflora telesiphe.* (m98)
	timareta	*Passiflora alata.* (x)

continued

Table 3.2 *Continued*

Genus	Species	Host plants
	wallacei	*Passiflora coccinea, glandulosa, quadriglandulosa, variolata,* and *vitifolia.* (x)
	xanthocles	*Passiflora eichleriana, nitida, oerstedii,* and *praeacuta.* (x, m80, r03)

Complete list of known host-plant records for heliconiine species. These records have been collected by many observers over many years and are likely to include errors and records of occasional 'accidental' hosts that are not major food plants. Better quantification of host use in the field is needed for more communities. Data collated by K. Kozak primarily from the Natural History Museum HOST database (Robinson et al., 2010), indicated with 'x', and the following: 0 = secondary host in Brown (1981); barreto85 = Menna-Barreto & Araújo (1985); bb75 = Brown & Benson (1975); bbg75 = Benson et al. (1975); dm10 = Dasmahapatra et al. (2010) (describes eratosignis, inferred to feed only on *Dilkea*); j98 = Jiggins & Davies (1998);. k05 = Kerpel & Moreira (2005); m80 = Mallet & Jackson (1980); m82 = Mallet & Longino (1982); m96 = Mugrabi-Oliveira & Moreira (1996); m98 = Knapp & Mallet (1998); md03 = MacDougal & Hansen (2003); p89 = Pemberton (1989); r03 = Reed (2003); romanovsky85 = Romanowsky, Gus, & Araújo (1985); s08 = Silva et al. (2008); sg78 = Gilbert & Smiley (1978); st78 = Smiley (1978a) 78; tol = tolweb.org; um04 = Ulmer & MacDougal (2004); v14 = Vargas et al. (2014).

heliconiines is to feed on older leaves, which generally rely more on structural rather than chemical defence (Coley, 1980). Thus, the genera *Dione, Eueides, Philaethria, Dryadula,* and *Agraulis* primarily feed on old expanded leaves of their *Passiflora* hosts (Benson et al., 1975; Mallet & Longino, 1982). *Heliconius doris* is somewhat intermediate, laying large rafts of eggs on old leaves of *Passiflora ambigua* and other species, but generally in the proximity of younger shoots (Benson et al., 1975).

In contrast, one of the defining characteristics of the *Heliconius* radiation (including the Neruda group) is the habit of feeding on young shoots or meristems (Benson et al., 1975; Gilbert, 1991). *Dryas iulia* is an outgroup species that also lays eggs on young shoots, as well as older leaves—this may be an independent origin of meristem feeding. Young shoots are generally more nutritious than older leaves, so that this feeding habit allows *Heliconius* larvae to develop very rapidly. Generally larval development takes about 2–3 weeks from egg to pupa (McMillan, Jiggins, & Mallet, 1997), but can be as little as 10 days (Longino, 1984), considerably faster than in species feeding on older leaves. Some species with gregarious larvae can move onto older leaves as fourth- or fifth-instar larvae, although this results in slower development (Longino, 1984). Unlike old leaves, young shoots are ephemeral resources that can be difficult to find. The traplining behaviour of *Heliconius* adults (see Chapter 5), whereby they regularly patrol the same host plants, allows them to efficiently make use of such shoots as they appear. The clear demarcation between old-leaf and young-shoot feeding probably allows co-existing species to use the same host-plant species while avoiding direct competition.

3.5 The evolution of host specialization

A challenge in evolutionary biology is to explain niche specialization (Futuyma & Moreno, 1988; Jaenike, 1990). Some species are highly specialist in their ecology, being restricted to a very narrow habitat range, such as insects that feed on a single host-plant species. Such specialization implies trade-offs—expertise in one task comes at a cost in terms of an ability to carry out more general tasks. On the other hand, some very successful species are highly generalist.

Why, then, do some species evolve specialization, whereas others can thrive without it? It has often been suggested that specialization is an evolutionary dead end, with a trend towards ever-increasing specialization over evolutionary time. In *Heliconius*, Spencer (1988) has argued that the Laparus, Neruda, and burneyi clades were more generalist, with some clades within *Heliconius* showing increased levels of specialization. However, phylogenetic analysis in other butterflies has provided little support for this hypothesis, with switches commonly occurring between specialists and generalists (Janz, Nyblom, & Nylin, 2001).

One way to study niche specialization is by comparing closely related species that differ in their patterns of niche exploitation. In *Heliconius*, host-plant specialization can be a purely behavioural adaptation, which may vary across a species range. This is best documented in *H. melpomene*, which feeds on only a single species in Central America (*P. menispermifolia* or *P. oerstedii*, depending on the locality), but whose larvae can be fed with equal success on a wide variety of host-plant species (although this may not be the case in some parts of the range; Claire Mérot, personal communication).

It has been suggested that genetic trade-offs in the ability to use different hosts might be a primary driver of specialization (Jaenike, 1990), but the ability of *H. melpomene* to feed on many hosts suggests a lack of biochemical adaptations to host plants (Smiley, 1978b). Host specialization is therefore almost entirely due to the oviposition preferences of the adult females.

So why do females reject host plants that are found in their local environment and are perfectly good hosts for their larvae? Several ecological factors have been suggested, including parasite pressure and interspecific competition (Smiley, 1978a). Naisbit carried out an experiment in which 120 paired first-instar larvae each of *H. melpomene* and *H. cydno* were placed on *P. menispermifolia* and other host species in the wild. Both butterfly species, but especially *H. melpomene*, showed slightly higher survival on *P. menispermifolia* as compared to other hosts (Merrill et al., 2013). This suggests that *P. menispermifolia* might be a better host in the wild. However, this doesn't address why there are different levels of specialization in the two species.

H. melpomene is more of a host-plant generalist in the Amazon and Guiana shield, where it is not in sympatry with *H. cydno*, suggesting that interspecific competition may be partly responsible for the change in strategy. Perhaps specialization is a recent adaptation to local conditions in Central America. Crossing experiments to investigate the genetic basis of host specialization suggest that the oviposition preference of specialist females is an autosomal, dominant trait, which may have a simple genetic basis (Merrill et al., 2013). This is consistent with specialization having arisen recently in evolutionary time.

This specific case study is consistent with patterns across the genus *Heliconius*, where there is a great deal of variation in the degree of host specialization. In fact, one of the best predictors of host range is geographic range—more widespread species have larger host ranges, perhaps driven by local adaptation to varying ecological conditions (Kozak, 2015). This suggests that the degree of specialization evolves rapidly and is perhaps shaped by local ecological selection pressures rather than constraints imposed by phylogeny. Furthermore, in a more comprehensive analysis of host use, there is no evidence that specialists evolve from generalists. Instead, it seems likely that generalist species evolve from specialist ones as frequently as the other way around (Kozak, 2015).

3.6 Larval gregariousness

Another strategy that varies between species is the degree of larval gregariousness. Some species such as *Heliconius erato* have highly cannibalistic larvae, and lay eggs singly or widely separated on a single shoot (Figure 3.2). Others, such as *H. melpomene* and *H. cydno*, are less cannibalistic, but lay no more than two or three eggs on a shoot. Solitary larvae are

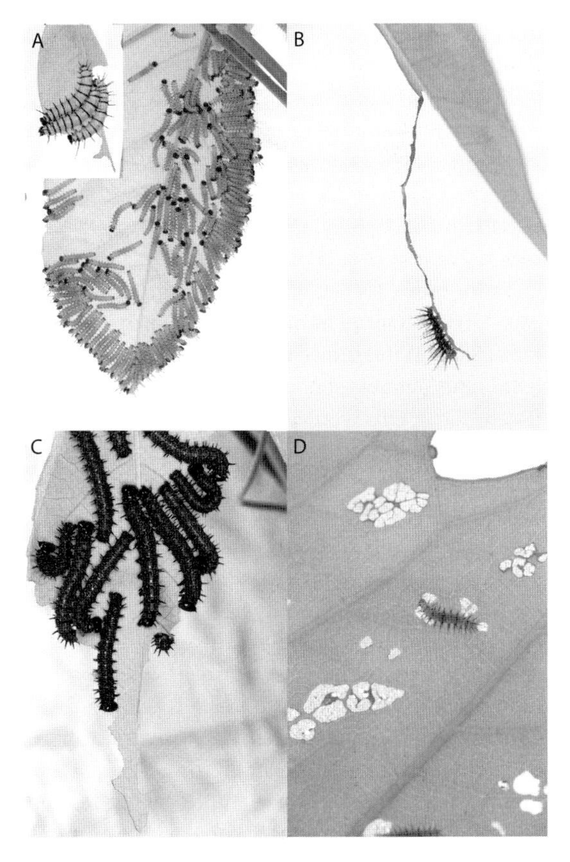

Figure 3.2 Caterpillar feeding strategies (see Plate 6)

(A) *Heliconius doris* caterpillars feed in a coordinated manner in large groups. Third-instar caterpillars are shown in the main image and fifth-instar caterpillars are shown inset. (B) Early-instar caterpillars of *Dryas iulia* cut a narrow strip of leaf and sit at the end of the hanging thread. This is a strategy to avoid the attention of ants patrolling the leaves. (C) Caterpillars of *Dione juno* are a crop pest on commercial *Passiflora edulis*. They are highly gregarious but do not feed in such a coordinated manner as *H. doris*. (D) Early-instar caterpillars of *Eueides aliphera* scrape the underside of leaves, leaving characteristic 'windows' of translucent cuticle in the leaf.

presumably an adaptation to the limited resources available—most light-demanding gap species of *Passiflora* produce small shoots that can support only one or a few larvae before the leaves harden and become inedible. Thus, there is almost a one-shoot, one-mature-larva rule, with these *Heliconius* species relying on rapid larval growth to reach maturity before the shoot becomes depleted.

At the other end of the spectrum, some species are highly gregarious—for example, clutch sizes of almost 200 eggs were recorded for *Heliconius hewitsoni* in Costa Rica (Reed, 2003). *Heliconius doris* is the most gregarious species; rafts of around 800 eggs are known (Benson et al., 1975), and there is one record of 1,204 pupae and pupating larvae found at the base of a *P. ambigua* vine in Costa Rica (Mallet, 1984).

Phylogenetic analysis suggests that gregariousness has arisen several times in the Heliconiini from an ancestor with solitary larvae (Beltran et al., 2007). In several species (*D. juno, H. doris, H. hewitsoni, H. sara, H. xanthocles*), multiple females have been seen to contribute to laying a single clutch (Mallet & Jackson, 1980; Turner, 1981; Longino, 1984; Reed, 2003). In the case of *Heliconius hewitsoni*, up to ten females laid on the same shoot, despite the availability of other suitable shoots in the immediate vicinity (Reed, 2003). This would seem to imply that there is an advantage to gregariousness among larvae, and this is supported by evidence that larval mortality is lower for larger egg batches (Longino, 1984).

Nonetheless, it has also been reported that the interactions between ovipositing females at the same shoot can be aggressive, suggesting that they are competing for a scarce resource rather than cooperating to increase group size. Females rarely lay more eggs on a shoot once it is occupied (i.e. after the first day; Longino, 1984). At the time of oviposition, shoots are generally tiny—only 1.5 cm in length on average. However, the laying females seem to know which shoots are the most promising, as clutch size is correlated with the subsequent growth rates (Reed, 2003). Thus, females have an impressive ability to assess the potential of shoots even before they begin their most rapid growth phase.

Gregarious larvae tend to feed on large canopy vines, often in closed canopy forest (Benson et al., 1975). These are the only *Passiflora* species that are capable of supporting large numbers of larvae on a single shoot. Gregariousness may therefore have evolved as a tactic to take advantage of the occasional sporadic availability of large, rapidly growing shoots of canopy vines. One of the unavoidable consequences of being gregarious is increased conspicuousness. Gregarious insects generally tend to be toxic and warningly coloured, and it is therefore interesting that among *Heliconius* species, gregarious and more host-specialist species, such as *H. sara* and *H. sapho*, tend to have higher concentrations of cyanogens as compared to more solitary species, such as *H. erato* and *H. melpomene* (Engler-Chaouat & Gilbert, 2007). Perhaps higher toxicity levels evolve to compensate for the increased probability of attack due to gregariousness. Alternatively, gregariousness might have arisen after increased toxicity, to enhance the warning signal through clustering of larvae.

Larval morphology and behaviour is also adapted to their sociality. Gregarious species generally have less spiny larvae, and the spines are softer, presumably so they don't spike their sisters. In the highly gregarious *Dione juno* and *H. doris*, the larvae have coordinated behaviour, eating entire leaves in coordinated lines and then moving on to new leaves or shoots as a group, and also resting together. Larval moulting and pupation are also coordinated, with the whole group ceasing to feed and aggregating before moulting together. In *Dione juno*, large egg clutches tend to become dispersed as smaller groups of larvae move apart to exploit different leaves or shoots, such that pupation generally occurs in groups of 3–10 (Muyshondt & Young, 1973). In contrast, *H. doris* pupates in large groups. There is also evidence for coordinated defence; for example, when threatened, the larvae emit a 'disagreeable odour' and make spasmodic movements as a group (Muyshondt & Young, 1973). These responses would presumably be ineffective if performed by a solitary individual, but are more likely to be effective when performed by a group. Such behaviours are likely to be favoured by kin selection acting on groups of siblings, although relatedness may be lower in some aggregations where different females cooperate to lay egg clusters.

3.7 Niche partitioning

In summary, there are several dimensions along which species differ in host-plant use (Benson, 1978): (1) old versus young leaves, (2) gregarious versus solitary larvae, which roughly equates to use of large versus small individual host plants, (3) varying degrees of host specialization, from complete host specialists to generalists that use most of the *Passiflora* species available, and (4) specialization on different host-plant species. Rather surprisingly, the last of these seems to be the most phylogenetically labile from the butterfly perspective. Over the *Heliconius* phylogeny, there has been a single major switch to meristem feeding, and between three and seven changes from solitary to gregarious larvae (Beltran et al., 2007).

In contrast, there are differences between generalist versus specialist strategies and multiple shifts between host-plant species even among populations of a single *Heliconius* species. Overall, the pattern is one in which the diversity of *Heliconius* in any locality is strongly influenced by the locally occurring *Passiflora* species, but the relationship is not one-to-one—there are fewer *Heliconius* species than *Passiflora* species. It has even been suggested that the diversity patterns of *Heliconius* might provide a model for predicting global insect biodiversity (Thomas, 1990).

The observational evidence therefore suggests that host-plant use is the primary axis of niche partitioning, which permits coexistence of heliconiine species. This is supported by the observation that reproductively isolated taxa that share a host-plant niche remain parapatric and do not appear capable of sympatric coexistence. For example, *Heliconius himera* and *H. erato* show no evidence for differences in host use, but are reproductively isolated. They replace one another over a narrow ecological transition zone in southern Ecuador (Jiggins, McMillan, Neukirchen, et al., 1996; Jiggins, McMillan, & Mallet, 1997). Similarly, *H. pachinus* is a parapatric replacement of *H. cydno*, and these species also share host-plant strategies.

Another line of evidence for niche competition is ecological release in the absence of a competitor. If competition constrains a species to a narrow niche, then removal of the competitor should lead to an increase in niche breadth. Again, there is some circumstantial evidence for this, with *H. melpomene* showing a much larger host range in the eastern part of its range, where it does not coexist with its close relatives *H. cydno* or *H. timareta*.

Nonetheless, there are a few cases of species that coexist and have very similar patterns of host use. *H. clysonymus* and *H. telesiphe* are Andean species, while *H. erato* is a close relative that is common in the lowlands. The three species overlap at 1,000–1,200 m where they largely share host plants. Anecdotally, it seems that the three species are never all common in the same site, with one of the three always considerably more abundant than the other two. This might suggest competitive interaction between these species, but has never been properly documented and might provide a useful arena in which to study interspecies competition. Unfortunately, there is no good experimental evidence for ecological host plant competition, or ecological release. This contrasts with some other systems such as *Anolis* lizards and Darwin's finches, where there is more direct experimental and observational evidence for the role of competition in structuring communities (Grant, 1999; Losos, 2011).

3.8 Evolution and coevolution, defence, and counter-attack

The heliconiines are relentless predators of their *Passiflora* host plants. To the *Heliconius* biologist, a leafless stick of *Passiflora* stem is a sure sign of recent herbivory by heliconiines (when you see that female butterfly fluttering determinedly around a spot in the understorey, look out for defoliated stems that may be all that is left of her *Passiflora* victim). Shoot occupancy in a study on Pipeline Road, Panama, varied from 4 to 23% depending on the *Passiflora* species, and the great majority of recorded herbivores were heliconiines (Naisbit, 2001; Merrill et al., 2013). Similar figures were obtained for a comparable community at Sirena (Mallet, 1984). In the case of *P. pittieri*, shoot occupancy varied considerably depending on the season, but was always high (20–80%), and growth of juvenile plants was severely limited by *H. hewitsoni* larvae (Longino, 1984). In one year, 107/489 shoots were completely consumed by larvae. In at least some populations, therefore, predation pressure by heliconiines is a very significant selective pressure on the growth of host plants. This has led to an elaborate evolutionary dance, with the host plants developing an array of defences against herbivory, most of which in turn have been overcome by one or more butterfly species.

3.9 Biochemical defences

One of the great perks of fieldwork in the Andes is the diversity of passion fruit—from the high-altitude taxo or curuba (*Passiflora mollisima*) to the ubiquitous maracuyá (*Passiflora edulis*), but the dark side to these fruit is that they can be poisonous if eaten unripe (Saenz, 1972). This is because *Passiflora* species contain a wide variety of biologically active molecules. Perhaps the most important of these are cyanogenic glycosides that hydrolyse to release toxic hydrogen cyanide (HCN), providing a potent defence from attack by herbivores (Figure 3.3).

Passiflora species also contain other classes of toxins, including flavonoids, phenols, and

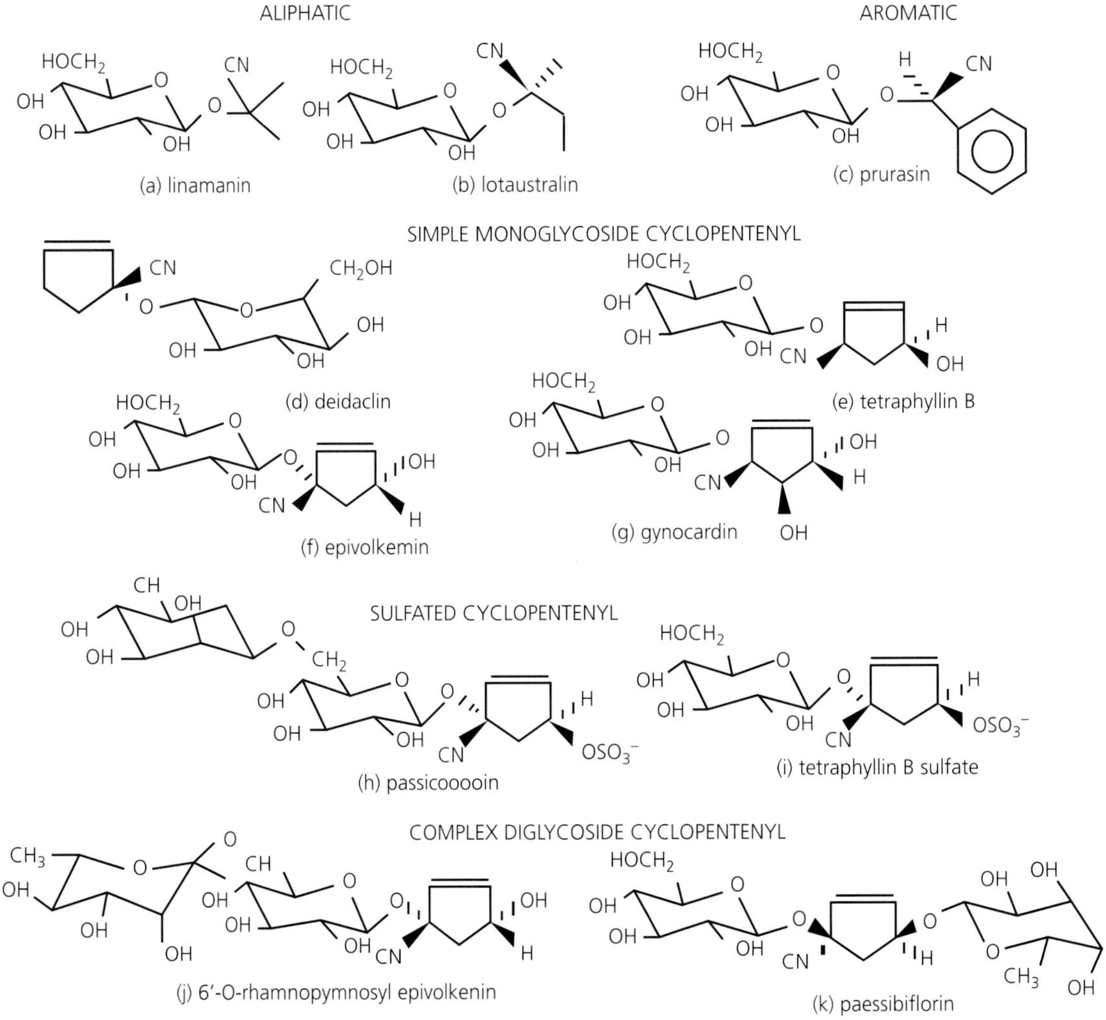

Figure 3.3 Cyanogenic compounds

Structural formulae of the major cyanogenic glycoside (cyanogen) classes found in Passifloraceae (Engler-Chaouat & Gilbert, 2007).

alkaloids (Dhawan, Dhawan, & Sharma, 2004). Native Americans have used *Passiflora* in traditional medicine, and there is current interest in the pharmacological properties of *Passiflora*-derived compounds that might be used for medicinal purposes (Dhawan et al., 2004).

Although cyanogenic compounds occur in many plant families, *Passiflora* have an unusual diversity of them. For example, Spencer and colleagues have described 29 distinct compounds in six major classes from *Passiflora* species (Seigler et al., 1982; Spencer & Seigler, 1984, 1985a, b, c; Spencer, Seigler, & Nahrstedt, 1986; Spencer, 1987, 1988). Most other cyanogenic genera of plants produce only one or two distinct compounds (Spencer, 1988). Single *Passiflora* species can contain up to four distinct compounds, and different species in the same taxonomic section of the genus can contain completely non-overlapping sets of compounds (Spencer, 1988). This diversity among closely related species is strongly suggestive of rapid evolution due to a coevolutionary arms race between plant and herbivore. Given the known ecological importance of *Heliconius* as *Passiflora* herbivores, it seems likely that they have played a major role in this process.

When a butterfly larva eats a passion vine leaf, there are two possible routes for the toxins—either safely sequester the chemicals in the butterfly's own tissue, or metabolize them. There is evidence for both. Engler and colleagues showed that all *Heliconius* species tested, when fed on plants containing simple monoglycoside cyclopentenyl (SMC) cyanogens (Figure 3.3), had these compounds present in the adult butterflies (Engler, Spencer, & Gilbert, 2000). The same butterfly species, when fed on plants without these SMCs, lacked the compounds in the adult (Engler-Chaouat & Gilbert, 2007). This implies that SMC cyanogens are directly sequestered by *Heliconius* from the host plants.

Evidence for metabolism of cyanogens comes from *H. sara*. A metabolic derivative of the cyanogen epivolkenin was identified in *H. sara* and named sarauriculatin, after the butterfly and its host, *P. auriculata*. This new compound was a cyclopentenyl cyanogenic glycoside that had undergone replacement of the nitrile group by a thiol (Engler et al., 2000). This metabolic change disarms the epivolkenin and prevents the enzymatic release of cyanide. In contrast, the burnet moth *Zygaena filipendulae*, which also feeds on cyanogenic host plants, uses a distinct mechanism whereby the enzyme β-cyanoalanine synthase detoxifies HCN to β-cyanoalanine, directly disarming the HCN (Zagrobelny et al., 2004; Zagrobelny & Møller, 2011). The distribution of different mechanisms for dealing with cyanogens across the insects remains poorly documented.

It is also known that butterflies can synthesize their own cyanogens. Nahrstedt and Davis (1983) fed radio-labelled amino acids, valine and isoleucine, to *Heliconius melpomene*, and showed that both larvae and adults used the amino acids as precursors for synthesis of the aliphatic cyanogens, linamarin and lotaustralin. In addition, Engler et al. (2000) showed that the young adults of 14 species of *Heliconius* all contained aliphatic cyanogens, whether or not this class of compounds were found in their host plants. The compounds are therefore synthesized in both the larval and adult stages.

More recent work in the burnet moth *Zygaena filipendulae* has worked out the biochemical mechanism for cyanogen synthesis. Two P450 enzymes (CYP405A2, CYP332A3) and a glycosyl transferase enzyme (UGT33A1) encode the complete biosynthetic pathway for cyanogenic glycosides (Jensen et al., 2011). This mechanism is remarkably similar and presumably convergent with that used in plants, with both involving two multifunctional P450 enzymes each catalysing unusual reactions and a glucosyl transferase, with the three enzymes acting in sequence. The *Heliconius* genome also contains orthologues of these genes, although their function remains unknown. However, given the similarity between known biosynthetic pathways in moths and plants, it seems highly likely that the butterflies use a similar mechanism.

There is another critical component to the defensive armour of the plants, as the glycosides need to be broken down (hydrolysed) in order for them to become biologically active toxins. Enzymes known as β-glucosidases carry out this reaction, and can be extremely specific in their action. Thus, Spencer (1987, 1988) has emphasized that the plant defence is more correctly seen as containing two complementary components—the glycoside (responsible for storage of the toxin) and its corresponding glucosidase enzyme (which activates its toxic effect). These are generally stored in different parts of the leaf tissue, so they come into contact with each other only during herbivore feeding.

Therefore, we might predict that the butterfly would not only evolve to deal with the cyanogens, but also to deactivate the plant glucosidases, and indeed this has been shown to be the case. β-glucosidase enzymes from *Heliconius* inhibit the same class of enzymes in the plant, either by binding competitively to the enzyme substrate or by deactivating the plant's β-glucosidase directly (Spencer, 1987, 1988). In an additional layer of complexity, plant tannins can have an inhibitory effect on butterfly β-glucosidase enzymes, suggesting that this class of plant compounds may have evolved in part to deactivate the butterfly defences (Spencer, 1988).

Patterns of biochemical host-plant specialization are at first sight puzzling. As described above, some butterflies are highly specialist but others are quite generalist. The usual explanation for ecological specialization invokes the concept of life history trade-offs. One cannot be a jack-of-all-trades without compromising the ability to exploit a particular niche relative to more specialist competitors. An intriguing hypothesis is that *Heliconius* face trade-offs in obtaining their chemical defence (Engler-Chaouat & Gilbert, 2007). Thus, specialist species such as *H. sara*, feeding on *P. auriculata*, and *H. sapho*, feeding on *P. pittieri*, rely more on sequestration of plant cyanogens. Their host plants are rich in a class of compounds known as simple monoglycoside cyclopentenyl (SMC) cyanogens, and these specialist species contain much higher concentrations of these compounds than generalist species feeding on the same plants.

Engler-Chaouat and Gilbert (2007) took advantage of variation between *Passiflora* species in their chemical composition. Thus the 'universal donor' host species, *Passiflora biflora*, contains almost no SMC cyanogens, and *H. sara* reared on this plant contained very low levels of cyanogens—apparently the high efficiency of sequestration of SMC cyanogens on its native host comes at the cost of a relative inability to synthesize cyanogens as an adult. In contrast, generalist species such as *H. cydno* rely more heavily on synthesis of their own cyanogens and contain mostly aliphatic cyanogens, irrespective of the host on which they are raised.

Thus, the specialists are probably much better protected overall due to their highly efficient sequestration, but this comes at the ecological cost of being unable to feed on many available host plants in their environment. Given the ecological and evolutionary importance of these trade-offs, a promising avenue for future research would be characterization of the genes and their enzyme products that are responsible for cyanogen synthesis and degradation.

Interestingly, *H. melpomene* behaves as if it were a biochemical generalist, with primarily synthesized cyanogens, and levels of SMC cyanogens no higher than generalist species when fed on its own host *P. oerstedii* (which does contain SMC compounds) (Engler-Chaouat & Gilbert, 2007). This supports the suggestion that *H. melpomene* has recently become a specialist as a result of ecological rather than biochemical selection pressures (see earlier).

There is also some evidence for within-species variation in levels of toxicity. One study of *H. erato* showed differences between populations in levels of cyanide production that were correlated with the cyanide content of their host plants (Hay-Roe & Nation, 2007). There is therefore the potential for host-plant use in generalist species to influence adult toxicity. A similar phenomenon in monarch butterflies leads to 'automimicry', whereby some individuals do not sequester toxic compounds and act as Batesian mimics of their conspecifics (Brower, Pough, & Meck, 1970). Future studies of inter-individual variation in host-plant use and toxicity will be needed to determine whether there is similar variation in *Heliconius*.

3.10 Physical defences

The chemical complexity of *Passiflora* suggests this is the main arena for the coevolutionary dance with *Heliconius*, but chemistry is by no means the only defence deployed by the plants. A striking example of a physical defence are the hooked trichomes on the leaves and stems of *Passiflora adenopoda* (Gilbert, 1971). Experiments have shown that most *Heliconius* species get hooked up on the trichomes and are unable to feed on this species (incidentally, the plant is easy to identify as it sticks to your clothes like Velcro). However, just as any evolutionary biologist would predict, eventually a herbivore will evolve to counter the defences.

Sure enough, *Heliconius charithonia* has done just that. *Heliconius charithonia* generally feeds on Decaloba species like its close relative *H. erato*, but unlike *H. erato*, it has somehow evolved to overcome the trichomes of *Passiflora adenopoda*, and in some areas this is its main host (Mendoza-Cuenca & Macías-Ordóñez, 2005). This is one of the best examples of reciprocal coevolution among *Heliconius* and their host plants, as there is evidence for evolutionary change in both parties in response to each other—first evolution of the host defence and then the evolution of a response by the herbivore, in this case specific to *H. charithonia*.

Another example is *Passiflora foetida*, which has glandular trichomes that give the plant a sticky feeling, and likely play a similar role in defence. This plant is not eaten by heliconines apart from *Agraulis vanillae* (Spencer, 1988). It has been suggested that these physical defences have evolved recently, after the heliconiines had overcome many layers of biochemical protection (Gilbert, 1971).

3.11 Psychological defences

Passiflora species also attempt to deceive their heliconiine herbivores in a number of ways. One of the most striking is by producing egg mimics—structures that superficially resemble heliconiine eggs. This provides a deterrent to oviposition in species where the larvae are cannibalistic or otherwise compete for host-plant resources. Females therefore avoid laying eggs in the presence of other larvae or eggs. Such egg mimics can be modified glands on the leaf blade (e.g. *P. biflora, P. auriculata*), stipules (*P. cyanea, P. ligularis*), or petiolar glands (*P. auriculata*—the Latin name of this species comes from the ear-like shape of these glands) (Figure 3.4).

Experiments have shown that the presence of egg mimics deters oviposition by females, but when they are made cryptic with green food dye, females are more likely to lay. This confirms that it is the visual appearance of the egg mimics that is acting as the deterrent (Williams & Gilbert, 1981). A similar deterrence was shown in shoots of *P. cyanea* with egg mimics, relative to experimental treatments where the mimic was removed (a control was also used in which the egg mimic structure was cut but not removed completely). Egg mimics, or indeed real eggs, do not prevent oviposition completely, but both reduce its probability and increase the time taken by females before oviposition (Williams & Gilbert, 1981). It has been suggested that egg mimics are primarily found on secondary or rarely used host species. The butterflies would readily evolve to discriminate egg mimics from real eggs if they were to occur on a primary host species (Benson, 1978).

Another striking feature of Passiflora species is their huge diversity of leaf shape (Figure 3.5). Coexisting species in any community commonly have very different leaf shapes, and those with similar leaf shape often differ in pubescence, reflectance, or other characteristics. It has been argued that this diversity may be an adaptation to visual predation by *Heliconius* species (Gilbert, 1975, 1982), which perhaps drives a form of negative frequency-dependent selection on leaf shape. I can see two ways in which this might work. First, if generalist heliconiines such as *H. cydno* learn the leaf shapes of local *Passiflora* species, then rarer leaf shapes might be at an advantage. Second, on an evolutionary timescale, if specialist heliconiine species showed genetic preferences for specific leaf shapes, this might favour the evolution of novel leaf shapes among *Passiflora* species, or the invasion of local communities by species with different leaf shapes. The existence of population differences in leaf shape—such as *P. oerstedii*, which has a tri-lobed and a single-lobed leaf form—perhaps represent an early stage in this process.

Alternatively, it has been proposed that leaf diversity may have evolved through mimicry, with leaf shapes being mimetic of other locally common vines or trees, as a form of crypsis (Gilbert, 1982). In support of this idea, it has been pointed out that many *Passiflora* species are named for similarity of leaf shape to other genera; e.g. *laurifolia, bauhiafolia, morifolia, capparidifolia, guazumaefolia* (Killip, 1938; Gilbert, 1975). However, in my own experience as a visual 'predator' of *Passiflora*, I find the distinctive leaf shapes a great aid to their discovery, so the mimicry hypothesis seems unlikely. By far the

Figure 3.4 Eggs and egg mimics (see Plate 7)

(A) Egg cluster of *H. doris* on *P. ambigua*; (B) eggs of *H. sara* on *P. auriculata*; (C) single egg of *H. erato* on *P. biflora*; glands on the petiole of (D) *P. quadrangularis* and (E) *Passiflora* spp. and on the leaf blade of (F) *P. auriculata* act as egg mimics to deter oviposition by *Heliconius* females.

hardest species to find are those with simple leaves (e.g. *P. pittieri*, *P. ambigua*).

There is now good evidence that leaf shape is used by female butterflies to find host plants. Leaf shape is known to be a cue in the searching behaviour of *Papilio* butterflies (Rausher, 1978), and *Heliconius* can also learn to associate shapes with nectar rewards (Gilbert, 1975; Dellaglio et al., 2016). Denise Dellaglio (2016) has also shown that individual *H. erato* females can learn to associate different leaf shapes with opportunities for egg-laying (see also Copp & Davenport, 1978; Kerpel & Moreira, 2005). Females trained on different artificial leaves were subsequently more likely to approach a shoot with

Figure 3.5 Diversity of leaf shape among coexisting *Passiflora* species

Leaf shapes of *Passiflora* species found in the Gamboa and Pipeline Road area in Panama. From top left clockwise: *Passiflora auriculata, P. menispermifolia, P. tryphostemmatoides, P. coriaceae, P. ambigua, P. biflora, P. vitifolia.*

the more familiar leaf shape. Learning in this way could therefore lead to formation of a 'search image' among the butterflies, which could favour species with rare and less well recognized leaf shapes. *H. erato* feeds on *P. auriculata, P. coriaceae*, and *P. biflora* in the Gamboa area, all with very different leaf shapes. It is tempting to speculate that the origin and maintenance of these species in this habitat is at least in part driven by the searching behaviour of the butterflies.

3.12 Biological defences: ants

Many *Passiflora* species recruit ants to their defence by means of extra-floral nectaries that provide a nectar reward. These are commonly located on the leaf blade, stipules, or petioles of the plant, and in the field a large proportion of *Passiflora* plants are patrolled by ants (Smiley, 1986; Leal et al., 2006). For example, Naisbit (2001) sampled 2,121 plants from eight *Passiflora* species on Pipeline Road, Panama, and found that 39% of plants had shoots that were patrolled by ants. It has been suggested that the vine growth form is particularly suited to attracting ants to nectaries, due to the large amount of contact between the vine and surrounding vegetation (Benson et al., 1975).

In return for their payment of nectar, the ants can be a significant source of mortality for eggs and larvae (Smiley, 1985a). For example, Smiley placed larvae on wild host plants and followed their survival and growth—9/60 (15%) larvae survived the first two days on ant-attended plants, whereas 22/66 (33%) survived on ant-free plants (Smiley, 1985b). Interestingly, this wasn't due to small size—levels of predation were similarly high for all larval instars tested (1st–3rd), but mortality was primarily during the first two days of the experiment, suggesting a difficult establishment phase (Smiley, 1985a). Similar mortality rates were recorded by Naisbit in Panama, with 26/120 (21%) of *H. cydno* and 20/120 (16%) of *H. melpomene* first-instar larvae surviving over two days, although in this case it wasn't clear how much of the mortality was due to ants (Naisbit, 2001). A slight caveat is that the high 'establishment phase' mortality in these experiments suggests that some of the mortality might be due to the experimental manipulation of placing larvae on wild hosts.

However, there is contradictory evidence regarding the effectiveness of ants in plant defence. A long-term study of mortality among *H. hewitsoni* larvae in Sirena also suggests that ants have an impact on larval mortality, but that this is relatively slight (Longino, 1984). There was higher early-instar mortality on shoots with dominant ants, but lower late-instar larval mortality (dominant ants are species that regularly visit the shoot, rather than being opportunistic visitors). Overall, shoots with dominant ants were somewhat less likely to be eaten by *H. hewitsoni* larvae. However, in a manipulation experiment, ant exclusion had little effect on survival of *H. hewitsoni* larvae, and no effect on growth or survival of *P. pittieri* host plants.

In contrast, exclusion of *H. hewitsoni* larvae led to dramatically improved plant growth. Indeed, *Heliconius* larvae are commonly seen coexisting with ants, which rarely attack late-instar larvae. The toxic chemicals in eggs and larvae and the presence of long spines is an effective defence against generalist predators, including ants—as Jack Longino said, '*H. hewitsoni* eggs and larvae are not good ant food', and the same is probably true of most *Heliconius* species (Longino, 1984).

Although the impact of ants may be slight overall, they can have a dramatic impact on the fate of individual plants. Once a plant has been colonized by a particular ant species, the dominant ant will generally remain in place for a long time (Longino, 1984; Smiley, 1986). Furthermore, ant species differ significantly in their impact on butterfly larval survival (Longino, 1984). A plant that is lucky enough to be colonized early in life by especially pugnacious ants is likely to have a stronger future.

In addition to attracting ants, *Passiflora* extrafloral nectaries may sustain other predators of heliconian larvae, such as the *Trichogramma* wasps that parasitize butterfly eggs (Gilbert, 1975), and larger predators such as vespid wasps. For example, Longino observed vespid wasps harvest an entire group of 26 first-instar larvae over a period of six hours (Longino, 1984). Other predators, such as pentatomid bugs (*Podisus aenescens* and *Oplomus nigripennis*), can also consume large numbers of larvae.

Other parasitoids include flies in the genus *Myothyriopsis* sp. (Tachinidae), which oviposit in larvae, and chalcidid wasps, which attack the pupae, eventually emerging to leave a perfectly round hole in the pupal case. At least in Sirena, both the chalcidid and *Myothyriopsis* are generalist, feeding on several *Heliconius* species (Longino, 1984).

Experiments with another nymphalid butterfly, *Eunica bechina*, have shown that the presence of fake rubber ants can deter egg-laying, demonstrating that adult females use visual cues to discriminate against ant-tended plants (Freitas & Oliveira, 1996). Although it has been commonly stated in the literature that ants deter oviposition by *Heliconius* (e.g. Benson, 1978), this has not been demonstrated quantitatively, and there are mixed reports. In one observation, a female *H. hewitsoni* was seen repeatedly attempting to lay eggs on a shoot attended by aggressive *Azteca* ants. Each time she was attacked by the ants, she leapt up into the air, before finally giving up after 30 minutes (Longino, 1984)! In contrast, however, most *H. hewitsoni* females were not deterred by the presence of less aggressive ant species (Longino, 1984), and larvae are commonly observed in the field coexisting with ants. It would clearly be interesting to repeat host-choice experiments involving ants with different *Heliconius* and *Passiflora* species.

The butterflies do have some evasive strategies that appear to have evolved to reduce ant predation. The common habit of laying eggs on the very tip of young tendrils or on dead tendrils seems likely to be such a strategy—these structures are less likely to be patrolled by ants. Similarly, the early-instar caterpillars of *Dryas iulia* often cut out a small strip of leaf on which they are feeding, which dries up and hangs down, and then rest at the end of this (Figure 3.2) (Mega & de Araújo, 2008). Presumably, pieces of dead vegetation such as this are also less likely to be visited by ants.

3.13 Passion flower life history

Attack by *Heliconius* herbivores has probably influenced many aspects of *Passiflora* life history. Perhaps the most striking and best documented is the growth habit of woody vines in the *Astrophea* subgenus. Species such as *P. pittieri* have a very sporadic growth pattern. They sit apparently dormant for long periods—up to 84 days in one case—before producing an extremely rapidly growing shoot that tries to grow towards the canopy before being found by *Heliconius* females (Longino, 1984). In their turn, the butterflies learn where all the vines are in their home range and visit daily to check for new growth. Their long lifespan and good spatial memory allow females to play a sit-and-wait game with the host plants, each hoping to outlast the other in a game of patience.

3.14 Phylogenetic patterns of host use

In their classic paper on coevolution, Ehrlich and Raven (1964) used *Heliconius* as an example to support their hypothesis that coevolutionary interactions between hosts and herbivores can promote diversification. Nonetheless, the relationship between host-plant use and diversification is fairly diffuse, at least within the Heliconiini.

The colonization of Passifloraceae occurred before the radiation of the Heliconiini. The Asian genus *Cethosia* is part of the Acraeini, sister clade to the Heliconiini, and also feeds on *Passiflora* (Wahlberg et al., 2009). Both *Cethosia* and *Acraea* have been shown to contain the same cyanogenic glycosides as are synthesized by *Heliconius* (Nahrstedt & Davis, 1983). Although *Passiflora* feeding and the biochemistry needed to synthesize cyanogens predated the radiation of the Heliconiini, it is unclear which came first: did the biochemistry required to feed on *Passiflora* facilitate synthesis of cyanogens by the butterfly, or vice versa? Neither innovation is unique in the Lepidoptera, as the burnet moths (Zygaenidae) also synthesize cyanogens (Davis & Nahrstedt, 1987), and many species feed on cyanogenic plants.

Molecular phylogenies of both *Heliconius* and *Passiflora* are now available (Yockteng & Nadot, 2004; Hansen et al., 2006; Beltran et al., 2007; Kozak et al., 2015), permitting a more systematic analysis of patterns of coevolution between these taxa (Kozak, 2015). Analyses before the development of phylogenetic comparative methods suggested some correspondence between the phylogenetic relationships of butterflies and their hosts. Thus, the melpomene/silvaniform clade primarily feed on the *Granadilla* subgenus; the wallacei clade on the *Distephana* subgenus; the erato clade on *Decaloba*; the most specialist are sapho clade species that feed on *Astrophea* and other large woody vines. This pattern is evident in patterns of host use in Panama, for example (Table 3.1).

However, analysis of all host records in the literature does not provide much support for this pattern—there are many host records for *H. melpomene* group species on *Decaloba* and *H. erato* on *Passiflora* (Table 3.2). Broadly, there seems to be little correspondence between *Heliconius* and *Passiflora* phylogeny. The current data on host use is far from perfect, and some of the records may be erroneous or represent rare hosts used only occasionally. A more quantitative and systematic survey of host use, especially in the poorly studied Amazon, would be extremely useful and might reveal patterns not evident in these data.

Nonetheless, quantitative analysis of dietary conservatism does show patterns across the major *Heliconius* clades (Figure 3.6). Krzystof Kozak (2015) used a measure of dietary conservatism that varies from 0 to 1; for example, all three Neruda species get a maximum conservatism score of 1 because they all feed on *Dilkea* spp. (the different species of *Dilkea* are generally not distinguished in the *Heliconius* literature).

In contrast, *Eueides procula* feeds on multiple *Decaloba* and *Passiflora,* and none of its preferences are shared with close relatives, so it has a conservatism score of 0. Across the *Heliconius* tree, erato and sapho clades show much more dietary conservatism than melpomene and silvaniform clades. The latter seem to happily feed on virtually all lineages of *Passiflora* and switch host plants with little apparent restriction, even between closely related species. Erato and sapho clade species, however, are much more likely to feed on *Passiflora* species similar to their close relatives.

The different strategies of cyanogen sequestration versus synthesis described by Engler may go some way to explaining this pattern—in the melpomene clade, synthesis is more important than sequestration, so these species are perhaps less constrained by the chemistry of their hosts and can therefore switch hosts more easily (Engler-Chaouat & Gilbert, 2007).

Consistent with these patterns of conservatism, there is broad evidence for codivergence. In other words, there is some correspondence between the branching patterns in the *Heliconius* and the *Passiflora* phylogenies—correspondence between the two phylogenies is significantly better than chance (Kozak, 2015), implying that interactions between butterflies and *Passiflora* are not entirely random. This is hardly surprising. It would be remarkable if the two clades were completely randomly distributed—this would be equivalent to every butterfly speciation event being associated with a completely random new choice of host plant(s). Even so, there are few consistent patterns. The best solution consists of many examples of a butterfly not diverging despite speciation of its host, and also multiple butterfly speciation events without host shifts.

In summary, although there is *some* correspondence between the butterflies and the plants they feed on, the overall pattern is one of remarkable flexibility. Butterflies readily switch between different host-plant lineages, and there is little phylogenetic evidence for coevolution. The lack of correspondence between host and butterfly phylogeny therefore provides little evidence for either the Ehrlich and Raven (1964) 'escape and radiate' hypothesis or alternative models of host-parasite cospeciation.

Nonetheless, trait-based studies show clear evidence for coevolution. There are specific defences clearly aimed at *Heliconius*, such as egg mimics,

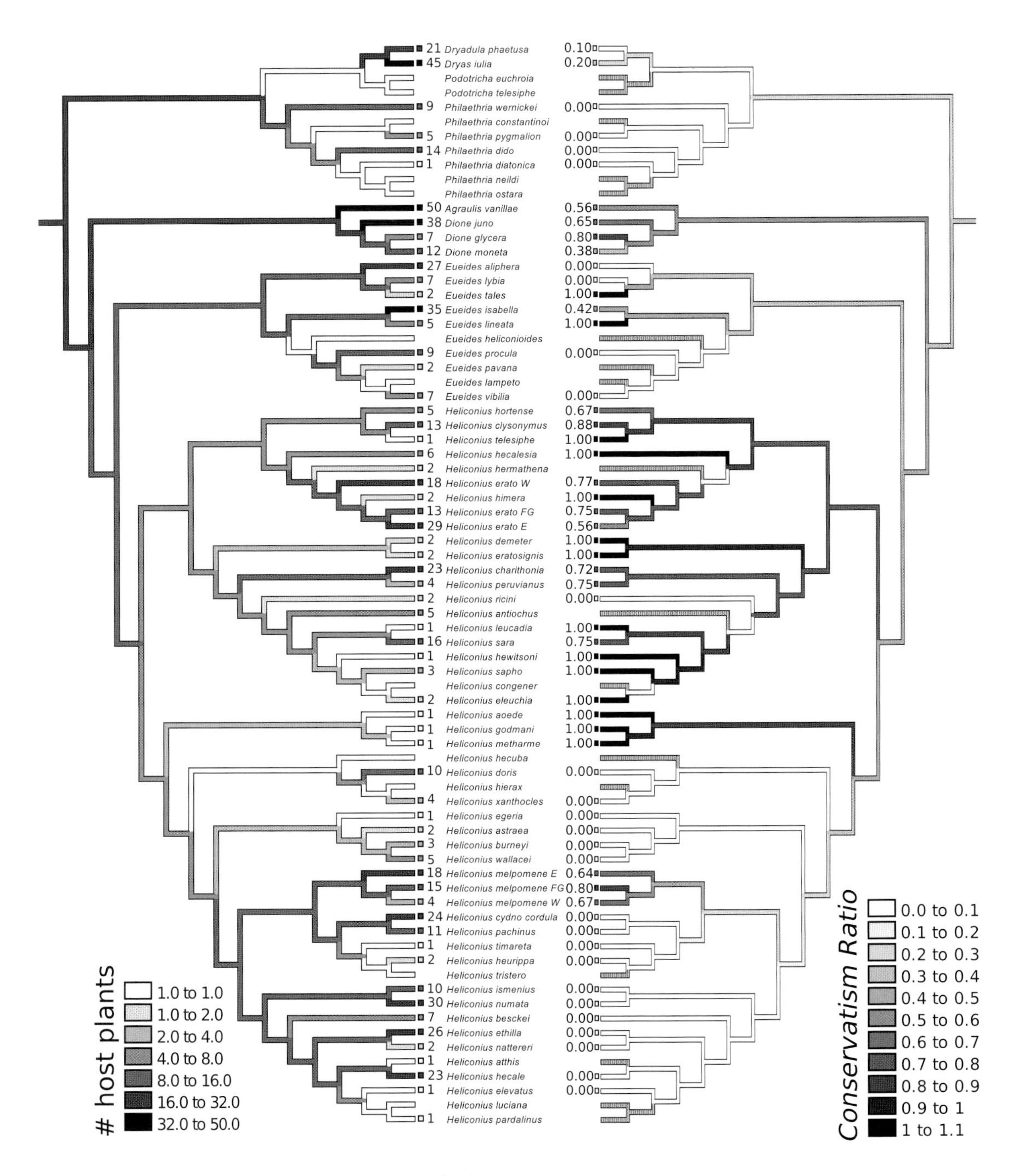

Figure 3.6 Diversity of host-plant specialization and extent of codivergence

(A) Number of host species used by different *Heliconius* species. There is a great deal of variation but no clear phylogenetic pattern in the number of *Passiflora* species attacked. In fact, geographic range area of the butterfly species is quite a good predictor of dietary range, suggesting that butterflies opportunistically exploit *Passiflora* species available across their range. (B) Conservatism in host use across the butterfly phylogeny. There is not a strong concordance between the phylogeny of the butterflies and their host plants, as measured by the conservatism ratio, although it is higher in the erato/sapho clade, indicating that phylogeny predicts host-plant preference most strongly in this group.

and evidence for reciprocal evolutionary change in both butterflies and hosts, such as the trichomes in *Passiflora adenopoda* and the evolution of strategies to overcome such defences in *H. charithonia*. It seems that trait-level coevolution does not necessarily translate into phylogenetic patterns of coevolution.

3.15 Future directions

The *Heliconius–Passiflora* interaction represents one of the best-studied tropical coevolutionary interactions, and provides excellent examples of coevolutionary processes. Still, most work to date on the ecology of this interaction has been carried out at just a few sites in Central America. There is a need for intensive study of the more diverse communities found in South America, especially in the Amazon basin. This will be logistically challenging and perhaps difficult to support financially, but would undoubtedly provide novel insights into niche partitioning and patterns of competition.

We are only just beginning to touch on the chemical and molecular mechanisms underlying the interaction of butterflies with their host plants. Perhaps the most promising area is the evolution of host chemistry. Studies in other tropical taxa suggest that defensive compounds can evolve extremely rapidly even between closely related species and can show parallel changes between lineages. It is likely that the apparent lack of coevolution at a phylogenetic level between butterfly and plant actually hides fast-evolving coevolutionary interactions at the level of host chemical defence. With recent advances in genomic studies of *Heliconius*, there is the potential for identification of genes involved in host-plant detoxification and cyanogen synthesis. There are already strong candidates for the latter from recent work on other insects, so this is a promising area for future work.

Inter-individual variation, both in the butterflies and the hosts, remains poorly understood, but holds the key to understanding the origin of coevolved traits such as egg mimics and chemical defence. In other butterfly species, individual variation in host quality can strongly influence female preference (Kuussaari et al., 2004). In *Heliconius*, we would like to know why butterflies select one particular host individual over another, and which cues are used. Particularly in generalist species such as *H. cydno* and *H. numata*, there is potential for behavioural flexibility in host use, either as a result of plastic or genetic variation. Is there evidence for individual-based alternative host strategies in these species? If they exist, cryptic 'host strategies' within generalist populations might provide insights into the early stages of host specialization. At a population level, it would be fruitful to look for evidence of local adaptation between host and herbivore, the so-called 'geographic mosaic of coevolution' that has provided important insight into coevolution in other species (Thompson, 1999).

There have been fairly frequent switches between a solitary and gregarious host-plant use strategy. This has involved parallel changes in larval morphology, feeding behaviour, and oviposition behaviour. It seems likely to have implications for larval immune systems, with much greater exposure to disease in gregarious caterpillars. This is therefore a promising area for investigation of parallel genetic changes associated with the evolution of gregariousness.

CHAPTER 4

The pollen: adult resources and life history evolution

Chapter summary

Heliconius adults have unusual feeding habits that have major implications for their life history and behaviour. Uniquely among butterflies, they systematically collect pollen and have evolved specialized behaviours to extract its amino acids. Many butterflies gain most of their nutrition from food the caterpillars eat, with adults limited to feeding mainly on sugar-rich nectar. However, pollen feeding in adult *Heliconius* has led to a shift from reliance on larval resources to a shorter larval development time and increased reliance on adult feeding. Compared to other butterflies, *Heliconius* have a much longer adult lifespan and more continuous reproduction. Pollen contributes nutrients to egg production in females and spermatophore production in males.

Heliconius saliva is rich in protease enzymes that must play a role in pollen digestion. There are also behavioural adaptations to pollen feeding; in particular, prolonged probing of flowers in response to the presence of pollen grains, and 'pollen processing' behaviour that involves release of saliva and repeated curling and uncurling of the proboscis to aid external digestion of pollen grains, before sucking up the rich juices. Surprisingly, however, it has been hard to identify any structures in the butterflies specifically adapted to pollen feeding. In contrast, the *Psiguria* vines that provide a major source of pollen for *Heliconius* are uniquely adapted to provide a reliable high quality pollen supply in return for pollination services. They produce regular flowers with large amounts of pollen that are visited by traplining *Heliconius* adults. There is evidence of niche partitioning among *Psiguria* vines according to flowering time during the day. This is an excellent example of coevolution between the plant and its pollinator.

4.1 Life history theory

Life history theory is a branch of evolutionary ecology that seeks to explain how organisms evolve to optimize their survival and reproduction in the face of diverse ecological challenges (Fabian & Flatt, 2012). Specifically, life history traits include factors such as the time animals take to grow, when they become mature, how many offspring—and of what size—they produce, and how long they live. Taken together, these patterns of growth, maturation, reproduction, and senescence make up an animal's life history.

There are trade-offs between these traits; for example, between longevity and early reproduction. High extrinsic mortality tends to favour reproduction at a younger age—if you are likely to be eaten by a predator, then it is better to get on with reproducing early to avoid missing out, but if you are safe from predators, then it may be better to live longer and invest in more high quality offspring. These intuitive predictions are supported both by theory and experimental evolution experiments, for example, in fruit flies (Stearns et al., 2000). *Heliconius* live long lives by butterfly standards, in part because their warning colours and chemical defences keep

The Ecology and Evolution of Heliconius Butterflies. Chris D. Jiggins, Oxford University Press (2017).

DOI 10.1093/acprof:oso/9780199566570.001.0001

them safe from predation, making it worthwhile to invest in long-lived tissues.

Another important axis of variation in life history among animals is the number of reproductive events. A female north Pacific giant octopus (*Enteroctopus dofleini*) lives for three to four years, after which it lays thousands of eggs in a single bout and then dies, a strategy known as semelparity. Similarly, many butterfly species live for just a few days or weeks and lay a large number of eggs in a short burst. In contrast, most *Heliconius* butterflies are highly iteroparous—in other words, they can live up to 6–8 months and continuously produce eggs daily throughout this lifespan. Life history theory predicts that low juvenile mortality but high adult mortality should favour semelparity. In contrast, a risky juvenile lifestyle, combined with a safe adult environment should favour iteroparity. *Heliconius* fit this prediction. Their high juvenile mortality at the hands of ants and parasitoids, but relatively risk-free adulthood, means that it makes sense to spread out reproduction over time. This acts as a bet-hedging strategy to maximize the chances of at least some offspring surviving.

Life history theory is sometimes characterized as resembling a pie (Fabian & Flatt, 2012). The pie represents the resources available to an individual during its lifespan, which can be partitioned in different ways; for example, towards current versus future reproduction, or current reproduction versus longevity. *Heliconius* have one unique adaptation that has broken this constraint and substantially enhanced the size of the pie itself, by making available a novel resource. This is pollen feeding, a key adaptation that influences all aspects of *Heliconius* ecology. Here I will first describe this behaviour and the morphological and behavioural adaptations that make it possible, and then investigate the life history implications of pollen feeding.

4.2 Pollen feeding by adult butterflies—a key innovation

Gilbert (1972) first recognized that adult *Heliconius* systematically collect pollen from flowers, providing a source of amino acids to adults. Unlike other butterflies visiting the same flowers, *Heliconius* accumulate large pollen loads on their proboscis. Pollen collection involves distinctive behaviours, including repeated probing of flowers with the proboscis and extended flower visits, with butterflies often spending several minutes at a single flower. After visiting one or more flowers, the butterfly perches to process the pollen, which is now collected in dry clumps along and under the proboscis.

Processing involves excreting saliva onto the pollen; the saliva contains proteases that help digest the pollen (Eberhard et al., 2007; Harpel et al., 2015). Repeated curling and uncurling of the proboscis serves to thoroughly mix the pollen with this liquid and agitate the mixture. The dry clumps of loose pollen become a single consolidated mass under the proboscis. The processing of pollen can last up to several hours, during which time the liquid is ingested, with the pollen-derived amino acids now in solution. The remaining solid matter is finally discarded. Ingeniously, therefore, pollen feeding provides access to a nutritious resource without the need to ingest solid material—which is obviously an impossibility for butterflies possessing only a proboscis. Most pollen processing occurs during the day immediately following flower visits, but pollen processing has also been observed at nocturnal roosts in *H. sara* females (Salcedo, 2010a).

Using artificial flowers, Gilbert (1972) demonstrated that adult *Heliconius erato* collected and processed pollen-sized glass beads in a similar manner to real pollen. Butterflies spent far longer at artificial flowers supplied with either pollen or glass beads, as compared to sucrose solution or pollen extracts. This demonstrates that the butterflies are responding to the physical presence of pollen, rather than accumulating pollen simply as a side effect of nectar collection. Furthermore, experiments with radio-labelled isotopes showed that nutrients are transferred from pollen to eggs (Gilbert, 1972). This transfer of nutrients is more than just a luxury: without pollen, females show drastically reduced egg production, teneral males are unlikely to mate, and lifespan is greatly reduced (Gilbert, 1972; Dunlap-Pianka, Boggs, & Gilbert, 1977; Jiggins, personal observation; O'Brien, Boggs, & Fogel, 2003).

All *Heliconius* species feed on pollen, including *Heliconius doris*. Related genera including *Eueides* and *Podotricha* do not pollen feed, and the behaviour is otherwise unknown among the butterflies.

The origin of this behaviour therefore represents a single evolutionary event that has been considered a 'key innovation' in the diversification of the genus (but see Penz, 1999 and Chapter 11). The one anomaly is the Neruda clade (*Heliconius aoede, H. metharme, H. godmani*), which do not pollen feed (Brown, 1981; Neil Rosser, personal communication). Molecular analyses clearly support these species as part of the genus *Heliconius*, suggesting that pollen feeding has been secondarily lost (Beltran et al., 2007; Kozak et al., 2015).

This phylogenetic distribution would seem to imply that some unique evolutionary innovation has permitted pollen feeding in this group but not others. It is therefore surprising to discover that there are no obvious morphological features unique to pollen-feeding butterflies. Instead, as far as we can tell, adaptation to pollen feeding has involved rather slight modification of existing structures. For example, pollen-feeding species have a longer proboscis and longer hair-like structures (known as bristle-shaped sensilla trichodea) on the proximal (upper) regions of the proboscis (Figure 4.1) (Krenn & Penz, 1998). These presumably facilitate extraction of pollen from deep flowers while keeping pollen attached to the proboscis. In contrast, some of the other structures originally proposed as being adapted for pollen feeding turn out to lie within the range of variation in non-pollen-feeding relatives (Krenn & Penz, 1998). The salivary glands of pollen-feeding species, which are presumably required to produce specialized saliva containing digestive enzymes (Eberhard et al., 2007), are also identical to those of non-pollen-feeding species (Eberhard & Krenn, 2003, 2005). Overall, therefore, it is striking how little morphological adaptation seems to have occurred.

There is some evidence of biochemical adaptation to pollen feeding. Recent proteomic analyses indicated a suite of proteins found in *Heliconius* saliva during pollen feeding with predominantly proteolytic activity (Harpel et al., 2015). Ten proteins likely to be involved in proteolysis included serine proteases, cysteine proteases, astacins, and a carboxypeptidase. This confirms colorimetric analysis showing strong proteolytic activity in *Heliconius* saliva (Eberhard et al., 2007). In contrast, the honeybee salivary proteome is dominated by enzymes involved in carbohydrate metabolism (Feng et al., 2013), as might be expected for a nectar-rich diet. Although this comparison suggests adaptation of the butterfly proteome to pollen feeding, a direct

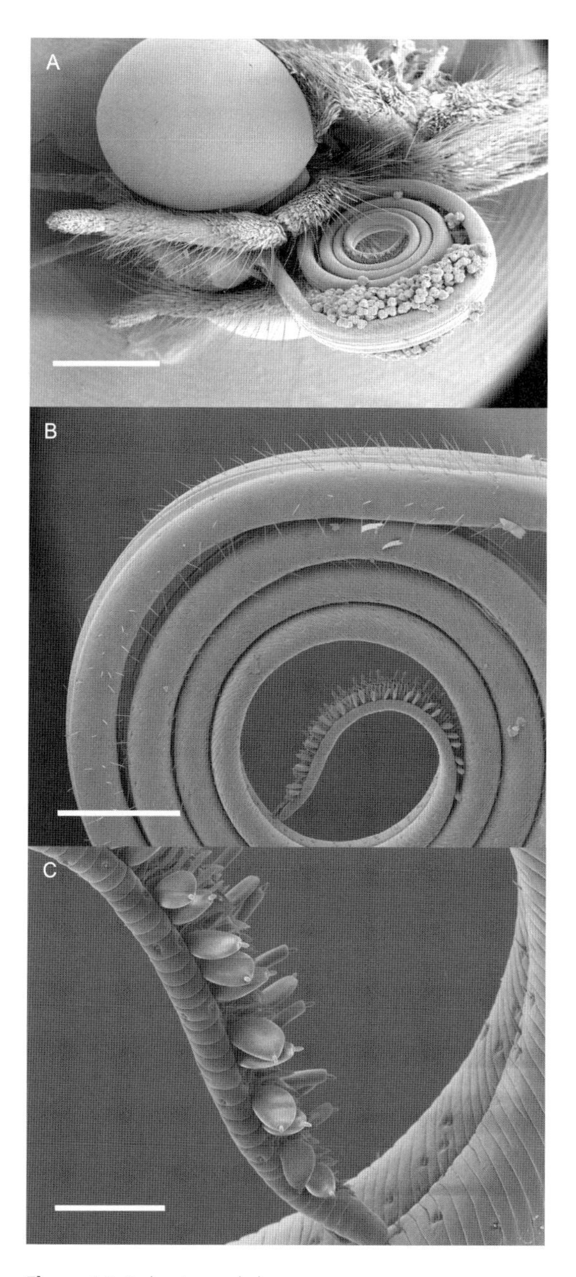

Figure 4.1 Proboscis morphology

(A) Head and mouthparts of *Heliconius ethilla* with left labial palpus removed and proboscis in resting position (scale bar = 1 mm). (B) Enlarged image of proboscis of *H. melpomene* showing proximal region of the proboscis with long bristle-shaped sensilla trichodea (scale bar = 500 μm). (C) Tip region of the proboscis in *H. melpomene* (scale bar = 100 μm). Photos courtesy of Harald Krenn.

comparison of the proteome of pollen-feeding versus non-pollen-feeding butterfly species is needed, perhaps combined with analysis of the molecular evolution of pollen-feeding associated enzymes, in order to test whether this dietary change has really led to significant adaptation.

There is also behavioural adaptation to pollen feeding. A study of ten species feeding on the generalist open corolla flowers of *Lantana camara* showed that pollen-feeding species spent longer, with more proboscis thrusting, at a single floret as compared to non-pollen-feeding species (Penz & Krenn, 2000). In fact, the non-pollen-feeding species actually removed more pollen, but this generally fell off the proboscis in dry clumps—in contrast to *Heliconius* species, which kept the pollen moist and stuck onto the proboscis (Figure 4.2). There is a need to extend these quantitative behavioural studies of pollen feeding to other flower species such as *Psiguria*, which are specifically adapted to *Heliconius* feeding. Overall, then, pollen feeding is an adaptation that has involved a suite of rather slight modifications to existing morphology and behaviour in the butterfly. Along with corresponding adaptations in the flowers that provide pollen (see later), these allow a more systematic exploitation of an existing floral resource—with far-reaching implications for all aspects of *Heliconius* biology.

Figure 4.2 Pollen-feeding behaviour (see Plate 8)

(A) *H. melpomene aglaope* processing a pollen load. Note the large-grained *Psiguria* pollen on the proboscis. (B) The first flower on an inflorescence of *Psiguria warcsewiczii* is relatively large, and (C) subsequent flowers get smaller. (D) A single inflorescence of *Psiguria bignoniaceae* with over 250 flower scars, indicating almost a year of continuous flower production.

4.3 Ecology of pollen feeding

Heliconius adults feed on a wide variety of flowers, and it is relatively simple to study pollen use. Pollen loads can be readily observed (or collected from wild butterflies without harm), pollen types identified, and the size of the pollen load recorded. Early studies of pollen use classified pollen by colour (Boggs, Smiley, & Gilbert, 1981; Cardoso, 2001), but later studies identified pollen types under a microscope (Murawski, 1986; Estrada & Jiggins, 2002). In Panama, there is a very useful guide, which makes pollen identification easier (Roubik & Moreno, 1991).

For example, from the pollen loads of seven species of butterfly, we found 32 different pollen types in Pipeline Road, Panama. Only a small proportion of these were common, the principal species being *Psychotria poeppigiana* (previously *Cephaelis tomentosa*), *Psiguria* spp. (probably mostly *P. warcsewiczii*), *Lantana camara*, *Tournefortia* sp., and *Manettia reclinata* (Estrada & Jiggins, 2002). Virtually all studies of pollen loads have shown significant differences in pollen use between species, suggesting a degree of niche partitioning in pollen use, similar to that seen for larval hosts. Although the details of these differences and their underlying causes are currently somewhat hard to interpret, there are some emergent patterns.

There seems to be a clear difference in pollen use between major clades of *Heliconius*. In Costa Rica, the silvaniform and melpomene clades used more white, large-grained pollen species such as *Psiguria* sp. and the erato clade more yellow, small-grained species such as *Lantana camara* (Boggs et al., 1981). In contrast, in a more open habitat site in Mexico, Marcio Cardoso showed that there was no difference between species in colour of the pollen collected, although the erato group species, *H. erato* and *H. charithonia*, collected larger pollen loads than the silvaniform *H. ismenius* (Cardoso, 2001).

In Panama, we also found that the melpomene and silvaniform clade species collected more white *Psiguria* pollen than the erato clade (Estrada & Jiggins, 2002). In fact, the proportion of *Psiguria* pollen collected was primarily explained by butterfly phylogeny, while most other pollen types varied more due to the microhabitat in which the butterfly was collected (except *Lantana*, which was influenced by both factors). This pattern suggests that most pollen types are collected largely according to their local availability, except for greater specialization on *Psiguria* pollen by the melpomene and silvaniform clades. *Psiguria* inflorescences open around dawn and pollen is rapidly depleted by visiting butterflies (Murawski, 1986). Melpomene and silvaniform clade species tend to be active slightly earlier, perhaps allowing them to compete more effectively for *Psiguria* pollen.

Insectary experiments also shed some light on these patterns. In a greenhouse containing *H. melpomene* and *H. erato*, the latter was shown to collect larger loads of *Lantana* pollen (Boggs et al., 1981). In addition, large-grained *P. warcsewiczi* flowers in our insectaries were visited by *H. erato* and *H. sapho*, but these species generally didn't collect pollen loads as large as those of *H. melpomene* or *H. cydno*, suggesting that the latter may be more efficient at collecting *Psiguria* pollen (Estrada & Jiggins, 2002).

Overall, the data suggest that melpomene/silvaniform clade species specialize on *Psiguria*, and erato/sapho clades on *Lantana*. It has been suggested that the *erato* group might represent the derived strategy, having evolved to use small-grained species of pollen and Decaloba larval host plants that are both common in open habitats (Smiley, 1985b). However, we have suggested that the direction of change may have been the reverse—i.e. the melpomene and silvaniform clades have largely driven coevolution with *Psiguria* (Estrada & Jiggins, 2002) through increased dependence on this single pollen source. In support of this hypothesis, *H. hecale* (a silvaniform species) was found to be more efficient at exploiting *Lantana* than *H. erato* in insectary experiments (Penz & Krenn, 2000).

As well as broad patterns across the butterfly phylogeny, closely related species occurring together also differ in pollen use. For example, *H. melpomene* used more *Lantana camara* and *H. cydno* more *Psychotria poeppigiana* (Estrada & Jiggins, 2002). This was not explained by preference for particular flowers—insectary choice experiments did not show any difference between the species in pollen use (unpublished data). The differences are more easily explained by the microhabitat use of the two species, with *H. melpomene* found in more open areas where *Lantana* is most common.

In addition to differences between closely related species, there are also differences between individuals. Older individuals tend to collect larger pollen loads, which might be due to greater nutritional demands, or to more efficient foraging gained from experience of the local environment (Boggs et al., 1981). Females tend to collect larger pollen loads than males, presumably due to the greater nutritional demands of egg production (Boggs et al., 1981; Cardoso, 2001; Mendoza-Cuenca & Macías-Ordóñez, 2005). Females are also more commonly found on flowers early in the morning and may gain a competitive advantage by foraging more efficiently when the *Psiguria* flowers first open.

4.4 Coevolution with Psiguria and Gurania vines

While pollen feeding may not have involved drastic modification of butterfly morphology, the same cannot be said for the vines that have coevolved with them. Two genera of vines in the cucumber family (Cucurbitaceae), *Psiguria* (around 16 species) and *Gurania* (around 40 species), provide large quantities of pollen to *Heliconius* and also exploit the butterflies for their own pollination. The two genera are easily distinguished, as *Gurania* flowers have a succulent, brightly coloured calyx tube and narrow, succulent yellow petals, while *Psiguria* have a green calyx and broad spreading colourful petals (Condon & Gilbert, 1990). *Psiguria* is most closely coevolved with *Heliconius*, while most *Gurania* probably rely to a greater degree on hummingbird pollination. Some *Gurania* flowers with long spiky petals and calyx lobes may have evolved to deter visitation by *Heliconius* in favour of hummingbirds (Condon & Gilbert, 1990).

Male *Psiguria* inflorescences are uniquely adapted as a pollen source. A single flower on each inflorescence opens almost every day, or in some species every 2–3 days. Each flower provides large quantities of large-grained pollen. Each inflorescence can last for several months or even years, providing a reliable source of pollen for long-lived *Heliconius* butterflies. One *P. umbrosa* plant grown in an insectary in Texas produced 10,000 flowers in a year, providing the equivalent of 20 g of pollen and 145 g of sucrose. A single inflorescence collected in the Carare Valley, Colombia, had 450 scars indicating past flowers and 35 buds, representing at least a year of continuous flowering (Gilbert, 1975).

When an inflorescence is young, the petals on each flower are large—up to 2 cm across on *P. warcsewiczi*, for example. This serves as an advertisement to attract butterflies. In subsequent flowers, however, the petals become smaller. The excellent spatial memory of long-lived *Heliconius* adults means that advertisement is no longer necessary. Having found the site, the butterflies can find the same inflorescence day after day without needing showy petals to guide them. Nectar production is also greater in earlier flowers produced on an inflorescence (Condon & Gilbert, 1990). Thus the vine avoids the cost of advertisement by maintaining loyal customers.

A significant problem faced by the vine is how to entice butterflies to visit female flowers and thereby ensure pollination, even though they provide no pollen reward. This is achieved primarily by mimicry, with female and male flowers having very similar appearance, and also by the monoecious, sex-changing nature of the plants (Condon & Gilbert, 1990). Thus, small vines produce male flowers, and the great majority of the flowers found in the forest are male (in one study the sex ratio was 32:1; Murawski & Gilbert, 1986). However, occasionally, a single branch on a large plant, or even a whole plant, will switch to producing female flowers.

Thus, female flowers are rare and produced by plants that are likely to have already been visited regularly by *Heliconius*. The butterflies probably visit female flowers in the hope of collecting pollen, but suffer little cost from doing so, as they are already visiting male flowers in the vicinity. Female flowers also generally provide more nectar than male flowers, perhaps compensating for the absence of pollen (Condon & Gilbert, 1990). Indeed, in Sirena, it has been shown that female and male flowers do not differ in visitation rates (Murawski & Gilbert, 1986), suggesting that *Heliconius* do not discriminate against female flowers.

Murawski studied *P. warcsewiczi* pollen dispersal using coloured dyes (Murawski & Gilbert, 1986). *Psiguria* flowers were visited mainly by five *Heliconius* species, but also less frequently by hummingbirds (primarily *Phaethornis longuemareus*). *Heliconius* dispersed pollen significantly further and to more individual plants than hummingbirds. Individual pollen movements were detected up to 350 m from the source flower; for the most dispersive species, *H. ismenius*, average pollen dispersal distance was 75 m (Murawski & Gilbert, 1986). Thus, for these low-density rainforest vines, mutualism with *Heliconius* provides an effective means of long-distance pollen dispersal.

Interestingly, the longest recorded pollen dispersal by an insect is from tropical fig trees, where a tightly evolved mutualism with fig wasps achieves individual pollen dispersal across distances over 10 km (Nason, Herre, & Hamrick, 1998). Similarly, mean pollen dispersal distance is around 200 m for animal-pollinated tropical tree species (Ward et al., 2005). Thus, coevolved pollination mutualisms are likely to be a general phenomenon in tropical rainforests, where individuals are typically found at low densities such that broadcast pollination methods such as wind are ineffective, but high species diversity offers many opportunities for the evolution of mutualisms.

In Trinidad, Gilbert noticed a *Heliconius* visiting a *Psiguria triphylla* flower in the early afternoon, while ignoring a nearby *Gurania spinulosa* flower that had been visited earlier in the day (Gilbert, 1975). This led to the suggestion that different *Psiguria* and *Gurania* species coexisting in the same habitat might release pollen and nectar at different times of day, thereby sharing the same pollinators. Observations on two sympatric species from Arima Pass support this speculation: *P. umbrosa* releases pollen around 5 a.m. with nectar production peaking by 11 a.m. whereas *P. triphylla* releases nectar at 9:30

and peaks nectar production at 6 p.m. Evidence for *Heliconius* consistently visiting particular flowers at different times of day suggest a link between circadian rhythm and spatial learning, such that they are adapted to exploit floral resources with different daily rhythms (Gilbert, 1975). Thus, 'circadian isolation' may contribute to reproductive isolation between *Psiguria* species, and perhaps make ecological coexistence easier.

Other flowers may be adapted to some extent for pollination by *Heliconius*, including some species in the families Rubiaceae and Boraginaceae (Gilbert, 1991), but the degree of coevolution is considerably less than in *Psiguria* and *Gurania*. In return for pollination services, all these plants provide heliconiines with the resources that permit them to sustain their high-investment lifestyle, involving a long lifespan and behavioural complexity.

4.5 Life history evolution—nectar, pollen, or larval resources?

The evolution of pollen-feeding behaviour has entailed a switch from reliance on larval feeding towards an increased use of adult-derived resources. In most butterflies, adults feed primarily on nectar. This was traditionally thought to contain little apart from sugars, but it has been recognized that the nectar of butterfly-pollinated plants in particular contains a broader range of amino acids and other nutrients than that used by more generalist pollinators (Baker & Baker, 1973). In addition, some butterflies supplement adult nutrition by feeding on other food sources such as bird droppings (Ray & Andrews, 1980), and the effectiveness of rotten fish or carnivore faeces as bait in butterfly traps (the smellier the better) suggests that many species may supplement adult nutrition by opportunistic feeding on protein sources such as rotting carcasses. Pollen feeding has taken this a step further, by providing a consistent and reliable source of amino acids to adult butterflies.

The influence of pollen feeding on *Heliconius* life history is dramatically illustrated by comparison with the related non-pollen-feeding species, *Dryas iulia* (Figure 4.3). Female *D. iulia* kept in insectaries in Texas lived 15–40 days, showed an early peak of egg production of 30–50 eggs/day that decreased by 21–28 days (Dunlap-Pianka et al., 1977). All females died a couple of days after cessation of egg production. In contrast, *Heliconius charithonia* lived for several months and maintained a consistent daily egg production of around 10 eggs/day throughout their lifespan. Individuals older than 70 days continued to produce 11–12 eggs/day.

However, when *H. charithonia* females were deprived of pollen and fed only on sucrose solution, their life history became similar to that of *D. iulia*. Pollen-deprived females lived 21–38 days, laying 70–330 eggs. They lost vigour around 20 days and soon stopped laying eggs. Hence, in *H. charithonia*, pollen is essential for both continued oviposition and viability beyond about one month (Dunlap-Pianka et al., 1977).

The development of female ovarioles also follows a different trajectory in pollen-feeding species. In *D. iulia*, the first stages of egg production (mitosis and pre-growth) cease in the latter stages of life, pre-empting the end of egg production. In *Heliconius*, by contrast, once the ovarioles have matured about a week after emergence, they have a similar structure throughout life, with a constant balance between eggs in early and late stages of generation and maturation. Hence, *D. iulia* and *Heliconius* illustrate the life history contrast of 'programmed senescence versus eternal youth' (Dunlap-Pianka et al., 1977).

Pollen feeding leads to a significant increase in the reliability and quality of adult nutrition. This led to the prediction that pollen-feeding species would need to lay down fewer resources as larvae (Boggs, 1981a). Carbon is generally readily available to butterflies as nectar sugars, but nitrogen is more commonly a limiting resource. Hence, a study of nitrogen resource allocation was carried out to test this prediction. As predicted, the most intensive pollen-feeding species, *H. cydno*, had the lowest proportion of abdominal nitrogen on emergence from the pupa; the more opportunistic pollen-feeder *H. charithonia* was intermediate; and the non-pollen-feeding *D. iulia* had the highest proportion of abdominal nitrogen (Boggs, 1981a). Nonetheless, studies of lifetime reproductive effort suggest that in one *Heliconius* species at least (*H. charithonia*), ongoing nutrient income from pollen feeding is less

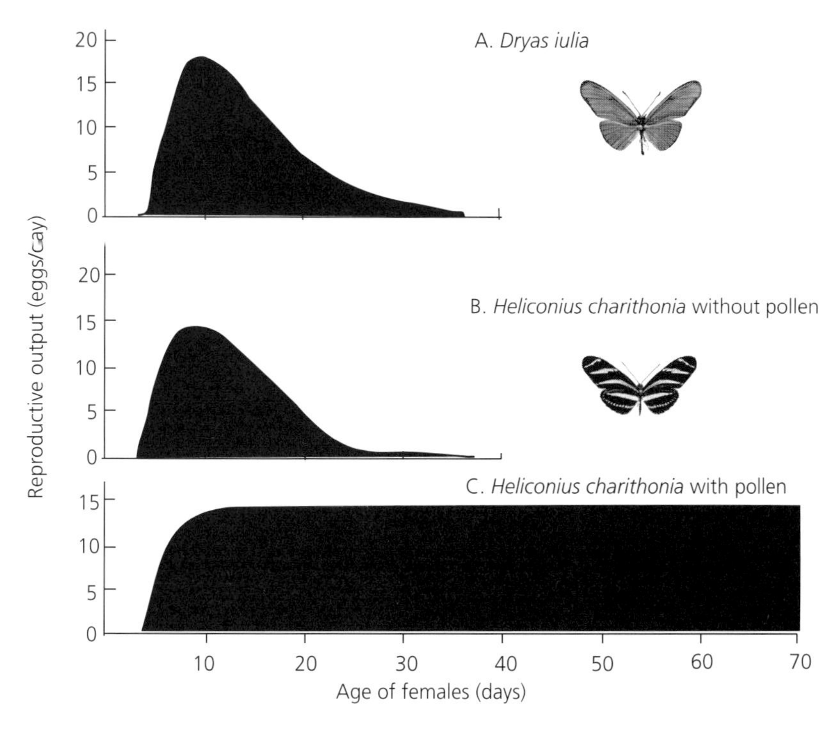

Figure 4.3 Pollen-feeding life history

Pollen feeding alters the life history of *Heliconius* species, permitting increased longevity and continuous egg production (Dunlap-Pianka et al., 1977). Daily oviposition patterns are shown as averages combining data from several females, with data smoothed by plotting a 2-day moving average. (A) *Dryas iulia*, a non-pollen-feeding species; (B) *Heliconius charithonia* reared without pollen; (C) *Heliconius charithonia* reared with pollen.

than reproductive output (Boggs, 1997). Body mass declined throughout adult lifespan, suggesting that reproduction continues to depend to some degree on stored reserves from the larval stage (Boggs, 1997). It isn't clear whether the same would also be true for more intensive pollen feeders such as *H. cydno*.

Furthermore, the *Heliconius* lifestyle involves a shift in the relative riskiness of larval and adult stages. As is probably the case in most butterflies, larvae incur high levels of predation and parasitism, but the *Heliconius* adult stage is characterized by a better quality of life. Pollen feeding provides nutrition, while distastefulness and warning colour reduce predation rates and the costs involved in evasive flight and crypsis (hiding is expensive in terms of lost opportunities incurred while pretending to be a dead leaf). Selection should therefore favour more rapid larval development (Gilbert, 1972). This prediction has not been tested explicitly, but *Heliconius* species that feed on young shoots generally develop faster as larvae than the related heliconiines (e.g. *Philaethria* and *Eueides*), whose larvae feed on older leaves and whose adults do not pollen feed.

O'Brien, Boggs, and Fogel (2003, 2004, 2005) have developed a clever technique to use carbon isotope ratios ($^{12}C/^{13}C$) to study carbon resource allocation in adult Lepidoptera, and quantify the contribution of different nutritional sources to egg production. Their method relies on the fact that cane sugar (from C4 photosynthesis) has a different isotopic ratio than beet sugar (from C3 photosynthesis). If the isotope ratio of the larval host plant is also known, then comparison of egg isotope ratios for females fed on beet sugar with those fed on cane sugar permits estimation of the relative contribution of larval versus adult diet to eggs.

Using this technique on four nectar-feeding species, including one heliconiine, *H. charithonia*, the contribution of adult nectar feeding to egg production was shown to increase to an asymptote at around ten days after emergence. Thus, the first few days of egg production use larval resources before a stable balance between depletion of larval and adult resources is achieved. In *H. charithonia* females that were deprived of pollen the proportion of carbon from the adult diet was estimated at around 55%, intermediate among the species studied.

There was an important distinction between essential and non-essential amino acids. The essential amino acids are those that cannot be synthesized by animals because the necessary enzymatic pathways have been lost during their evolution. Of 20 amino acids, 9 are essential in all animals; arginine is essential in insects but not humans; and 2 further amino acids can be synthesized only from other essential precursors (O'Brien et al., 2005). Thus, 12 amino acids were considered 'essential' for butterflies.

In all butterfly species studied, essential amino acids were derived primarily from larval food sources, whereas non-essential amino acids were more likely to be derived from carbon acquired during adult nectar feeding (O'Brien et al., 2005). Thus, for most butterflies, adult reproduction is potentially limited by essential amino acids collected during the larval stage.

Of course, in *Heliconius* there is a third major source of nutrition—pollen feeding—that can potentially overcome the amino acid limitation. This was studied by pollen-feeding adult *H. charithonia* with C4 maize pollen, which is isotopically distinct from the larval C3 *Passiflora* diet. Comparison of pollen-fed and pollen-deprived butterflies allows an estimation of the contribution of pollen feeding to egg production. Around 17% of the carbon in essential amino acids was derived from pollen feeding, but there was no detectable contribution to non-essential amino acids (O'Brien et al., 2003). This seems a surprisingly low figure given the dramatic life history implications of pollen feeding, and may reflect the fact that maize pollen is not a natural food source, and may therefore be nutritionally poor. Unfortunately, the methods employed rely on using pollen from C4 plants, which are not natural food sources for *Heliconius*. Additionally, these methods measure carbon contributions, while pollen feeding might be important primarily as a source of nitrogen. Nonetheless, the research of O'Brien and colleagues represents the first attempt to quantify the relative contributions of larval host plant, nectar, and pollen to reproduction in *Heliconius*.

4.6 Future directions

The 'key innovations' that characterize *Heliconius*—notably pollen feeding, but also feeding on young shoots, longevity of adults, behavioural complexity, and adult warning colour—have all evolved only once in this group of butterflies, making it difficult to tease apart cause and effect. Did pollen feeding and improved adult nutrition permit shortening of the larval stage and a switch to meristem feeding, or vice versa? Some of these questions may be impossible to answer, but there is plenty of scope for investigating the fitness consequences of the traits described here.

For example, pollen availability likely fluctuates between seasons and between localities across species ranges, as well as between closely related species. Variation in resource availability must have implications for the butterflies in terms of resource allocation and life history investment in reproduction versus survival, including, for example, investment in cyanogenic defensive compounds. Patterns of local and individual adaptation could therefore provide a rich arena for testing life history allocation theory, in a similar manner as has been applied in other animals including butterflies (Hanski, Saastamoinen, & Ovaskainen, 2006).

Heliconius saliva is rich in proteases, but it is not clear whether this represents an adaptation specifically to pollen feeding. A first step would be to make a direct comparison with non-pollen-feeding relatives such as *Dryas iulia* to see whether proteolytic activity is greater in *Heliconius*. Similarly, gene expression studies could determine whether protease enzymes are specifically expressed in salivary glands, or whether they are enzymes also used in other contexts. A molecular approach could therefore offer insight into pollen-feeding adaptations, and might identify vital innovations that morphological analyses have failed to identify.

The change in life history in *Heliconius* is a classic shift from a 'live fast, die young' life history towards increased longevity. There is currently great interest in the evolution of ageing. The contrast between short-lived genera such as *Dryas* and the longer-lived *Heliconius*, or between species within *Heliconius* with differing life histories, could provide a fruitful area for investigating the genetic basis for longevity.

Finally, while broadly the interaction of *Psiguria* and *Heliconius* provides an excellent example of coevolution, there are many open questions regarding the pollination ecology of *Heliconius* and their floral partners. How important are butterflies as compared to hummingbirds for the pollination of different *Psiguria* and *Gurania* species? Do butterflies with greater cognitive abilities make better pollinators, which might indicate coevolutionary interactions driving increased brain size in the butterflies? Is there any evidence for a 'geographic mosaic' of local adaptation in coevolutionary interactions between butterflies and flowers?

CHAPTER 5

Roosts and traplines: patterns of dispersal and movement

Chapter summary

The scale of dispersal is an important parameter for understanding diversification and population dynamics. In mark-recapture experiments, *Heliconius* show strong site fidelity, commonly returning to the same location, suggesting low dispersal rates. However, experiments have shown that there is an early dispersal phase prior to establishment of a home range, during which individuals can move large distances. Although dispersal estimated from a release-recapture experiment was ~296 m, indirect estimates from hybrid zones suggest between 2 and 4 km dispersal per generation. This is consistent with genetic data that show little evidence for local population structuring. Another unusual behaviour is gregarious roosting, in which individuals aggregate at night. Several hypotheses have been proposed to explain this behaviour, including group defence against predation and information sharing. There is experimental evidence that aggregation acts to enhance the protective effect of warning colours against predation and so has likely evolved primarily to afford protection from predators, although it probably also has other social functions.

Patterns of dispersal and gene flow across a species range are fundamental to understanding a wide range of evolutionary processes. If individuals do not disperse readily, it can lead to local clusters of related individuals, which may promote the evolution of cooperative behaviours through kin selection. Low levels of dispersal can also facilitate local adaptation, allowing populations to become genetically specialized to their local environment. However, low dispersal can also lead to inbreeding and a lower overall effective population size, meaning that populations are genetically less variable.

In contrast, more dispersive species are less likely to become locally adapted, unless selection is strong enough to overcome the mixing of genes caused by high gene flow. High dispersal can be associated with a higher overall effective population size and greater genetic variation. Species that harbour more genetic variation might be better able to adapt to a changing environment. The patterns of gene flow across a species range depend on the summed influence of decisions made by individual animals about whether to disperse or to remain and reproduce in their natal range. The behavioural tendency towards dispersal is therefore an important evolutionary parameter.

At the end of the previous chapter, I summarized the life history implications of pollen-feeding behaviour. This unusual resource allows *Heliconius* to shift to a long-lived, high-quality lifestyle. Much of this chapter concerns the behavioural changes that have evolved in association with that shift in life history strategy. These include the tendency to dispersal, as well as some unusual *Heliconius* behaviours such as traplining and fidelity to gregarious nocturnal roosts.

The Ecology and Evolution of Heliconius Butterflies. Chris D. Jiggins, Oxford University Press (2017).

DOI 10.1093/acprof:oso/9780199566570.001.0001

5.1 Longevity and site fidelity

Nick, Goliath, and Higgins were Caracas regulars, hanging out every night under the same tree, week after week for almost two months. Something was afoot. Well, actually they weren't Venezuelan Mafiosi but three butterflies (*H. charithonia*) returning to a communal roost and watched by the pioneering tropical biologist William Beebe. And, yes, he really did give them names (Beebe, 1950). This was one of the first observations that gave a hint of both the longevity and lack of dispersal that characterizes *Heliconius* populations. A couple of decades later, the first serious mark-recapture studies were carried out (Turner, 1971a; Ehrlich & Gilbert, 1973; Cook, Thomason, & Young, 1976). John Turner marked *Heliconius erato* in Trinidad, using numbers now rather than names (Turner writes—tongue firmly in cheek—that 'Beebe, within this much smaller experiment, could afford to name, rather than number, his butterflies', 1971a, p. 29). *Heliconius* are robust and can be handled without harm, and also have convenient bands on the forewing on which numbers can be written with a marker pen. Wild individuals can be captured, numbered, and released, and subsequently the location and date of recapture recorded. The degree of wing wear can also be used as an indicator of the age of an individual (Ehrlich & Gilbert, 1973). In this way, Turner documented high site fidelity, with individuals showing little movement between adjacent habitat patches just tens of metres apart, and adult longevity, with an adult lifespan estimated from recapture data of 100 days, and a maximum observed lifespan of 68 days (Turner, 1971a).

A two-year study of *H. ethilla*, also in Trinidad, came to very similar conclusions (Ehrlich & Gilbert, 1973). One individual lived for 161 days, and the average lifespan was 50 days. It seems likely therefore that the average lifespan of adults in this population is somewhere around 2–3 months. A study of *H. charithonia* in Costa Rica also documented broadly similar patterns (Cook et al., 1976). Ehrlich and Gilbert also confirmed that long-lived individuals remained reproductively active: mating was observed between tethered females and old wild males, and when older females were brought into captivity, they continued to lay eggs.

The site fidelity of adult *Heliconius* observed in these early mark-recapture experiments led to the suggestion that local populations might consist of closely related individuals (Benson, 1971). This could lead to kin selection, which might favour the origin of novel warning colour patterns (see Chapter 7), and could also favour cooperative traits such as information sharing about the whereabouts of local resources. In the late 1970s, *Heliconius* were often cited as a likely example of kin selection.

5.2 An early dispersal phase

However, an anecdotal observation by Turner (1971a) suggested *Heliconius* adults might actually be more dispersive than was supposed: a single freshly emerged individual moved across his entire study site in just a few hours, in marked contrast to the lack of movement among established adults. This was later confirmed in an exhaustive experiment by Mallet (1986b), who marked both wild and reared butterflies as they emerged from pupae on, or adjacent to, their natural host plants. In total, 108 freshly emerged *Heliconius erato* adults were marked, of which 49 were subsequently recaptured. The dispersal of these teneral individuals was compared to 'control' individuals captured as adults in the same area, demonstrating that the average dispersal distance (σ) of the former (266 m) was considerably larger than the latter (132 m) (Figure 5.1). If we assume that all butterflies have an initial dispersal phase, followed by movement around an established home range, these data give an estimate of per-generation dispersal of 296 m (Mallet, 1986b)—twice the value estimated from standard mark-recapture data.

Even this is likely to be an underestimate, however. One pupal-release individual moved 1,000 m before first recapture and a further 820 m before the second recapture, crossing the entire study area. Many of the released individuals that were never recaptured probably moved outside the study area completely. Much larger estimates of dispersal obtained from genetic studies (2.6 km in *H. erato* and 3.7 km in *H. melpomene*; see Chapter 10) support the suggestion that per-generation dispersal is actually significantly higher than 296 m (Mallet et al., 1990).

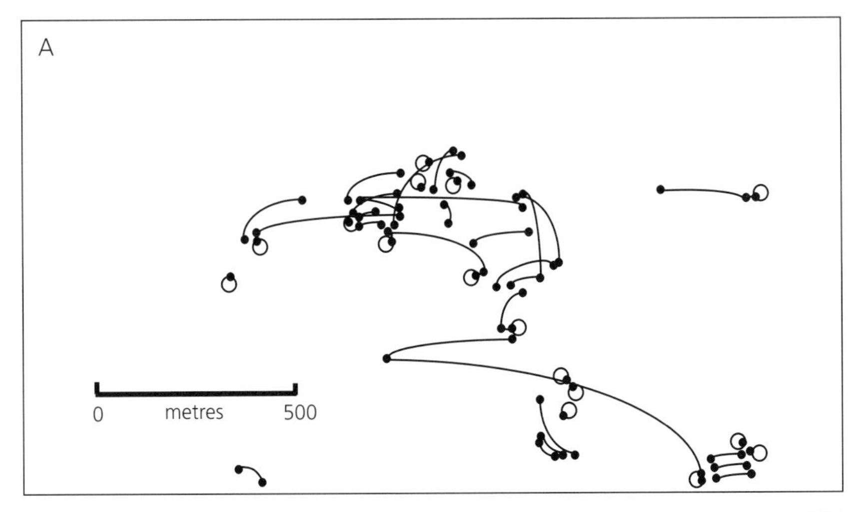

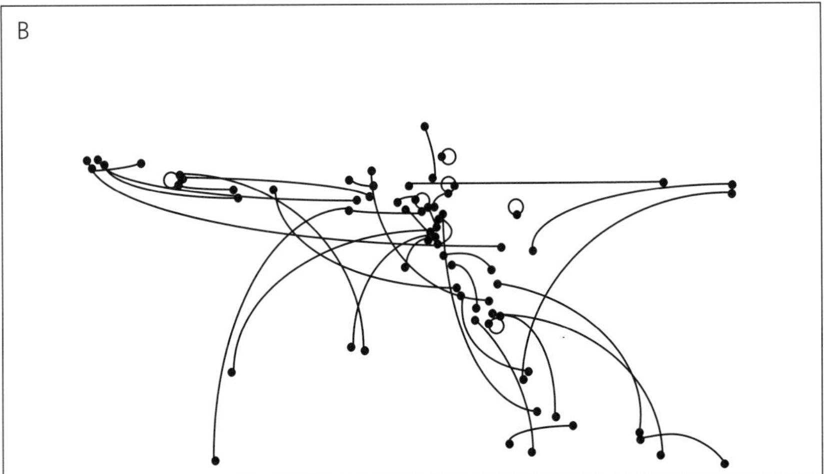

Figure 5.1 Dispersal behaviour of adult *Heliconius erato*

Movements of field-captured and pupal-released *H. erato* indicate the greater dispersal seen in recently eclosed adults as compared to established individuals. (A) Movements of 49 field-captured *H. erato* between first and second capture within the Sirena study site. Movements are represented by lines connecting dots at the sites of capture. A circle connecting a dot to itself represents a recapture at the site of first capture. (B) Movements of 49 pupal-released *H. erato* between eclosion and first capture (Mallet, 1986b).

In summary, *Heliconius* adults have a pre-reproductive dispersal phase, in which adults move before setting up home. Local populations of individuals identified in standard mark-recapture methods are therefore probably not closely related to one another compared to randomly chosen individuals from the regional population. This is also supported by genetic studies that show high variability among individuals but little genetic differentiation between local populations (Jiggins, McMillan, King, et al., 1997; Flanagan et al., 2004). Analysis of whole-genome data in *H. melpomene* similarly shows that although there is regional differentiation between populations on the scale of hundreds of kilometres, there is little differentiation on a local scale (Martin et al., 2016). Kin selection is therefore not a likely explanation for apparently cooperative behaviour in *Heliconius* adults. Furthermore, knowledge of the dispersal ability and extent of local population structure in a species is important for understanding hybrid zones and the origins of local adaptation.

5.3 Why so many males?

One of the first things that *Heliconius* biologists notice in the field is a tendency to collect more males than females. There is no evidence for a biased primary sex ratio—larvae reared in the laboratory produce equal numbers of the two sexes—so the difference must be either due to differential survival in the wild or, more likely, differences in behaviour.

Indeed, the most strongly male-biased samples are obtained by the lazy field biologist (this was first pointed out to me by James Mallet after I returned from my first three-month field trip with a highly male-biased sample). The field worker who stumbles out of bed for a long breakfast, starts off in the field after 9 a.m. and then sits by a sunny spot waiting for passing butterflies, will catch mostly males. Females tend to feed earlier in the day (Gilbert, 1975), and subsequently are more active in the forest understorey or undergrowth (depending on the species) where they search for host plants. Males move longer distances, both in their initial dispersal phase and once home ranges are established (Mallet, 1986b; Mendoza-Cuenca & Macías-Ordóñez, 2005), spend more time visiting flowers, but tend to collect smaller pollen loads (Gilbert, 1975), and may even visit different species of flowers (Mendoza-Cuenca & Macías-Ordóñez, 2005). In *H. numata*, males tend to fly higher than females (Joron, 2005).

One explanation for male-biased captures is therefore greater availability of males to the observer—they are simply more likely to be flying out in the open during the middle of the day. Nonetheless, observations at gregarious roosts are also male-biased, suggesting that the phenomenon is not entirely due to diurnal behaviour (Mallet, 1986a). It has also been suggested that females, due to behavioural dimorphism, might be more susceptible to predator attack and therefore suffer higher mortality (Ohsaki, 1995); if true, this would also contribute to the biased sex ratio.

5.4 Home-range behaviour

Heliconius adults demonstrate an extraordinary ability to learn their way around their habitat. The most detailed studies of foraging behaviour have come from long-term studies at Sirena, on the Osa Peninsula in Costa Rica, by Gilbert and others (Gilbert, 1984; Mallet, 1986b; Murawski & Gilbert, 1986). Marked individuals regularly visited the same flowers and host plants, and each can be found in a unique home range. As can be seen in Figure 5.2, no two butterflies shared exactly the same home range, and individuals differed considerably in the size and shape of their home range. One individual *H. hewitsoni* regularly moved 1.2 km from her roost to a host plant to lay eggs (Mallet, 1986a), although this is probably at the extreme end of the spectrum—most individuals move around 100 m each day. These home ranges are distinct from territories—they are not 'defended' against intruders and overlap considerably between individuals. The route followed by an individual through the home range is known as a 'trapline', similar to the daily path followed by hunters visiting their traps. In the case of *Heliconius*, a trapline will include flowers such as *Psiguria* as well as larval host plants. The latter are visited primarily by females for oviposition, but also by males searching for mates (see later).

Fidelity to a local home range provides a marked contrast to the behaviour of other warningly coloured tropical butterflies, notably Danaiines and Ithomiines, which are decidedly nomadic or even migratory. Anecdotally, I have found that it is possible to mark large numbers of ithomiine butterflies, with extremely low recapture rates.

It has been suggested that the behaviour of *Heliconius* is an adaptation to maximize use of sparse but reliable resources (Gilbert, 1975). Pollen sources such as *Psiguria* and growing shoots of *Passiflora* both occur at extremely low density in naturally forested areas (although *Passiflora* density can be high in disturbed habitats). However, once located, *Psiguria* inflorescences can be relied upon to provide a regular food source. To a lesser extent, *Passiflora* plants (unlike *Psiguria*, they are after all trying to avoid interaction with *Heliconius*) can be relied upon to produce new shoots eventually. Thus, regularly patrolling the same home range is an efficient means of exploiting these resources. By comparison, more nomadic butterfly species probably have commoner and/or less stable resources, although this hasn't been tested explicitly (Mallet et al., 1987).

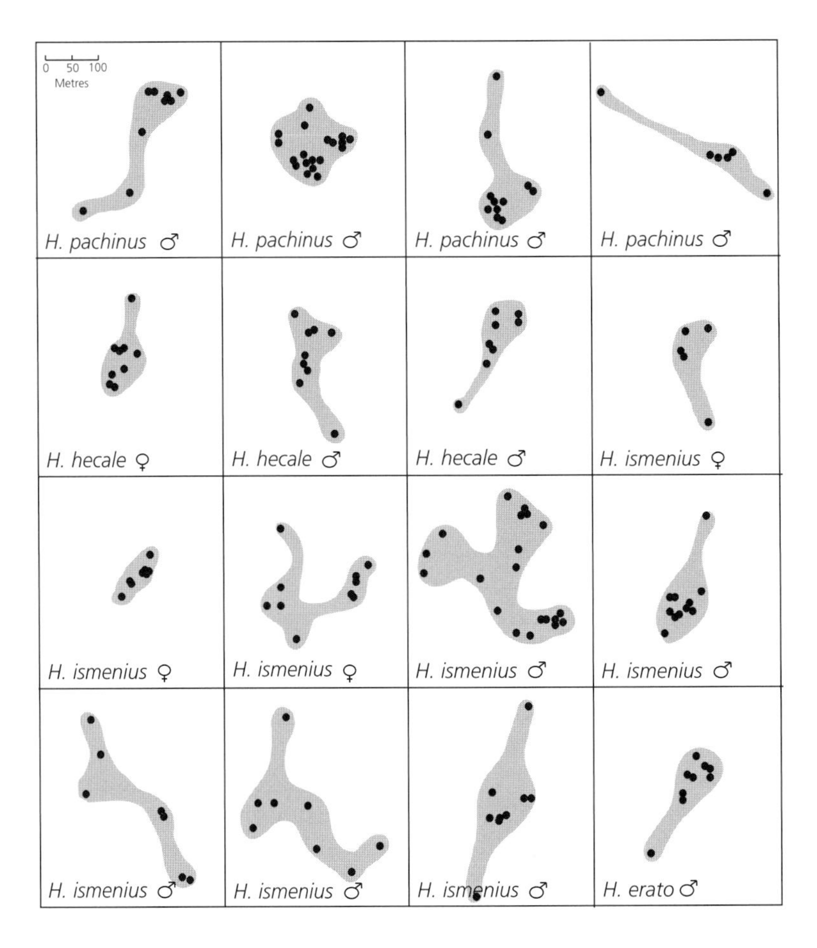

Figure 5.2 Estimated home ranges of individual *Heliconius* butterflies

The grey area is shown for clarity to connect observations and indicate approximate extent of the home range. Dots represent either capture locations of butterflies or the location of plants containing dye-marked pollen that was observed on the same butterfly (Murawski & Gilbert, 1986).

5.5 Gregarious roosting

During the day, each individual sets off on its own trapline, but at night *Heliconius* adults come together to roost gregariously (Figure 5.3). Aggregation of individuals at night-time roosts was first documented by W. H. Edwards in 1881, and discussed by many early biologists (Jones, 1930; Poulton, 1931a, b; Moss, 1933; Young & Thomason, 1975; Edwards, 1881; Waller & Gilbert, 1982). In Sirena, *H. erato* begin to aggregate about 3:30–4:00 p.m. depending on the weather (earlier if it is overcast), and perch, wings up, on leaves around the roost site (Mallet, 1984). By 4 p.m. individuals begin to approach the actual roost site, generally dead

Figure 5.3 Roosting behaviour of *H. erato* (see Plate 9)

One female is shown fanning a group of four individuals already roosting together. This photograph was taken in an insectary in Gamboa.

twigs or vines some 1–2 m above the ground, and start to hang, wings down, from the twigs.

Exact roosting times can depend on the local light environment, with individuals at darker forest sites arriving earlier and leaving later (Finkbeiner, 2014). As subsequent individuals arrive, they will often flutter around those already present, 'fanning' them with their wings. Commonly, this involves a characteristic to-and-fro hovering, which is similar to courtship behaviour. They may also clutch at the legs or wings of individuals already on the roost. The disturbance is vigorously opposed by the roosted individual, which 'lashes out violently with its wings' (Mallet, 1984; see also Salcedo, 2011a). Occasionally, such behaviour displaces the roosted individual, who will then join in the jostling for a spot to hang from (indeed, it all looks very annoying, a bit like your spouse pushing you out of bed before settling into the best spot on the mattress).

However, the usurper does not necessarily take over the empty spot, but instead will often continue to hover around other individuals or roost on a nearby twig, so the behaviour cannot readily be explained by competition for a particular roosting position. Eventually, all the individuals will settle down on the roost, hanging in a fairly tight cluster on the same or adjacent twigs (this varies considerably between species—see later).

Intriguingly, there is a sex-specific wing posture among the hanging butterflies. Roosting females draw back their forewings so the band on the forewing is largely obscured by the hindwing. Males keep their forewings extended and more exposed. The female posture is maintained only while it is light—after dark, females relax their wings into the same posture as males, but if a light is shone on females, the posture is slowly reassumed (Mallet, 1984). Females may also adopt this posture during mating and egg-laying, and it likely serves to reduce the unwanted attention of males (Mallet, 1986a).

This hypothesis is supported by experimental evidence showing that dead butterflies with the forewing band blacked out are less likely to be approached by potential roostmates (Mallet, 1986a), and by a failure to roost gregariously among butterflies whose wing patterns are completely blacked out (Salcedo, 2010b). Colour is clearly an important cue for gregarious behaviour. Nonetheless, in hybrid zones, individuals with different wing patterns roost together, suggesting that colour is not critical for roost aggregation (Finkbeiner, 2014; Jiggins, personal observation).

In one of the earliest experimental studies of *Heliconius* behaviour, Jones swapped the twig used for a gregarious roost with another from nearby, and showed that individuals learn the location of the roost, rather than being attracted to a particular twig—this implies the use of visual rather than chemical cues to locate the roost site (Jones, 1930). The experiment has since been repeated with similar results in Colombia (Mallet, 1986a). Nonetheless, when individuals are aggregating, there is evidence that they use chemical as well as visual cues in conspecific interactions at roosts (Salcedo, 2010b; Finkbeiner, 2014). In summary, the site is first discovered by use of landmarks, and gregarious behaviour is then initiated by means of both visual (wing pattern) and chemical cues.

Roosting behaviour is quite variable between species. The erato and sapho clades tend to be more gregarious, with *H. sara* and *H. charithonia* particularly so. In these clades, around 80% of individuals roost gregariously; average roost size is 3–4 (Mallet, 1984, 1986a). *H. charithonia* will commonly aggregate so closely that individuals are hanging from one another's wings and legs. In contrast, the melpomene and silvaniform clades are less gregarious, with around 30–60% of individuals roosting gregariously and an average roost size of 1.3–2.5 (Mallet, 1984). These figures are primarily derived from observations at Sirena, so are likely to vary considerably across species and populations. Roost height also varies considerably, with a tendency for co-mimetic species to roost together at similar heights (Mallet & Gilbert, 1995) (Figure 5.4).

Studies suggest that roost membership can be highly constant (Mallet, 1986a; Turner, 1971a; Waller & Gilbert, 1982), but this varies between studies (Cook et al., 1976). For example, in a 3-week study of an *H. charithonia* roost in Mexico, 24 butterflies were observed returning at least once to the roost, and roost members spent on average 11.4 nights on the roost (Waller & Gilbert, 1982). However, individuals show distinct 'personalities' in their roosting behaviour (Finkbeiner, Briscoe, & Reed, 2012; Mallet, 1986a). Some return faithfully to the same

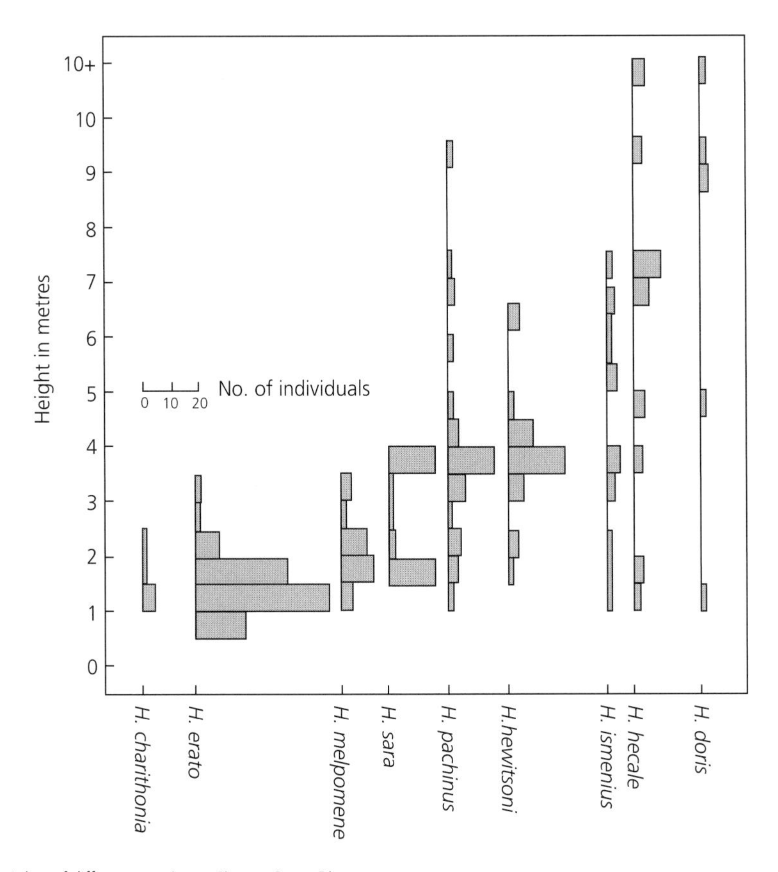

Figure 5.4 Roost heights of different species at Sirena, Costa Rica

The number of individuals found roosting at different heights for coexisting species. Note similarity in roost heights between co-mimic pairs of species, *H. erato* and *H. melpomene; H. pachinus* and *H. hewitsoni.*

site night after night, while others consistently switch between one or more roosting sites. The establishment of new roosts begins with regular site fidelity by a single individual, who is later joined by others (Finkbeiner, 2014).

One of the mysteries of *Heliconius* biology is why individuals should take part in gregarious roosting behaviour. One suggestion is that there might be a benefit to gregariousness associated with warning colour signalling or avoidance of predation (Mallet, 1986a; Turner, 1975a). Early in the morning, butterflies generally perch with their wings open in sunspots near roosts, while they raise body temperature sufficiently before foraging. It can be a striking sight to see ten or more brightly coloured *Heliconius* adults perched in a small area, catching the early morning sun. This is the time of day foraging birds are most active.

Thus, the butterflies are still cold and slow moving at the same time of day that birds are actively foraging—it is probably then that the risk of predator attack is highest. The benefit of grouping to dilute predation risk has been termed the predator satiation hypothesis, stated most succinctly by John Turner (1975a, p. 222): 'A predator which encounters one of them and experiments with tasting it will immediately encounter all the others and leave them alone, whereas if the prey were dispersed a number of them might fall victim to several different predators'.

It also seems likely that the warning signal itself is enhanced by gregariousness, by effectively sending a stronger 'group' signal to predators. A mass of brightly coloured butterflies is more memorable than a single individual, and there is evidence that after predators have experienced a distasteful butterfly, the mere sight of additional individuals with the same pattern reinforces the memory of the pattern (a process known as secondary reinforcement; Thorpe, 1956). Both the role of colour in cueing aggregation behaviour (see earlier), and the tendency for co-mimic species to roost together are consistent with this hypothesis (Mallet & Gilbert, 1995).

An alternative suggestion, termed the conspecific cueing hypothesis, suggests that naive individuals might choose to sleep near established roosts, as these sites are more likely to be safe from predation than a twig chosen at random. Analysis of *H. sara* roost sites support the idea that these are indeed safe localities, with roost sites being dryer, darker, and better protected from rain than other nearby locations (Salcedo, 2010b).

A single observation of a predation event on an *H. charithonia* roost in Tamaulipas, Mexico, demonstrates that predation does occur at roosts (Mallet, 1984): a bird flew through the roost in an explosion of butterflies, leaving a single female dead on the ground. However, intensive observations of *H. erato* roosts in Sirena showed very high levels of roost disturbance, mainly by birds in the early morning, but no actual predation—each butterfly was disturbed on average every 4.3 mornings (Mallet, 1984). Similarly, video recordings of nocturnal roosts showed occasional disturbance by bats but little evidence of predation by visual predators such as birds in the early morning (Salcedo, 2011b). Nonetheless, actual predation rates on distasteful and warningly coloured butterflies are likely to be low, so the lack of observations of predation does not preclude its importance in the evolution of roosting behaviour.

Given the rarity of direct observations of predation on live butterflies, Susan Finkbeiner set out to test the predation hypothesis using artificial models of *Heliconius erato demophoon* in Panama and Costa Rica. The models were printed and matched to real butterflies using computer models of bird vision (Finkbeiner et al., 2012). A clever addition was the use of 3-OHK, the natural *Heliconius* yellow pigment, to mimic the UV reflectance of the real butterflies. Predation events were then observed on the plasticine bodies, or paper wings that had been dipped in wax to record the imprint of bird attacks. The results clearly showed a benefit to gregariousness. Solitary individuals were considerably more likely to be attacked (21%) than an individual in a group (3%).

One benefit of group living is a dilution effect: if predators encounter groups at a similar rate to single individuals, and become quickly satiated, then individuals within groups would benefit from a reduced probability of predation. However, the 'per group' attack rate was lower on groups than on solitary butterflies, suggesting that gregariousness enhances the protective effect of warning colour. Furthermore, there was some evidence for an optimal group size, with groups of five showing lower attack rates than groups of two or ten (Finkbeiner et al., 2012).

It would be interesting to further explore these findings to confirm that there is an optimal group size and, if so, whether it varies between species or between habitats. Research to date used models of the ventral wing surface of perching butterflies—it would also be useful to carry out experiments with models of the dorsal surface, which is more brightly coloured and probably subject to stronger predation selection. Although individuals roost with only the ventral surfaces exposed, predation likely occurs in the early morning when they are basking with their wings open and dorsal surfaces exposed.

A second possible benefit of group living is information sharing. One study of roosting behaviour noted that almost half the *H. charithonia* individuals from one roost fed at a single large *Psiguria* vine near the roost site (Waller & Gilbert, 1982). Indeed, nearly all the individuals observed at that pollen source came from the same roost. Such observations led to the suggestion that individuals learn the location of resources in the forest from other roost members, perhaps by following them from the roost (Gilbert, 1975, 1984).

However, observations on the home ranges of *H. erato* roost members do not support this suggestion. Mallet (1984, 1986a) observed that butterflies leave roosts individually with no evidence of following behaviour, and commonly had very different feeding stations and home ranges. Obviously, roosts are by definition part of an individual's home range, so there was some overlap between individuals in their diurnal range also, but these ranges were in some cases virtually non-overlapping (Mallet, 1984, 1986a). If roosts were information centres, roostmates should have diurnal home ranges that are more similar to one another than to other butterflies in the local area, but this does not seem to be the case.

Similarly, Finkbeiner followed butterflies from roosts and, of 74 observations, never saw two butterflies leave the roost together and subsequently feed on the same flowers. She did occasionally observe new recruits arriving at the roost site with an established visitor, suggesting that following may play a role in roost recruitment (Finkbeiner et al., 2012). Nonetheless, she also observed multiple individuals from several different roosts following the same traplines between flowers, suggesting that following may be important in finding flowers but unrelated to roosting behaviour (Finkbeiner, 2014). Thus, there is little support for information sharing as a major factor in roost establishment, although it may still be important in some populations.

Evolutionarily, there is also difficulty in explaining information sharing, as there is presumably a cost due to increased competition—why should experienced individuals give up their valuable local knowledge to competitors? The concept of information sharing originally came from vertebrate species where resources were hard to find but locally very abundant, such that there would be little cost from competition (Ward & Zahavi, 1973). However, this is clearly not the case in *Heliconius*. If the beneficiaries were close relatives, it could be a genuinely altruistic trait driven by kin selection, but high dispersal and low relatedness within local populations makes such an explanation unlikely (see earlier). Alternatively, some form of reciprocity may lead to information sharing: perhaps the immediate cost of increased competition from sharing information is balanced by the potential benefit gained from finding new resources. Of course, it may be that there is no adaptive explanation for information sharing, and, if it does occur, then it is a side effect of a gregarious behaviour that has evolved for other reasons.

It is also possible that gregarious roosting has some other sexual or social function. There are occasional records of mating at a roost—a pair of *H. erato* was found in the morning adjacent to a roost by Salcedo in Gamboa, Panama, and Finkbeiner observed a pair in copula arriving at a roost (Finkbeiner, 2014). However, such behaviour is unusual: Mallet never saw mating at roosts in Sirena after many hours of observation.

Even so, the 'fanning' behaviour described earlier is similar to that seen during courtship (see later), and although it occurs between all combinations of males and females, there is a tendency for males to fan females more commonly than would be expected by chance (Mallet, 1986a). Therefore, although it is unlikely that the primary function of roosting is mating, it is probably a derived form of courtship behaviour (Crane, 1957b). The function of fanning may be species recognition, as conspecifics tend to roost together, although multi-species roosts are also common, especially between co-mimics.

In summary, there is good evidence supporting the protective effect of group living against predators, but it does not clearly explain the social behaviour observed at roosts, nor substantial variation in gregariousness between species. A full explanation will need to account not only for the gregarious behaviour itself, but also why there are such differences in the degree of gregariousness between species, and why individuals should apparently invest so much effort in fanning their roostmates.

5.6 Where have all the butterflies gone?

Amazingly, butterflies can remember where they have been captured and avoid such dangerous locations for a few days (Mallet et al., 1987). During extensive mark-recapture experiments in Costa Rica, it was noticed that individuals seemed less likely to return to a site after they had been captured with a net. This observation was tested in an experiment in which the probability of 'recapture' was compared after days on which butterflies were captured by net, and those on which they were identified by sight (Mallet et al., 1987). *Psiguria* flowers were staked out by the experimenters every other day for a period of seven days. On the first day, all butterflies were captured by net, but on the third and fifth day, they were identified by sight, using binoculars. On the seventh day, they were again captured with a net.

Butterflies were more likely to return to a flower when they had not been captured. Even after two days, only 17% of butterflies returned to the same flower after net capture, compared to 51% after visual identification. Although there was a trend in the expected direction after 4 days, it was not significant, suggesting the effect lasts 2–4 days. Importantly, observations of the same butterflies at their nocturnal roosts showed that they hadn't abandoned the area altogether—they were still found sleeping in the same place. Thus, the effect seems to be specific avoidance of the spot where they had been captured.

There are obvious reasons such behaviour might be adaptive in a natural context. Jacamars are specialist predators of flying insects and commonly implicated as *Heliconius* predators (Chai, 1986). They often return to the same perch site, generally adjacent to a forest clearing, from which to sally forth and capture their prey. A *Heliconius* that had escaped capture from such a predator would be wise to avoid the site for a few days. Alternatively, the net capture in the experiment described would have prevented the butterfly from feeding, so the response could have been simply due to avoidance of a flower that had not provided pollen. Whatever the origin of the behaviour, it clearly implies a complex learnt response by a butterfly. As the authors state, 'Insect behaviour could be vastly more complex and subtle than is catered for in most ecological studies' (Mallet et al., 1987, p. 384).

5.7 Future directions

The scale and patterns of dispersal remain poorly documented across *Heliconius* species. However, high-resolution genomic data could be used on large population samples to track patterns of movement and population differentiation in wild populations in order to address dispersal ability across multiple species. In particular, it would be interesting to compare patterns seen in polymorphic species such as *H. doris* with highly geographically differentiated species such as *H. erato* and *H. melpomene*, and with broadly monomorphic species such as *H. charithonia*. If local population structuring and movement have played a role in population differentiation (see Chapter 10), we might expect different patterns of dispersal and population structure in these species.

Although considerable progress has been made in understanding the function of gregarious roosts, there remain outstanding questions. Notably, the suggestion that there is an optimum intermediate group size for a nocturnal roost is intriguing but needs to be tested directly. The considerable variation in roost size between species might offer a means to investigate how group size is determined. Better understanding of *Heliconius* chemical ecology and pheromones may offer some insight into the function of the fanning behaviour seen at roosts. Finally, traplining behaviour might prove to be a tractable system for studying spatial learning in insects (see also Chapter 6). In turn, such understanding could have implications for economically important insect species such as crop pests or disease vectors, as well as providing insight into insect neurobiology in general (Lihoreau et al., 2013).

CHAPTER 6

Brains, sex, and learning: behaviour, sexual and social selection

Chapter summary

The behaviour of *Heliconius* butterflies is adapted for exploiting their complex forest environment. *Heliconius* have large and highly plastic brains, associated with their spatial learning of traplines and resources in the forest. The visual systems of the heliconiines are also highly developed and show a broad range of spectral sensitivity and colour discrimination. Analysis of visual opsin genes has shown a duplication of the UV-sensitive gene in *Heliconius* relative to other heliconiines, potentially giving *Heliconius* the ability to distinguish colours in the UV spectrum. Olfactory and gustatory senses are less well studied, but expression of gustatory receptor genes in female legs is likely associated with their role in host-plant discrimination. The diversity of genes associated with olfactory and gustatory senses is similar to that seen in moths.

Sexual selection is an important force in speciation and diversification. *Heliconius* show a diversity of mating behaviours including the unusual habit of pupal mating, restricted to the erato and sapho clades, which essentially precludes the opportunity for female choice. However, we remain largely ignorant of the frequency of pupal mating in the wild. When mating occurs between adults, colour is important for mate-finding and signalling. Chemical communication is also important, but much less well studied. There is evidence for alternative mating strategies in some species, in which larger males compete for females at pupae while smaller males adopt alternative strategies, either patrolling for females or defending territories. Another unusual adaptation is the transfer of an anti-aphrodisiac pheromone from males to females during mating. The anti-aphrodisiac reduces sexual harassment of mated females, but may come at a cost to females of lost mating opportunities. Conflict between the diverging interests of males and females has resulted in the rapid evolution of anti-aphrodisiac pheromone composition.

6.1 Learning

Anyone who has watched *Heliconius* in the forest will realize that these are not your average butterflies. A female flies into a clearing and heads straight for the tiny *Psiguria* flower hidden among the leaves, then swoops down to the juicy new shoot of a *Passiflora* seedling on the forest floor. They often seem to know exactly where they are going.

Implicit in much of the behaviour described in the previous chapters is a remarkable learning ability. If individuals are to follow their own distinct traplines through the forest, remain faithful to a characteristic home range and sleep every night on the same twig, they must be able to learn reliable spatial cues in order to orient through an extraordinarily complex environment. Indeed, anyone who has reared *Heliconius* in an insectary will also be aware of their ability to learn about aspects of their environment, such as the location of food resources and the boundaries of the enclosure. Many of these behaviours are time-dependent, such as foraging on

The Ecology and Evolution of Heliconius Butterflies. Chris D. Jiggins, Oxford University Press (2017).

DOI 10.1093/acprof:oso/9780199566570.001.0001

particular *Psiguria* flowers, suggesting that memory is closely tied to the circadian clock.

It is increasingly recognized that learning plays an important role in the lives of insects—even those with relatively simple nervous systems such as a fruit fly larva (Dukas, 2008). However, although learning ability has been repeatedly noted in *Heliconius*, there have been relatively few studies of learning, since the colour learning experiments conducted in the 1970s (Swihart & Swihart, 1970; Swihart, 1971; see also Dellagio et al., 2016). Nonetheless, with their long lifespan and learning of complex routes through the forest, *Heliconius* must be at the upper end of insect learning ability.

As noted, the bold assertion that *Heliconius* are some of the brainiest insects has not been well tested by experimental studies of learning ability. However, recent work on brain morphology in *Heliconius* does suggest that they have a rather unusual brain. The mushroom body, a structure associated with learning in insects, is proportionately the largest of any Lepidoptera that has been tested to date (Figure 6.1) (Sivinski, 1989; Montgomery et al., 2016). The monarch butterfly, famous for its long-distance migration and ability to use a sun compass, also has a larger mushroom body than most Lepidoptera, but it is still smaller than that of *Heliconius hecale* (Heinze & Reppert, 2012; Montgomery et al., 2016).

Plasticity in brain morphology can also tell us about the potential for learning and plasticity in behaviour. The mushroom body of some bees and ants can increase in volume between 20 and 30% from emergence to full maturity, associated with a period of learning through foraging and social experience (Gronenberg, Heeren, & Hölldobler, 1996; Fahrbach et al., 1998; Jones et al., 2013). *Heliconius* brains show comparable if not greater increases during post-eclosion development: 38% for the calyx and 34% for the lobe system in *H. erato*; 28% for the calyx and 24% for the lobe system in *H. hecale* (Montgomery et al., 2016). A similar size difference is also found between laboratory-raised and wild individuals in *H. hecale*, implying strong experience-dependent plasticity in brain development. These are exciting discoveries. What we need to do next is to link brain morphology with the remarkable array of ecological interactions in order to understand how the unusual *Heliconius* brain facilitates their unusual lifestyle.

Social learning—learning from other individuals—has not been commonly documented in insects, aside from the famous waggle dance in honeybees and a few other cases in Hymenoptera (Dukas, 2008). It seems likely that *Heliconius* also show social learning. This has been suggested by the information-sharing hypothesis, proposed to explain gregarious roosting, and anecdotally occurs

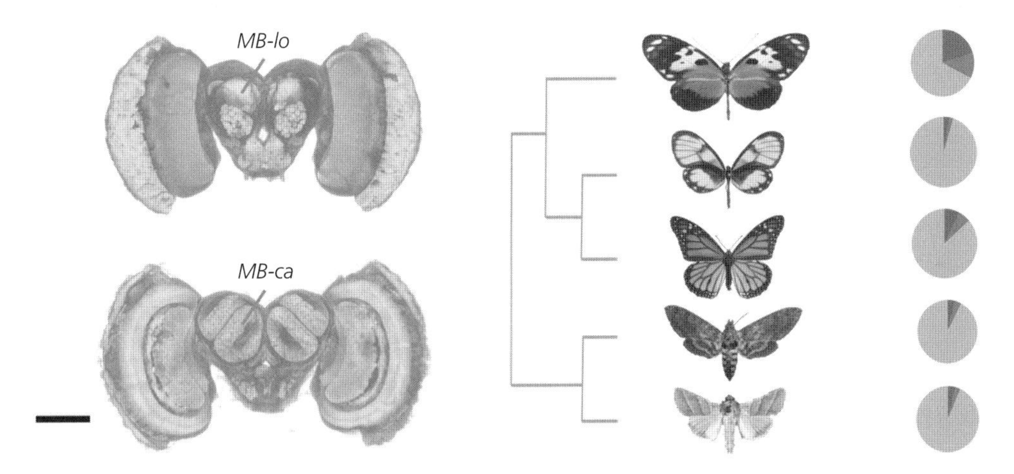

Figure 6.1 Expansion of mushroom bodies in *Heliconius* butterflies (see Plate 10)

(A) 3D surface rendering of the brain of *H. hecale* (shown left, scale = 2.5 mm) viewed from the anterior (top) and posterior (bottom) indicating the mushroom body lobes (MB-lo) and calyx (MB-ca). Scale = 500 μm. (B) Pie charts show the proportion of the midbrain occupied by MB-calyx (dark red) and MB-lobe + peduncules (light red) in *H. hecale, Godyris zavaleta, Danaus plexippus, Manduca sexta*, and *Heliothis virescens* (top to bottom) (Merrill et al., 2015; Montgomery et al., 2016).

in insectary cages. Individuals seem to learn the location of food resources far more readily when experienced individuals are present (Jiggins, personal observation). Careful observation of foraging in the wild suggests that individuals follow one another during trapline foraging, perhaps to rapidly learn where local resources are located without having to find them independently (Pinheiro, 2009; Finkbeiner, 2014). Although social learning and behavioural plasticity are exciting areas for future *Heliconius* research, most of this chapter will focus on better-studied areas of sensory ecology and sexual behaviour, starting with how *Heliconius* sense their environment.

6.2 Sensory ecology: vision and colour

Evolutionary biologists have long focused on the obvious morphological and behavioural traits that differ between individuals and the extent to which such variation results from natural and sexual selection. More recently, there has been a realization that the sensory systems of animals are also closely adapted to their environments and shaped by the need to find food and mates (Endler & Basolo, 1998).

For example, the spectral sensitivity of some firefly species has been shown to be closely matched to the wavelengths of light produced by their bioluminescence organs (Cronin et al., 2000; Briscoe & Chittka, 2001), presumably driven by selection for mate-finding. Similarly, in a comparison of two Lycaenid butterflies, *Lycaena heteronea*, a species with blue wing patterns, had much greater expression of blue photopigments in the ventral portion of the eye than did *Lycaena rubida*, a species without blue wing patterns (Bernard & Remington, 1991; Sison-Mangus et al., 2006). These examples show that both ecological and sexual selection play important roles in shaping insect visual systems.

Similarly, *Heliconius* use their visual system to both find mates and food, and the large heads of adults represents a large investment in the visual neuropile (Gilbert, 1975), implying selective pressures for increased visual sensitivity. *Heliconius* adults can perceive colours across a wide range, from ultraviolet at 300 nm up to red at perhaps as long as 700 nm (Crane, 1955; Swihart, 1967a, 1972; Briscoe et al., 2010; Bybee et al., 2012).

Christine Swihart showed that *Heliconius charithonia* can be trained to coloured flowers (Swihart & Swihart, 1970; Swihart, 1971). When presented with an array of differently coloured artificial flowers, naive butterflies showed a bimodal response, with peaks in the orange/red and the blue/green regions of the colour spectrum. However, they could be trained to prefer either yellow or green flowers by providing artificial flowers of either colour for two days. Trained butterflies visited yellow flowers 50% of the time compared to 9% for untrained individuals, while the effect was even more striking for green flowers, where training increased visits from 2% to 55% (Swihart & Swihart, 1970).

Subsequent experiments showed that even fairly slight differences in hue can be distinguished in the yellow region of the colour spectrum, and that individuals can be trained to more complex artificial flowers with multiple colours (Swihart, 1971). These experiments demonstrate the existence of both 'true' colour vision in *Heliconius* and the ability to remember an association between colour and nectar reward. The memory lasted overnight, as all experiments were carried out on the morning following the training period. Furthermore, the ability of butterflies to become trained to almost any part of the visual spectrum suggests that strong behavioural responses to particular colours, such as during courtship (see later), result from the way in which neurones are connected, rather than being an intrinsic phenomenon of the visual system (Swihart & Gordon, 1971).

The adaptive value of colour associative learning for foraging has been demonstrated in a series of experiments on *Agraulis vanillae* (Weiss, 1995). The flowers of *Lantana camara* are initially yellow, but turn red after a single day. Only the yellow flowers produce nectar, but the red flowers are retained on the flat-topped inflorescences, presumably to enhance their attractiveness. *Agraulis vanillae* had no initial preference for yellow, but rapidly learnt to associate yellow with nectar reward, and subsequently showed a strong preference for yellow over red flowers. This preference was learnt in fewer than ten flower visits and seemed to be enhanced by further visits. This is a rate of learning similar to that of honeybees, which require one to five visits to learn preferences (Menzel, 1993).

Visual systems use visual receptor molecules to convert photons of light into neuronal signals. The visual receptors consist of a light-sensitive chromophore, which in butterflies is *11-cis-3-hydroxretinal*, bound to an opsin protein. Colour vision requires the comparison of inputs from at least two photosensitive pigments that are sensitive to different wavelengths of light. Changes in the amino acid sequence of the opsin protein can alter the spectral sensitivity of the receptor, allowing animals possessing multiple opsin genes to detect colours, provided these are expressed in distinct photoreceptor cells.

A common mechanism for the evolutionary diversification of visual systems is through gene duplication and subsequent diversification of opsin genes. Some insect species possess large numbers of distinct opsin genes (Futahashi et al., 2015), but among the Lepidoptera most species have just three. These are known as ultraviolet- (S), blue- (M), and green/yellow- (L) sensitive opsins, but an additional red-sensitive opsin has arisen in several lineages (Briscoe & Chittka, 2001). *Heliconius* do not show any unusual patterns of evolution in the M opsin (Hsu et al., 2001), but show some novel adaptations at the UV and long wavelength ends of the spectrum.

Heliconius are unusual because there has been a duplication of the UV-sensitive opsin to give four opsins: UV1 (S1), UV2 (S2), blue, and green, with peak spectral sensitivities at 360, 390, 470, and 550–560 nm, respectively. After duplication, one of these copies underwent a burst of rapid molecular evolution, with evidence for an accelerated rate of amino acid substitutions (Briscoe et al., 2010). This is a common pattern seen after duplication, as one of the gene copies takes on a new function (a process known as neo-functionalization). Furthermore, in *H. erato*, at least, the extra copy is expressed only in females, suggesting a role in female-specific functions (McCulloch, Osorio, & Briscoe, 2016). Duplication of the UV opsin potentially gives *Heliconius* the ability to detect subtle differences in UV 'colours' that are not detectable by other organisms in their environment. In particular, their likely avian predators have a single UV opsin, so can detect UV light but may not be able to distinguish UV colours.

Heliconius are also unusual in the long wavelength end of the spectrum. In bees, moths, and butterflies, each ommatidium has six (or seven) receptors expressing L opsin, and two receptors expressing either both M opsin, both S opsin, or one M and one S opsin (Zaccardi et al., 2006a, b). However, in addition to the photopigments, insect eyes possess filtering pigments.

In *Heliconius*, there is evidence that the pigments provide another more unusual method of colour perception. In principle, photoreceptors expressing the same photopigment could be sensitive to different wavelengths due to the presence of different filtering pigments—this mechanism would not extend the overall spectral sensitivity of the organism, but would rather allow colour detection over a wider range of spectra. It is fairly easy to study the distribution of filtering pigments by shining light into the eye and observing the colour of the reflected light. In *H. erato*, there are two kinds of ommatidia that reflect either yellow or red light due to differences in the distribution of the filtering pigments (Zaccardi et al., 2006b). Furthermore, behavioural experiments and electrophysiology confirm that *H. erato* can distinguish colours in the long wavelength range (i.e. 590 nm, 620 nm, and 640 nm), but there is no evidence that there are two opsin pigments sensitive across these wavelengths (Zaccardi et al., 2006b). *Heliconius* are therefore the first animals known to use filtering pigments in order to generate true colour vision from photoreceptors that express exactly the same opsin gene (Zaccardi et al., 2006b). This means—in *H. erato*, at least—that there are five visual sensitivity peaks, with λ_{max} = 360 nm, 390 nm, 470 nm, ~ 560 nm, and ~ 600 nm (McCulloch et al., 2016).

Does the unusual visual system of *Heliconius* represent an ecological or sexual adaptation? Enhanced sensitivity to colours in the red spectrum likely plays a role in detection of mates, flowers, or both, although this has not been explicitly demonstrated. Another intriguing possibility is that UV signals might represent a 'cryptic' channel of communication that can be seen by butterflies but not birds. This could be especially important for species involved in mimicry, where colour pattern similarities between species might lead to confusion in mating (Estrada & Jiggins, 2008). Recently, methods have been developed to test perceptual differences between colours perceived by different animal visual systems (Vorobyev & Osorio, 1998). The approach is based on modelling the known spectral absorbance of the visual receptor molecules in an organism, and comparing the stimulation of photoreceptors by different colours.

The first application of this approach in *Heliconius* compared the perception of the yellow colours on the wings of *Heliconius* and related genera such as *Eueides* and *Dryas*. *Heliconius* yellow colour, produced by the pigment 3-OHK, has a marked UV reflectance peak around 300–350 nm. Visual modelling suggests that *Heliconius* can distinguish this from non-*Heliconius* yellows better than bird visual systems (Bybee et al., 2012). This indicates that the *Heliconius* visual system is 'tuned' to detect *Heliconius* patterns. However, the real challenge is to distinguish between co-mimic species within *Heliconius* that all use the same 3-OHK yellow pigment. There does seem to be more variation between different yellows in *Heliconius* than in other butterflies (Briscoe et al., 2010), but signalling between *Heliconius* co-mimics has not been explicitly tested.

Nonetheless, Violaine Llaurens and colleagues compared *Heliconius numata* with its *Melinaea* co-mimics (Llaurens, Joron, & Théry, 2014). Again, the most striking result was found for the yellow pigments. The study used the contrast between yellow and adjacent black regions, rather than the yellow colour alone. This analysis similarly indicated that *Heliconius* might be better than birds at distinguishing conspecifics from co-mimics, suggesting that there may indeed be 'cryptic' channels of communication for the butterflies to detect differences between co-mimics that are not detectable by birds.

Even so, all of these studies must be regarded as somewhat preliminary. The visual system modelling has not been tested with behavioural experiments in *Heliconius*, so, although the results are suggestive, they do not prove that the butterflies distinguish UV signals. Similarly, we know little about the visual systems of the likely avian predators of *Heliconius* in the tropics—the visual models used are largely based on behavioural studies of blue tits and peafowl. More behavioural experiments with butterflies and with tropical insect predators are needed.

Insects, including butterflies, are able to detect polarized light. Light from the sun is depolarized, but becomes partially polarized as a result of scattering in the atmosphere. Many naturally occurring objects also polarize light, including reflectance from water and many shiny biological surfaces. It has been argued that polarized light in forested environments might provide a useful source of signalling information in a complex light environment (Douglas et al., 2007). Butterflies that fly in forests, including those *Heliconius* species with iridescent colours, often have polarized reflectance patterns (Douglas et al., 2007). We have shown that *H. cydno* are more attracted to moving models with their polarized patterns intact than to models displayed under a depolarizing filter (Sweeney, Jiggins, & Johnsen, 2003), confirming that adult butterflies respond to polarized signals and probably use them in signalling between individuals. Better methods are now available for generating controlled polarized signals, however, and further work is needed to confirm the ecological importance of polarized signals.

6.3 Sensory ecology: scent

Chemical communication plays a role at least as important as vision in the lives of *Heliconius*, but has been less well studied. For example, naive *H. melpomene* significantly preferred artificial flowers spiked with the floral scent of *Lantana camara*, a preferred flower in the wild, over those with the scent of vegetative parts of the same plant (Andersson & Dobson, 2003a). After learning to associate yellow with nectar rewards, however, colour was a much stronger stimulus. Yellow flowers with vegetative scent were preferred over green flowers spiked with floral scent (Andersson & Dobson, 2003a).

The antennae are the main organs for detection of chemical cues, and the sensitivity of butterflies to different compounds can be measured by electroantennographic detection (EAD)—measurements of the neural activity of the antenna. This can be combined with separation of compounds using gas chromatography (GC). Thus antennal responses to the individual components of complex scents, such as those produced by a flower, can be measured in a single experiment (the whole procedure earns the acronym GC-EAD).

The floral scent of *Lantana* is remarkably complex, containing over 65 compounds. *H. melpomene* antennae responded strongly to many compounds, including several benzenoids, monoterpenoids, and irregular terpenoids in the floral scents of plants including *Lantana camara* (Andersson & Dobson, 2003b). Some of these were unique indicators of flowers (i.e. they were not present in vegetative parts of flowering plants).

Intriguingly, two of the compounds identified as components of floral scent that elicited strong antennal responses have also been shown to play a role in communication between *Heliconius* pupae and adults. Linalool and linalool oxide are produced by pupae; they permit males to distinguish the pupal sex (see later for discussion of pupal mating) (Estrada et al., 2010). This suggests that compounds initially used as cues in foraging may be adopted for intraspecific communication. Chemical communication also plays an important role in sexual selection among adult *Heliconius*, as discussed in section 6.5 on mating behaviour.

Chemical cues are also used in oviposition. Tethered *Agraulis vanillae* females flew towards the smell of *Passiflora* vines (Copp & Davenport, 1978). When looking for a site to lay eggs, females will spend extended periods of time flying around the plant and 'probing with the antennae and proboscis as well as tapping with the forelegs' before deciding to lay an egg (Benson et al., 1975, p. 666). In particular, the reduced forelegs are used in 'drumming' the plant, presumably investigating its chemical and physical characteristics. The forelegs show sexual dimorphism: females have a set of cuticular spines, each associated with four gustatory sensillae (hairs used in taste perception) that are not present in males (Figure 6.2) (Renou, 1983; Briscoe et al., 2013). It seems likely that the forelegs are used as drumsticks in

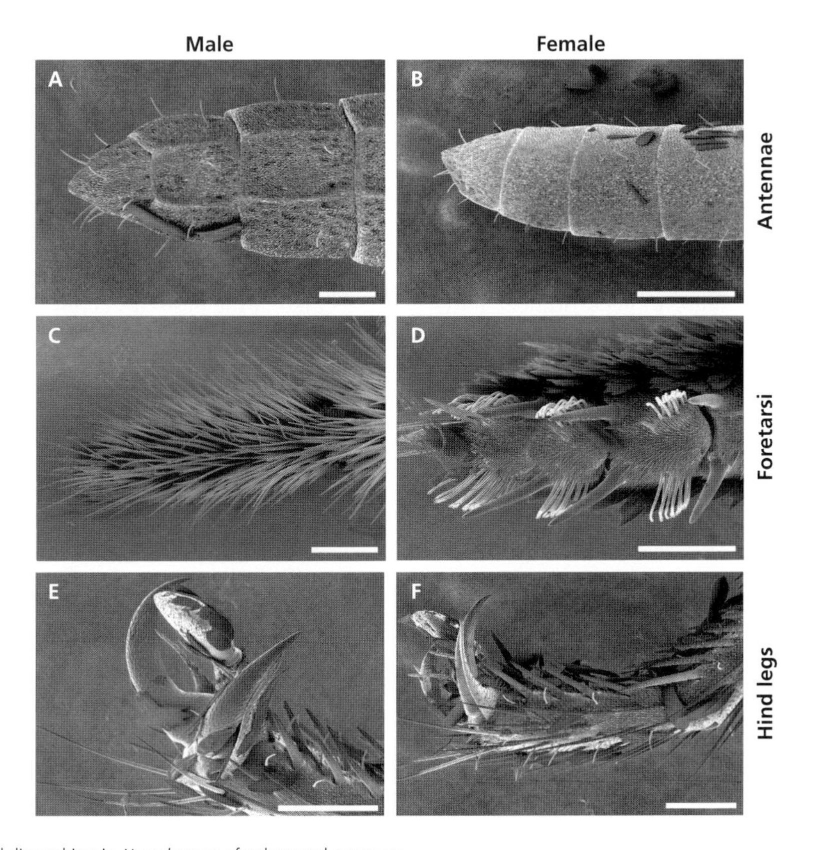

Figure 6.2 Sexual dimorphism in *H. melpomene* forelegs and antennae

Scanning electron micrographs of adult antennae and legs showing a sexual dimorphism in gustatory (trichoid) sensilla. Antennal tips of a male (A) and female (B) antenna. Foreleg foretarsi of a male (C) and a female (D). Four pairs of clumped taste sensilla are each found associated with a pair of cuticular spines on each female foot (only three are shown). Hindlegs of an adult male (E) and a female (F) showing individual gustatory sensilla (small white hairs).

order to puncture and taste host plants to assess their suitability for oviposition. *Passiflora* vines are commonly entangled with other plants, and chemical cues must play a key role in determining which leaves belong to which plant, and in separating the young shoots that are suitable for oviposition from older leaves. Unfortunately, nothing is known of the chemical cues used in this process, but given the importance of oviposition preferences, it would be an excellent area for future work.

To a much greater extent than in visual systems, gene duplication and diversification play a crucial role in the evolution of chemosensory systems. Genomic approaches are beginning to document the diversity of genes involved in chemosensation in *Heliconius* and other insects. Perhaps surprisingly, there is a similar diversity of chemosensory genes in *Heliconius* as in the moth *Bombyx mori*—despite our initial expectation that the switch to a diurnal lifestyle in the evolution of the butterflies might have led to an increased dependence on visual rather than chemical signals (The Heliconius Genome Consortium, 2012). Comparison of a few well-annotated insect genomes suggests that the major chemosensory gene families show a reduction in number relative to Hymenoptera and Coleoptera, but appear to be similar across the more closely related Diptera and Lepidoptera. Nonetheless, this apparent stability in gene number hides considerable turnover—for example, 64% of the gustatory receptors (GR) in *Heliconius* have arisen through gene duplication since the divergence from *Danaus*, while a larger proportion of the odorant receptors (ORs) show a 1–1 orthology relationship across *Heliconius*, *Danaus*, and *Bombyx* (Table 6.1).

Analysis of the expression patterns of these genes gives some insight into their possible function. In particular, 17 GRs were found to be expressed in the legs only of females (Briscoe et al., 2013). This parallels sexual dimorphism in taste hairs also on the forelegs (Renou, 1983; Briscoe et al., 2013; Figure 6.2), suggesting that the genes are likely involved in oviposition decision-making. In other insects, plant cyanogens can act as phagostimulants and as cues in choosing plants on which to lay eggs, and this might prove a fruitful area for future investigation of host selection in *Heliconius* (Gleadow & Woodrow, 2002).

In summary, there is considerable evidence for great genetic diversity and rapid evolution in the chemosensory system of *Heliconius*, but the work is in its early stages. We remain largely ignorant of the function of any of the chemosensory genes in regulating the ecology of the butterflies—whether it be detecting host plants or assessing possible mates.

6.4 Sensory ecology: hearing

Heliconius erato can hear. In an insectary, adults respond to sound by fluttering their wings. In a series of slightly gruesome experiments, Stewart Swihart (1967b) removed parts of the anatomy until the response disappeared. By a process of elimination, the hearing organs of *H. erato* were shown to be located at the base of the hindwing (Figure 6.3). The speed of the response to sound was very rapid—around 30 msec—so a likely explanation for the adaptive value of hearing is that it is used to escape predators rather than as a means of communication. There is, however, evidence for sound production by *Heliconius*—pupae occasionally

Table 6.1 Diversity of chemosensory genes across insect genomes

Gene family	Heliconius	Danaus	Bombyx	Drosophila	Apis	Tribolium
Odorant Binding Proteins (OBP)	43	35	46	52	21	49
Odorant Receptors (OR)	68	66	69	62	163	299
Gustatory Receptors (GR)	72	57	69	73	10	220
Chemosensory Proteins (CSP)	34	34	24	4	6	20
Ionotropic Receptors (IR)	31	27	25	66	18	23

Data shown for *H. melpomene, D. plexippus, Bombyx mori, Drosophila melanogaster, Apis mellifera,* and *Tribolium castanaeum*. Data from Briscoe et al. (2013), van Schooten et al. (2016) and Sánchez-Gracia et al. (2011). Not including pseudogenes.

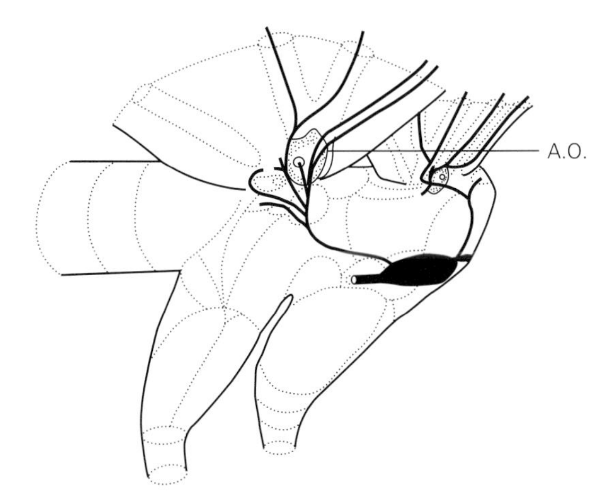

Figure 6.3 Diagram of the location of the hearing organ in *H. erato*

The thorax of *H. erato* is shown with the location of auditory organs (A.O.). Neural elements associated with these organs are indicated by heavy black lines.

make audible squeaking sounds (Gilbert, 1975), and female *H. cydno* have been recorded making clicking sounds during interactions with members of their own species and with other species (Hay-Roe & Mankin, 2004). The function of these sounds remains unknown.

In summary, *Heliconius* have highly developed vision, a complex chemosensory system, and a large brain for processing this information. The ecological complexity of the rainforest environment and the behavioural adaptations that *Heliconius* show to living in this environment likely play the major roles in shaping this system, but another important factor is finding and selecting a mate—sexual selection. In the remainder of this chapter, I will discuss mating and sexual selection in *Heliconius*.

6.5 Sexual selection and mating behaviour

Sexual selection is a form of natural selection caused by variation between individuals in their reproductive success. This can take several different forms, including competition between males for mating opportunities and female choice of males. Sexual selection is perhaps the most important cause of sexual dimorphism and male ornamentation, but *Heliconius* are unusual in that they lack obvious sexual dimorphism in wing pattern. Their appearance is primarily adapted for mimicry and defence (see Chapter 7). Although wing patterns do play an important role in mate selection, strong sexual dimorphism in chemical cues suggest that these are likely to be under stronger sexual selection.

Males of the majority of *Heliconius* species (Neruda, Laparus, burneyi, melpomene, and silvaniform clades) patrol home ranges and mate females as they encounter them after emergence. Males are also known to visit host plants (Mallet & Jackson, 1980), and probably mate females very soon after emergence. Indeed, it is extremely rare to find an unmated female of any *Heliconius* species in the wild, suggesting that mating is rapid and probably often occurs near the host plant (mated females can be recognized by their distinctive odour—see later) (Walters et al., 2012).

In some species found in tall primary forest, notably *H. numata, H. cydno*, and *H. nattereri*, males adopt a behaviour known as 'promenading' (Brown, 1970; Brown & Benson, 1974; Joron, 2005). This is a vigorous flight—often alternating between gliding and flapping—generally at a height of 3–10 m, but sometimes much higher in tall lowland forest, well above the height of the undergrowth. The males are presumably searching for females, who are generally to be found flying more slowly and lower down, where they are searching the understorey for host plants (Joron, 2005). Males can be attracted down to net height by vigorous waving of coloured rags (generally red, but white works well in the case of *H. cydno*). Males will sometimes fold their wings back and dramatically drop out of the sky from a great height in response to such a stimulus. The combination of colour and movement seems to be most effective, suggesting that the males are looking for fluttering females.

The suggestion that visual cues play a prominent role in finding females is supported by insectary experiments. By manipulating colour cues using painted females or paper models, Crane (1955) showed that red or orange colours were most likely to elicit courtship behaviour by *H. erato hydara* males. A combination of colour (hue) and movement were necessary to stimulate the initiation of courtship. Emsley (1970) also noticed that the yellow-and-red-banded species *H. besckei* was attracted to a race of *H. erato* with a similar combination of colours, but not to a race of *H. melpomene* kept in the same insectary that lacked a yellow band.

We have followed up on these early studies with choice tests involving fluttering models, made either from dissected female wings or printed paper (Jiggins, Naisbit, et al., 2001; Jiggins, Estrada, & Rodrigues, 2004). In both cases, males were attracted to the models, often initiating the preliminary stages of fluttering courtship (see later). When given a choice, males almost invariably showed a preference for their own wing pattern over that of conspecific races or other closely related species.

These results demonstrate that, as wing patterns diverge between populations (see Chapter 10), mating behaviour evolves in tandem. Novel wing patterns may arise primarily for signalling to predators, but they are also then used as cues for courtship. It seems most likely that the preferences arise as a result of selection for efficient mate-finding: movement and colour are the main cues by which patrolling males locate females.

There has been considerable excitement about the potential role of learning in mate selection, both in insects and vertebrates (Verzijden et al., 2012). The plasticity inherent in learnt mate preferences might contribute to rapid speciation. However, there is no evidence that learning plays a role in mating behaviour in *Heliconius*, at least in the context of colour cues. Isolating males from exposure to red, either in other butterflies or on themselves, had no effect on their courtship response (males were allowed to emerge in the dark before painting out their red bands) (Crane, 1955). Exposure to females of a different wing pattern race for several days prior to testing also had no effect (Jiggins, Estrada, & Rodrigues, 2004). Hybrid males that had their own patterns blacked out before exposure to light retained differences in preference associated with their wing patterns, indicating that self-matching is also not occurring (Merrill et al., 2011b).

Thus, all the evidence to date suggests that *Heliconius* mating preferences are genetically determined, in marked contrast to the extensive evidence for learning in foraging behaviours. Perhaps it isn't surprising that foraging, which involves interacting with a constantly changing environment, is more plastic than courtship. Other insects, including other butterflies, do show evidence for learnt mating preferences, however (Verzijden et al., 2012; Westerman et al., 2012). It is interesting that *Bicyclus* is highly variable in wing pattern, with distinct phenotypic morphs in wet and dry seasons, whereas *Heliconius* are generally monomorphic in wing patterns. Perhaps species with more variable phenotypes are more likely to evolve learning mechanisms to control mate selection.

In contrast to females, males need to mature before they can mate, and generally will not engage in courtship until at least 2–3 days (depending on the species) after emergence from the pupa. Once males have located a female, they begin an elaborate courtship (Figure 6.4) (Crane, 1955, 1957b; Klein & Araújo, 2010). If the female is flying, the male will chase her vigorously, sometimes forcing her to

Figure 6.4 Mating behaviour and courtship sequence (see Plate 11)

A courting pair of *H. ismenius* are shown at various stages in the courtship. (A) The male hovers over the female fanning his wings; (B) the male lands beside or even on top of the female; (C) the female can reject males by raising her abdomen and opening her pheromone glands; (D) the male attempts to mate by landing beside the female and bending his abdomen towards her; (E) during mating the male and female can stay attached for several hours. Note that panel E shows a different pair of individuals.

perch by flying above her and towards the vegetation. The chasing will sometimes involve mutual spiralling and circling.

Once the female is perched, the male begins hovering up and down behind or above the female, fanning her vigorously with his wings. If the female flies off, the male will take chase again, often initiating a rapid chasing flight until the female perches once again. (In a crowded insectary, the male will sometimes lose track of the female he was chasing and end up chasing another individual, which, frustratingly, may be another male.) During fanning, the male often adopts a distinctive hovering flight in which the fore- and hindwings separate, and the male exposes his androconial patches. The fanning has long been thought to blow androconial pheromones over the female (Crane, 1955, reports detecting a flowery scent from the male at this stage, although I have never been able to smell it).

The androconia are patches of scales in the overlapping region of the fore- and hindwing, with a distinct greyish sheen and marked sexual dimorphism in colour. In male *H. melpomene*, they have an unusual brush-like morphology on the hindwing not seen in the females, which is presumably associated with pheromone production, release, or both (Figure 6.5). Chemical analyses show that a bouquet of cuticular hydrocarbons can be detected in mature males about a week after eclosion, but they are not present in females or in other wing regions (Mérot et al., 2015; Vanjari et al., 2016). The composition of such chemicals differs even among closely related *Heliconius* species. The differences in chemical composition suggest that the signals may be under strong sexual selection, although this hypothesis remains to be demonstrated.

During fanning behaviour, the male is often above and facing in the same direction as the female, but may also fly around the female and face her from the front (Crane, 1957b, calls this 'disoriented fanning', but it seems to me a normal part of courtship behaviour rather than an 'irrelevant' behaviour). The female often adopts a characteristic rejection behaviour, in which she raises her abdomen vertically and everts her scent glands (note that Crane considered this part of the normal courtship sequence, but it is more likely a rejection

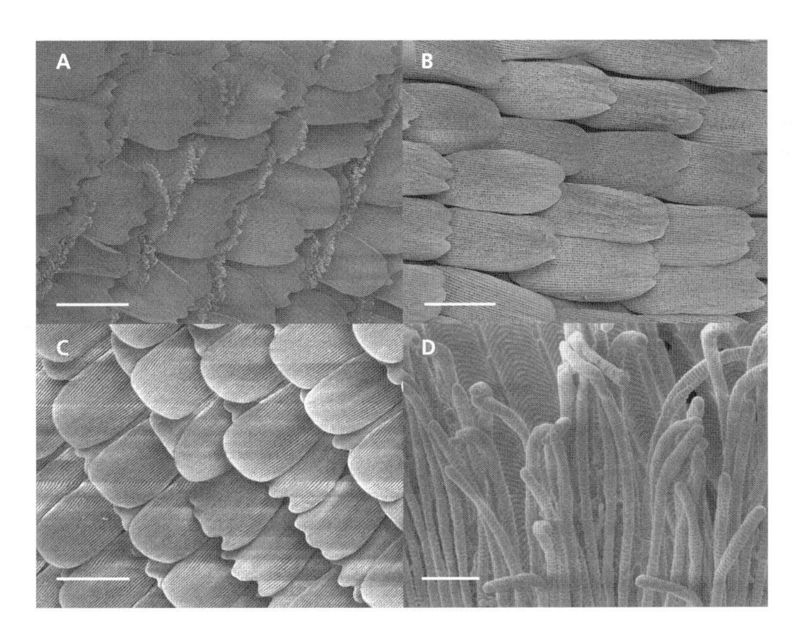

Figure 6.5 Structure of androconial scales on the wing

Scanning electron microscope image of putative androconial scales of *H. melpomene*. Scales in the region of the hindwing vein Sc + R1 and forewing vein 1A are shown. (A) Male hindwing; (B) male forewing; (C) female hindwing; (D) magnified view of brush-like structures of the androconial scales. Scale bars indicate 50 μm (A–C) and 2 μm (D).

behaviour—see also Mallet, 1986a; Klein & Araújo, 2010). If she is already mated, the glands will smell strongly of 'anti-aphrodisiac' pheromone (see later) and serve to deter the male unless he is highly motivated. At this stage, she may also flutter her hindwings while the forewings remain closed, which presumably serves to waft the unsexy pheromones towards the male.

6.6 Male wing pheromones

Fritz Müller (1912) was the first to describe scales on the wings of male *Heliconius* and speculate about a possible role in pheromone release. As noted previously, these androconial scales lie on the hindwing in the region of the wing that overlaps with the forewing. Researchers have begun to identify compounds found only in this wing region that are candidate pheromones (Mérot et al., 2015; Vanjari et al., 2016). Experiments with male wing models perfumed with wing extracts show that females alter their behaviour in response to this pheromone (Vanjari et al., 2016). As in other butterflies (Nieberding et al., 2008, 2012; Bacquet et al., 2015), it seems likely that the male sex pheromone is involved in conveying information used in sexual selection and female choice. Claire Mérot and colleagues have also performed experiments in which females were perfumed with the pheromone extracts of heterospecifics. Male courtship intensity was found to decrease, suggesting that a female-derived pheromone is also involved in triggering male courtship (Mérot et al., 2015).

If the female is receptive, the male will perch beside her, facing in the same direction, and bend his abdomen at right angles to the body. The male searches for the female abdomen between her hindwings with his claspers, and often mating happens rapidly at this stage. The female rejection behaviour described earlier prevents mating at this stage—if her abdomen is raised, it is impossible for the male to make contact with his own claspers. Once mating occurs, the male moves slowly around to face in the opposite direction to the female, their abdomens now joined. During mating, the male may take flight, with the female hanging below. Mating itself can last from an hour up to several hours or even overnight, depending on the status of the male—depleted males that have mated recently often stay in copula for much longer.

This courtship sequence is highly variable in duration (Klein & Araújo, 2010). If a virgin female is receptive, or still recently emerged from the pupa and unable to resist, the male may mate almost immediately without any of the preliminary behaviour. On the other hand, if the male is persistent and the female unreceptive, courtship may continue for many hours. This is certainly true in the insectary, where the female cannot escape, but repeated, persistent, and apparently unsuccessful courtship is also seen in the wild, with an ongoing sequence of chasing, fluttering, and attempted mating. In the insectary, such extended chasing and courtship sequences almost never result in mating, and presumably—given very low remating rates of females (see later)—this is also true in the wild. The only attempt so far to quantify courtship, by Klein and Araújo (2010) in *Heliconius erato*, demonstrated great variability in the courtship sequence and considerable investment in chasing made by males that never result in mating.

The behaviours are similar in all heliconiines, although there is undoubtedly variation between species. To date only one attempt has been made to characterize interspecific variation in courtship behaviour (Crane, 1957b).

6.7 Sexual behaviour—pupal mating

Some *Heliconius* are renowned for a dramatic mating behaviour in which males mate females as they are eclosing from the pupa (Figure 6.6) (Gilbert, 1975, 1976; Deinert, Longino, & Gilbert, 1994). Edwards (1881, p. 211) first reported males of *H. charithonia* sitting on pupae. After speculating about other functions—'Did they come to protect the chrysalis or to assist the butterfly to come forth, or was there anything of sexual desire?'—he correctly concludes that the sitting behaviour was associated with mating.

It is now known that males search larval host plants for fifth-instar larvae, which they closely inspect, returning regularly after the larvae have pupated (Longino, 1984). Experiments with *H. charithonia* have shown that males use odours produced by herbivore-damaged plants to help locate larvae

Figure 6.6 The behavioural sequence of pupal mating in *H. charithonia* (see Plate 12)

Three males are shown contesting for a position on a female pupa (A), one in flight and two perched on the pupa. As the female begins to emerge, the male inserts his abdomen into the pupal case (B and C). In the inset (B), one male attempts to mate (left) while a second male is perched on the pupa (right). Finally, as the female ecloses from the pupa, she is already mating (D). Photos courtesy of Andrei Sourakov.

(Estrada et al., 2010). The males sit on pupae the day before eclosion, often in groups of two, three, or four, and finally compete to mate with the females as they emerge. In some species, such as *H. hewitsoni* and *H. charithonia*, the males may actually insert their abdomen into the pupal case and mate with the female before she emerges (Gilbert, 1991). In other cases, females may be mated as they eclose. Just under half of *Heliconius* species mate in this way, and the behaviour seems to have a single origin in the genus, being confined to a monophyletic group consisting of the erato and sapho clades (Beltran et al., 2007).

The behaviour has been best studied in *Heliconius hewitsoni* at Sirena, Costa Rica. Like several species in the sapho clade, *H. hewitsoni* larvae feed in large gregarious groups on rapidly growing shoots of a single Astrophea subgenus host, *Passiflora pittieri* (Deinert et al., 1994; Deinert, 2003). Consequently, pupae are often found in large aggregations, which likely serves to enhance the intensity of mate competition.

Males locate young pupae and initially visit both male and female pupae. However, a few days before eclosion, they are able to distinguish pupal sex, and from then on concentrate on female pupae. The chemical cues used to distinguish males from females have been identified (in *H. erato*) as the compounds linalool and linalool oxide, found in male and female pupae, respectively (Estrada et al., 2010). At this stage, males begin to sit on the pupa, folding their wings forward to cover the pupa and prevent other males from gaining a purchase. There is considerable competition between males attempting to guard a pupa. For example, of 27 pupal mating events studied, 235 males were recorded and each pupa was visited by nearly 9 males on average (Deinert et al., 1994).

At this stage, the larger males—in particular, those with longer wings—had an advantage in competing for a position on the pupa. Sitting males were significantly larger than those that failed to gain a seat. Just before eclosion, one or two males puncture the cuticle of the female pupa, inserting up to three abdominal segments. Mating takes place as the female begins to emerge from the pupa; females mate only once.

At this final stage, there is an advantage for smaller males—in particular, those with a shorter abdomen—presumably they are more agile in puncturing the cuticle. Once on the pupa, males that succeed in mating are significantly smaller than those that failed to mate. Overall, however, there is no evidence for size selection—among all males visiting pupae, there was no significant difference in size between mated and unmated males. The two stages of competition balance out, with selection for larger males at the sitting stage balanced by selection for smaller males during mating.

As noted, long wings are favoured initially and short abdomens later. These two characteristics are obviously strongly correlated—larger males tend to be larger in all directions. But one would predict that pupal mating selection might favour a change in allometry, such that the ratio of wing length to abdomen length would increase. There is some evidence that this is the case, as pupal mating group species do indeed have significantly larger wing:body size ratios (Deinert et al., 1994). However, there is the problem that pupal mating has arisen only once, so it is difficult to determine whether this change is a direct result of pupal mating selection. Nonetheless, species with alternative mating strategies might provide suitable variation to test for selection on the allometry of body size (see later) (Hernandez & Benson, 1998; Mendoza-Cuenca & Macías-Ordóñez, 2010).

6.8 Does pupal mating involve sexual conflict?

It is easy to envisage a scenario for the evolution of pupal mating behaviour. Males are already endowed with the ability to locate specific sites in the forest, and have the genetic machinery necessary to find *Passiflora* host plants, as a result of natural selection on females. Patrolling these plants would increase the chances of finding newly emerged virgin females, and hence increase the chances of mating (see later—remating rates in all *Heliconius* are low, so the best chance for mating is to find a virgin female). This is essentially the strategy taken by 'non-pupal mating' *Heliconius*, where males are also occasionally seen fluttering around host plants. It would then be a small step to actually locate pupae, and defend them against other males with similar intentions.

But what about the female? This behaviour does not seem to be in the interest of the female. First, there is likely a direct cost to pupal mating: eclosion is an extremely delicate moment in the life of a butterfly, and presumably one doesn't really want to emerge into the world to find a gang of males competing violently to take advantage of your situation. Until the wings are fully extended and hardened, the eclosing adult is highly vulnerable to disturbance. Indeed, it has occasionally been observed in the insectary that male competition for a female leads to the female falling to the floor during eclosion, which is invariably fatal (Gilbert, personal communication). A high cost indeed. Second, there is likely to be a more subtle cost in terms of lost opportunities for female choice. Females have absolutely no ability to reject males and choose their mates, and therefore lose out on the opportunity to select the 'best' mate. The optimal mate for her may not necessarily be the male that can most effectively compete for a position on the pupa.

Even so, there may be advantages to the female. First, if population densities are very low it may be hard to find a mate, so pupal mating ensures that a mate is found early in life. Second, males make a significant contribution to female nutrition via the spermatophore (see later)—so it may be important to maximize early reproductive opportunities by getting mated quickly to obtain this nutrient boost. The fact that pupal sex is clearly signalled, and that the chemical compounds involved are presumably fairly easy to synthesize, has led to the suggestion that the female is complicit in sending this signal—the argument being that if female pupae wanted to smell like males, it would be easy for them to do so (Estrada et al., 2010).

Although pupal mating has been observed in many erato and sapho group species, both in the insectary and in the wild, it remains unclear what proportion of matings occur this way in natural populations. It is an important question, as it will determine the potential for female choice, which, as recognized by Darwin, can be a critically important force in driving sexual selection.

In *H. hewitsoni*, it seems clear that most pupae are found by males and pupal mated, although field studies tend to be carried out on high-density populations where pupal mating may be more likely. It is also common in *H. charithonia*, and I have seen it occur regularly in insectary populations of *H. sara*. In contrast, in *H. erato*—although pupal mating does occur in the insectary (Gilbert, 1976)—it has not been well documented in the wild. In the dispersal study described earlier (Mallet, 1986b), female pupae were placed into the wild near their natural host plants but were never pupal mated. Insectary populations of *H. erato* do not always pupal mate (McMillan et al., 1997), and males show similar courtship behaviour to that seen in non-pupal mating species, which leads to mating with adult females in insectary conditions (Crane, 1955). It has been argued that males need to identify larvae before pupation in order to trigger pupal mating behaviour, which might explain the lack of observations of the behaviour (Gilbert personal communication), but more recent experiments on *H. erato* in Panama, in which larvae were placed on host plants in the field, also showed that most females mated after eclosion rather than as pupae (Brodie et al., 2015).

In summary, it seems likely that the traditional dichotomy between pupal mating and non-pupal mating species may be more of a continuum. In the species traditionally considered as non-pupal mating, such as the melpomene and silvaniform groups, males may patrol host plants and mostly mate females soon after emergence (cf. Estrada & Gilbert, 2010). In species such as *H. erato*, males do defend pupae, but commonly mate after female emergence. Finally, species in the sapho clade are more strictly pupal mating and both defend pupae and commonly mate females during eclosion (Longino, 1984; Deinert, 2003). Nonetheless, mating behaviour is poorly studied in the wild across most species, and the extent to which females might be able to exert mate choice remains an open question.

6.9 Alternative mating strategies—territoriality and patrolling

A major area of interest in studies of sexual selection is the evolution of alternative mating behaviours. In some cases, species can show genetically distinct sexual phenotypes. One example is the white-throated sparrows found in North America, with tan and white-striped morphs that show different levels of aggression and distinct mating preferences (Horton, Moore, & Maney, 2014).

In other species, there are environmentally determined phenotypes, which are often condition-dependent. For example, some beetle species have short and long-horned male morphs that either sneak matings or compete directly for females. These morphs develop as a result of a non-linear allometric relationship between body size and horn length—below a certain threshold body size, it is no longer worthwhile for males to develop a horn and try to fight to defend a tunnel entrance (Moczek & Emlen, 2000). In *Heliconius*, there is also evidence for different mating phenotypes that result from the intensity of male-male competition induced by pupal mating.

In erato and sapho clade species, males that are unlikely to succeed in competition for pupal matings may adopt alternative strategies. Seitz (1913) originally described, in *Heliconius* and *Eueides*, 'the habit of flying for hours or half days at a time up and down for a certain distance, turning sharply around at a certain point and returning the same way' (cited by Benson, Haddad, & Cardoso, 1989, p. 34). This represents truly territorial behaviour quite distinct from the patrolling of home ranges and 'promenading' described earlier. Males of several species (*Heliconius sara, H. leucadia*, and *E. tales* in Brazil; Benson et al., 1989; *H. sapho* on Pipeline Road, Panama; Jiggins, personal observation) defend light gaps, flying to and fro repeatedly along an area of 10–15 m, commonly along a forest opening created by a trail (Figure 6.7).

Benson and colleagues (1989) describe the remarkable defensive behaviour of *H. sara* males in Brazil, whereby intruder and resident males drop to the ground together, circling rapidly around one another at a height of only 5–15 cm above ground. Generally this circling ends with the resident getting underneath the intruder and chasing him up with zig-zagging flight, sometimes dashing at the intruder to chase him up and away. In this way the intruder is ejected from the territory, sometimes with the chase moving up to 40 m away before the resident abandons the pursuit. Changes in the territory holder were generally associated with extended battles such as this, although expulsion of the resident male was never observed directly. In some cases, there was no contest: the intruder flew in a leisurely manner out of the territory, generally with the resident following some 0.5 m below the height of the intruder. It was suggested that this gentle retreat is a form of appeasement behaviour by the intruder, signalling that he poses no threat to the resident (Benson et al., 1989).

The rather surprising fact about these territorial males is that they are on average smaller than males caught away from territories (Benson & Hernandez, 1991; Hernandez & Benson, 1998). The difference represents around 3% of wing length. Smaller males spend more time defending a territory, and there is less tendency for disputes to escalate in which the

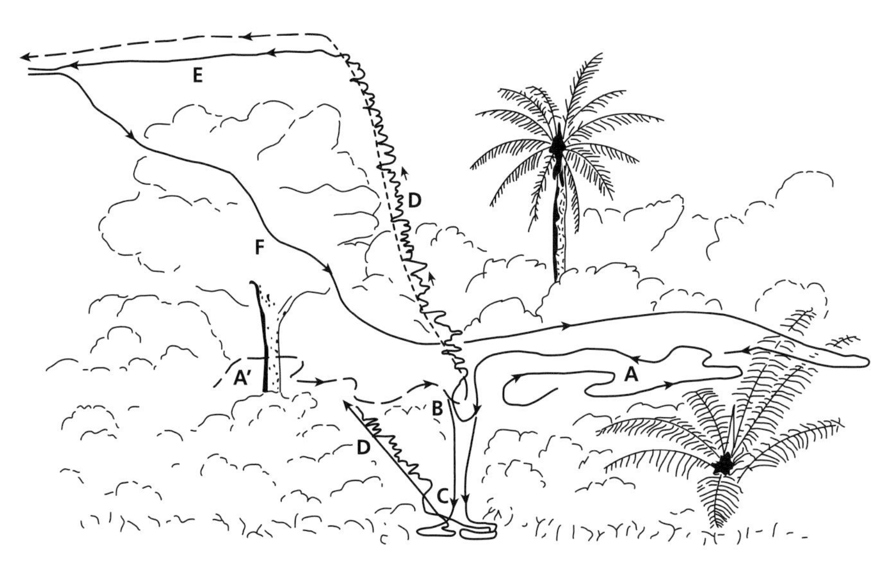

Figure 6.7 Territorial dispute between males of *H. sara*

Typical territorial interaction in *Heliconius sara* at Serra dos Carajás, Para, Brazil. Resident *H. sara* (solid line) promenades (A) or perches on territory while intruder (broken line, segment length approximately proportional to intruder velocity) patrols along forest margin (A'). One or both butterflies attack, with resident attempting to get below invader (B) and both butterflies sometimes diving to ground where they circle adjacent to each other (C). Intruder starts flying slowly upwards with resident darting back and forth below and behind it (D) until reaching tree-top level, where invader may dash away with resident in pursuit (E). Resident returns to territory where it flies briskly over core and peripheral areas as it resumes promenading.

resident is much smaller than the intruder. In other words, where the size difference is greater, the intruder is less likely to challenge for the territory. Usually, one would expect that larger males would be the strongest, and therefore better able to defend a territory, but this does not appear to be the case. So what is going on? It is possible that smaller males are actually better at defending territories, perhaps having greater agility. However, this seems unlikely—large size is generally a result of better larval nutrition, and is associated with greater body mass and presumably also stronger flight muscles.

An alternative suggestion is that defence of a territory is a second-best alternative strategy for small males (Hernandez & Benson, 1998). Pupal mating is commonly observed in *Heliconius sara*. The work on *H. hewitsoni* described earlier suggests that larger males are more competitive, at least in the initial 'sitting' stage of defending pupae, and may therefore be more likely to obtain a mating. If smaller males have little chance in pupal mating contests, perhaps they defend a territory as the 'best of a bad job'. This is also known as a paradoxical strategy, where the small males would have more to gain or less to lose from territory defence. Defending a territory may be their only chance of obtaining a mating with the occasional female that was not mated on emergence from the pupa (Hernandez & Benson, 1998), so it is worth risking all in a territorial contest.

In *H. charithonia*, another mating strategy has been termed 'patrolling' (Mendoza-Cuenca & Macías-Ordóñez, 2005). Smaller males flew farther and visited more flowers; presumably they were adopting a strategy of searching widely for unmated females. This patrolling strategy is more similar to that adopted by males in non-pupal mating species.

In an impressive long-term study, Luis Mendoza-Cuenca and Rogelio Macías-Ordóñez followed male mating success over several years and provided strong support for the hypothesis that there are two mating strategies (Mendoza-Cuenca & Macías-Ordóñez, 2010). The distribution of male size was bimodal, with smaller males patrolling and never visiting areas where pupae were located. Wild pupal matings were always by larger males. Competition was intense, with up to 11 males visiting a single pupa and up to 5 sitting on a pupa at one time. A set of 50 females was released as pupae, none of which were pupal mated. The virgin females that emerged were chased by patrollers, with five observed matings. All 22 females that were recaptured had been mated as adults. This clearly shows that adult mating does occur, and supports the idea that the distinct class of patroller males dominate these matings. Nonetheless, it remains unclear what proportion of females are adult mated, and therefore what the payoff is for the patroller males. It is also possible that these strategies are genetically controlled, although that remains to be demonstrated.

Mendoza-Cuenca compared two populations to gain some insight into the conditions that favour the mixed mating strategy (Mendoza-Cuenca & Macías-Ordóñez, 2010). In a second population at Zimpizahua, female emergence was more synchronous, and there was no evidence for a dual strategy. Instead, all males competed for pupal mating opportunities. Because more females were available at the same time, it was argued that the intensity of male competition was less intense and there was weaker selection favouring the alternative strategy. Overall, therefore, there is intriguing evidence for alternative mating strategies among males in pupal mating species, but no such evidence in non-pupal mating species.

6.10 The male contribution: toxins, food, and unsexy pheromones

The dynamics of sexual selection are strongly influenced by the patterns of investment of males and females in reproductive effort. As recognized by Darwin (1871), females are generally coy, more choosy in selecting mates, whereas males are more ardent and undiscerning. Bateman (1948) recognized that it is the greater variance between males in reproductive success that drives this pattern, such that males benefit more than females from multiple mating. Variance in male fitness is sometimes therefore considered a useful measure of the 'opportunity for sexual selection' on males (Parker & Birkhead, 2013).

More recently, research has focused on the importance of female choice in sexual selection, and in particular the importance of multiple mating by females (known as polyandry) (Parker & Birkhead, 2013). Many more species are polyandrous than was previously recognized, and females can benefit from multiple mating both indirectly (by producing more genetically diverse offspring) (Fedorka & Mousseau, 2002; Tregenza & Wedell, 2002) and directly (from nuptual gifts or by gaining extra paternal care for their offspring). In *Heliconius*, males make a significant investment in mating through the transfer of a nutrient-rich spermatophore, but polyandry is rare.

Many male insects transfer a spermatophore during mating that not only contains sperm for fertilization, but also represents a male contribution to reproductive effort (Khalifa, 1949). In *Heliconius*, the spermatophore is so large and tough that it can often be seen and felt in the female abdomen after mating. If male larvae or pupae are injected with radio-labelled amino acids and then allowed to emerge and mate, the females that they mate with lay eggs containing radioactivity (Boggs, 1981b; Boggs & Gilbert, 1979). Even unfertilized eggs contain labelled carbon, indicating that this transfer of material is not simply due to fertilization. Thus, nutrients used by a female in egg production come in part from the father. The radio-labelled carbon is found throughout the body of the female within a week of mating, indicating that the nutrients from the male are also used in general female nutrition as well as egg production (Boggs & Gilbert, 1979). The transfer of nutrients is a general phenomenon in butterflies and has been documented in *D. iulia, H. charithonia, H. ethilla*, and many other butterfly species (Boggs, 1981b; Boggs & Gilbert, 1979).

There is also evidence in several *Heliconius* species for transfer of cyanogens in male spermatophores, which presumably contribute to chemical protection of the female and her eggs (Cardoso & Gilbert, 2007). A similar transfer occurs in the burnet moth, where there is also evidence that females may reject males that do not have a sufficiently high cyanogen content. Painting of cyanogens onto unattractive males can make them more acceptable to females—a particularly toxic form of sexual attraction (Zagrobelny & Møller, 2011). It remains unknown whether similar signalling happens in *Heliconius*.

In pupal mating species, the spermatophore is gradually degraded by the female—after about two weeks, all that remains is an 'orange oily substance' in *H. charithonia* (Boggs, 1981b), although a later study in *H. erato* suggests that degradation generally takes over three weeks, but also that it is highly variable (Walters et al., 2012). There is evidence that distention of the bursa copulatrix acts as a disincentive for further matings in butterflies, so one reason for males to transfer a large spermatophore is to ensure paternity for a longer period before the female remates.

Unsurprisingly, there is evidence that spermatophore production is costly to males. In *D. iulia* males, there was a correlation between time since last mating and both mating duration and spermatophore size—a longer recovery period after a first mating means that males can subsequently mate more quickly and transfer a larger spermatophore (Boggs, 1981b). The spermatophore produced by the male is presumably replenished gradually after mating, and when remating occurs before replenishment is complete, the male mates for longer while he attempts to reallocate resources for transfer to his mate. Although Boggs found no correlation between time since last mating and mating duration for *H. charithonia* males (Boggs, 1981b), in my experience, males of *H. erato* and *H. melpomene* also tend to mate for much longer in a second mating after they have recently mated.

Further evidence for the costly nature of mating to males is the observation that male *H. cydno* that have more lifetime matings also collect more pollen (Boggs, 1990). Overall, therefore, there is strong evidence that spermatophore production is costly, and that replacement of the resources lost during mating takes several days.

From the female perspective, there is considerable evidence for the benefit of nutrients transferred during mating. Female *H. cydno* that remate more often tend to collect less pollen. This suggests that there is a nutritional balance between income from mating versus that from feeding—mating more often and getting more spermatophores means that feeding is less important (Boggs, 1990). This would also imply that there is some cost to extra feeding, perhaps in terms of lost opportunities to search for oviposition sites or increased predation.

The work of Boggs (1990) also raises the intriguing possibility that there might be polymorphism for female remating, with some individuals choosing to enhance their nutrition by remating whereas others prefer to feed. In *H. charithonia*, remating is far less frequent, so females apparently do not have the option of multiple mating to replenish nutrition. Instead, female *H. charithonia* greatly increase feeding rate 15–20 days after mating—this is roughly the time at which the spermatophore has become completely degraded and when the male-contributed nutrients in eggs decreases (Boggs, 1990). Therefore, it seems likely that female *H. charithonia* have to increase feeding to compensate for the reduction in male-derived nutrients. Unfortunately, there is as yet no study of lifetime fecundity that directly measures the benefits of multiple mating in *Heliconius*.

This brings us to remating rate. How often do females actually remate in the wild? This has important implications for the action of sexual selection. If females are pupal mated and never have an opportunity to remate, then there can be no female choice, and the only arena for sexual selection is inter-male competition for pupal matings. However, if females mate multiple times as adults, there is an opportunity for female as well as male choice and sperm competition, both of which play an important role in sexual selection in other insects.

The accepted wisdom among *Heliconius* biologists is that females of pupal mating species in the erato and sapho groups mate only once, whereas the non-pupal mating melpomene, silvaniform, and Neruda/burneyi/Laparus groups mate multiple times. Thus, for example, in the insectary populations described earlier, *H. charithonia* mated on average only once (1.0±1.3; x±SD), while *H. cydno* mated on average 2.8 times (2.8±2.8; x±SD) (Boggs, 1990). Data on spermatophore counts from wild populations largely confirm this. The adult mating species showed a pattern consistent with a constant but low probability of remating with female age. In the most heavily sampled species, *H. melpomene*, around half of the females in the two oldest age classes, as judged by wing wear ('worn' and 'ragged'), were remated at least once. There was less information for other species in the melpomene and silvaniform clades, but patterns were broadly similar (Walters et al., 2012).

In contrast, of 251 females sampled of *H. erato*, only three were double mated. These were all very young individuals as judged by wing wear, suggesting that they had been remated during or soon after eclosion. Perhaps pupal mating sometimes involves remating, or perhaps females that are pupal mated by an unsatisfactory male may occasionally choose to remate. Nonetheless, no fresh spermatophores were observed in older females, nor were older females ever found with two spermatophores.

Therefore, despite the confounding factor of spermatophore degradation, it seems that *H. erato*—and most likely also the other species in the erato and sapho clades—are largely monandrous. In contrast, *H. melpomene* shows a low but consistent probability of remating. Sampling was comprehensive for *H. erato* and *H. melpomene*, but only a few females were sampled for many other species, so there may be further undocumented variation in mating rate. One species in particular that merits further investigation is *H. doris*, which readily remates in insectary cultures (Liz Evans, personal communication), and for which four of five females showed more than one spermatophore (Walters et al., 2012). A single *H. doris* female has been found with seven spermatophores in Peru (Claire Mérot, personal communication).

6.11 Both males and females are choosy

The large investment made by males at mating means that in general both sexes are choosy during mating. Experiments involving closely related species and races show that males are reluctant to court butterflies with wing patterns different to their own or with divergent pheromones (see Chapter 11). However, females are choosy also, and commonly reject males. Therefore, the dichotomy of ardent males and choosy females is less marked in *Heliconius*, where both sexes make a large investment in contributing nutrients to reproduction.

6.12 Anti-aphrodisiac pheromones

In addition to nutrients and toxic compounds, females also receive a chemical chastity belt during mating. In virgin females, the extrusible glands at the end of the abdomen are small and white. After mating they expand, become bright yellow, and smell strongly when extruded. In *H. erato*, in particular, the smell (said to be similar to witch-hazel) can be extremely powerful—sometimes in the forest the whiff of an *H. erato* female can be detected before sighting the butterfly.

Gilbert (1976) originally recognized that the scent comes from the male at mating. Different races of *H. erato*, with distinct odours, were mated to one another (Gilbert, 1976). In particular, *H. chestertonii*, a genetically divergent relative of *H. erato* from the Cauca Valley, Colombia, has a 'more fragrant' odour. Females mated to this race had the characteristic smell of the male rather than that of their own race (Gilbert, 1976). Although this experiment was rather subjective, relying on the experimenter's ability to detect odour differences, later chemical analysis confirmed the result (Schulz et al., 2008).

H. melpomene is said to smell of fried rice (Turner, 1975b)! This complex odour bouquet contains primarily the volatile compound β-*ocimene*, with many other esters and alcohols, and is produced by males (Schulz et al., 2008). The bouquet is absent from virgin females. The transfer to females during mating was confirmed by feeding males with radio-labelled glucose, which was incorporated into β-*ocimene* in the male and then transferred to the female. The mechanism of transfer is unknown, but there is a 'pouch lined with glands' in the male claspers that may represent the site of synthesis (Eltringham, 1925; Gilbert, 1976). The female has tiny 'stink clubs' adjacent to the large abdominal glands, and both glands and clubs likely play a role in receiving, storing, and disseminating the pheromone (Crane, 1955; Gilbert, 1976). In particular, the glands fill out and become yellow after mating, suggesting a probable role in storage of the pheromone.

The 'anti-aphrodisiac' nature of the pheromone was originally suggested based on the observation that pupal mating aggregations of ardent *H. erato* males would disperse if a smelly female was brought close. Such males are otherwise extremely difficult to dislodge. More recently, once the pheromone was identified in *H. melpomene* as β-*ocimene*, bioassays could be carried out with the compound directly applied to females (Schulz et al.,

2008). Virgin females that were otherwise receptive became much less attractive to males when painted with β-*ocimene*. Furthermore, the effect lasted longer when β-*ocimene* was mixed with low volatility esters, suggesting that the esters in the natural odour bouquet act to slow evaporation of the pheromone. In *H. melpomene* and related species, males also smell quite strongly, so it seems possible that the anti-aphrodisiac pheromone acts by making females smell more male-like (Schulz et al., 2008).

6.13 Multimodal signals and aposematism

It was originally suggested that the characteristic *Heliconius* smell was an anti-predator mechanism (Crane, 1955). It seems likely that these chemicals also serve this additional function, especially as *Heliconius* are particularly pungent compared to other butterflies. Indeed, poison and sex are inextricably linked in many butterflies, as distasteful compounds are quite often also used as sex pheromones (Boppré, 1984). The odour bouquet of *H. melpomene* contains three pyrazines that are general warning odours of chemically defended insects and are known to deter rats and birds (Schulz et al., 2008). Thus, the smelly chemical signals probably act to deter predation and to enhance the learning of warning colours by predators (Lindström, Rowe, & Guilford, 2001). It seems likely that the complex mixtures of chemicals found in analysis of pheromone glands in males and females have multiple functions. A challenge for the future will be to associate particular functions with particular chemical compounds.

6.14 Anti-aphrodisiacs as an arena for sexual conflict

The evolutionary advantage to donor males (the signaller) of transferring the unsexy pheromone is clear—they prevent or reduce remating in the female and ensure their own paternity of her offspring. However, for the female and for subsequent courting males (receiver males), there is potential for conflicting interests. In the short term, females clearly benefit from reducing harassment by courting males, which likely has costs. When approached by males, mated females raise their abdomen and evert their yellow pheromone glands, releasing the pheromone.

Because males cannot force copulations on adult females, such behaviour in the interest of both parties: receiver males avoid wasting time on fruitless courtship of unreceptive females, and females reduce harassment costs that likely include increased wing wear, spend less time in avoidance behaviour, and perhaps decrease their exposure to predation. Mating itself may also be costly, as a mating pair are easy to capture and might be vulnerable to predation.

However, as time goes by and the resources in the original spermatophore are used up, there is the potential for diverging interests. The signaller male would prefer the female not to remate, so that his sperm can continue to fertilize her eggs. However, the female may benefit from multiple mating because each spermatophore is a valuable resource that contributes to female reproduction and nutrition. Receiver males may therefore benefit from persisting in courtship of mated females even in the presence of the pheromone, just in case they can find a receptive individual that is ready to remate. Although the pheromone is a strong deterrent to courtship, males do indeed persist in protracted courtship of mated females. Although the intensity of the pheromone signal likely declines as the compounds transferred by the donor male are used up, it has been argued that the female is unable to control the release of the pheromone, and therefore may not be able to honestly signal her receptivity (Andersson, Borg-Karlson, & Wiklund, 2004).

An arena for evolutionary conflict is thereby created. Donor males should be under selection to evolve novel signals that increase deterrence, either by persisting in the female for longer or by lowering the threshold for deterrence of receiver males. In contrast, receiver males will be under selection to persist in courtship as soon as there is any chance of a female being receptive. Females might be under selection to increase the honesty of the pheromone as a signal of their receptivity, perhaps by degrading the pheromone once they are ready to remate.

The conflict is therefore primarily due to male-male competition, mediated by a pheromone that is released from females. The conflict model therefore predicts that there should be rapid evolution of the pheromone composition, and that it should be exaggerated in species with higher remating rates, where there is greater potential for sexual conflict.

This prediction is supported by the analysis of compounds across species. Even to our own sense of smell, there are quite marked differences in odour between species (I rather like the sweet smell of *H. sapho*). Chemical analysis has shown that there is considerable divergence between species in the composition of male and female abdominal chemicals (Ross et al., 2001; Schulz et al., 2007; Estrada et al., 2011). Furthermore, the rate of pheromone evolution is faster in the melpomene clade than in erato and sapho clades, consistent with the prediction of faster evolution associated with higher rates of remating (Estrada et al., 2011). Even closely related species in the non-pupal mating clade, such as *H. melpomene* and *H. cydno*, or *H. numata* and *H. ismenius*, are highly divergent in their pheromone composition (Estrada et al., 2011). In contrast, major compounds are shared across quite divergent species such as *H. sapho, H. charithonia, H. sara*, and *H. hewitsoni* in the pupal mating clade. Nonetheless, we need to know more about the function of different chemicals as anti-aphrodisiacs, attractant pheromones, or defensive compounds before we can fully understand why they are evolving so rapidly.

In addition to the pheromone compounds, males transfer a variety of proteins in the spermatophore. In other insects, these seminal fluid proteins are known to mediate sperm competition and sexual conflict, in some cases increasing short-term female fecundity at a cost of reducing longevity (Chapman et al., 1995; Wigby & Chapman, 2005). This benefits the male but at a cost to the female. These proteins are commonly some of the fastest-evolving loci in the genome, thought to be a result of intense diversifying selection due to sexual conflict (Swanson & Vacquier, 2002).

Heliconius are interesting in this context because low remating rates mean that there is relatively little opportunity for sperm competition, providing a contrast to other insects in which seminal fluid proteins have been studied. Proteomic analysis of male accessory glands in *H. melpomene* and *H. erato* has identified 51 candidate seminal fluid proteins, 40 of which are found in both species (Walters & Harrison, 2010). Ten are serine proteases (chymotrypsins), which are commonly found in seminal fluid proteins across insects and mammals. These proteins evolve faster than randomly chosen control genes (i.e. they have higher average ratios of non-synonymous to synonymous substitutions, dN/dS).

Two proteins had dN/dS ratios significantly > 1, which is generally taken to provide evidence for positive selection. (Most coding sequences in the genome have $dN/dS < 1$ due to purifying selection removing deleterious amino acid substitutions, while $dN/dS = 1$ is consistent with neutral evolution.) The pupal mating species in the erato and sapho clades were expected to show slower rates of evolution because of reduced sexual conflict in the absence of multiple mating, but in fact this group showed faster rates (Walters & Harrison, 2011). Perhaps the seminal fluid proteins are not needed in the absence of multiple mating and are therefore released from purifying selection. Further work on the actual function of the proteins is needed to follow up on these results.

6.15 Future directions

Heliconius are an excellent system for studying learning and behavioural plasticity, which remains poorly understood in insects. One challenge is to develop assays of spatial learning, which is clearly important in the wild but has proven difficult to study in controlled conditions. Studies of brain architecture are in their infancy, but *Heliconius* are interesting subjects for studying the influence of learning on brain architecture, as has been demonstrated in other butterflies (Snell-Rood, Papaj, & Gronenberg, 2009; Montgomery et al., 2016). Do butterflies that learn to navigate more complex environments develop larger brains?

The visual systems of the heliconiines are becoming well understood, but we still do not fully understand the behavioural implications of the discovery of duplicated UV opsins. Behavioural assays to test for true colour vision in the UV range would be particularly valuable, as well as electrophysiological studies of the diversity of receptor types in *Heliconius* eyes.

A benefit of multiple mating in terms of increased fecundity has not been convincingly demonstrated in *Heliconius* as it has in other butterflies (Wiklund et al., 1993). Indeed, the high quality of adult food resources may mean that the benefits of multiple mating are more marginal in than in other species, which perhaps explains the relatively low rate of remating. Understanding the dynamics of sexual conflict will require a better estimation of the costs and benefits of remating.

It is also intriguing that in the erato and sapho clade species, the females smell far more strongly than males, whereas in the melpomene clade the males can be just as smelly as females. It would be interesting to investigate the possibility that this finding reflects selection for females to more rapidly degrade the pheromone in order to signal receptivity to remating—it is the melpomene clade species that are more likely to remate. Finally, intraspecific variation in pheromones, which was originally exploited in their identification (Gilbert, 1976), might prove a useful system in which to test the predictions of the sexual conflict model and to understand the fitness consequences of individual variation in pheromone composition.

Finally, it is exciting to speculate about how genomics can contribute to our understanding of sensory systems. While opsin evolution is relatively well understood, our understanding of chemosensory systems lags well behind. Which genes are responsible for detecting those chemicals involved in mating, host-plant selection, etc. and what are the selection pressures that have driven rapid gene duplication among chemosensory gene families? This will require cellular in vitro analyses of receptor molecule sensitivities in order to understand which molecules are being detected. Patterns of sex- and age-specific expression will help refine which molecules should be studied. It seems likely that these receptors play a role in species-specific differences in mate- and host-plant recognition, and ultimately, therefore, a role in speciation.

CHAPTER 7

Beware! Warning colour and mimicry

Chapter summary

This chapter reviews the evidence for visual predation pressure as a selective force on *Heliconius* wing pattern evolution. There is good evidence that predation pressure drives the evolution of warning colour and mimicry. This comes in part from experimental evidence that bird predators learn to avoid *Heliconius* patterns with experience. The experiments also demonstrate variation between species in their toxicity, suggesting a gradual increase in toxicity in moving from the outgroup heliconiine genera to *Heliconius*. Natural selection can be demonstrated using experiments to study the survival of butterflies in the wild, as well as by indirect methods that take advantage of natural variation in hybrid zones. These methods are broadly in agreement and suggest that mimicry selection can be extremely strong, of the order of 10–30%. This implies a large cost to individuals that carry the 'wrong' wing pattern in a particular habitat. Recent work therefore supports the assertion by Bates (1862) that mimicry represents an excellent example of natural selection in action.

Müllerian mimicry theory predicts monomorphism, but there is considerable diversity in wing patterns, with several species polymorphic across much of their geographic range. The best-supported explanation for such polymorphism is spatially variable selection due to heterogeneous distribution of mimicry rings. This hypothesis has experimental support from mark-recapture experiments in *H. cydno* and field studies in *H. numata*. Alternatively, it has been suggested that variation in the degree of palatability might lead some Müllerian mimics to impose a cost on their more toxic co-mimics, so-called quasi-Batesian mimicry. If true, it might lead to selection for polymorphism, although the hypothesis remains controversial and untested.

7.1 Warning colour

Heliconius are attractive subjects for entomologists in part due to their bright and memorable wing patterns. The colours employed by *Heliconius* are similar to those used on road signs or warning notices: bold and clearly defined blocks of red, yellow, or white on a sharply contrasting black background. Darwin originally thought that bright colours in the animal world primarily functioned in attracting mates, and he was puzzled by brightly coloured caterpillars and other larval stages, which have no need for sexual signals (see more details in Ruxton, Sherratt, & Speed, 2004). It was Wallace who realized that the bright colours of caterpillars had evolved to warn predators that they were toxic, bad tasting, or otherwise unsuitable to eat (Wallace, 1867). The aesthetic appeal of *Heliconius* patterns is therefore no evolutionary accident. The wing patterns have evolved to be recognizable to predators with visual systems rather similar to our own. It is now recognized that warning colour—perhaps more correctly known as aposematism—is widespread in the natural world, and involves organisms signalling their unsuitability as prey.

Warningly coloured butterflies also have distinct behavioural and morphological characteristics.

The Ecology and Evolution of Heliconius Butterflies. Chris D. Jiggins, Oxford University Press (2017).

DOI 10.1093/acprof:oso/9780199566570.001.0001

Distasteful butterflies tend to fly slowly and directly, showing off their warning patterns, in contrast to the fast zig-zagging flight of palatable species. These flight characteristics are associated with long slim bodies, lower body temperatures (Srygley & Chai, 1990), and often long, relatively narrow wings. It has been shown that the slow flight of aposematic species is more energetically costly than the rapid flight of palatable species, perhaps implying a cost to the evolution of warning colour (Srygley, 2004). Warningly coloured species also tend to be tough and resistant to attack, as any butterfly collector who has tried to squeeze the thorax of a *Heliconius* knows. This is presumably an adaptation to survive attacks from predators after having been taste-rejected. Palatable species are unlikely to be released once caught, but distasteful prey may get a second chance, so it is worth investing in resilience.

The ability to fly at lower temperatures and relatively free from the threat of predation means that warningly coloured species can use a broader range of ecological habitats (Speed, Brockhurst, & Ruxton, 2010). They are able to fly in open exposed areas and for more hours of the day. Notably, *Heliconius* are often active in rainy weather and early in the morning, when most butterflies are still at rest. This is presumably because, by relying on warning colour rather than fast flight to escape predation, they can afford to be flying at lower ambient temperatures and levels of solar radiation. It seems likely therefore that the metabolic costs of maintaining chemical defences are outweighed by the competitive ecological advantage of a broader habitat range. Thus, the evolution of warning colour is a major evolutionary innovation that has an impact on all aspects of the biology of a species.

Aposematism commonly involves multimodal signalling, involving the combined effects of sounds, smells, and tastes—the rather pungent odour of adult *Heliconius* may be an example. The characteristic slow flight of aposematic butterflies, as well as being a result of selection to reduce investment in escape ability, likely also represents part of the signal that birds associate with unpalatability. Signalling in several dimensions has been shown to enhance learning in predators (Rowe & Guilford, 1996).

7.2 Mimicry

Observations on *Heliconius* and other neotropical butterflies prompted nineteenth-century biologists to formulate the theory of mimicry. Henry Walter Bates (1862) originally recognized that rare, palatable species might gradually evolve to resemble distasteful, warningly coloured species. The palatable species is termed the mimic, and the distasteful species, the model. Such similarity would confer an obvious advantage to the individual of the mimic species, if even occasionally predators confuse mimic with model and therefore avoid attacking the edible prey. Hence, Batesian mimicry, as it is now termed, can evolve gradually, as small increases in similarity will be favoured by natural selection.

Bates also recognized, however, that there were large families of butterflies that were apparently all unpalatable, but showed considerable similarity between species, most notably the Heliconiini and Ithomiini. He recognized that rare and somewhat unpalatable species might become mimics, but he did not have a clear explanation for the evolution of similarity between two common species. He hypothesized that the similarity was due to adaptation to similar local conditions rather than convergence due to mimicry. Wallace (1871, p. 85) subsequently suggested that even closely related species might not have similar palatability—their 'distasteful secretion is not produced alike by all members of the family'—and therefore even members of the same family might be Batesian mimics of one another.

It was left to Fritz Müller (1879), however, to recognize that mimicry does not have to involve one species being rare or palatable. Müller was influenced by his studies of two genera, *Thyridia* and *Ituna*, that closely resembled each other and were initially thought to be close relatives. However, Müller recognized them as being members of different families, then known as Danaidae and Ithomiidae, respectively, both of which are considered distasteful. Moreover, in some localities one was most abundant, while this relationship was reversed in others, such that it was unclear which might be the model and which the mimic.

This led Müller to recognize that if one unpalatable species were learnt by predators, there would be a benefit to another unpalatable species in sharing the wing pattern. When two unpalatable species occur together and look similar, they share the cost of teaching local predators. If a bird needs to sample five butterflies to learn that a pattern is best avoided, then the probability of any individual being sampled decreases as the pattern becomes more abundant. So if the local population of species A has 50 individuals, then 5/50 (10%) of the butterflies will be sampled each generation to teach the local predators. If species B, also with a population size of 50, evolves to mimic species A, then both species benefit, as only 5/100 (5%) of the butterflies will be sampled. Stated in this way, Müllerian mimicry is therefore a form of mutualism, whereby all parties involved benefit (but see later, quasi-Batesian mimicry).

7.3 Are Heliconius unpalatable?

It has long been assumed that *Heliconius* are unpalatable to predators, but for a long time this was based on very little evidence. The nineteenth-century explorer and biologist, Thomas Belt, noticed that a pair of foraging birds collected many butterflies and dragonflies, but never included *Heliconius* in their catch. He also demonstrated that *Heliconius* were unpalatable to a tame white-faced monkey—an interesting observation that is probably not especially relevant, as monkeys are unlikely to be major butterfly predators in the wild (Brower, Brower, & Collins, 1963; Belt, 1874).

The first experimental studies of *Heliconius* palatability were carried out by Lincoln Brower and colleagues, with caged silver-beaked tanagers in Trinidad (Brower et al., 1963). These are brightly coloured, omnivorous birds that feed on fruit and insects. Frozen heliconiine butterflies were presented to the tanagers, alternated with edible species as controls to ensure the birds were hungry. While virtually all the edible butterflies were eaten, less than 1% of *Heliconius* were eaten. The exception was *H. doris*, 11% of which were eaten, and the outgroup genera *Dryas iulia* and *Agraulis vanillae*, 25% of which were eaten. This provided the first evidence suggesting a gradual increase in unpalatability during the evolutionary history of the genus *Heliconius*.

Tanagers are probably not major natural predators of butterflies. Instead, specialist flying-insect predators are more likely to be the selective force responsible for *Heliconius* mimicry. In particular, forest-dwelling jacamars are flying-insect predators par excellence (Chai, 1986; Mallet & Barton, 1989a; Langham, 2004). Up to 30% of the prey of jacamars may consist of adult lepidopterans (Tobias, 2002; Langham, 2006). Jacamars look a bit like slender kingfishers, to whom they are related. They regularly perch in the same spot at the edge of a forest clearing, from where they spot flying insects and sally out to catch them, often following in pursuit if their prey tries to escape. Their long slender bill is ideal for both snatching flying prey from the air, and subsequently manipulating it for consumption. The jacamar often then returns to the same perch, and appears to taste its prey before consumption (Langham, 2004). When unpalatable prey are rejected after tasting, the bird vigorously shakes its head and wipes its bill—it all looks fairly unpleasant but presumably is better than vomiting up the crop contents after eating a nasty butterfly, as has been observed in jays (Brower et al., 1968; Langham, 2006).

Unfortunately, experimental work with jacamars is not straightforward. They will eat only live flying insects. Peng Chai hand-reared three jacamars from nestlings, and also captured adult birds, laboriously feeding them with wild-caught dragonflies. When the birds were presented with butterflies in a large experimental cage, there was variation in the palatability of different species, similar to that demonstrated by Brower. *Dryas* and *Agraulis* were the most palatable heliconiines (75% and 92% eaten, respectively), with *Dione* and *Eueides* intermediate (20–50%), and *Heliconius* the least palatable (0–10%). The most palatable *Heliconius* species was *H. melpomene* with 2/20 individuals eaten. Interestingly, in contrast to the results of Brower, *H. doris* was rejected at a similar rate to other *Heliconius* species, with none of the 35 butterflies tested being eaten (Chai & Srygley, 1990; Srygley & Chai, 1990).

Another large family of insectivorous birds found in the neotropics is the tyranid flycatchers (Tyrannidae), some of which are sallying predators of flying insects, similar to jacamars. However, tyranids are more common in open areas than in forests. One such species is the kingbird (*Tyrannus melancholicus*), a bright yellow and grey bird, commonly seen perching in conspicuous spots throughout the neotropics. This species was studied by Carlos Pinheiro in the city of Carajás in Brazil (Pinheiro, 1996).

The urban kingbirds were so tame that butterflies could be released close to their perch, and large numbers of individuals tested for their palatability. The rather surprising result of these experiments was that a very large number of *Heliconius* were eaten. Of 58 butterflies from eight species, 49 (84%) were eaten. Nonetheless, seven individuals (12%) were taste-rejected, a higher proportion than in *Eueides* (7%) or other heliconiine genera (2%). It is a little hard to know what to make of these results. The kingbirds were naive predators with respect to *Heliconius*—they hadn't previously encountered the species in the city environment—so their responses were unaffected by previous experience.

It isn't clear, therefore, whether such a large proportion of *Heliconius* would be eaten by more experienced birds. Whatever the case, these results support the general result of increasing unpalatability in moving from outgroup heliconiines to *Eueides* to *Heliconius*. They also highlight the fact that there is a diversity of natural predators, some of which may not be strongly averse to *Heliconius*.

The results of these palatability tests with natural predators are in general agreement with analysis of the chemical composition of the species involved (Engler-Chaouat & Gilbert, 2007). *Dryas* and *Agraulis* have both lower cyanide concentrations, and in all cases are less unpalatable than *Heliconius*. *H. doris* has a much lower cyanide concentration relative to other *Heliconius*, and was also more palatable in Brower's experiments. However, chemical analysis also demonstrates that the cyanogen composition of sapho clade species was much higher than other *Heliconius*, but there is little support for this from the palatability data. In fact, Brower showed that *Heliconius sara* was actually more palatable than other *Heliconius* (apart from *H. doris*), the reverse of what would be predicted (Brower et al., 1963). Further tests with sapho clade species are needed to determine whether they really are as nasty as their high cyanide concentration would seem to suggest.

It should also be noted that, relative to other butterflies, *Heliconius* are somewhat intermediate in their nastiness. In particular, the ithomiine butterflies use pyrrolizidine alkaloids for defence rather than the cyanogenic glycosides used by *Heliconius*, and are thought to be far more toxic (Boppré, 1984; Brown et al., 1991; Chai, 1996). Experiments with predators confirm that ithomiine butterflies are generally more strongly avoided than *Heliconius* (Chai, 1986, but see Arias et al., 2015). Consistent with this result, *Heliconius* are far more nifty flyers than most ithomiines. Although individuals from both taxa tend to glide along rather slowly when undisturbed, *Heliconius* can accelerate rapidly when threatened. In contrast, many ithomiines have completely lost the ability for fast, evasive flight. These observations suggest that there is indeed a 'palatability spectrum' along which *Heliconius* are somewhat intermediate and ithomiines are towards the more distasteful end.

7.4 Experimental evidence for mimicry—experiments with predators

Experiments with predators confirm that *Heliconius* are unpalatable—they are commonly either taste-rejected or avoided entirely by predators. The two other requirements for Müllerian mimicry are that predators (1) should develop an aversion to attacking unpalatable species with experience, by associating their appearance with the bad taste, and (2) should generalize between species that are similar in appearance (co-mimics).

Experiments with predators support both tenets. In Brower's experiments described earlier (Brower et al., 1963), tanagers were presented with a series of individuals of the same *Heliconius* species, alternating with an edible butterfly. In all cases, avoidance of the *Heliconius* wing pattern increased with experience, generally within only two to three trials. Furthermore, when a co-mimic butterfly was presented subsequently, the tanagers were far less likely to eat it than when the same species was presented to a naive bird. Similarly, Peng Chai (1986, 1996) presented naive jacamars with a wide variety of butterflies, and showed that *Heliconius* species were increasingly and rapidly rejected with experience. The only exception was *H. charithonia*, which was generally acceptable to all the jacamars. He also painted edible butterflies, such as *Dryas iulia*, with a variety of colours of toxic paint. The jacamars rapidly learnt to associate the nasty experience with the colour and avoid future attacks (Chai, 1988). Interestingly, bright colours typically associated with warning (red, yellow) were learnt more rapidly than colours more associated with camouflage (green, brown). This result supports the hypothesis that bright colours provide a more effective aposematic signal than muted colours.

Gary Langham (2004) conducted by far the largest set of experiments with jacamars. In the llanos (savannahs) of western Venezuela, he was able to capture 80 rufous-tailed jacamars (*Galbula ruficauda*) over three field seasons. These were each presented with butterflies in cage trials. As all the birds were captured in the wild, they were assumed to have had previous experience with the local *Heliconius* fauna, in particular the commonest local mimicry pair, *H. erato hydara* and *H. melpomene melpomene*.

In each trial, the bird was presented with an unmanipulated *H. erato hydara* ('local morph'), and two 'novel morphs' consisting of *H. erato hydara* individuals with the forewing band either blacked out or made brighter red with a marker pen. As predicted, none of the birds attacked the local morph, but 29/80 jacamars attacked a novel morph, with both red and black morphs being attacked (albeit with more attacks on the black morph). These experiments clearly show that the local predators avoided the *Heliconius* pattern with which they had previous experience, but were more likely to attack a novel pattern. Furthermore, the red morph was only 15% brighter in the forewing band than the local morph, so the experiment also demonstrates that jacamars could distinguish relatively minor differences in pattern—an important prerequisite for the evolution of the extremely exact mimicry seen between *Heliconius* species.

Langham's experiments also offer some intriguing insight into predator psychology. A total of 46 of the same jacamars were recaptured in a subsequent field season and retested (Langham, 2004). Although most birds never attacked any of the butterflies presented, those individuals that attacked a novel morph the first time round were similarly likely to attack in the second trial. This suggests that some individual birds have adventurous personalities and are more likely to attack novel phenotypes, while others avoid all '*Heliconius*-like' butterflies. Intriguingly, the birds most likely to try novel morphs were older (at least, they had longer tails, which is probably an indication of age) (Langham, 2006), the reverse of what might have been predicted. Instead of young, naive predators being most likely to attack novel forms, it seems that older individuals, perhaps more confident of their discriminatory powers, are more willing to take a chance.

Similarly, Peng Chai had one female jacamar that ate a much larger proportion of *Heliconius* than all other individuals (24 of 49; 49%), suggesting individual variation in acceptability within predator species, possibly due to nutritional status (Chai, 1986, 1996). Whatever the reason, this is exactly the kind of behaviour that would lead to the gradual evolution of perfect mimicry. Poor mimics would initially benefit by avoiding attacks from conservative jacamars, while perfection of mimicry would be favoured by continued sampling of slightly different forms by the more adventurous birds.

There was no evidence that the time between the two trials was related to the probability of attack the second time around—so the birds did not learn the novel morph and then gradually forget it. Perhaps the simplest explanation is that a single trial was simply not enough experience for the predator to learn in the first place.

7.5 Experimental evidence for mimicry—experiments with butterflies

Thus, experiments with birds support the basic tenets of mimicry theory: predators learn to associate bright wing patterns with distasteful prey, and can also generalize between different butterfly species. The prediction is therefore that butterflies with unusual wing patterns, which have not been previously experienced by local predators, should be attacked more frequently. A fairly easy way to test this is to manipulate the patterns of wild butterflies to see if it affects their survival. *Heliconius* are robust, and pretty resilient to manipulation, so their patterns can be altered using marker pens.

Woody Benson (1972) carried out just such an experiment in Costa Rica (Figure 7.1, panel A), in which wild *Heliconius erato* were captured and their

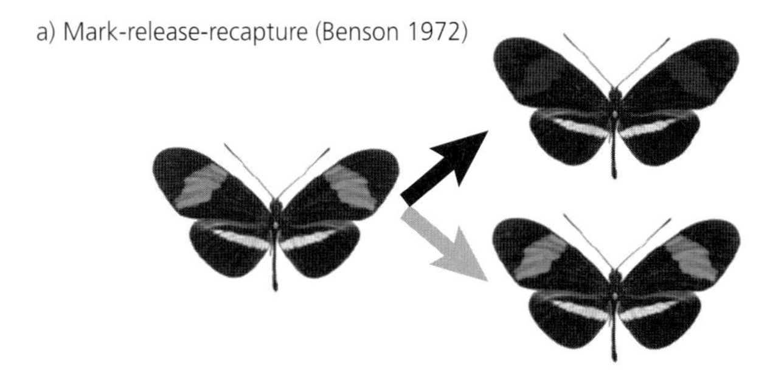

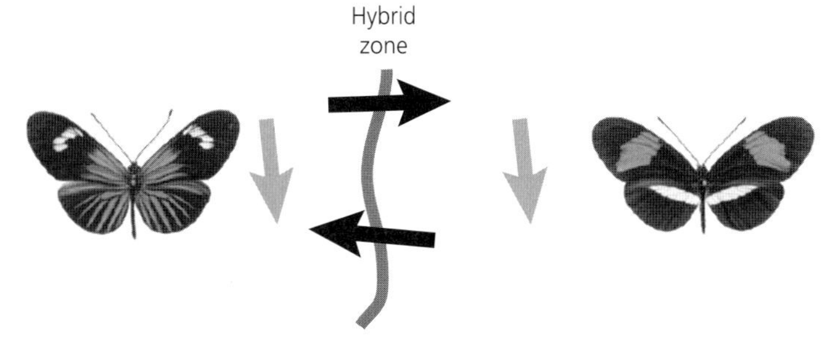

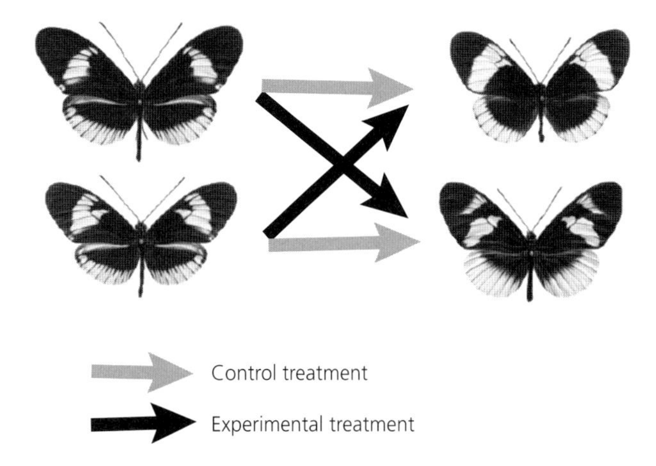

Figure 7.1 Summary of field tests of mimicry selection in *Heliconius* (see Plate 13)

(A) Benson blacked out the forewing band of *Heliconius erato* in Costa Rica, with control butterflies similarly manipulated without a colour change; (B) Mallet moved butterflies across a wing pattern hybrid zone in Peru, with control butterflies being moved a similar distance but within the same wing pattern range; (C) Kapan moved polymorphic *Heliconius cydno* adults between populations with different morph frequencies, demonstrating that the locally abundant morph showed higher fitness.

red forewing bands blacked out. Other individuals were left unaltered in appearance, but with a similar amount of black ink applied to the surrounding wing regions, controlling for any adverse effect of the ink treatment. Benson was able to take advantage of the gregarious roosting behaviour of the butterflies in order to maximize the rate at which butterflies were re-sighted after manipulation. Over two years, a total of 73 butterflies were marked.

The 'blacked out' treatment butterflies showed reduced survival—on average, their survival time on the roost was significantly less than for controls with unaltered wing patterns. The blacked out butterflies also had higher levels of beak mark damage, suggesting that their reduced survival was indeed due to bird attacks. The use of beak marks as a measure of predation is a little problematic; the beak-marked butterflies have escaped, so beak marks don't actually measure the important predation events from the perspective of natural selection. Nonetheless, the significant effect of manipulating the wing pattern on both survival and probability of beak marking provides good evidence for selective predation on wing patterns.

An alternative approach is to take advantage of the natural variation in wing patterns between races of *Heliconius*. Across natural hybrid zones, variants of the same species can have remarkably different wing patterns. In order to measure selection across one such zone, James Mallet spent a year driving a battered Land Cruiser around the highlands of north-eastern Peru (Mallet & Barton, 1989a). (To add to the challenge, it was in the mid-1980s, at the height of guerrilla activity by the Shining Path, and Peruvian involvement in the international cocaine trade.) The butterflies in the mountains had a pattern very similar in appearance to those studied in Costa Rica by Benson, while those in the lowlands look very different (see Figure 7.1, panel B). Mallet moved individuals to sites where the local butterflies had a different appearance (foreign patterns) and released them alongside control individuals that were moved between sites with similar patterns (local patterns). He then returned to the release sites and recaptured butterflies, estimating their survival.

As predicted, butterflies with foreign patterns showed lower survival than those with local patterns. In particular, there was an initial establishment phase during which the foreign patterns suffered significantly reduced survival. The estimated selection pressure against individuals with the 'wrong' colour pattern was a remarkable 52%—in other words, butterflies with a wing pattern not recognized by local predators were about half as likely to survive. This provides evidence for the strength of natural selection acting on wing pattern in these butterflies.

7.6 The paradox of Müllerian mimicry

Overall, therefore, there is good evidence both from experiments with predators in cages, and from survival of butterflies in the wild, for the selection pressure that drives mimicry. Mimicry selection is rather unusual, as it favours whichever pattern is most common in the local environment. This is known as positive frequency-dependent selection—the higher the local abundance of the pattern, the more it is favoured. In fact, it is probably more correctly termed number-dependent—the fitness of a pattern is dependent on the number of individuals in the local environment that possess that pattern (Joron & Mallet, 1998). This kind of selection promotes monomorphism and convergence, and therefore the ultimate equilibrium state would seem to be for every butterfly to converge on the same pattern. Very clearly—as a glance through the plates in this book will confirm—this is not the case, which raises what has been termed the 'paradox of Müllerian mimicry'—diversity.

Across the neotropics, warningly coloured butterflies have a startling diversity of wing patterns. I will describe how this diversity is distributed and then attempt to explain how it can be squared with mimicry theory. The first striking observation is that within any one locality, coexisting butterflies fall into diverse groups of species that share the same wing pattern, which we term 'mimicry rings'. In the Amazon basin, up to nine heliconiine species may share the same pattern. However, a number of such mimicry rings can be found flying together (for example, see Figure 7.2). The number of coexisting mimicry rings that include *Heliconius* species does not generally exceed about six, but there can be many more coexisting mimicry rings if the ithomiine butterflies are also included. The second observation is that the pattern of the mimicry rings varies geographically, 'as if at the touch of an enchanter's wand', in the memorable phrase of H. W. Bates (quoted in Müller 1879, p. xxix). Most famously, the species *H. melpomene* and *H. erato* vary in parallel across their geographic range, with nearly 30 named races in each species. In summary, contrary to the prediction of mimicry theory, there is considerable pattern diversity both within and between geographic locations.

7.7 Diversity between mimicry rings at a single locality

The coexistence of different wing patterns in a single locality is most readily explained by ecological segregation between different species. In Panama,

Figure 7.2 Sympatric mimicry rings (see Plate 14)

Coexisting mimetic species found in central Panama with upper (ventral; above) and underside (dorsal; below) wing patterns. Top row: *H. hecale melicerta, H. ismenius boulleti, H. sara magdalena*; second row: *H. cydno chioneus, H. sapho sapho, H. doris obscurus*; third row: *H. erato demophoon, H. melpomene rosina, H. hecalesia longarena.*

for example, we collected mimetic *Heliconius* along Pipeline Road, a transect that runs from open habitat through to closed canopy forest. The red, black, and yellow pattern of *H. melpomene* and *H. erato* was most common in the more open areas, whereas the black and white pattern of *H. cydno* and *H. sapho* was more common deeper into the forest (Estrada & Jiggins, 2002).

Such ecological segregation could contribute to stable maintenance of mimicry patterns in two ways. First, each mimicry ring might be primarily exposed to different suites of predators. In Panama, the main predators of flying insects in open areas are tyranid flycatchers such as kingbirds, while butterflies flying in the forest interior are more likely to be attacked by forest specialist birds such as jacamars. Thus, different patterns might be adapted to signal to different predators. This may also be combined with the adaptation of patterns to the local light environment, such that particular patterns might be more contrasting or more memorable in particular microhabitats, although this has not been particularly well documented. A second explanation is that the habitat itself might be a part of the signal. Thus, predators might show context-dependent learning of wing patterns, and recognize 'black and white in the forest' and 'red and black in the sun' as distinct patterns to avoid. Even if the same predators experienced both patterns, therefore, habitat differences could act to stabilize the coexistence of the two wing pattern forms.

Flight behaviour can also contribute to segregation between mimicry rings in a similar manner. These same Panamanian species also differ in flight: *H. cydno* and *H. sapho* have an asymmetric wingbeat, with a swift downstroke and more gentle upstroke, whereas *H. melpomene* and *H. erato* have a more symmetric wingbeat, with an even upstroke and downstroke (Srygley, 1999). In flight, this gives the former species a characteristically floating flight, whereas the latter are more even, determined flyers. In this case, wing pattern is actually a better predictor of flight pattern than is phylogeny. This convergence in behaviour must also facilitate learning of distinct mimicry rings by predators.

How ecologically segregated are *Heliconius* mimicry patterns? One of the first studies of ecological segregation between patterns was by Christine Papageorgis (1975), working at several sites in eastern Peru. She showed vertical segregation in the height at which ithomiine and heliconiine mimicry rings fly, with 'blue'-patterned species such as *H. sara* in the highest forest strata, and the transparent ithomiine species at the lowest levels. Subsequent work has confirmed that vertical segregation in flight is certainly important among mimicry rings of ithomiine butterflies, which show marked differences in flight height even over a scale of just a few metres above the forest floor (Beccaloni, 1997; Elias et al., 2008).

However, there has been scepticism about the importance of vertical segregation in the flight of adult *Heliconius*, as individuals are commonly observed to range widely through the forest. Indeed, species such as *H. sara* that are supposed to be at the highest levels in the Papageorgis data, are commonly seen 'mud-puddling' or visiting flowers near ground level. High-flying species such as *H. burneyi* will swoop down into light gaps in response to fluttering red rags, and are mimetic with much lower-flying species such as *H. erato*. This intuition from field observations is supported by data on the height at which adult *Heliconius* visit *Psiguria* flowers, which showed no evidence for habitat segregation (Mallet & Gilbert, 1995).

Instead, vertical segregation in roosting height may be more important (Figure 5.4). *Heliconius* species tend to have characteristic roost heights in the forest, and co-mimic species tend to roost together at similar heights (Mallet & Gilbert, 1995). Thus, in Costa Rica, *H. melpomene* and *H. erato* roost at a median height of 1–2 m, while *H. hewitsoni*, *H. sara*, and *H. pachinus* roost at a median height of about 4 m.

At first it seems puzzling: why should visual mimicry be associated with nocturnal behaviour, at a time of day when wing patterns cannot be seen? However, it makes sense in the context of roosting behaviour, whereby butterflies begin to aggregate before sunset and remain aggregated in the morning in the vicinity of the roost. Butterflies leave a roost site and perch on nearby leaves with their wings open to catch the warmth of the morning sun. Once body temperatures have risen, they leave one by one to forage. Thus, at precisely the time when forest birds are most active early in the morning, co-mimic butterflies are aggregated and basking in the sun, showing off their aposematic wing patterns. It seems likely that selection for the extraordinarily precise mimicry between *Heliconius* species occurs mainly at this time of day. In summary, fine-scale niche segregation between mimicry rings likely contributes to their stable coexistence.

7.8 Other ecological functions of wing patterns

Mimicry theory requires only that wing patterns are memorable, and therefore the details of the pattern may be arbitrary. A bit like fashion, it may not matter too much what you look like, as long as you look the same as everyone else. However, in some cases there may be reasons that particular patterns are adaptive in particular habitats. This could be for thermoregulation (Heinrich, 1993), or for signalling in different light environments (Endler & Basolo, 1998). Extensive work on *Colias* butterflies has shown that darker wing patterns are associated with high altitudes where they are thought to be better at capturing solar radiation, giving a thermal advantage (Watt, 1968). In *Heliconius*, there are examples of darker patterns associated with higher altitude, such as the orange and black *H. numata bicoloratus*, which is found in the Andean habitats alongside its ithomiine co-mimics. At lower altitudes, *H. numata* tends to have more pale patterns with larger areas of yellow. However, thermoregulation seems unlikely to explain the majority of variation in wing patterns among the heliconiines.

It is increasingly realized that signals are adapted to maximize their transmission in particular habitats (Endler & Basolo, 1998). In *Heliconius*, there is evidence for certain patterns being associated with particular light environments. For example, the bold red and black 'postman' patterns of *H. erato* and *H. melpomene* tend to be associated with savannah and open habitats, while the more complex 'rayed' patterns are more often found in the Amazonian forests (Endler, 1982). This is especially striking where the postman races of both *H. melpomene* and *H. erato* follow the northern coast of South America as far as the mouth of the Amazon.

In French Guiana, there is good evidence that the hybrid zone is associated with a habitat transition (Endler, 1982; Blum, 2008). It seems that this narrow strip of postman-patterned butterflies is maintained by some ecological advantage in the open gallery forest found along the coast. The advantage could be related to signalling in open habitats, although this hypothesis has not been tested directly. Nonetheless, habitat associations of particular wing pattern forms are not strict, and hybrid zones between such forms are not necessarily associated with any habitat discontinuity (Mallet, 1993).

It has even been suggested that *Heliconius* patterns are cryptic, and are adaptations for disruptive coloration in dappled forest light (Papageorgis, 1975). However, I have yet to meet a field biologist who has watched *Heliconius* adults in the wild who accepts this idea (although many *Heliconius* are more cryptic on the dorsal side, presumably to reduce visibility when resting). In summary, the null hypothesis is that *Heliconius* wing patterns are simply arbitrary signals of distastefulness. However, there is some evidence that they are locally adapted for signalling in particular environments. There is, however, considerable scope to revisit the signalling properties of *Heliconius* patterns in the light of recent developments in the sensory ecology of butterfly and bird visual systems (see also Chapter 6).

7.9 Geographic races and hybrid zones

Bates first noticed that the dominant *Heliconius* patterns vary from place to place (Bates, 1862). He originally described the divergent forms mostly as distinct species, but they are mostly now recognized as races or subspecies (Figure 7.3a-d). Thus, most famously, the species *H. melpomene* (Figure 7.3a) and *H. erato* (Figure 7.3b) vary in parallel across their geographic range, with nearly 30 distinct races in each species. Here I will describe this diversity and its maintenance, leaving the origins of wing pattern races for Chapter 10.

For many years, biologists were puzzled by the fact that some populations of species such as *Heliconius melpomene* are highly polymorphic. This is not expected in Müllerian mimicry, where selection favours monomorphism. In particular, a region of western French Guiana was renowned with collectors for the diversity of forms that could be collected (many of which were given Latin names, adding greatly to taxonomic confusion). John Turner used laboratory hybridization (partly conducted in the insectaries of Phillip Sheppard and Cyril Clarke in Liverpool) and collections from across the region to establish that this was in fact a stable hybrid zone (Turner, 1971b). The polymorphic populations represent fairly narrow zones of contact between geographic forms that are stable and monomorphic over larger geographic areas. The fact that illustrations of these hybrids were known from Dutch collections from the eighteenth century, made in what was then the colony of Surinam, established the fact that hybridization was a stable phenomenon that had lasted as long as '2,000 generations' (Turner, 1971b).

Now it is recognized that across their entire range, species such as *H. melpomene* and *H. erato* are broken up into a patchwork of different wing patterns that extend for several hundred kilometres. The geographic races are separated by hybrid zones and are considered conspecific as they mate freely where they meet. Some of these hybrid zones are extremely narrow, but others are much broader regions of intermingling between forms.

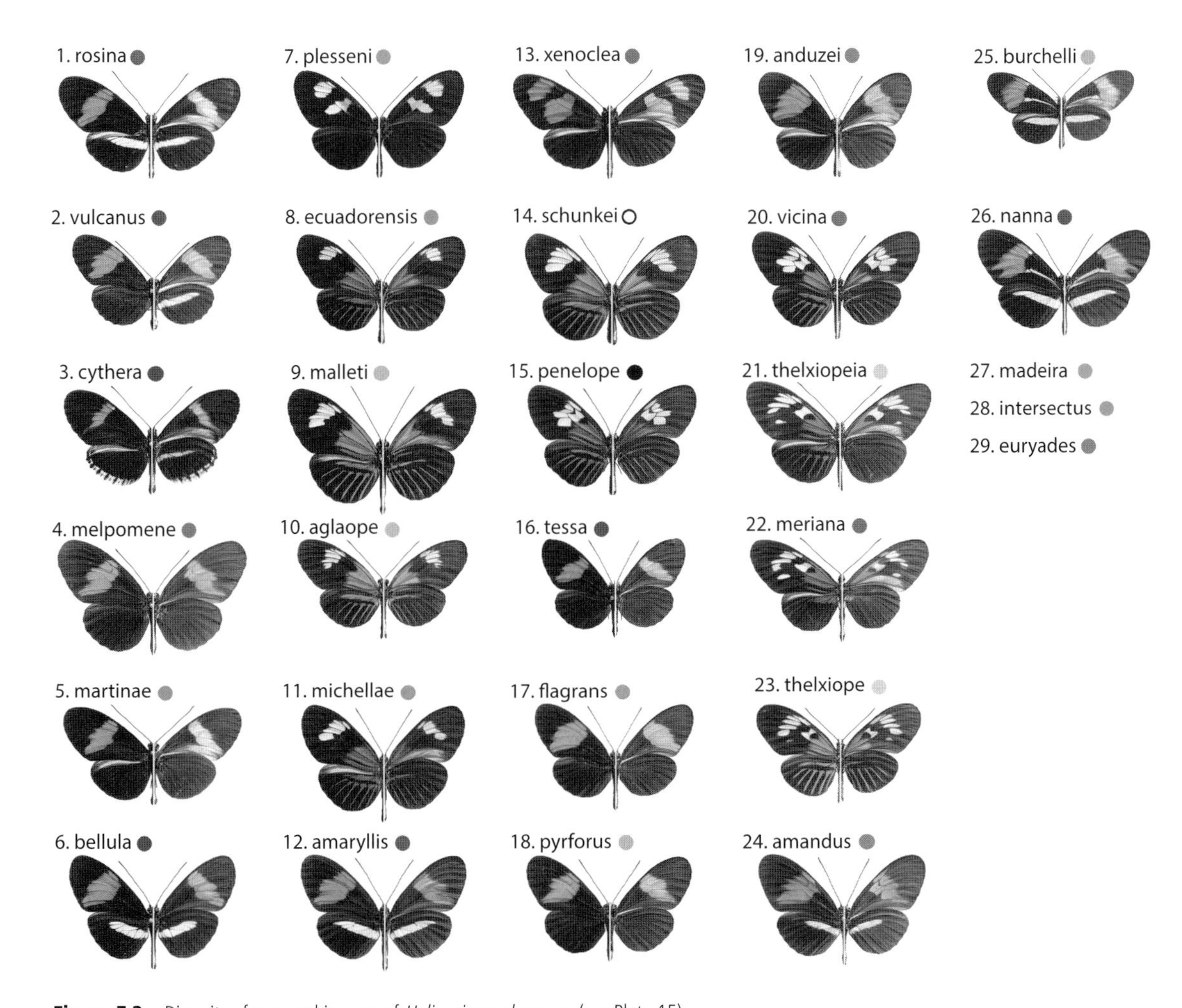

Figure 7.3a Diversity of geographic races of *Heliconius melpomene* (see Plate 15)

Three forms lack images: *madeira* is similar to *thelxiope; intersectus* is the Isla de Marajó form with a distal yellow band otherwise similar to *thelxiopeia; euryades* is a postman form similar to *melpomene*. Note that *H. m. martinae* is polymorphic for the hindwing yellow band, with some forms looking very similar to *H. m. rosina*. Butterflies are shown with dorsal (left) and ventral (right) images.

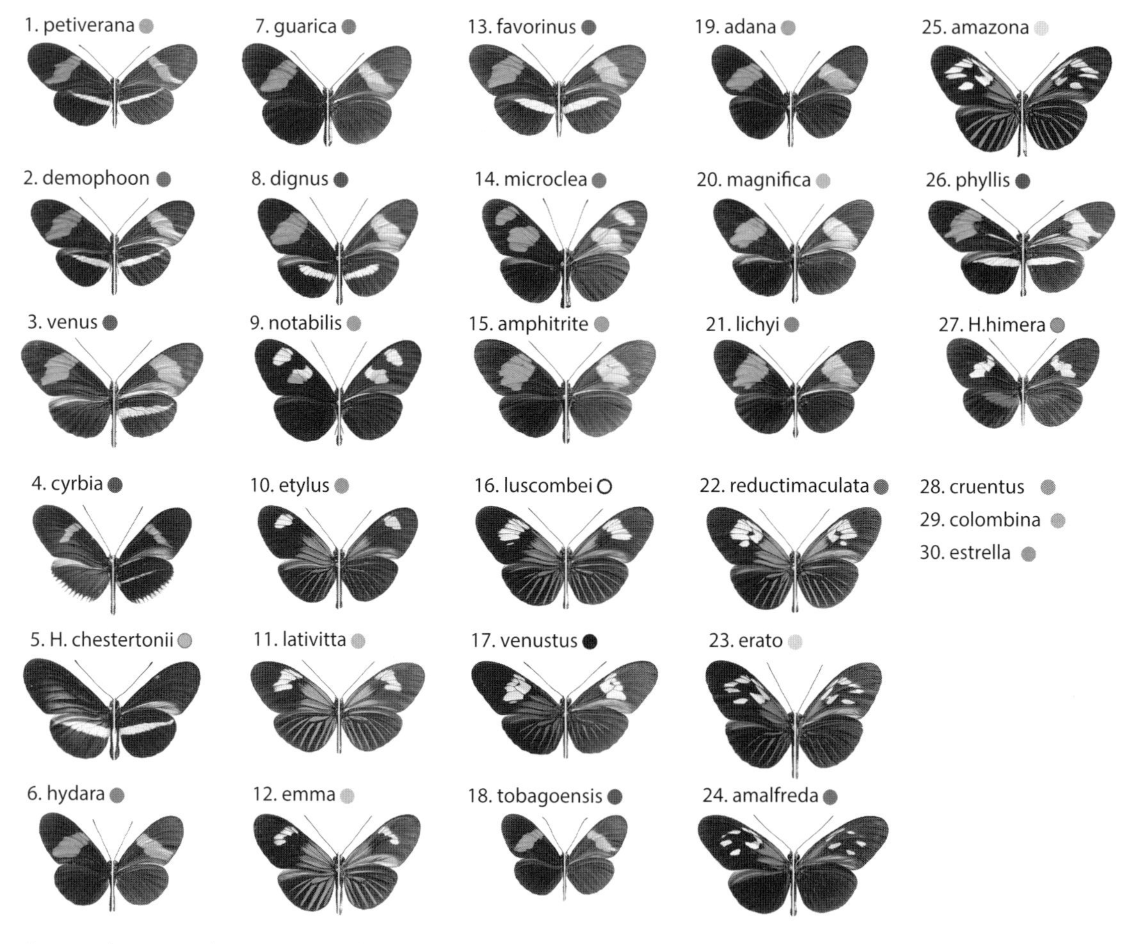

Figure 7.3b Diversity of geographic races of *Heliconius erato* (see Plate 16)

Three forms lack images: *cruentus* is similar to and somewhat intermediate between *petiverana* and *demophoon* with a broad and slightly shorter hindwing yellow bar; *colombina* is similar to *demophoon; estrella* is the Isla de Marajó form with a distal yellow band otherwise similar to *erato*. Butterflies are shown with dorsal (left) and ventral (right) images. *Heliconius chestertonii and H. himera* are geographic replacements of *H. erato*, that are considered to be distinct species.

Such hybrid zones have been termed 'natural laboratories' for the study of evolution and speciation (Barton & Hewitt, 1989), because they allow us to study the interaction between genetically divergent populations in the wild.

The stability of a warning colour hybrid zone is explained by the same force of frequency-dependent natural selection that leads to mimicry. If a butterfly flies across a hybrid zone from its place of birth into the range of a different wing pattern form, it is moving into an area where its own wing pattern is in a minority and not recognized by predators. It is therefore expected to suffer from a higher chance of attack. Thus, the positive frequency-dependent selection on wing pattern that leads to mimicry also acts to stabilize hybrid zones and maintain diversity.

7.10 Indirect estimates of selection

The dynamic nature of hybrid zones allows us to measure natural selection in the wild, complementing direct estimates from mark-recapture experiments described earlier. Hybrid zones between divergent races of *Heliconius* are a dynamic balance between natural selection and gene flow, known

Figure 7.3c Diversity of geographic races of *Heliconius hecale* (see Plate 17)

Three forms lack images: *paraensis* is similar to *paulus; barcanti* is similar to *clearii; zeus* was described from Bolivian specimens and is similar to *felix*; the distributions for these forms are combined. Note that *H. hecale ithaca* also has a form with a black hindwing (not shown) that is mimetic with *H. numata messene*. Butterflies are shown with dorsal (left) and ventral (right) images.

as 'tension zones' (Figure 7.5). Selection tends to make the cline narrower, as individuals are eliminated from the extremes of the hybrid zone. Dispersal broadens the cline by moving genotypes across the zone. If these two forces are held roughly constant, then the cline will remain in a stable state. Inferences from clines in *Heliconius* are made much easier by the fact that *Heliconius* wing patterns have relatively simple genetic control, so the genotypes of individuals can be inferred from their phenotype (see Chapter 8). Thus, collections of butterflies from across wing pattern hybrid zones can provide detailed information on the spatial distribution of genotypes at multiple loci under selection.

One easily measured parameter from a hybrid zone is the width of the cline, which is approximately proportional to gene flow (σ) divided by the square root of selection ($\sqrt{s}$). Gene flow also causes non-random associations between alleles at different loci, known as linkage disequilibrium, which can also be estimated from field collections. With estimates of these two parameters from wild populations, it is possible to estimate both selection and dispersal (Mallet and Barton, 1989b).

Figure 7.3d Diversity of forms of *Heliconius numata* (see Plate 18)

The distinction between geographic races and within-species polymorphism somewhat breaks down in *H. numata*. Many of the taxa are sympatric forms controlled by alternate alleles at the *P* supergene, but for simplicity I consider all forms as subspecies. Butterflies are shown with dorsal (left) and ventral (right) images.

Mallet worked with the theoretical population geneticist Nick Barton to apply these methods to the Tarapoto hybrid zone, the same zone that he studied using release-recapture experiments. The clines were narrow in *H. erato*—8.5 km for the co-dominant *D* locus that controls red patterns, and ~10 km for each of the other two dominant loci (known as *Sd* and *Cr*). Combined with estimates of disequilibria between loci, these cline widths imply selection of around 0.23 per locus on each of the three colour pattern loci, and a dispersal distance (σ) of 2.6 km per generation (Mallet et al., 1990). There is a hybrid zone in exactly the same place in the co-mimic species *H. melpomene*, which has slightly broader clines (11–13 km), but gives similar estimates for the strength of selection and a slightly larger dispersal distance (σ) of 3.7 km per generation (Mallet et al., 1990).

Later re-sampling of the same hybrid zone by Neil Rosser and colleagues showed that the zone in both species was extremely stable in both width and location between 1986 and 2011 (Figure 7.5) (Rosser et al., 2014). This provides further evidence that the zone is maintained by selection—the alternative

Figure 7.4 The Amazonian dennis-ray mimicry ring (see Plate 19)

First row: *H. burneyi huebneri*, *H. aoede auca*, *H. xanthocles zamora*; second row: *H. timareta timareta* f. *timareta*, *H. doris doris*, *H. demeter ucayalensis*; third row: *H. melpomene malleti*, *H. egeria homogena*, *H. erato emma*; fourth row: *H. elevatus pseudocupidineus*, *Eueides heliconioides eanes*, *E. tales calathus*; bottom row: *Chetone phyleis*, a pericopine moth. These butterflies are from populations in both Ecuador and Peru and do not all occur in the same locality.

hypothesis of secondary contact between two divergent and previously isolated populations would predict that the zone should decay over time, becoming gradually wider. Rosser et al. (2014) also estimated independent selection coefficients for each locus from the locus-specific cline widths, giving $S \approx 0.38$ for the *D* locus, $S \approx 0.17$ for *Cr*, and $S \approx 0.15$ for *Sd* in *H. erato*. In summary, these data show that natural selection on mimicry is strong.

Remarkably, the shape and width of the clines in these analyses fits extremely well with known phenotypic effects of these loci in *H. erato* (Mallet & Barton, 1989b; Mallet et al., 1990). Thus, the red (*D*) locus cline is narrower than the other two loci and

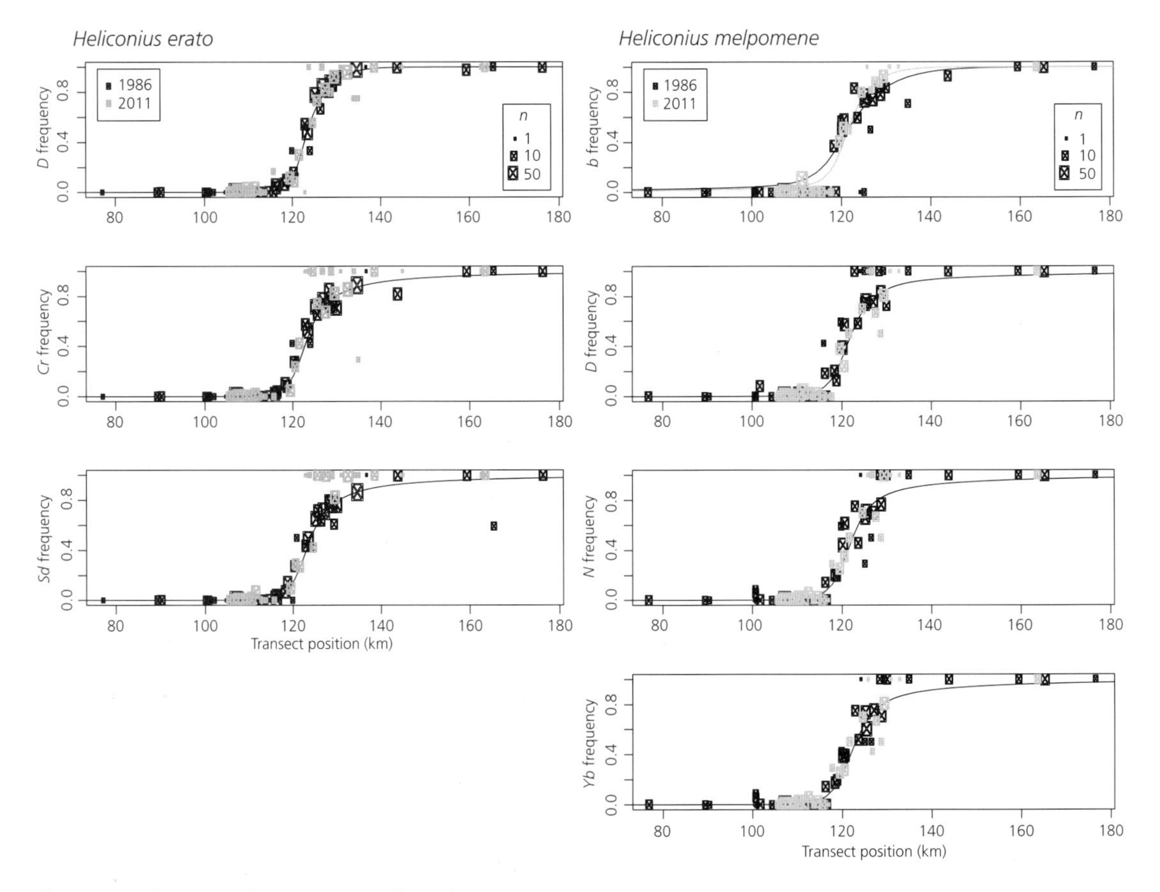

Figure 7.5 Colour pattern clines in *H. erato* and *H. melpomene* in Peru

The wing pattern cline that has been best studied lies in north-eastern Peru, near the town of Tarapoto. The width of these clines was used to estimate the strength of selection on wing pattern alleles. The loci correspond to phenotypic effects of the red locus, *optix* (*D* and *b*), the yellow locus, *cortex* (*Cr*, *Yb* and *N*), and the shape locus, *WntA* (*Sd*). Allele frequencies for 1986 are shown in black, those for 2011 in grey, showing no evidence for cline movement over 25 years. The size of datapoints has been $\log_{10}$ scaled in proportion to the number of alleles sampled; the legends in the bottom right of the top panel show this expressed as *n* butterflies sampled. For details, see Rosser, Dasmahapatra, and Mallet (2014).

has a correspondingly larger selection coefficient. This finding is consistent with the larger influence of the red locus on appearance—it controls all the red patches on fore- and hindwings and influences a larger wing area than other loci. Visual selection is therefore expected to be stronger on this locus, for the simple reason that it has a larger effect.

Cline shape is also informative. The two dominant loci fit best to an asymmetric shape, which is exactly as expected under dominance. Recessive alleles can move into the range of the other population by gene flow, largely unimpeded by selection because they are hidden in heterozygote individuals. Only when they occasionally find themselves in a homozygote butterfly are they selected against. In contrast, on the dominant side of the cline, selection is more effective at removing immigrant alleles, as they are all expressed and exposed to selection. These observations strongly support the assertion that visual mimicry is the selective force in these hybrid zones—they are hard to explain otherwise.

As noted earlier, the estimates of selection based on indirect inferences from the clines give locus-specific cline widths of $S \approx 0.38$ for the *D* locus, $S \approx 0.17$ for *Cr*, and $S \approx 0.15$ for *Sd* in *H. erato*, remarkably similar to estimates from the release-recapture experiment described previously: 0.33, 0.15, and 0.15 (Rosser et al., 2014). Considering the assumptions that underlie both analyses, these highly similar estimates clearly support the general assertion that mimicry selection in the wild is strong and readily measurable.

In contrast, the estimate of dispersal from the cline study is considerably greater than that made from direct observation. When Mallet released pupae and observed the dispersal distance of the young adult butterflies that emerged, he calculated a dispersal of 266 m (σ), almost an order of magnitude lower than the estimated 2.7 km for *H. erato* from the cline width (Mallet, 1986b; Mallet et al., 1990). It seems likely that the higher estimate is closer to the truth, mainly because ecological estimates of dispersal are always limited by the size of the study area and the loss of individuals that fly out of the area. However, it is also possible that dispersal differs between the habitats studied in Costa Rica and Peru.

These analyses of the *Heliconius* hybrid zones are especially powerful because of the unusually clear genotype-phenotype relationship in *Heliconius*, such that the exact genotypes of individual butterflies can be estimated from their phenotype alone. Molecular genetic analyses mean that the basis for the genotype-phenotype relationship is now better understood in many other species. These cline-fitting methods could therefore be more widely applicable, where molecular data can be used to determine exact genotypes of loci under selection. For example, in *Culex* mosquitoes, a similar approach was used to investigate patterns of selection and migration of a locus for insecticide resistance along the French Riviera (Lenormand et al., 1999). These also led to very high estimates of selection: $s = 0.33$ in the presence of insecticide.

7.11 Comparison of dispersal estimates from assumptions of selection versus drift

The estimates of selection are derived from the assumption of an equilibrium between dispersal and selection, at a locus (or loci) under selection. An alternative method of estimating dispersal comes from assuming a dispersal-drift equilibrium, often using the Wright-Fisher island model (Whitlock & McCauley, 1999). Any neutral locus in the genome is under some balance between drift, which gradually leads to populations becoming differentiated, and dispersal, which homogenizes such differences. This equilibrium can also therefore be used to infer dispersal (or, more correctly, effective gene flow) between populations. This method has been far more widely used, primarily because any putatively neutral genetic marker can be used.

However, drift is a much weaker force than selection, so populations reach equilibrium much more slowly under drift-mutation than selection-migration. For example, unless the effective population size is very small, it would take a long time for two populations that were completely isolated to reach the 'equilibrium' state of complete fixation of genetic differences that would be expected under zero gene flow (Whitlock & McCauley, 1999). Estimating dispersal from selected loci therefore offers an alternative approach that can be more robust to perturbation (Mallet, 1990).

7.12 Polymorphism in Müllerian mimicry

Polymorphism in a few *Heliconius* species is not easily explained as a result of hybridization. Notably, *H. numata, H. ismenius, H. doris*, and some populations of *H. cydno* show polymorphism that is apparently stable across broad geographic regions. This seems to contradict a basic prediction of mimicry theory: positive frequency-dependent selection on Müllerian mimics favours monomorphism. The contradiction has been considered an 'evolutionary paradox' and remains one of the outstanding problems in the biology of mimicry (Joron & Mallet, 1998; Mallet & Joron, 1999). Polymorphism in mimicry therefore represents a specific case of the more general problem of how variation is maintained in natural populations, which intrigued evolutionary biologists throughout the twentieth century.

The best studied and most remarkable polymorphic species is *H. numata* (Figures 7.3d and 7.6). This is an elegant slow flying species that 'promenades' in open forest areas (see Chapter 6). It is distributed across the Amazon basin and Guiana shield. Throughout its range, it is highly polymorphic with as many as seven distinct forms found in any one locality. Each is a mimic of sympatric species of

Figure 7.6 Mimicry between *Melinaea* species and *Heliconius numata* forms (see Plate 20)

First row: *Melinaea idae vespertina* and *H. ismenius metaphorus*; second row: *Melinaea satevis cydon* and *H. numata arcuella*; third row: *Melinaea marsaeus mothone* and *H. n. bicoloratus*; fourth row: *M. menophilus cocana* and *H. n. euphrasius*; fifth row: *M. menophilus menophilus* and *H. n. tarapotensis*. All photos by the author except *M. menophilus cocana*, courtesy of Keith Willmott.

ithomiine butterflies, primarily in the genus *Melinaea*. It is a host-plant generalist that lays its eggs on many species of *Passiflora* in forest gaps and open areas near tall forest. The extent of wing pattern polymorphism in this species and its close relatives caused a great deal of taxonomic confusion, until it was resolved through careful analysis by Keith Brown and Woody Benson in the 1970s (Brown & Benson, 1974). Eggs were raised from wild females and many different wing pattern forms, which had often previously been given species status, were shown to be conspecific morphs.

Mathieu Joron has since studied this species in north-eastern Peru, where at least six distinct forms coexist (Joron et al., 1999). Across a small region of around 30 × 30 km, Joron regularly visited 11 study sites to collect *H. numata* forms and their models. He showed that there was striking heterogeneity in the distribution of the different *Melinaea* species, with any particular site dominated by a different wing pattern. The larvae of *Melinaea* feed on large canopy epiphytes such as *Markea* and *Juanulloa*, and local abundance of these resources may drive their very patchy distribution. In contrast, *H. numata* was widespread across all study sites with no genetic evidence for population structuring. *H. numata* wing patterns are therefore being pushed in different directions by natural selection, caused by the spatial distribution of *Melinaea*. However, ongoing migration prevents fixation of any particular pattern. This is known as a selection-migration equilibrium, in which polymorphism is maintained by patchiness in the mimicry environment.

These descriptive studies of *H. numata* are complemented by a manipulative experiment on another polymorphic species, *H. cydno*, in western Ecuador that provides further support for the spatial heterogeneity hypothesis. Here, *H. cydno* mimics two alternative models, *H. sapho* and *H. eleuchia*, which have single white and double yellow forewing bands, respectively. The polymorphism is controlled by two genetic loci that regulate the white/yellow and solid/split band phenotypes.

Durrell Kapan (2001) identified sites in western Ecuador in which either *H. sapho* or *H. eleuchia* was the dominant model. As in the *H. numata* study, the frequency of different *H. cydno* morphs correlated with the abundance of their respective models. With the help of a cocktail fridge and a tube of superglue, Durrell captured *H. cydno* individuals and translocated them between sites (Figure 7.1, panel C). The fridge was used to keep the butterflies cool and happy during transport; the superglue was used to 'castrate' the butterflies and prevent them from breeding in their new localities. He then monitored their survival after release, both by recapture and by re-sighting using binoculars. The butterflies with wing patterns matching the locally most abundant 'model' species showed the highest survival, as predicted by mimicry theory. This provides experimental support for the spatial variation in mimicry selection hypothesized for *H. numata*.

Overall, therefore, the empirical studies of polymorphic Müllerian mimics suggest that the scale on which the mimicry environment varies is the critical factor in determining whether populations are polymorphic or monomorphic. In *H. erato* and *H. melpomene*, the mimicry environment varies on a scale of hundreds of kilometres, much larger than the scale of dispersal of individual butterflies, so populations are monomorphic except for narrow hybrid zones. In contrast, in *H. numata* and some populations of *H. cydno*, the mimicry environment varies on a scale of just a few kilometres. This heterogeneity is more fine-grained than the dispersal distance of individual butterflies, which leads to a stable polymorphism, maintained by a balance between selection and migration.

Nonetheless, there are some cases of polymorphism in which variable selection across the landscape does not quite make sense as an explanation. *Heliconius doris* has red, blue, green, and yellow forms that are broadly distributed across its range. The red form is similar to the Amazonian dennis-ray mimicry ring, and also to species such as *H. clysonymus* and *H. hortense*. It has also been suggested that this form mimics butterflies in the genus *Parides*, and for this reason is more abundant in Mexico (Mallet, 1999). The blue form is mimetic with the *H. sara* mimicry ring and is most common in South America, while the rarer yellow morph is most common in Costa Rica and western Panama where it is similar to *H. hewitsoni* and *H. pachinus*. These correlations perhaps indicate a spatially variable mimetic environment, but they are very broad-brush, and most populations of *H. doris* are both polymorphic and not especially accurate mimics.

One possibility is that *H. doris* is not particularly unpalatable (Brower et al., 1963) and acts as a Batesian mimic—in which case polymorphism may be favoured as a means of preventing the palatable mimic becoming too frequent relative to its models. However, this explanation does not explain the presence of completely non-mimetic morphs in some parts of the range, such as the red morph in lowland Central America. A second possibility is that the unusual biology of this species may favour polymorphism. It has highly gregarious larvae and can reach locally high abundance, as large numbers of adults tend to emerge synchronously in a local area. Perhaps this saturates the local predators leading to weak selection against polymorphism (see later for number-dependent mimicry theory, which supports this assertion).

Another case is the population of *H. timareta* in the Puyo region of Ecuador, which has an unusual polymorphism involving the dennis and ray patterns (Mallet, 1999). This was long thought to be an isolated population with a largely non-mimetic polymorphism. However, we later discovered nearby lowland populations of *H. timareta* that are monomorphic for the dennis-ray patterns, suggesting that this polymorphism can perhaps be explained, at least in part, as a result of hybridization between a lowland and an upland population (Nadeau et al., 2014).

Overall, therefore, polymorphism in wing pattern is unusual in *Heliconius* outside of narrow hybrid zones. The few examples of stable polymorphisms that do exist are explained partly by local heterogeneity in the mimicry environment, but this does not provide a universal explanation. I shall return to mimicry polymorphism in the context of the genetic control of wing patterns (Chapter 8).

7.12.1 Sexual dimorphism

Many butterflies show sexual dimorphism in wing pattern. For example, species such as *Papilio dardanus* have Batesian mimic females, but males with typical yellow and black patterns. Polymorphism is expected under classic Batesian mimicry theory, and such sexual dimorphism provides one means of achieving it. However, as with polymorphism, it is harder to explain sexual dimorphism in Müllerian mimicry. Although there are some Müllerian mimics that show sexual dimorphism, it is rare in *Heliconius* (Figure 7.7).

The only two species across the Heliconiini that show striking dimorphism in wing pattern are *Eueides pavana* and *Heliconius nattereri* (Turner, 1968b). The latter is a rare species restricted to the Atlantic forests of south-eastern Brazil, in which the females are orange and yellow, mimetic with *Heliconius numata ethra* and *H. ethilla* (Brown, 1972a). In contrast, males are black and yellow and non-mimetic, or possibly rather poor mimics of *H. charithonia*. Females fly in the forest while males patrol in more open areas, so the dimorphism may be related to habitat use (Brown, 1972a).

Other species show rather slight sexual dimorphism. For example, in *Heliconius demeter*, males show an unusual fusion of the rayed hindwing pattern, and sometimes rather fuzzy forewing yellow bands, while females are much better dennis-ray mimics (Figure 7.7) (this dimorphism seems to be absent in the recently recognized sister species *H. eratosignis*). In some silvaniforms—notably, some populations of *Heliconius hecale*—females have duller patterns than males.

There are two complementary explanations for dimorphism. First, mimicry selection may differ between males and females due to behavioural differences between the sexes. In particular, females may be more strongly predated due to their slow flight and need to search for host plants. Analysis of beak marks on museum specimens of a variety of butterflies supports this assertion (Ohsaki, 1995). This may explain why females are better mimics than males in both *H. demeter* and *H. nattereri*, simply because selection for mimicry is stronger in females. Sexual selection is the second explanation. Males, females, or both may use wing patterns in choosing mates, leading to divergent selection

Figure 7.7 Examples of sexual dimorphism in *Heliconius* (see Plate 21)

Not many species of *Heliconius* show sexual dimorphism. Top two rows: Several silvaniform species such as *H. hecale quitalena* show dimorphism in the brightness of their wing patterns. Middle two rows: *H. demeter demeter* shows much more accurate mimicry in females as compared to males. Bottom two rows: *H. nattereri* is the most sexually dimorphic species in the genus *Heliconius*. Dorsal (above) and ventral (below) images are shown for all specimens.

on wing pattern. It seems likely that the dramatic sexual difference in wing pattern in *H. nattereri* and the brighter patterns of *H. hecale* males is driven by female choice.

7.13 The dynamics of mimicry—theory

So far I have considered mimicry from a natural history perspective, assuming a clear dichotomy between unpalatable Müllerian mimics and palatable Batesian mimics. However, the reality is considerably more complex. As is evident from the experiments with bird predators described earlier, there is a spectrum of palatability, with some species considerably more unpleasant than others. In addition, there is a great diversity of predators. It seems likely that avian predators are the main selective force driving wing pattern evolution, with puffbirds, motmots, jacamars, and many species of tyranid flycatchers all likely to be candidate butterfly predators.

In addition, however, bats, lizards, geckos, mantids, and spiders are all known to consume *Heliconius* either in the insectary or in the wild, and could also play a role in the evolution of visual mimicry and other forms of defence. Even for a single predator, attack rates likely vary during the year depending on availability of alternative prey and levels of hunger. However, remarkably little is known of the details of these interactions—which species are the major predators, what are their visual systems and what is their psychology of learning and forgetting?

A body of theory considers the influence of predator psychology on mimicry dynamics. Mathematical models based on a variety of assumptions about predator psychology challenge the traditional dichotomy between Müllerian and Batesian mimicry (for a summary see Speed & Turner, 1999). The classic formulation of mimicry theory describes Batesian mimicry as parasitic—mimics increase the attack rates on their models—whereas Müllerian mimicry is mutualistic—the presence of mimics always benefits their models. The newer models suggest that mildly unpalatable mimics can also be parasitic and increase the attack rate on their model. This has been termed 'quasi-Batesian mimicry' (Speed, 1993).

Huheey (1976) proposed the first mathematical model of quasi-Batesian mimicry. He assumed that predators sample prey and learn about their unpalatability, then over time forget the experience. Huheey's model had the strange feature that the rate of forgetting was dependent not only on the toxicity of the prey, but also on the local frequency of the different prey species. If we imagine a less toxic mimic evolving similarity to an existing toxic model species, the frequency-dependent forgetting rule led to the conclusion that the mimic would increase the rate at which predators forget the shared pattern, and hence increase attack rates on the model species. Thus, Müllerian mimicry becomes a parasitic relationship similar to Batesian mimicry, in which the toxic model species suffers a cost from the presence of the less unpalatable mimic—even though the less unpalatable mimic on its own would also act to deter predation. Huheey has argued that this may explain wing pattern polymorphism and diversity in *Heliconius* (Huheey, 1976, 1988).

Huheey's model has been heavily criticized (Benson, 1977; Turner, 1984; Speed, 1999), not least because of the unusual frequency-dependent forgetting rule. However, more realistic models of predator psychology can yield similar results. Indeed, Speed and Turner (1999) reviewed a range of different assumptions about predator learning and forgetting that produce broadly similar outcomes and support quasi-Batesian mimicry (Speed & Turner, 1999).

Nonetheless, this conclusion has been contested. Mallet and Joron extended Müller's original number-dependent theory to include variation in palatability between model and mimic, and this model supports the traditional view of Müllerian mimicry (Joron & Mallet, 1998; Mallet & Joron, 1999). They assume that predators need to sample a fixed number of prey in order to learn to avoid an unpalatable pattern. Less unpalatable prey may therefore require a larger number of individuals to be sampled before aversion is learnt, but the number attacked will always tend towards zero. In addition to providing support for the traditional view of Müllerian mimicry, this theory leads to the interesting prediction that the shape of the frequency-dependence curve is strongly dependent on population size (Figure 7.8). At large population sizes, a very small

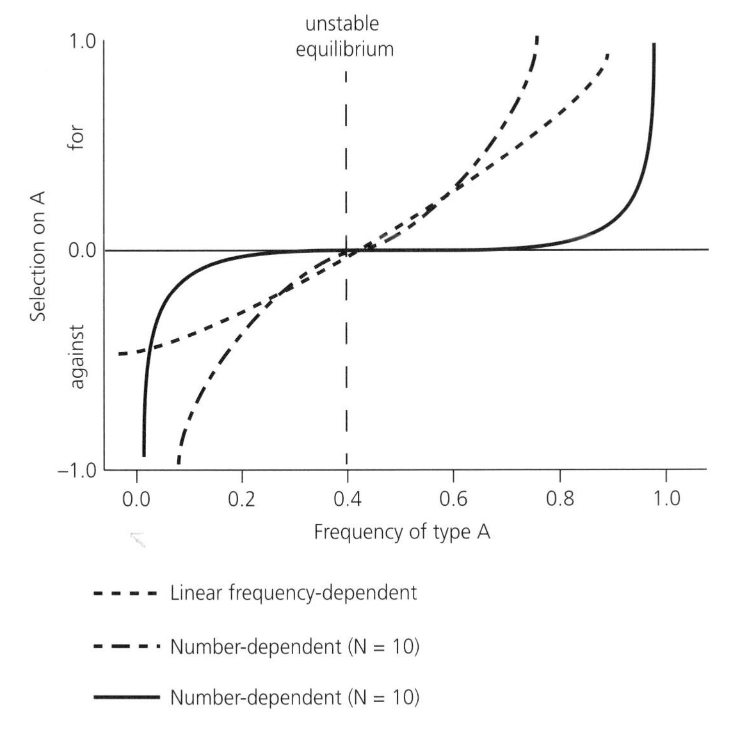

Figure 7.8 The shape of mimicry selection

Number- and frequency-dependent selection in mimicry and aposematism. Assume two patterns in the population, a novel pattern (A) and the wild type (a). The number-dependent theory assumes predators need to consume a constant number of prey in order to learn, and predicts a 'trough' of weak selection at intermediate frequencies. Selection primarily acts to remove rare variants. Under frequency dependence, there is no such trough of weak selection. The selection pressures on A and a have been chosen such that A has 1.5 times greater fitness than a, giving an equilibrium frequency of 0.4 (Mallet & Joron, 1999).

proportion of the total number individuals need to be sampled for predator learning, so selection is very weak except on extremely rare forms. Mallet and Joron (1999) state that selection is 'weak in a broad basin of intermediate frequencies, allowing opportunities for polymorphisms and genetic drift' (p. 201). Empirical studies using artificial models broadly support this non-linear frequency-dependent selection (Chouteau, Arias, & Joron, 2016). This might explain how several morphs are maintained in species such as *H. doris*, because these are effectively neutral with respect to one another at high population densities.

The foregoing arguments may seem rather arcane, but they have implications for our understanding of polymorphism and diversity in mimicry. Both approaches claim to provide explanations for the existence of polymorphism and geographic diversity, and virtually all of the papers presenting these models have extensively discussed *Heliconius* in support of one or another hypothesis.

Nevertheless, I believe that both approaches make some assumptions that are unrealistic. First, the inherent assumption of most models supporting quasi-Batesian mimicry is frequency dependence, such that predators attack prey depending on their frequency in the local population. This leads to the slightly absurd idea that as prey become highly abundant, predators should continue to consume a fixed proportion of available prey—presumably stuffing themselves with larger and larger numbers of toxic prey. As emphasized by Mallet and Joron, this seems improbable; some form of density (or number) dependence needs to be included in the models (Mallet & Joron, 1999).

Another difference of opinion concerns the asymptotic attack frequency—the proportion of prey that will be attacked once a predator has gained full experience of the prey. Mallet and Joron argue that for all unpalatable prey this should be 0: any unpalatable prey will eventually teach a predator complete avoidance (Mallet & Joron, 1999). In contrast, Speed and others have argued that a prey can be unpalatable and yet still have an appreciable asymptotic attack frequency (Speed & Turner, 1999). In other words, a prey can be unpalatable and teach predators increased avoidance, but yet still suffer some constant probability of being attacked. This might be because a small dose of a toxin can be tolerated; that the predator continues to try prey just in case they prove profitable; or that there is a constant probability of forgetting what predators have experienced.

There is good evidence that predators balance nutritional needs against toxin accumulation and can evolve to include small amounts of toxic prey in their diet (Skelhorn & Rowe, 2007; Halpin, Skelhorn, & Rowe, 2014). Herein lies the key to understanding quasi-Batesian mimicry: if constant attack probability varies depending on unpalatability, then the presence of a more palatable mimic could teach predators to associate the pattern with a less toxic experience, and in turn lead to a higher asymptotic attack frequency on the model.

The key requirement is therefore a non-zero attack rate on mildly unpalatable prey. This seems plausible given the experiments with predators described earlier in Section 7.4. Predators do indeed attack Müllerian mimics at surprisingly high rates, and their attack probability varies across butterfly species (Brower et al., 1963; Chai, 1986, 1996; Sargent, 1995; Pinheiro, 1996). The experiments therefore suggest that predators do indeed continue to attack these butterflies with a non-zero asymptotic probability. The fact that *Heliconius* retain their ability for rapid escape flight, unlike other aposematic butterflies such as the ithomiines, also suggests that they are prey for at least some predators and may lie in the zone of intermediate palatability.

However, what is really needed is experimental evidence on the behaviour of predators in response to differing prey types over extended time periods. One type of evidence comes from the so-called novel world. This is an experimental set-up in Finland that uses wild great tits attracted into an experimental arena, in which unnatural prey items are presented. These are generally flakes of almond glued between two pieces of paper; arbitrary patterns drawn on the paper can be made cryptic or conspicuous. Prey palatability can be varied using different concentrations of chloroquine.

There have been a long series of experiments using this system to address quasi-Batesian mimicry and other related problems. Some have varied model/mimic frequency at a constant density (Lindström et al., 2006; Ihalainen et al., 2008); others have varied density (Rowland et al., 2007). Some experiments support traditional mutualistic Müllerian mimicry (Lindström et al., 2004, 2006; Rowland et al., 2007), whereas others support parasitic quasi-Batesian mimicry (Ihalainen et al., 2008; Rowland et al., 2010). Undoubtedly, under some circumstances the addition of mildly unpalatable mimics to a nastier model does result in mutualistic benefit for both sides (Rowland et al., 2007), but that is perhaps unsurprising, as virtually all theoretical models predict mutualistic Müllerian mimicry in some areas of parameter space (Speed & Turner, 1999).

On the other hand, the experiments provide strong support for stable intermediate attack frequencies on mildly unpalatable prey, which is a key necessity for quasi-Batesian mimicry (although it is hard to know how often real prey would fall into this intermediate zone in the spectrum of palatability). One experiment in which mildly unpalatable models were kept at a constant density—but varied in the frequency of either mimics or non-mimics—demonstrated a cost to the toxic model of the increasing presence of the mimic, and therefore seemed to support quasi-Batesian mimicry (Rowland et al., 2010). This scenario would seem to be a reasonable approximation to an unpalatable species evolving towards mimicry of a more toxic model. A similar experiment using pastry baits and garden birds gave comparable results (Speed et al., 2000).

In summary, quasi-Batesian mimicry is plausible, but it remains unclear whether it is likely or frequent. The relative abundance of species may have a much stronger influence on the costs and benefits of mimics than relative palatability (Joron & Mallet, 1998). If quasi-Batesian mimicry does occur, it might contribute to explaining polymorphism in species such as *H. doris*. However, as outlined earlier, we probably do not need to invoke quasi-Batesian mimicry to explain polymorphism, so the existence of species such as *H. doris* cannot by itself support quasi-Batesian mimicry. The extensive literature on quasi-Batesian mimicry mainly serves to highlight how ignorant we are of predator psychology and behaviour. Much of the evidence comes from European garden birds, which are easy to study, but hardly a good proxy for the astute insectivores of the tropics.

7.14 Early stages

A common question is whether there is mimicry in early life stages. Many heliconiine larvae are aposematic. The spines and bright white background of species in the melpomene and silvaniform clades clearly stand out against the host plant. The behaviour of the caterpillars also serves to advertise unpalatability, as most species rest on exposed parts of the host plant during the day. Nonetheless, the pace of evolution in larval coloration is much slower than for adult wing patterns, with little within-species variation. The larvae of closely related species can generally be distinguished, but differences can be slight. It is not clear why the larvae do not show the rapid diversification of pattern seen in adults. There are, however, a few cases of larval mimicry, typically among species that share a host plant. In Costa Rica, for example, the larvae of *H. sara*, *H. hewitsoni*, and *Eueides vibilia* all look very similar (Mallet & Longino, 1982), whereas in the Amazon, *Eueides tales* has apparently converged to mimic *Heliconius numata* larvae, with which it shares a host plant (Brown & Holzinger, 1973).

In contrast, the pupae are cryptic, mostly looking like dead leaves. They show some phenotypic plasticity in colour, with the pupal case being darker against darker backgrounds. Experiments in *H. erato* have shown that the light environment during the early prepupal stage influences the final colour of the pupal case, implying that the pupa can detect the brightness of its surroundings at this stage (Ferreira, Garcia, & de Araújo, 2006). In many species, the pupae also have shiny patches on the ventral surface that may act as a form of disruptive colouration to break up the outline of the pupa.

7.15 Future directions

Despite the overwhelming evidence for strong selection on the warning colour patterns of *Heliconius*, a fundamental piece of natural history information that is still lacking is an understanding of exactly which predators attack these butterflies in the wild. One potential approach is to use DNA barcoding of predator faeces to identify which predator species are consuming which butterfly species. We also need to know how predators perceive and learn patterns, and how this varies between seasons and between predator species. It seems likely that the intensity of mimicry selection varies across seasons, as predators become more or less satiated with edible prey. Such a relaxation of selection might provide insights into how new patterns arise and potentially also into the maintenance of polymorphic patterns.

Specifically, it would be useful to determine whether *Heliconius* species do actually lie in the zone of intermediate palatability in which predators would consume them at some non-zero rate. A first step would be to more extensively survey variation in the cyanogenic content of wild butterflies. This could be complemented with experiments with real predators to investigate the extent to which they can tolerate ingestion of small quantities of these toxins. Work on the comparative palatability of the *H. numata* mimicry ring offers one approach using European birds, although it would be even better to use tropical predators (Arias et al., 2015). Such studies will offer real insights into the dynamics of mimicry in the tropics.

CHAPTER 8

Genes on the wing: colour pattern genetics

Chapter summary

Heliconius wing patterns are an adaptive trait under strong selection in the wild. They are also amenable to genetic studies and have been the focus of evolutionary genetic analysis for many years. Early genetic studies characterized a large number of Mendelian loci with large effects on wing pattern elements in crossing experiments. The recent application of molecular genetic markers has consolidated these studies and led to recognition that a huge range of allelic variation at just four major loci controls patterns across most of the *Heliconius* radiation. Some loci consist of tightly linked components that control different aspects of the phenotype and can be separated by occasional recombination. More recent quantitative analyses have also identified minor effect loci that influence the expression of these major loci.

Population genetic analysis has confirmed that different races of *H. melpomene* and *H. erato* differ little except at the major wing patterning loci. Indeed, the regions identified from crosses studying wing patterns are associated with the most divergent regions in the genome between races. However, these genetic studies provide little evidence for recent rapid selection (selective sweeps), suggesting that the variants are relatively ancient.

Studies of a single locus polymorphism in *Heliconius numata* provide an example of a 'supergene' in which a single major locus controls segregation of a variable phenotype. This supports Turner's 'sieve' hypothesis for the evolution of supergenes, whereby sequential linked mutations arise at the same locus. In addition, inversion polymorphisms are associated with wing pattern variation, which reduce recombination across the supergene locus. This finding provides direct evidence that the architecture and organization of genomes can be shaped by natural selection. There is evidence that patterns of dominance of the alleles at this locus have also been shaped by natural selection. Mimicry therefore provides a case study of how natural selection shapes the genetic control of adaptive variation.

On the wall in my office in Cambridge is a framed plate from the book on mimicry by Reginald Punnet (1915). The plate was uncovered during the renovation of one of the offices in the Department of Zoology, and offered to me, presumably as the local butterfly geneticist. Punnet worked in Cambridge in the Department of Genetics and at Gonville and Caius College, where he studied polymorphic mimicry in butterflies.

The plate shows variation in the wing patterns of *Papilio polytes*, an Asian swallowtail butterfly. Punnet showed that in species such as this, just one or a few loci could control dramatic differences between individuals in their wing patterns. Punnet and colleagues interpreted this as evidence that major mutations contributed to the direction of evolution (hence this school of thought became known as the mutationists). These discoveries seemed to

The Ecology and Evolution of Heliconius Butterflies. Chris D. Jiggins, Oxford University Press (2017).

DOI 10.1093/acprof:oso/9780199566570.001.0001

contradict gradual Darwinian evolution and were even portrayed as a challenge to the importance of natural selection. The genetic basis for pattern variation in mimetic butterflies therefore became an iconic example cited by early evolutionary geneticists in arguments over the mechanisms of evolution.

In defence of Darwinism, and to counter the arguments made by Punnet and others, Ronald Fisher devoted an entire chapter to mimicry in his book *The Genetical Theory of Natural Selection* (Fisher, 1930). Fisher was also based in Cambridge and, curiously, also a fellow of Gonville and Caius College. He argued that the loci of major effect discovered by Punnet could evolve through many small steps. Their large effect was not the result of a single major mutation, but rather the outcome of a long process of gradual evolution. Indeed, Fisher recognized that in polymorphic populations, the evolution of a single gene taking control of wing pattern might be directly favoured by natural selection. Such major genes later became known as 'supergenes'. We are now in a position to test alternative hypotheses for the evolution of mimicry by directly studying the changes in DNA sequences that regulate different wing patterns.

Indeed, a major research effort in evolutionary biology is devoted to determining the molecular changes in DNA sequences that control adaptive phenotypic changes at the level of the organism. By identifying the number and identity of genes controlling traits, and the relative contribution of individual mutations to changes in the appearance of an organism, we can address questions debated by Fisher, Punnet, and others, including the importance of large versus small mutations in evolution. It is exciting to be in a position to address these long-standing questions with modern techniques.

The earliest genetic studies of mimicry mainly focused on the genus *Papilio*, but *Heliconius* began to catch up in the 1950s and 1960s. Once again, pioneering work was begun by William Beebe, who in 1954 travelled to Surinam and visited a collecting locality on the Maroni River famous for its diversity of *Heliconius* (now recognized as a stable hybrid zone; see Chapter 7). He transported adult butterflies back to his research station in Trinidad and raised offspring from males and females of known phenotype. The resulting publication is surprising for its apparent lack of any genetic inference made from these crosses (Beebe, 1955); nevertheless, this work provided the first hint at the underlying genetic control of wing pattern variation in *Heliconius*.

Already evident in Beebe's experiments—and confirmed by subsequent work—is the finding that loci of major effect play an important role in controlling *Heliconius* patterns (Figure 8.1). Subsequently, Emsley used crosses between *H. erato* and *H. melpomene* from Ecuador and Trinidad to infer genetic control and linkage patterns of major patterning loci (Emsley, 1964).

However, the greatest advances in understanding were made by a large consortium of collaborators, including Cyril Clarke and Philip Sheppard working in Liverpool, Keith Brown and Woody Benson in Brazil, Mike Singer in Texas, and Sheppard's student, John Turner. A series of publications from these efforts culminated in a wonderful, if impenetrable, paper published in the *Philosophical Transactions of the Royal Society* in 1985 (Sheppard et al., 1985). The paper (mainly written by John Turner, after Philip Sheppard's death at an early age) reported experiments in which numerous races were crossed within the two species, *H. erato* and *H. melpomene*, collected from different parts of South America. In general, brood sizes were rather small, and often the full genotypic ancestry of the individuals involved was not known, so at times the results can be difficult to follow. Nonetheless, the genetic scheme outlined in this work has held up and provided a foundation for more recent molecular analysis.

Later experiments used crosses between individuals of known ancestry with larger brood sizes and 'test crosses' to investigate specific genetic hypotheses (Mallet, 1989; Jiggins & McMillan, 1997). This approach eventually led to the identification of the genes controlling wing patterns at a molecular level. Our understanding of wing pattern genetics has made major advances, so I will not attempt to review the old literature but will instead organize this chapter by genetic loci and just touch on how our understanding has evolved over time. The first half of the chapter is descriptive; in the

Figure 8.1 Phenotypes from a hybrid zone in eastern Ecuador (see Plate 22)

There are three parental races that contribute variation to the hybrid zone, pictured here along the top row: *H. melpomene plesseni, H. m. malleti*, and *H. m. ecuadorensis*. Three major loci control the wing patterns: the red locus, *optix*, controls red/orange pattern elements; the shape locus, *WntA*, controls the shape of the forewing band (two spots or one); the yellow locus, *cortex*, produces the yellow forewing band. The butterfly hybrids are all from the Neukirchen collection. Scale bar is 1 cm.

second half, I will return to some of the evolutionary questions posed by Fisher and others, and consider how *Heliconius* patterns can begin to provide some answers.

8.1 Background versus pattern

In the description of wing pattern genetics that follows, I describe the red, orange, yellow, and white coloured regions as the 'pattern elements' that are

Table 8.1 Homology of wing pattern loci

Species	Locus	Phenotypic effect	Reference
Red locus - Optix – LG18			
H. melpomene	D	Dennis patch	1
	B	Red FW band	1
	R	HW rays	1
	M	Yellow FW band	2
H. erato	Y	Yellow/red FW band	1
	D	Dennis patch	1
	R	HW rays	1
	Wh	White in FW	1
H. cydno	Br	Brown cydno 'C'	3
H. pachinus/heurippa	G	Red HW spots	3,4
H. hecale	HhBr	HW orange/black	5
H. ismenius	HiBr	HW orange/black	5
Yellow locus - cortex – LG15			
H. melpomene/cydno	Yb	Yellow HW bar	1,3
	N	Yellow FW band	1,3
	Sb	HW white margin	3,5
	Vf	Pale ventral FW band	3
H. erato	Cr	Cream rectangles	1
H. hecale	HhN	FW submarginal spots	6
H. ismenius	HiN	FW submarginal spots	6
H. ismenius	FSpot	FW subapical spots	6
H. ismenius	HSpot	HW marginal spots	6
H. numata	P	All pattern variants	7
Shape locus - WntA – LG10			
H. melpomene/cydno	Ac	FW band shape	1,3
	C	Broken FW band	1
	S	Shortens FW band	1,8
H. erato	Sd	FW band shape	1,9
	Sd	HW bar	1,9,10
	St	Split FW band	1,9
	Ly	Broken FW band	1,9
	Yl	Yellow FW line	1,11
H. hecale	HhAc	Yellow FW band	6
H. ismenius	HiAc	Yellow FW band	6

continued

Table 8.1 *Continued*

Species	Locus	Phenotypic effect	Reference
Colour locus - LG 1			
H. melpomene/cydno	K	FW band colour (yellow/white)	3,12
	Khw	HW margin colour (yellow/white)	13
LG13			
H. melpomene	Unnamed	FW band shape	14
H. erato	Ro	Rounded FW band	15
Unknown			
H. melpomene	Or	Orange/red switch	1
H. cydno	L/Wo	Forewing white spots	16
H. cydno/pachinus	Ps	Pachinus 'shutter'	17
H. cydno	Fs	Forewing 'shutter'	17
H. cydno	Cs	Cydno 'shutter'	17

A summary of previously described wing patterning loci and their homology to the recently identified major effect genes. HW and FW refer to hindwing and forewing respectively. Notes: [1] Sheppard et al. (1985). [2] The *M* locus interacts with *N* to influence the forewing yellow band in *H. melpomene* (Mallet 1989). Unpublished work (Baxter and Mallet pers. Comm.) indicates that *M* is an effect of the *optix* locus. [3] Naisbit, Jiggins, & Mallet (2003). [4] Mavarez et al. (2006). [5] Linares (1996). [6] Huber et al. (2015). [7] The *P* supergene locus in *H. numata* controls all aspects of phenotype. The locus is homologous to *Yb* although it seems likely that the supergene includes several functional loci (Joron et al., 2006). [8] Nijhout, Wray, & Gilbert (1990). [9] Papa et al. (2013). [10] Mallet (1989). [11] Sheppard et al., (1985) infer that *Yl* and *Sd* are linked, but that *Yl* and *Ly* segregate independently. *Sd* and *Ly* are now known to be the same locus, so it is unclear whether *Yl* is unlinked. Further crosses of Brazilian forms would be needed to test this. [12] Kronforst et al. (2006a). [13] Joron et al. (2006). [14] Baxter, Johnston, & Jiggins (2009). [15] The *Ro* locus was mapped to linkage group 13 by means of a hybrid zone association study (Nadeau et al., 2014). [16] L and Wo are linked loci that control forewing white elements in *H. cydno* and may be homologous to *Ac* (Linares, 1996). [17] *Ps*, *Fs* and *Cs* from Nijhout, Wray, & Gilbert (1990) are included for completeness but patterns of segregation and linkage are not known. These may be effects of the *WntA* locus.

expressed against an otherwise black background. It has been proposed that we should invert this logic and instead consider the black regions as 'pattern' against a brightly coloured background (Nijhout, 1991; Gilbert, 2003). I find that this makes the segregation of patterns in crosses much more difficult to describe, not least because there are several possible background colours. Even authors who promote this viewpoint tend to flip between the two perspectives, such that, for example, Papa et al. (2013) assert that 'natural variation of the size and shape of the bands and bars elements in the fore- and hindwing were consistent with the concept that regions of black scales genetically define the position and size of these elements', but in the following paragraph they consider segregation of 'red and orange pattern elements' (p. 2). The background/pattern distinction is arbitrary and I suspect will become meaningless once we understand the patterns better from a developmental perspective. For the moment, it is easier to retain the more intuitive approach used by John Turner and others; I will return to the question of the 'default' or 'background' state of wing patterns in the next chapter.

A summary of previously described wing patterning loci and their homology to the recently identified major effect genes is given in Table 8.1.

8.2 Phenotypic effects of major loci: the red locus optix

Much like Punnet and his crosses in the genus *Papilio*, the most striking aspect of *Heliconius* wing pattern genetics is that a few major loci control large phenotypic changes (Figure 8.2). This major locus

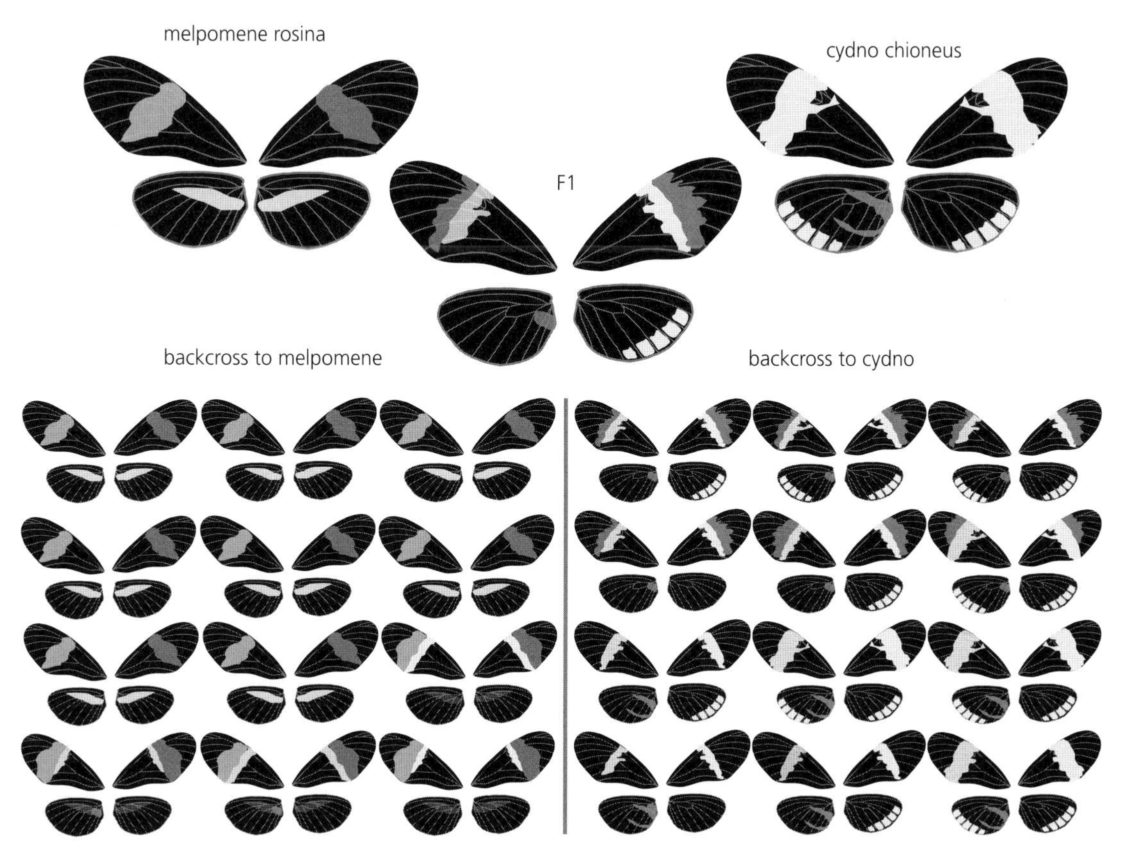

Figure 8.2 Inheritance of wing pattern variation in an interspecies cross (see Plate 23)

Hybrid phenotypes for a cross between *H. melpomene rosina* and *H. cydno chioneus* from Panama. The backcross families are generated by crossing an F_1 male back to either a female *H. melpomene rosina* (left) or a male *H. c. chioneus* (right). The main pattern loci with alleles that segregate in this cross are the red locus *optix*, controlling the red forewing band; the yellow locus *cortex*, with alleles that control the yellow/white forewing band, yellow hindwing bar, and white margin; the shape locus *WntA*, which controls the hourglass-shaped forewing band element; and the colour locus *K*, which alters the colour of the forewing band from yellow to white. Drawn from crosses described in Naisbit et al. (2003). Both ventral (left) and dorsal (right) patterns are shown.

control of adaptive traits is an emerging pattern in other organisms, but studies of butterflies provided some of the first clear examples (Nadeau & Jiggins, 2010). The locus that is best understood at a molecular level and has perhaps the largest phenotypic effect controls red patterns. The existing nomenclature for these loci is highly confusing and is summarized in Table 8.1, but I will here adopt a simplified scheme using both descriptive names and gene names. I will therefore term this the 'red locus', or *optix*.

Variants at this locus likely represent regulatory switches controlling expression of the gene *optix* (Reed et al., 2011), which in turn controls variation in both *H. melpomene* and *H. erato* red and orange pattern elements. Even to William Beebe, it must have been evident that a major genetic locus was controlling these patterns. In his *H. erato* 'Brood C', a female parent with 'red forewing band only' produced two offspring 'like [the] parent' and two with both a 'broken red forewing band' and 'red radiations on fore and hindwing', thus indicating that at least the red ray pattern was inherited in an all-or-nothing manner, in this case inherited from the father of the brood.

In fact, the red patterns can be divided into three main elements: the red forewing band, the red ray pattern on the hindwing, and the basal patch on the forewing. The latter is known as the 'dennis' patch, after an individual butterfly that William Beebe named Dennis the Menace (note that this is the likeable American cartoon character Dennis the Menace, rather than his altogether more menacing British namesake). These three elements are controlled by allelic variants at the *optix* locus with the presence of red patterns being dominant over their absence. Hence, heterozygote hybrids between butterflies with a red forewing band and those with red dennis-ray patterns express both phenotypes.

8.3 The red locus is homologous between species

Early crossing experiments clearly showed that *H. melpomene* red phenotypes were inherited in a similar manner to *H. erato*, as allelic variants at a single major locus. Early workers therefore speculated that there was homology between the loci controlling red pattern elements in the two species (Turner, 1981; Nijhout, 1991). Nonetheless, significant differences in the appearance and inheritance of patterns in the two species led Mallet to suggest that they might not be homologous loci (Mallet, 1991). The *H. erato* ray pattern is constructed from red pigment entirely deposited between the wing veins in the hindwing. In contrast, the *H. melpomene* rays form a 'nail-head' shape that abuts a band across the hindwing (see Figure 7.4). Unlike the pattern in *H. erato*, this band crosses wing veins, suggesting a distinct developmental origin that is not vein-dependent. *Eueides tales* represents a third phenotypically distinct form of hindwing rays, in which the orange pigment that forms the rays is deposited on the veins rather than between the veins as in all other species.

Since it is impossible to cross *Heliconius erato* and *H. melpomene* with each other, testing for the homology of loci between the species required linked genetic markers. Once these became available, it turned out that despite the differences, there is a remarkable degree of homology between species in the loci used to control similar patterns. Homology of the red locus between the two species was first demonstrated at the level of chromosomes (Joron et al., 2006; Kronforst, Kapan, & Gilbert, 2006b). Subsequently Simon Baxter obtained sequence from AFLP markers for the red locus in *H. melpomene*, and demonstrated at a fine scale that the region was homologous in *H. erato*, definitively establishing homology between the two species (Baxter et al., 2008).

8.4 Other phenotypic effects of the red locus

The main phenotypic effect of the red locus is in placement of red pattern elements on the wing. However, the locus also controls the switch between red and other colours. For example, in crosses between *H. erato* races with red and those with yellow forewing bands, alternate alleles control the switch between red and yellow. Expression of the two colours is co-dominant, with heterozygotes typically showing a mixture of yellow and red scales in the forewing band.

The red locus also controls the switch between yellow and white in the forewing of the race *H. e. notabilis*, an effect previously ascribed to a locus named *Wh* (Sheppard et al., 1985; Papa et al., 2013). This contrasts with the situation in *H. melpomene*, where a separate unlinked locus controls yellow patches in the forewing (the yellow locus, *cortex*). In the *H. melpomene* clade, there is strong epistasis such that when yellow forewing alleles are combined with red band *optix* alleles, the combination produces distinct stripes of proximal yellow and distal red scales, with a sharp boundary between the two. This seems to imply a developmental interaction between two genes such that, at least in *H. melpomene*, they are never expressed together (Figure 8.3).

It is perhaps not so surprising that convergent, similar patterns in mimic species are controlled by the same genetic mechanism. But what about other types of patterns? Historically, most of the genetic studies have been carried out in *H. melpomene* and *H. erato*, but it turns out that even very different patterns in other species can be controlled by the same genetic loci. For example, the *optix* locus controls red/orange pattern elements in silvaniform butterflies. Crosses between two races of *H. hecale* from Panama and Venezuela, *H. hecale melicerta*

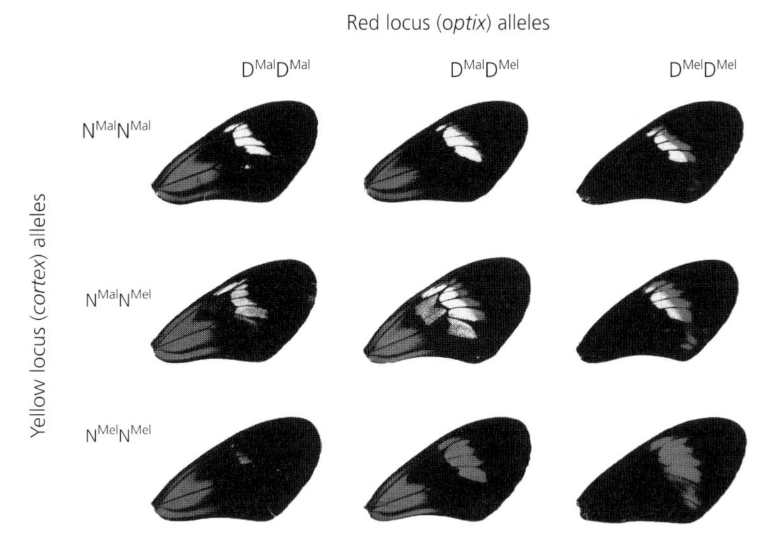

Figure 8.3 Epistatic interaction between the yellow and red loci (see Plate 24)

In crosses between the red forewing banded postman races and the yellow forewing banded dennis-ray races of *H. melpomene* and relatives, there is an epistatic interaction between alleles at the yellow (*N* refers to the effect of this locus on the forewing band) and red (*D*) wing patterning loci. For example, the wings shown are from a cross between *H. m. malleti* (top left) and *H. m. melpomene* (bottom right). The full red band in the forewing is produced only in the absence of yellow alleles at the yellow locus. In hybrid genotypes, the presence of yellow band alleles at the yellow locus pushes the red band distally, to produce the characteristic double yellow and red band (top right). This is similar to the phenotype of the hybrid species, *H. heurippa*.

and *H. h. clearei*, have shown that the loss of orange patterns on the fore- and hindwing of *H. h. clearei* is also controlled by this locus (Huber et al., 2015).

Similarly, in *H. ismenius*, variation in a black hindwing band on an otherwise orange pattern differs between the Central American races *boulleti* and *telchinia*, and also maps to the same region (Huber et al., 2015). The brown C-shaped pattern on the ventral hindwing of *H. cydno*, which is replaced by red spots in crosses with *H. melpomene*, is also controlled by a linked locus known as *Br* (Naisbit et al., 2003; Chamberlain et al., 2011). In summary, in every species so far investigated genetically, the red locus has major phenotypic effects wherever there is variation in red and orange pattern elements.

8.5 Recombination within the red locus

Historically, different patches on the wing were typically ascribed to the influence of distinct loci. This was a reasonable approach, but even in the older literature it was already clear that many of these distinct 'loci' were actually so tightly linked as to effectively be inherited as a single locus. Until now, I have considered the red patterning loci—known as *B*, *D*, and *R* in *H. melpomene* and *D*, *R*, and *Y* in *H. erato*—as diverse phenotypic effects of a single locus.

Nonetheless, it is clear that the red locus contains distinct, tightly linked elements (Figure 8.4). Evidence for recombination between these elements comes mainly from rare natural putative recombinants. In *H. erato*, a single individual with ray but not dennis was collected in a Peruvian hybrid zone, and similar individuals are known from Bolivia (Mallet, 1989). Similarly, in *H. melpomene*, Mallet caught one individual with dennis but not ray among 903 sampled in the Peru hybrid zone, while a single individual with the reverse phenotype—ray without dennis—occurred in a laboratory-reared brood (Mallet, 1989). Two individuals with dennis but not ray were also captured in the Ecuador hybrid zone near Puyo (P. Salazar and Jiggins, unpublished).

In both clades, there are also established races that appear to have recombinant genotypes. Thus, *H. e. amalfreda* and *H. m. meriana* are mimetic races from the Guianas that have dennis but not ray, while *H. timareta timareta f. contigua* is a form with ray but not dennis. Recent molecular analysis has confirmed that these phenotypes are indeed recombinants between tightly linked elements, rather than novel mutations (Wallbank et al., 2016). Thus, the red locus consists of at least three very tightly linked elements that independently control different patches of red on the wing.

Linkage mapping suggests that the *Br* locus, which controls a brown forceps-shaped mark on the underside of the hindwing in *H. cydno*, is ~20 cM away from *optix* on linkage group 18 (Naisbit et al., 2003; Chamberlain et al., 2011). This might seem to suggest that this pattern is controlled by a different gene, but in fact *optix* is expressed in association with this phenotype in a similar manner to other red elements (Reed et al., 2011). It is possible that a distantly linked region on the same chromosome regulates the expression of *optix*, although alternative explanations are either that this locus is incorrectly mapped perhaps due to epistasis, or that there is a second wing patterning gene on this chromosome that acts upstream of *optix*. The latter seems unlikely given the lack of any evidence for such a locus from any other mapping or developmental studies to date.

8.6 Phenotypic effects of major loci: the yellow locus *cortex*

This second major locus is similar in many ways to the red locus—it consists of tightly linked elements that also control different patches of similar colours—yellow and white in this case. The yellow locus is located on linkage group 15 and is associated with expression differences in the gene *cortex* (Nadeau et al., 2016). The effects of this locus include factors previously known as *Yb*, *Sb*, and *N* in *H. melpomene*, and *Cr* in *H. erato*. The *Yb* locus is named in *H. melpomene* for its effect in placing the hindwing yellow band (Mallet, 1986c; Sheppard et al., 1985).

Alleles that produce a yellow band are recessive to the absence of the band, although heterozygotes typically show an alteration in scale morphology in the band region that can be seen in altered reflectance in the otherwise black hindwing. Another allele at the same locus produces a band only on the underside of the hindwing and is present in the west Colombian race *H. m. venustus*. The patterns of inheritance in *H. erato* are virtually identical to those in *H. melpomene* in the Central American and Colombian races (Mallet, 1986c). A distinct but tightly linked element (*Sb*) at this locus controls the white hindwing margin found in the west Ecuador races *H. e. cyrbia* and *H. m. cythera* (Ferguson et al., 2010a; Jiggins & McMillan, 1997).

Many of the coloured patches on *Heliconius* wings are controlled in this very simple one-allele-makes-one-phenotype manner. However, there are also more complex interaction effects between the major loci. For example, in east Andean populations of *H. erato*, the yellow hindwing bar results from the joint effects of the yellow locus and the shape locus (see later – respectively termed *Cr* and *Sd* in *H. erato*). Thus, in Peruvian *H. e. favorinus*, recessive alleles at both loci are required for full expression of the hindwing bar (Mallet, 1989). In contrast, in Central American *H. erato*, a very similar bar results from just a single recessive allele at the yellow locus, much like in *H. melpomene*. The difference in inheritance patterns between populations could be either because the yellow bar has originated twice in *H. erato*, or because differences in the genetic background led to different patterns in the two populations (Mallet, 1989).

As outlined earlier, genetic control of the yellow forewing also differs between the species. The yellow locus places the yellow forewing band in *H. melpomene*, but is strongly epistatic with the red locus such that yellow and red are never mixed in the forewing of *H. melpomene* hybrids (Figure 8.3). These genetic interactions indicate that the genes at the major loci interact with one another at a molecular level during wing development, but how this occurs is not yet understood.

The yellow locus also has major effects in other species across the genus. In *H. cydno*, *H. pachinus*, and *H. heurippa*, loci homologous to *N*, *Yb*, and *Sb* all control variation in the forewing band, hindwing bar, and hindwing margin, respectively (Linares, 1996; Linares, 1997; Mavarez et al., 2006; Naisbit et al., 2003). White scales on the underside of the red forewing band in crosses between *H. cydno* and *H. melpomene* are also controlled by the yellow locus (known as *Vf*) (Naisbit et al., 2003). In crosses in the silvaniform taxa, *H. hecale* and *H. ismenius*, alleles at this locus similarly control placement of yellow spots on the forewing and hindwing (Huber et al., 2015). In contrast to *H. melpomene*, however, this locus has little influence on the broad yellow forewing band, whose presence is instead controlled by the shape locus (see later) (Huber et al., 2015). The yellow locus also corresponds to the 'supergene' *P* in *Heliconius numata*. I describe that species in more detail later (Joron et al., 2006).

8.7 Recombination at the yellow locus

As at the red locus, there is evidence for rare recombination events between tightly linked loci at the yellow locus. Thus, for example, *Yb* and *Sb* were mapped to within ~1 cm of each other, with two recombinant phenotypes identified in 175 individuals (Ferguson et al., 2010a). Previously it has been suggested that rare recombinant genotypes observed in the wild or in mapping families might represent rare de novo mutations, casting doubt on the hypothesis of tightly linked elements at these major effect loci. However, molecular genotyping has confirmed that recombinant phenotypes are indeed associated with recombination at this locus rather than mutation (Ferguson et al., 2010a). Similarly, in crosses between *H. melpomene rosina* and *H. c. chioneus*, we observed two recombinant phenotypes between *N* and *Yb*, giving an estimate of 4.3% recombination (Naisbit et al., 2003). These crosses also suggest a possible gene order of *N-Sb-Vf-Yb*.

In summary, there is a remarkable similarity between the red and yellow loci. Both consist of a set of tightly linked genetic elements that control major phenotypic changes. Each locus controls pattern elements with broadly similar phenotypic effects. Patterns of dominance are also quite predictable, with alleles for red elements dominant, and those for yellow or white elements recessive, giving a dominance series of red > black > white > yellow. In each case, these loci most likely represent tightly linked *cis*-regulatory elements of the same protein-coding gene, as I will outline in the next chapter; linkage is most likely a result of genetic architecture rather than a phenotype that has been favoured by selection (see later).

8.8 Phenotypic effects of major loci: the shape locus *WntA*

The third major locus is located on linkage group 10 and primarily controls the shape of the forewing band. This locus has been known as *Ac* in *H. melpomene* and *H. cydno*, *Sd* in *H. erato* (Table 8.1), and likely results from variation in expression of the gene *WntA* (Martin et al., 2012). This locus controls different aspects of the shape of the forewing band (Figure 8.4). In the Ecuadorean *H. m. plesseni*, this locus produces the 'split' forewing band—the largely recessive *H. m. plesseni* allele expresses the more proximal of the two white patches of this form, and also influences the shape of the more distal patch (Salazar, 2012). In crosses between *H. melpomene rosina* and *H. cydno chioneus*, a recessive allele places a white triangle that extends the forewing band into a small white hourglass shape in *H. cydno* (Naisbit et al., 2003). Similar phenotypes are seen in crosses within *H. melpomene*, with a homologous yellow element expressed in the forewing yellow band of Amazonian races (Sheppard et al., 1985). *L*, which places white patches on the forewing of *H. cydno weymeri f. weymeri*, may also correspond to this locus (Linares, 1996).

In *H. erato*, Papa et al. (2013) used linked molecular markers to show that variation previously attributed to three loci, *St*, *Sd*, and *Ly*, all maps to the same genomic location, corresponding to the shape locus. All these loci influence the shape of forewing band elements. In some cases, the phenotypic effects of this locus are extremely similar to those seen in *H. melpomene*. Thus, for example, in *H. e. notabilis*—which is mimetic with *H. m. plesseni*—*Sd* also acts to generate the split forewing band phenotype (Salazar, 2012). In the Amazonian forms such as *H. e. erato*, the allele at this locus also generates the broken yellow forewing band, while a major shift in position of the forewing bands between *H. himera* and *H. e. etylus* is also caused by the respective allelic variants at this locus (Sheppard et al., 1985; Papa et al., 2013).

In common with the yellow locus, alleles for yellow or white patterns are typically recessive to those for black. However, unlike the yellow and red loci, there is no clear evidence at this locus for recombination between linked elements. Sheppard et al. (1985) did report 3% recombination between *Yl* and *St*, although the status of *Yl* remains a little unclear and the finding needs to be confirmed. None of the other crosses to date report recombination at this locus. This may be because the relevant crosses have not been carried out in order to test for recombination between the effects of *St*, *Sd*, and *Ly*. Analysis of genome sequencing data should allow

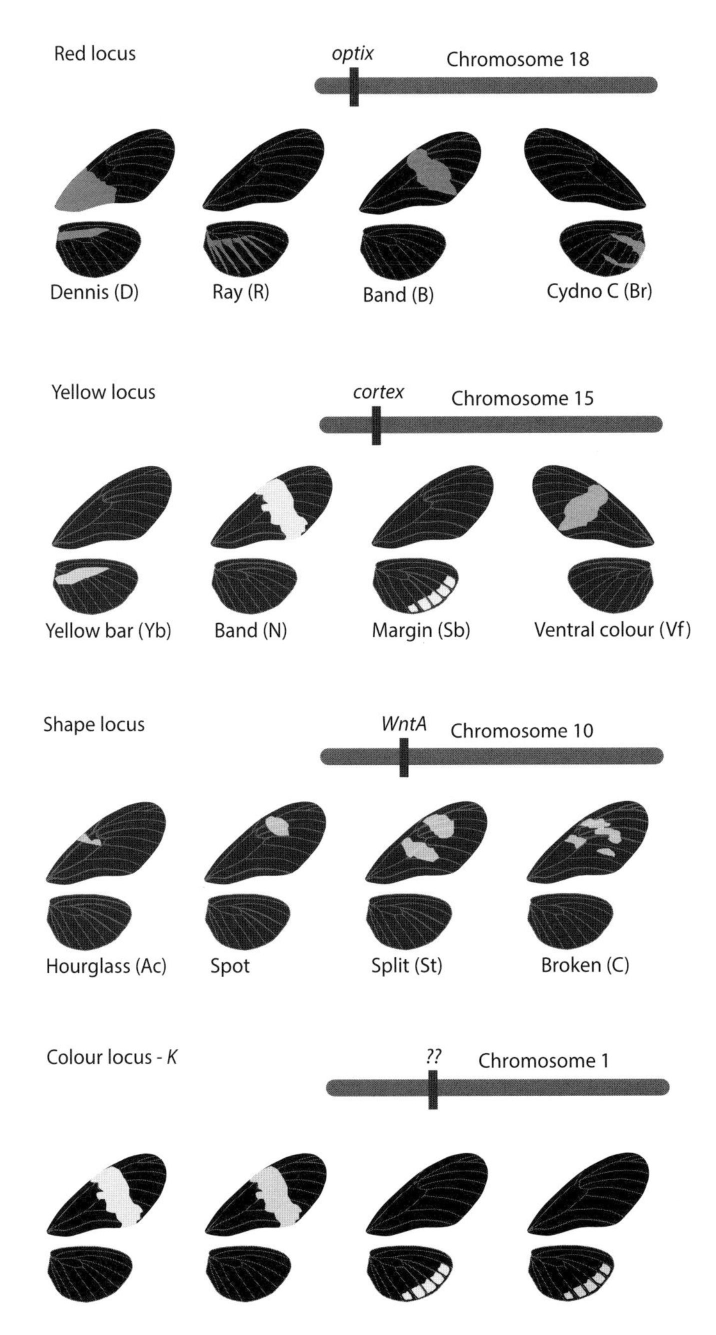

Figure 8.4 Summary of major wing patterning genes (see Plate 25)

Major effects of four loci on wing patterns in the *H. melpomene* clade. Dorsal wing surfaces are depicted, except for wings in reverse orientation that show two ventral phenotypes. Allelic variants at the red and yellow loci, *optix* and *cortex*, control the presence/absence of the pattern elements depicted, except where the yellow locus alters the colour of the ventral red band from red/orange to pink (known as *Vf*). In contrast, the shape locus *WntA* has alternate alleles controlling different forewing band shape phenotypes, so rather than the presence/absence of elements, this locus controls the shape of elements whose presence is controlled by the other loci. The colour locus, *K*, alters patterns from white to yellow.

identification of the genetic modules associated with particular patterns at this locus and determine whether the *WntA* locus shows a similar structure of linked elements associated with specific phenotypic effects as the other major loci.

Again, the shape locus *WntA* has major effects in other species across the genus. In the silvaniform species, *H. hecale* and *H. ismenius*, it is primarily implicated in controlling the extent of the yellow forewing band, which is broken into spots in some races (*H. h. zuleika* and *H. i. telchinea*) and forms a complete band in others (*H. h. melicerta* and *H. i. boulleti*) (Huber et al., 2015). These effects are very similar to the phenotypic effects of the shape locus in generating the broken forewing yellow band in Amazonian races of *H. erato* and *H. melpomene*.

8.9 Phenotypic effects of major loci: the colour locus K

Another locus located on chromosome 1 controls the colour change between yellow and white pigments in *H. melpomene*, *H. cydno*, and *H. pachinus*. As this locus has not yet been definitively associated with a gene, I will continue to use the older nomenclature and call this locus *K*, or the 'colour locus'. This locus therefore differs from all those described previously in that it influences solely colour, with no effect on pattern. Most strikingly, the locus controls a polymorphism of yellow and white forms in *H. cydno alithea* in western Ecuador (there is also polymorphism at the *WntA* locus in the same populations influencing band shape). The colour locus is located on linkage group 1, and is linked, although not closely, with the gene *wingless* (Kronforst et al., 2006a). This locus also segregates in crosses between *H. melpomene rosina* and *H. cydno chioneus*, in which the *H. melpomene K* allele turns the white *H. cydno* forewing yellow (Naisbit et al., 2003).

There is some evidence for a separate locus that has the same phenotypic effect on the hindwing margin, named *Khw* (Joron et al., 2006). This was mapped to about 20 cM away from *wingless*, so appears to be a separate locus, although again they might represent different *cis*-regulatory elements of the same gene. In natural populations, some races, such as *H. c. cydnides*, have a yellow forewing and a white hindwing margin, perhaps representing the independent effects of *K* and *Khw*. At both loci, white alleles are dominant to yellow alleles.

8.10 Other loci

There are a large number of minor effect loci described in the older literature, but in most cases in which these have been studied more recently, they have been found to represent allelic effects of the major loci described earlier. Nonetheless, some of these loci are likely to be distinct. For example, a locus named *Or* described in both *H. melpomene* and *H. erato* controls the switch between red and orange colours (Sheppard et al., 1985). 'Postman' races typically have a bright red forewing band, whereas Amazonian forms have orange *dennis* and *ray* patterns. Crosses between the two often show variation in the colour of elements controlled by the *optix* locus, varying from orange to red. More recent crosses have not mapped the *Or* locus, however; in my experience, it seems to be a continuous quantitative trait. This may be in part because phenotypes are hard to characterize, as red pigment fades to orange with time (see next chapter for further discussion of pigment chemistry).

Another locus that has been better characterized is *Ro*, which generates a rounded forewing band phenotype such as that seen in *H. e. notabilis*. This segregates as a Mendelian locus in large mapping crosses (Salazar, 2012; Papa et al., 2013), and has been mapped to linkage group 13 using a genetic association study (Nadeau et al., 2014).

Other traits show more quantitative continuous variation and have been poorly characterized in genetic analysis to date. Some of the most interesting and beautiful are the iridescent blue and green colours that result from structural variation in the wing scales. These traits vary continuously and are also difficult to quantify (Jiggins & McMillan, 1997).

8.11 Quantitative analysis

Historically, most analysis of *Heliconius* genetics has relied on the scoring of presence/absence of major pattern elements. This has served us well in identifying loci of major effect, but in order to objectively quantify the contribution of different loci to phenotypic variation, quantitative trait locus mapping (QTL) of complete wing pattern variation is needed. QTL analyses take advantage of variation between individuals in a mapping family, usually generated by making hybrids in the laboratory between two strains or species. Offspring are scored for the segregation of genetic markers, and then tested for associations between quantitative variation in traits of interest and inheritance of alleles across the genome. The result is a QTL map showing peaks of association between genomic regions and the variable phenotype.

Relatively few such QTL studies have been carried out in *Heliconius*. The first studies involved measuring individual quantitative traits that varied the expression of major pattern elements. Thus, for example, Simon Baxter measured variation in the red forewing band shape in a cross between *H. m. cythera* and *H. m. melpomene* (Baxter et al., 2009). Six chromosomes were found to influence band shape, two of which contained the major pattern loci *optix* and *WntA*. A major QTL on linkage group 13 was identified that likely represents a novel pattern locus of large effect. This may represent a homologous locus to *Ro* described in *H. erato* that is located on the same chromosome.

A much more comprehensive QTL analysis was carried out by Papa et al. (2013), using a large set of crosses between *H. e. notabilis* and *H. himera*. This quantitative analysis confirmed the subjective finding from generations of earlier researchers, that major loci control segregation of most of the wing variation in crosses. For example, an additive model showed that the red locus controlled 87% of variance in the amount of white versus yellow in the forewing, while the amount of red was best described by an epistatic model in which the red locus explained ~56% of the variance. For the extent of red, the next smallest QTL explained 15% and 10% of the variance, respectively, while for white/yellow there was a sex chromosome effect that explained 10% of the variance.

The sizes of the two forewing spots showed a less skewed distribution of effect sizes, and were controlled by several QTL of moderate effect (>5%), some as large in effect as the major shape locus (*WntA*). For example, four QTL together explained 63% of the variance in the 'big spot', one of which was the shape locus. Several of these QTL were located on chromosomes not previously implicated in *Heliconius* wing pattern variation. The spot shape analysis therefore suggests a less skewed, more quantitative genetic architecture. Nonetheless, this cross was specifically chosen as exhibiting a large amount of quantitative variation, so the overall variance explained across the complete set of *H. erato* crosses described by Papa et al. (2013) is strongly dominated by the large effect loci. These QTL analyses of specific wing pattern traits nonetheless still fail to capture and quantify both segregation of the presence and absence of major pattern elements in the same analysis as quantitative variation in the expression of those traits.

More recently, analytical methods have been developed that capture all of the variation in colour and pattern into a single principal component analysis (PCA) (Le Poul et al., 2014; Huber et al., 2015), permitting a more comprehensive and unbiased analysis. The method first combines colours into a few different pigment classes (yellow, white, orange, etc.). The wings of different individuals are then aligned and warped to coincide with one another. Finally, pigment variation across all pixels of hindwings and forewings is combined into a single PCA. This approach, named 'colour pattern modelling', allows all of the variation across the wing to be combined into a single analysis.

Barbara Hüber used this approach to analyse broods of *H. hecale* and *H. ismenius* for QTL analysis. Since nothing was known previously about the genetics of these species, it was surprising to discover that all the significant QTL identified corresponded to the existing major wing patterning loci. Minor QTL did not pass the statistical significance threshold, although some of these additional loci would likely become significant with larger sample sizes.

In summary, quantitative analyses of wing patterning strongly support the view that most variation is controlled by a handful of major effect loci that are deployed repeatedly to control a huge diversity of patterns. The expression of major loci is modified by a number of minor effect loci, although the actual distribution of effect sizes remains to be quantified. There is a clear need for studies that combine large mapping families with objective methods for pattern analysis to properly characterize the effect-size distribution of wing patterning variants.

8.12 Sex-limited and sex chromosome effects

Many mimetic butterflies show strong sexual dimorphism, but in general there is very little dimorphism in *Heliconius* (but see Chapter 7). Mimicry selection seems to act similarly on both sexes. Nonetheless, it is fairly common to observe some sexual dimorphism in the expression of patterns in hybrids. For example, in *H. himera* × *H. e. cyrbia* crosses, some families showed a significant effect of sex on the relative expression of yellow and red scales in the forewing band (Jiggins & McMillan, 1997). In a QTL study of *H. numata*, after the major patterning locus *P*, some of the largest effect sizes resulted from segregation of the sex chromosome (Jones et al., 2012). These sex-limited effects therefore appear in hybrids, but are typically not seen in pure races. This suggests that natural selection has acted to canalize sex-limited expression of wing patterns in order to perfect mimicry.

In most crosses, sex-specific effects are likely due to sex-limited expression of autosomal wing patterning factors. There is little evidence for sex linkage of wing patterning genes. Theory predicts that loci involved in species divergence might accumulate disproportionately on the sex chromosomes (Qvarnstrom & Bailey, 2008). The Lepidoptera are female-heterogametic: females have divergent sex chromosomes (known as Z and W), unlike mammals, in which males are heterogametic (X and Y). Recessive alleles are more effectively exposed to selection on the Z (or X) chromosome, because one sex has only a single copy of that chromosome. Consequently, divergence due to positive selection is expected to be faster on the sex chromosome than on the autosomes (i.e. non-sex chromosomes) (Charlesworth, Coyne, & Barton, 1987).

In general, there is evidence in Lepidoptera that genes differing between species are indeed often found on the sex chromosomes (Sperling, 1994; Prowell, 1998). In the light of these theoretical expectations and empirical data in other Lepidoptera, the lack of sex linkage of wing patterns in *Heliconius* is perhaps surprising. However, theory also predicts that under some circumstances, genes with sex-specific functions are expected to accumulate on the sex chromosomes (Ellegren & Parsch, 2007). Conversely, genes that control functions where sexual monomorphism is favoured—as is the case of *Heliconius* mimicry—may be more likely to reside on the autosomes, which are equally represented in the cells of both sexes. So the fact that similar mimicry patterns are favoured in both sexes may also explain why *Heliconius* wing pattern genes tend not to be found on the sex chromosomes.

8.13 Non-genetic effects and plasticity

There has been considerable interest in the role of phenotypic plasticity in evolution. It has been proposed that plasticity can promote evolutionary novelty, for example, by allowing populations to explore new phenotypes without genetic change (Pfennig et al., 2010; Moczek et al., 2011). However, there is little or no evidence for phenotypic plasticity in the expression of *Heliconius* wing patterns. First, most of the variation in wing pattern among hybrid butterflies can be explained by genetic variation at just a handful of major loci. Second, in the wild—apart from wing pattern races that are genetically divergent—there is very little phenotypic variation in wing pattern among individuals occurring across a wide range of altitudes and habitats. Some pigment colours do fade with age, or among stressed individuals raised in poor conditions, but this is not adaptive plasticity. In summary, although plasticity may play a role in many aspects of *Heliconius* biology, such as learning of spatial information, there is no evidence that it plays a role in wing pattern evolution.

8.14 Summary of wing pattern genetics

It is now clear that the rather complex genetic scheme described by Sheppard et al. (1985) and reviewed by Nijhout (1991) is in fact far simpler than ever imagined (Figure 8.4). Virtually all the major pattern variation described by the early crosses results from allelic variants at just a handful of loci whose identity at the genetic level is now well established (Table 8.1). This major advance in our understanding came about through development of molecular markers, which allowed homology of loci to be established across different species and between crosses within the same species (Table 8.1). In this way, distinct variants described from early crosses were recognized as resulting from allelic variants at the same major loci.

Surprisingly, the same major loci act across most of the *Heliconius* radiation that has been studied. The fine mapping conducted by Barbara Huber on *Heliconius ismenius* and *H. hecale* indicates that the same four major loci control most of the wing patterning variation in these species (Huber et al., 2015). This study used a whole-genome mapping approach that did not depend on 'candidate genes' identified in previous studies, so provides unbiased evidence that the same genes are repeatedly implicated in pattern evolution across species. Remarkably, the more we study pattern variation in *Heliconius*, the simpler it all seems to become (Table 8.1).

Alongside the consolidation of the major loci, quantitative analyses have identified additional loci across multiple chromosomes that influence the expression of major elements. Effect sizes of these loci are commonly small compared to those of the major genes (Papa et al., 2013; Huber et al., 2015). However, studies are needed that combine the large brood sizes used by Papa et al. (2013) with the sophisticated analysis of colour pattern and genome-based markers used by Huber et al. (2015), in order to characterize minor effect loci and quantify their relative effects. I will now consider the theoretical background for the genetic basis of adaptation and the implications of this work for our understanding of adaptation more broadly.

8.15 A distribution of effect sizes?

In countering Punnet's (1915) evidence for major mutations in evolution, Fisher (1930) argued that mutations with a large effect on the organism will almost always be deleterious. Organisms are typically fairly well adapted in multiple dimensions, so any major change will tend to mess things up.

If we imagine a population that lies some distance from an optimum phenotype (often depicted as a peak in an adaptive landscape), mutations will arise randomly; those that move the phenotype closer to the peak will be favoured by selection. Allen Orr has shown that under such a process, we can expect the accumulation of an exponential distribution of mutational effect sizes (Orr, 1998, 2005). Early in the process, there is a high likelihood of mutations that move the population a large distance relative to the optimum. The reason is that—although individually large effect mutations are highly unlikely—over many steps in the 'adaptive walk', there is a cumulative probability that at least one large effect mutation will be fixed. Later in the process, as the population nears the optimum phenotype, smaller effect mutations are more probable; these act to 'fine-tune' the adaptation.

Hence, across the whole process of the adaptive walk, an exponential distribution of effect sizes is expected, with a few substitutions of major effect on the phenotype and many substitutions of more minor effect. Note that 'major' here is entirely relative to the initial distance to the optimum. Adaptation of populations that start further from the optimum are expected to involve larger absolute effect sizes than those that start close to the optimum. Although this theory has been criticized, in part as being too abstracted from biological reality (Rockman, 2012), it does provide one of the few general and quantitative predictions for adaptive evolution.

The theory developed by Orr and others hypothesized a population evolving towards a single adaptive peak. However, the frequency-dependent nature of mimicry and warning colour means that these traits may not evolve in this way (Figure 8.5). If a population of butterflies has a bright warning colour pattern (hereafter the 'mimic'), predators will learn this pattern and the population will generally be well protected from predation. There may be other butterfly species locally that are more abundant or more toxic (the 'model'), and therefore have a better-protected wing patterns, so our mimic species would gain in fitness by evolving mimicry of the model pattern.

However, an individual mimic that deviates from the rest of the population is likely to be selected against, even if it becomes slightly more similar to the model. In order for gradual evolution towards the model to be possible, the two patterns would have to be so similar that predators occasionally generalize between them (Turner, 1981). However, many current *Heliconius* patterns are sufficiently different from one another that gradual convergence seems unlikely. In our adaptive landscape metaphor, there is a valley of low fitness between the model and mimic that would prevent gradual evolution of mimicry.

The difficulty can be overcome if a single mutation causes a large change, sufficient to induce enough similarity to the model in one step that overall fitness is increased (Figure 8.5). This initial mutation is unlikely to produce a perfect mimic, so subsequent mutations will then be needed to perfect the phenotype.

This argument was first outlined in detail by Nicholson (1927) and subsequently termed the 'Nicholson two-step model' by John Turner (Turner, 1977, 1984, 1987). We might therefore expect mimicry to have a different genetic architecture to traits evolving under a single-peak-climbing model (Baxter et al., 2009).

The major locus control of *Heliconius* patterns seems to fit with the predictions of the Nicholson two-step model (Turner, 1981; Baxter et al., 2009; Papa et al., 2013; Huber et al., 2015). There are divergent alleles at four major loci, with small effect loci modifying the expression of these patterns. The major effect changes controlled by Mendelian loci would therefore represent the first step in the evolution of a new mimicry pattern. It seems unlikely that a single mutation would generate perfect mimicry ('hopeful monsters') (Turner, 1981), so the subsequent perfection of the pattern would result from mutations of small effect, perhaps corresponding to the minor effect QTL detected in crosses—the second step of the two-step model.

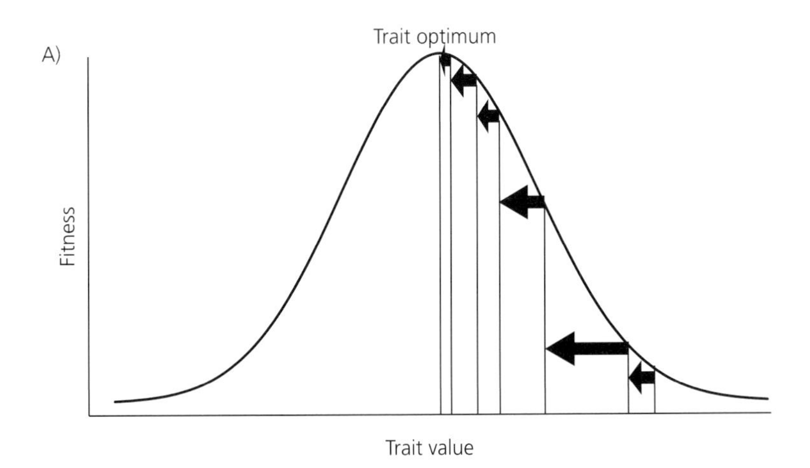

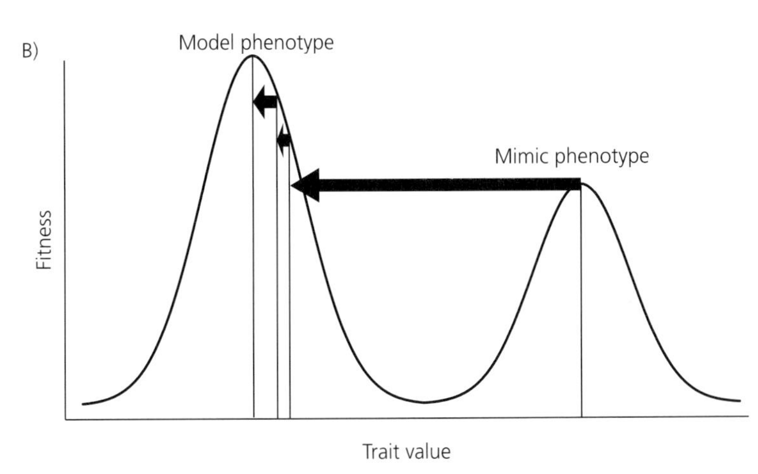

Figure 8.5 Adaptive landscapes in warning colour (see Plate 26)

The curve represents the fitness of an individual at a range of phenotypic values. (A) An adaptive walk towards a single fitness peak. Arrows show possible mutations fixed as a population evolves towards the fitness peak. Fixation of large effect mutations is more likely early in the adaptive walk, with an exponential distribution of mutational effects expected overall. (B) Warning colour and mimicry are expected to show an especially rugged adaptive landscape, shown here with distinct peaks of fitness for the model and the mimic species. In order for the mimic to evolve similarity to the model wing pattern, a large mutational step may be required to produce sufficient initial similarity between the species.

This is an appealing interpretation that fits empirical data with theory, but there are a number of reasons to be sceptical. First, many races within *H. erato* and *H. melpomene* differ at several unlinked major effect loci. For example, hybrid zones in Peru and Ecuador between races of *H. melpomene* and *H. erato* differ in 2–3 major loci (Mallet, 1989; Salazar, 2012; Nadeau et al., 2014). It isn't clear whether a substitution at just one of these loci would be sufficient to gain enough mimetic similarity to provide protection, while the population 'waited' for a subsequent mutation at the second locus. Turner (1977) has acknowledged this difficulty and suggested that there may be multiple rounds of 'two-step' evolution, or that changes at just one of the loci would be sufficient to confer a fitness advantage. To address the question we would need to understand the influence of phenotypic variation on fitness and the extent to which predators perceive and generalize between patterns (see Chapter 7). Experiments could potentially be designed, either with captive predators or in the field, in order to test the shape of the protection curve around extant mimicry rings.

Another mismatch between the theory and empirical data is that the data from crossing experiments refer to the phenotypic effects of genetic loci, not separate mutations (Baxter et al., 2009). As pointed out by Fisher (1930), and more recently in dissection of major effect QTL in other organisms (Stam & Laurie, 1996; Linnen et al., 2013), major effect loci can result from accumulation of many mutations at a single locus. Unfortunately, genetic analyses in *Heliconius* have not yet reached the stage of mapping genotype to phenotype for individual mutations.

However, it seems likely that single large effect genetic loci harbour many mutations corresponding to adaptive steps towards the peak. Testing the two-step model therefore becomes a much more challenging problem of separating the order and effect size of individual mutations at a single locus, which is still a distant goal. Nonetheless, there are at least some cases where mimicry can arise through a single step: hybridization with a related species and subsequent introgression can introduce an already well-adapted large effect allele. This represents a clear case of single-step, 'major effect' evolution, so there certainly are at least *some* cases in which large steps are involved (see also Chapter 10) (The Heliconius Genome Consortium, 2012).

Overall, therefore, we can conclude that the 'rugged' adaptive landscape of mimicry likely favours adaptation via large steps as described under two-step theory, which might provide part of the explanation for the major effect loci involved in *Heliconius* mimicry. These theoretical predictions also explain the existence of small effect alleles that modify the expression of major pattern elements.

However, two-step theory is only part of the explanation for major loci. It doesn't explain why repeated mutations involved in such a wide variety of different phenotypes all map to the same genetic locus, or the existence of complex alleles at these major loci that must involve the fixation of multiple sequential mutations. In order to become such large effect and multipurpose loci, the same genomic regions are repeatedly targeted by evolution. They have become 'hotspots' for evolution—I will address the reasons for this in the next chapter.

8.16 Supergenes and polymorphism

The broad picture of wing pattern genetics outlined earlier seems to apply to most *Heliconius*, but there is one species in the genus that has a very different system of wing pattern inheritance: *H. numata*. As we have seen in Chapter 7, mimicry patterns in *H. numata* are polymorphic, with different morphs mimetic with different species mostly in the genus *Melinaea*. In order for this dramatic polymorphism to be stable, a system is needed to switch between alternate forms without producing intermediates.

We can think about the problem in the context of another ubiquitous polymorphism in the animal world: sex. Males and females perform different reproductive roles and differ in all sorts of ways, including behaviour, morphology, size, etc. It is extremely important to ensure each individual is unambiguously either one sex or the other, as intermediates with indeterminate sex would not be successful in either role. In most animals, sex determination is controlled by a genetic system that has evolved to regulate this binary decision. For example, in mammals, differentiated sex chromosomes ensure that individuals are either male, with XY, or female, with XX chromosomes.

Similarly, in polymorphic butterfly mimics, a genetic system has evolved that reliably switches between alternate forms. In *Heliconius numata*, dramatic differences between different wing patterns are controlled by a single genetic locus, with several alternate alleles. If human height were controlled this way, we would all be either 1 m 60 cm, 1 m 70 cm, or 1 m 80 cm tall, but never anything in between (Wade, 2011). Such loci are known as 'supergenes'.

We defined the term supergene as 'a genetic architecture involving multiple linked functional genetic elements that allows switching between discrete, complex phenotypes maintained in a stable local polymorphism' (Thompson & Jiggins, 2014, p. 3). In *H. numata*, wing variation is controlled by such a supergene that has been termed *P* (the name *P* was coined by Keith Brown and Woody Benson after the mythical Pushmi-Pullyu, a two-headed antelope discovered by Doctor Dolittle; Brown & Benson, 1974). Genetic control of the phenotype results from an allelic series in which each allele at a single locus produces one of the phenotypic forms, each mimicking a different *Melinaea* species.

In order to understand why selection favoured the evolution of a supergene in *H. numata*, we can look at another species that has a 'stable local polymorphism' but has not evolved a supergene. The *H. cydno alithea* populations in western Ecuador are in a stable local polymorphism, but genetic variation is controlled by two independent loci (the shape and colour loci, *WntA* and *K*). With two alleles at each of two loci, there are four possible phenotypes, two of which are close mimics of *H. sapho* and *H. eleuchia*, respectively (Figure 8.6). However, the other two phenotypes are slightly less than perfect mimics. Every generation, mating between individuals will produce more of these less well adapted forms. The evolution of a single gene switching between the alternatives should be favoured by selection in order to avoid the production of these less fit individuals. However, this has not yet occurred, perhaps because the fitness differences between morphs are relatively slight (the shape locus has a relatively small effect on phenotype – Figure 8.6), or perhaps due to a lack of suitable genetic variation. This polymorphism in *H. cydno* therefore illustrates the evolutionary problem that is solved by evolution of a supergene—if alleles at the shape and colour loci were to become tightly linked in the correct combinations, all individuals would be excellent mimics of either *H. sapho* or *H. eleuchia*.

There are two major characteristics of the *Heliconius numata* supergene that maintain an integrated phenotype. First, a lack of recombination—in other words, all aspects of the phenotype are inherited as a single non-recombining locus. Two hypotheses have been proposed to explain how a single locus could control different aspects of a phenotype. Evolutionary geneticists have typically hypothesized tight linkage between multiple genes. In contrast, some developmental biologists have suggested that a single gene might evolve to regulate multiple aspects of the phenotype (Thompson & Jiggins, 2014). As we shall see, aspects of both of these hypotheses are likely to be correct. The second major characteristic is dominance—alternate alleles need to show complete dominance relationships such that heterozygote genotypes develop the wing pattern of one or other parent. I will discuss these characteristics in turn.

The first clue into the genetic origins of the *H. numata* supergene came from comparative mapping between *H. numata* and *H. melpomene*. Mathieu Joron and I were studying these two species in parallel in Edinburgh, and were surprised to discover that the *P* supergene is genetically homologous to the yellow (*cortex*) locus in *H. melpomene* (Joron et al., 2006). This doesn't seem all that surprising now, given the subsequent demonstration of homology in wing patterning between many different *Heliconius* species, but at the time it was the first demonstration that very different-looking wing pattern phenotypes could be controlled by the same locus. The *H. melpomene* genetic architecture of three to four major loci is clearly ancestral because it is shared by all other species in the genus that have been studied—not just *H. melpomene* and *H. erato* but also silvaniform species *H. hecale* and *H. ismenius* that are closely related to *H. numata* (Huber et al., 2015). The single locus *H. numata* architecture has therefore come about through one locus taking over control of all aspects of pattern variation (Jones et al., 2012).

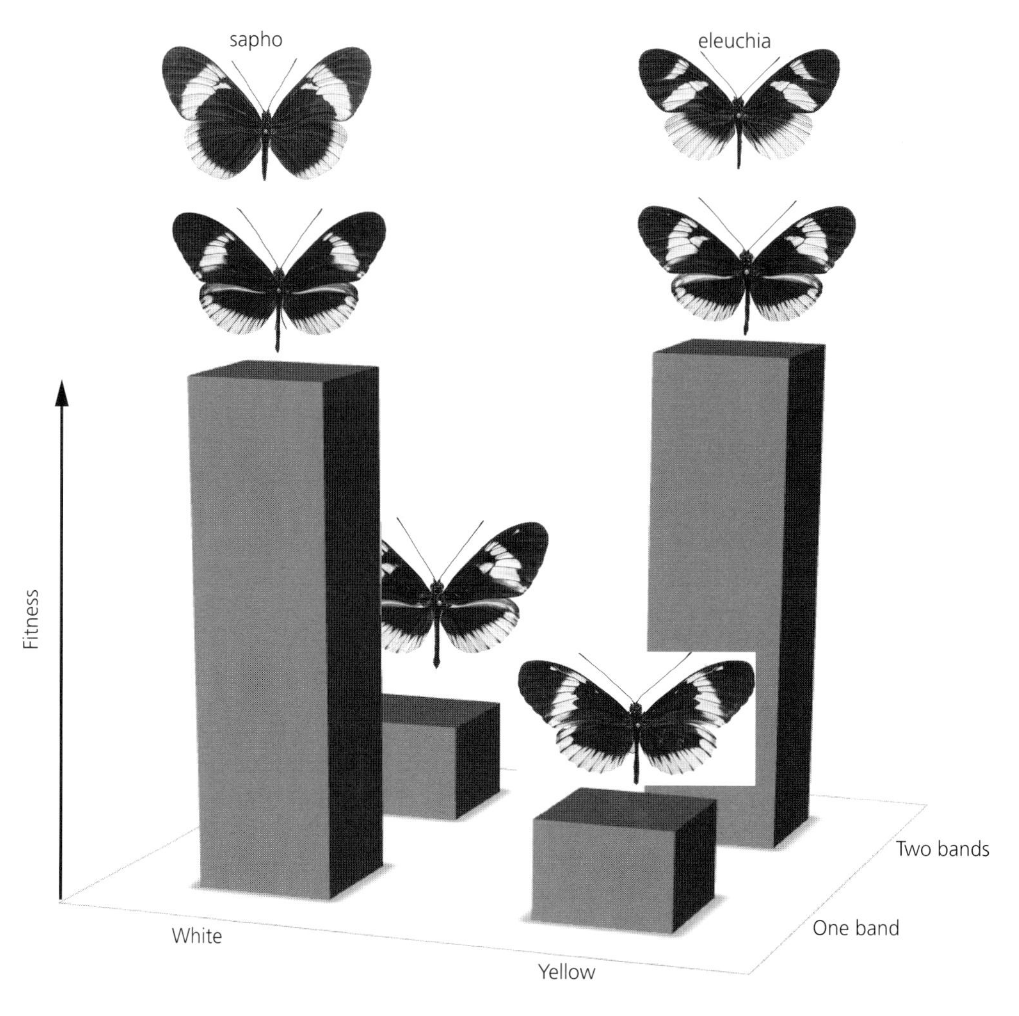

Figure 8.6 A two-locus adaptive landscape (see Plate 27)

In western Ecuador, *H. cydno alithea* is polymorphic (four forms shown below) and mimics *H. sapho* and *H. eleuchia* (top). Variation at the shape locus, *WntA*, acts to facilitate mimicry: although *H. cydno* never has two distinct bands, the *WntA* pattern element gives the appearance of a broken band that mimics *H. eleuchia*. The colour locus, *K*, controls the switch between yellow and white. Since there are two loci involved in this mimicry polymorphism, some of the phenotypes produced are less effective mimics. Recombination between the unlinked *K* and *WntA* loci produces these less fit combinations every generation. Note that fitness values are hypothetical—the joint fitness effects of the two loci have not been estimated in the wild.

It is worth noting that the red and yellow pattern loci in *H. melpomene* and *H. erato* are similar in some ways to our definition of a supergene: they have 'multiple linked functional genetic elements that [allow] switching between discrete, complex phenotypes' (Thompson & Jiggins, 2014, p. 3). However, these are not 'maintained in a stable local polymorphism' and there is no selection for increasing linkage between these elements as they are typically not variable within populations. Recombination generates intermediate phenotypes only where different races meet in hybrid zones. The zones are narrow, so selection to reduce recombination is weak. I would therefore not use the term supergene to describe the *optix* and *WntA* loci.

The structure of colour pattern loci in *Heliconius* is nonetheless somewhat pre-adapted for the evolution of a supergene—the *cortex* locus already has multiple linked elements controlling major aspects of the pattern. These linked elements are most likely different regulatory modules that control expression of a single protein-coding gene, so blurring the distinction between the 'multiple gene' and 'single regulatory gene' hypotheses that have been proposed to explain the existence of supergenes.

There are several hypotheses to explain the gradual evolution of tightly linked elements in a supergene. A long-standing hypothesis is that alleles located in different regions of the genome might be translocated into tight linkage (Turner, 1967b). Later, this hypothesis was considered theoretically unlikely, mainly on the grounds that it requires maintenance of a multilocus polymorphism prior to translocation (Charlesworth & Charlesworth, 1976). Furthermore, genome sequences for multiple *Heliconius* species have revealed no evidence for long-range movement of genes. The gene content of the *cortex* locus region is similar in all *Heliconius*. Other known wing patterning loci (*optix, WntA*) have also not changed their gene content in *H. numata*, even though aspects of pattern normally controlled by those loci, such as orange patches, are now controlled by *P*. The *P* locus has therefore evolved to take over aspects of pattern variation normally controlled by genes on different chromosomes, rather than by moving those genes into linkage.

The second hypothesis is that sequential mutations might arise in tight linkage with the polymorphic locus and be favoured by selection (Charlesworth & Charlesworth, 1976; Turner, 1977). Mutations that improve one mimetic form are likely to make things worse for other forms. However, if a new mutation is tightly linked at the *P* locus, then it will always be inherited with the alleles with which it is coadapted. This process has become known as 'Turner's sieve', because it involves sieving of the genetic variation that arises in order to select only linked variants (Charlesworth & Charlesworth, 1976; Turner, 1977, 1978).

It seems likely that such a sieve has been at work on the *P* locus. In *H. melpomene*, several linked elements within the yellow locus (e.g. *Yb*, *Sb*, and *N*) control variation in similar aspects of the phenotype to those controlled by *P* (Joron et al., 2006), and there is phenotypic evidence from *H. numata* for occasional recombinants at *P* (Brown, 1976; Joron et al., 1999). This suggests that *P* consists of multiple linked elements that have arisen through multiple sequential mutations. Of course, this is similar to what we see in all *Heliconius* species, but in *H. numata*, 'Turner's sieve' imposes additional selection for these mutations to be tightly linked.

Once linked elements have arisen, theory predicts that selection can act to further reduce recombination between them (Turner, 1967b; Charlesworth & Charlesworth, 1976). Such a reduction in recombination will reduce the production of maladapted recombinant genotypes and increase the fitness of coadapted alleles (Kirkpatrick & Barton, 2006). The idea that selection should reduce recombination around supergene loci is a long-standing hypothesis, but it lacked direct support for some time.

Mathieu Joron and his group have now used genomic analysis to demonstrate the process in *H. numata* (Joron et al., 2011). There are large genomic inversions across some 400 kb of the genome that segregate in polymorphic populations around the *P* locus (Figure 8.7). Alternate gene arrangements are fully associated with wing pattern phenotypes in natural populations. Furthermore, there are strong genetic associations between variable sites ('linkage disequilibrium') in natural populations across the locus, supporting the hypothesis that recombination is reduced in natural populations. Effectively, there is a large block of about 400 kb of DNA sequence that is inherited in complete association with different wing pattern forms (Joron et al., 2011).

Similar inversions have been seen in complex polymorphisms in other species—notably, a behavioural and plumage polymorphism in the white-throated sparrow (Huynh, Maney, & Thomas, 2011), a social polymorphism in fire ants (Wang et al., 2013), and a behavioural polymorphism in the ruff, a wading bird (Küpper et al., 2015; Lamichhaney et al., 2015b; see also Thompson & Jiggins, 2014). In all cases, inversions lock together a large part of one chromosome, which is therefore inherited as a single supergene. Perhaps most similar to the *Heliconius numata* case is *Papilio polytes*, in which a very localized inversion around the *Dsx* gene controls a wing pattern mimicry polymorphism (Kunte et al., 2014; Nishikawa et al., 2015). These examples all suggest that the evolution of inversions to reduce recombination between coadapted alleles may be a fairly common phenomenon.

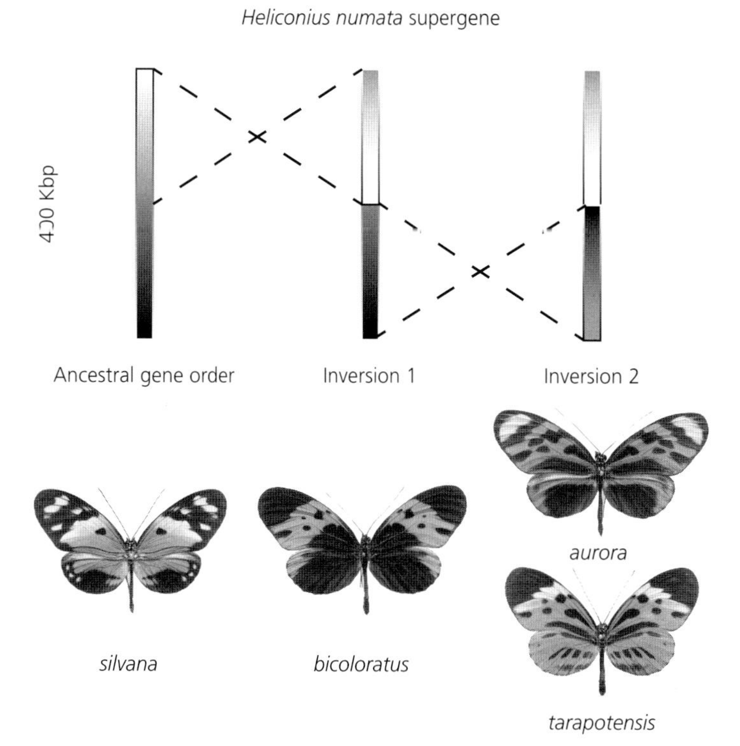

Figure 8.7 Structural variation associated with the *Heliconius numata* supergene (see Plate 28)

At least two genetic inversions are associated with the *H. numata* supergene. The ancestral gene order, which matches that in *H. melpomene* and *H. erato*, is shown on the left and is associated with ancestral phenotypes such as *H. n. silvana*. Two sequentially derived inversions are associated with more dominant alleles and are shown middle and right. Redrawn from Joron et al. (2011).

8.17 Accumulation of deleterious mutations

Inversions are a neat genetic solution to the problem of maintaining associations between alternate alleles at linked loci. In heterozygote individuals with different gene orders, chromosomal pairing in meiosis is disrupted, and hence crossing over reduced in the inverted gene region. In contrast, in homozygote individuals, gene order is the same on both chromosomes, and recombination can occur normally. This is important, as a complete cessation of recombination generally leads to a build-up of deleterious mutations. Selection is much more efficient at removing harmful mutations when there is recombination. Chromosomes that do not recombine, such as our own Y-chromosome, tend to accumulate deleterious alleles and degenerate over time.

Although inversions permit recombination in homozygotes, supergenes may not be immune from genetic degeneration. In particular, dominant alleles are likely to be found mostly in heterozygote individuals. For example, in a two-allele system, with the dominant morph forming 25% of the population, dominant homozygotes will form only 2% of individuals (where p is the recessive allele, $p^2 = 0.75$, so $p = 0.87$ and $q = 0.13$, hence $q^2 = 0.02$). This fraction will be even lower if dominant morphs are rarer or in a system with more alleles.

Thus there may be little opportunity for recombination among dominant alleles, because they are rarely found as homozygotes during meiosis. Furthermore, selection against the accumulation of deleterious mutations in the dominant allele is likely to be weak, because deleterious mutations are typically recessive. (Beware of the potential confusion here: the phenotypic effect on colour pattern is dominant, but the deleterious cost of the mutation is recessive.) The deleterious alleles are therefore mostly masked because they are found in heterozygotes.

Is there any evidence for degeneration of dominant supergene alleles? White-throated sparrows provide one of the best examples. The dominant white morph has reduced viability as a homozygote, and rarely occurs in the wild due to assortative mating between white and tan morphs (Tuttle et al., 2016). The ruff shows similar effects (Küpper et al., 2015). There is limited evidence for similar inviability of dominant alleles in butterfly supergenes, including *H. numata* and *P. dardanus* (Thompson et al., 2014; Joron and Chouteau, personal communication), although it remains to be well documented.

8.18 The evolution of dominance

The second aspect of a supergene that is required to ensure perfect inheritance of alternate morphs is a strong pattern of dominance (Llaurens, Joron, & Billiard, 2015). Alternate alleles at *P* show complete dominance, with an allelic series of dominance between morphs. Hence, in the Tarapoto region of Peru, the *silvana* pattern is the bottom recessive, while *bicoloratus* is the top dominant. Other alleles found in the same region form a strict dominance series (Brown, 1976; Joron et al., 2011; Le Poul et al., 2014).

There is one exception to this pattern of dominance. Heterozygotes between the *arcuella* and *tarapotensis* alleles are not similar to either homozygote. Instead they are named as a distinct morph *timaeus* that mimics ithomiine species such as *Melinaea menophilus hicetas* and *Athyrtis mechanitis* (Figure 7.3d) (Le Poul et al., 2014). Remarkably, as an alternative to dominance, a heterozygote genotype therefore appears to have been stabilized because it is an effective mimic.

In other *Heliconius*, there are predictable rules for dominance. Red/orange pattern elements are generally dominant over black, while yellow/white pattern elements are recessive. White is typically dominant over yellow. So, for example, in crosses between red-banded and dennis-ray butterflies in both *H. melpomene* and *H. erato*, hybrids have both the red bands and the dennis-ray elements from each parent (Figure 8.1). Although individual elements show complete dominance, heterozygotes express several elements inherited from each parent and are therefore intermediate. These 'rules' of dominance presumably result from the developmental pathways that produce the patterns. In contrast, in *H. numata* crosses, black is generally dominant over orange, and orange over yellow, but all these patterns are occasionally reversed: both yellow and orange can be dominant over black. Furthermore, co-dominance of different pattern elements, producing intermediate heterozygotes, is never seen among sympatric morphs (except for *timaeus* mentioned earlier). The complete dominance of alleles across the entire wing surface in *H. numata* therefore represents a derived state that apparently overturns typical rules of inheritance seen in other species. Dominance has been optimized by natural selection.

These patterns of dominance could be controlled by mutations within the supergene itself (e.g. Turner's sieve; see earlier), or could be due to unlinked loci acting to control dominance at *P*. There is evidence for both processes. Crosses between different geographic populations show breakdown of dominance, providing evidence for unlinked modifiers (Turner, 1977; Le Poul et al., 2014). However, there is also good evidence for evolution of dominance at the *P* locus itself. The evidence comes from comparing patterns at different *P* alleles. Two alleles, *illustris* and *silvana*, carry the ancestral gene order similar to that seen in other *Heliconius* species such as *H. melpomene*. The remaining derived alleles carry a rearranged gene order, with at least one inversion. Patterns of dominance *between* derived and ancestral alleles show unusual patterns of dominance in which the typical dominance patterns are overruled. In contrast, *among* derived alleles, patterns of dominance follow the typical colour hierarchy seen in other *Heliconius* (Le Poul et al., 2014). These patterns suggest that dominance is a property of the alleles themselves, and perhaps that a special mechanism of regulating dominance has evolved between derived and ancestral alleles.

We do not yet understand how dominance is regulated in *H. numata*, but some clues as to how this might work come from self-incompatibility loci in plants. These loci prevent pollen tubes from growing in selfing crosses: if a pollen grain lands on a female flower that shares an S-allele, it fails to grow and fertilize the flower, preventing self-fertilization. In some systems, S-alleles also show dominance in a manner very similar to that seen at *P* in *H. numata*. For example, in *Arabidopsis*, there is a class of pollen-recessive haplotypes that are not expressed in heterozygote genotypes with pollen-dominant alleles. It has been shown that a small RNA molecule (siRNA) encoded in the dominant allele targets the recessive allele and prevents expression, perhaps by inducing transcript degradation (Tarutani et al., 2010). This demonstrates a possible mechanism for dominance of *P* alleles, which could be encoded at the *P* locus.

In summary, in a long series of elegant studies of *H. numata* led by Mathieu Joron, we have seen how the organization of the genome and patterns of dominance can be shaped by natural selection. Genome organization is often viewed as resulting from random neutral processes, and dominance as an inevitable result of cellular processes—this work has therefore shown how these fundamental genetic processes can also represent adaptations.

8.19 From making butterflies cross to finding genes

So far I have considered the wing pattern genes as genetic factors, without considering which genes they are or how they act at a molecular level. I will discuss the identity and action of these major genes in the next chapter when I describe the development of a wing pattern. Here I will briefly summarize the methods we have used to go from genetic loci segregating in crosses towards identifying specific genes.

Finding genes that control adaptive traits in wild populations is a major area of evolutionary biology, and *Heliconius* are a group in which we have achieved this goal. It was a major research effort over 15 years, which coincided with a period during which genomic technologies underwent a revolution, such that the technical approaches that became available were considerably more powerful than when we started out.

The efforts to identify wing patterning genes in *Heliconius* have been a collaboration between my group working on *H. melpomene* (Jiggins, Mavárez, et al., 2005; Joron et al., 2006; Baxter et al., 2008, 2010); Owen McMillan, working in parallel with *H. erato* (Tobler et al., 2005; Kapan et al., 2006; Papa et al., 2008a; Counterman et al., 2010); and Marcus Kronforst working with *H. cydno/pachinus* (Kronforst et al., 2006b; Chamberlain et al., 2011). Later this effort was joined by Robert Reed, whose group pioneered developmental approaches to testing wing pattern evolution (Reed et al., 2011; Martin et al., 2012). When we began the crossing experiments around 2002, there seemed little prospect for identification of the specific genes involved in pattern specification (or at least we didn't admit to this ambitious goal in grant proposals). Rather, the immediate aim was to test for homology in the genetic basis of patterns between mimetic species.

The first step was to develop markers of known DNA sequence around the wing pattern loci. Originally, this was first achieved using laborious sequencing of AFLP markers by Margarita Beltran and Simon Baxter (Joron et al., 2006; Baxter et al., 2008), but later genome-wide sequence-based markers became available for linkage mapping (e.g. RAD sequencing) (Huber et al., 2015). The next step was to use the markers to identify the genome region implicated. Again, this is fairly straightforward now that a reference genome is available for *H. melpomene*, but originally it involved sequencing bacterial artificial chromosome (BAC) clones one by one across the target region (Baxter et al., 2008; Papa et al., 2008a).

Although more laborious to obtain than whole-genome sequence, these BAC sequences are of high quality and completeness, and still provide useful reference sequence. The genome sequence for the target region was then used for fine mapping, in order to delineate the precise region implicated (Baxter et al., 2010; Counterman et al., 2010; Ferguson et al., 2010a). Unfortunately, the power of this approach is limited by the size of families available. With hundreds of individual offspring used for mapping, the regions identified typically contained many candidate genes. The recombination rate in *H. melpomene* is estimated at 180 kb/Cm. Genome-wide, we expect about a gene every 20 kb (12,000–15,000 genes in a 270 Mb genome), so mapping with 100 individuals might be expected to localize a region containing approximately five genes, while 500–1,000 individuals would be needed to narrow down a single gene, which is at the upper limit of what is feasible in *Heliconius*. In reality, gene density is highly variable, so these numbers are only an approximation, and the linkage mapping approach identified single candidate genes in some cases but not others.

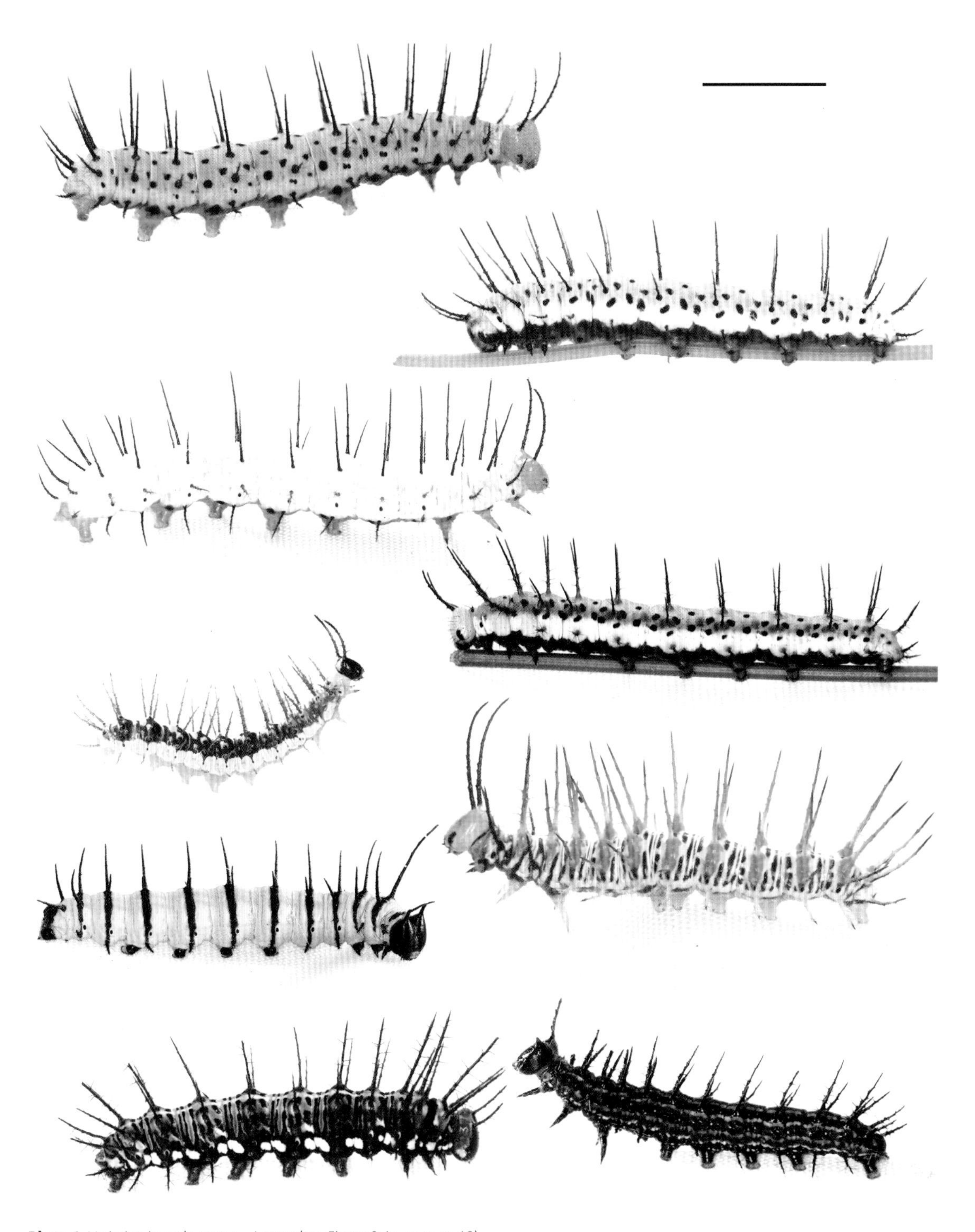

Plate 1 Variation in early stages—Larvae (see Figure 2.1a on page 12)

From top: *H. cydno; H. charithonia; H. hecale; H. erato; Eueides aliphera; Philaethria dido; Heliconius doris; Agraulis vanillae; Dryas iulia* (bottom left). All specimens were raised and photographed in Gamboa, Panama. Images are not precisely scaled but scale bar indicates approximately 1 cm.

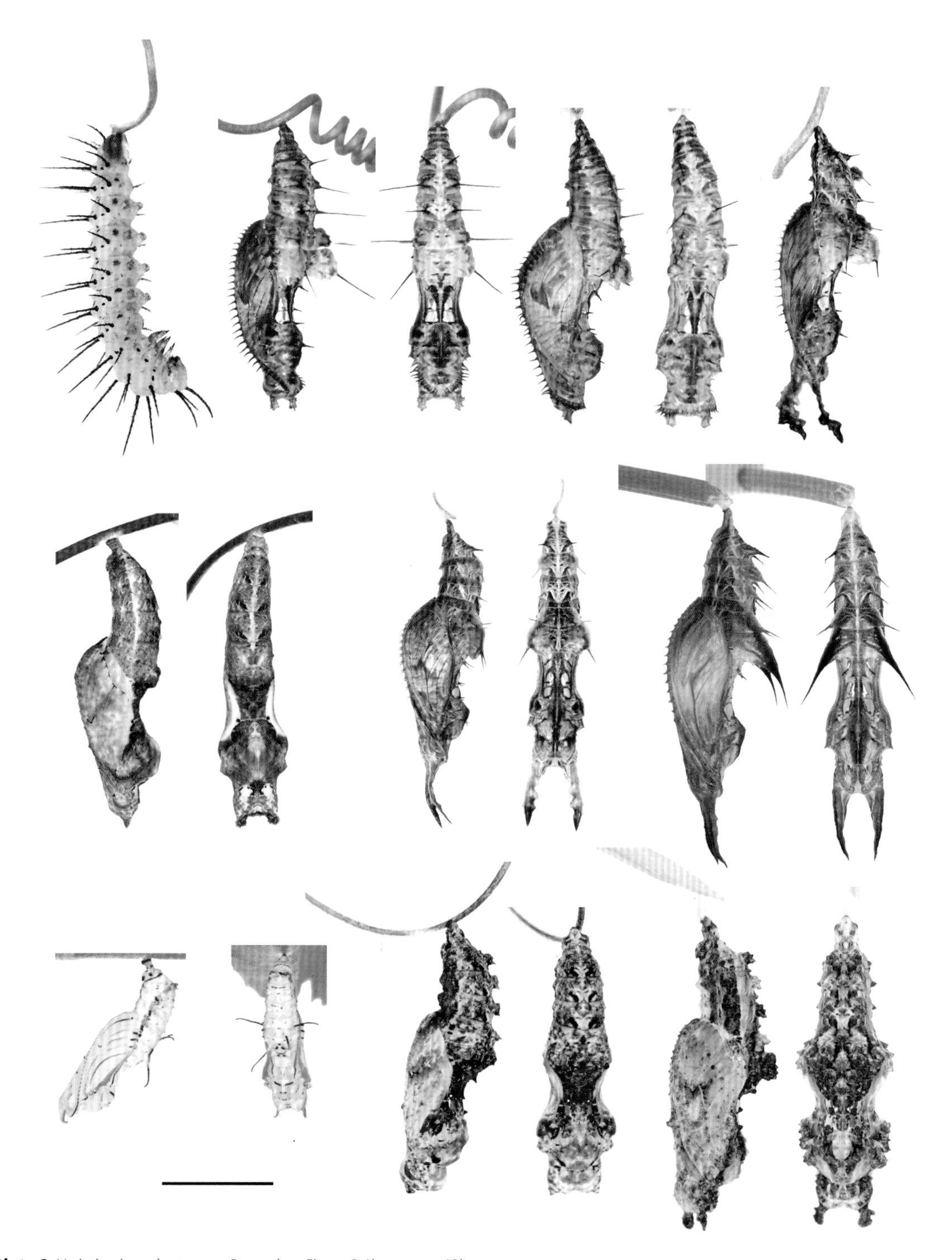

Plate 2 Variation in early stages—Pupae (see Figure 2.1b on page 13)

First row: *Heliconius cydno* prepupa; *H. cydno* pupa (side, front); *H. hecale* pupa (side, front); *H. charithonia*. Second row: *Agraulis vanillae* (side, front); *H. erato* (side, front); *H. hecalesia* (side, front). Third row: *Eueides aliphera* (side, front); *Dryas iulia* (side, front); *Philaethria dido* (side, front). All specimens were raised and photographed in Gamboa, Panama. Images are not precisely scaled but scale bar indicates approximately 1 cm.

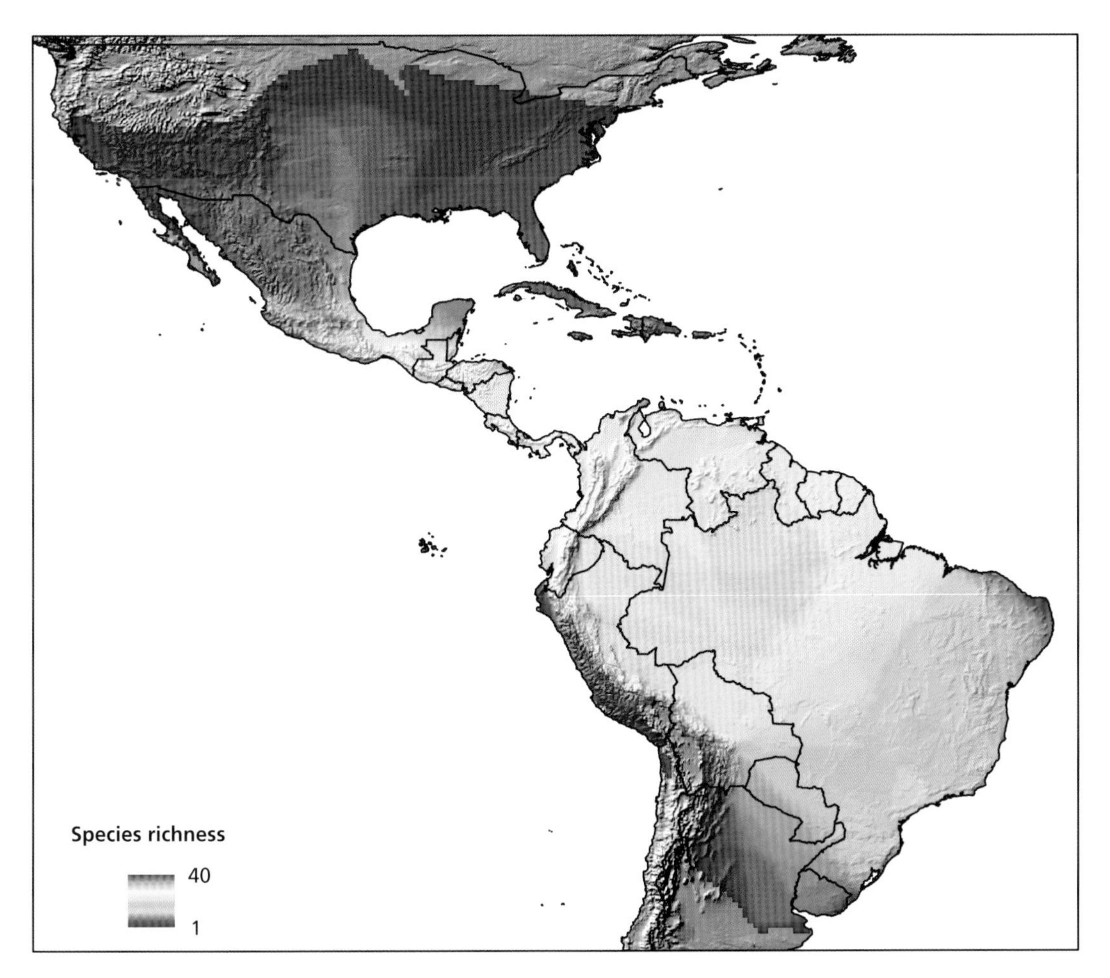

Plate 3 Patterns of species richness (see Figure 2.2 on page 15)

The diversity of heliconiine species reaches a peak in the upper Amazon, where lowland Amazonian taxa overlap with Andean species. This region corresponds to peak diversity in many taxa. Image courtesy of Neil Rosser.

Plate 4 Main wing pattern forms (see Figure 2.5 on page 20)

Representative taxa are shown for major wing patterning forms with upper (ventral; above) and underside (dorsal; below) wing patterns. Note that this is not a definitive classification of pattern diversity, but rather is intended as a rough guide to the terms that will be used throughout the book to refer to different types of patterns. Row 1, Postman: *H. ricini insulanus, H. erato hydara*; Dennis-ray: *H. xanthocles napoensis*; Sara: *H. leucadia pseudorhea*; Row 2, Cydno: *H. cydno cydnides, H. eleuchia eleusinus*, Metharme: *H. metharme metharme*, Nattereri: *H. nattereri*; Row 3, Tiger: *H. numata silvana, H. pardalinus tithoreides, H. numata aristiona*, Hecuba: *H. hecuba hecuba*; Row 4, Dryas: *Dryas iulia alcionea*, Agraulis: *Agraulis vanillae maculosa*, Philaethria: *Philaethria dido dido*, Eueides: *Eueides lybia olympia*. Note that the genus *Eueides* contains a large diversity of wing patterns including forms with postman, dennis-ray, tiger, and dryas patterns. Scale bar = 1 cm.

Plate 5 Oviposition behaviour (see Figure 3.1 on page 28)

Heliconius charithonia laying on *Passiflora biflora* and *Heliconius melpomene melpomene* laying on *Passiflora menispermifolia*. Both photographs were taken in the insectary in Gamboa, Panama.

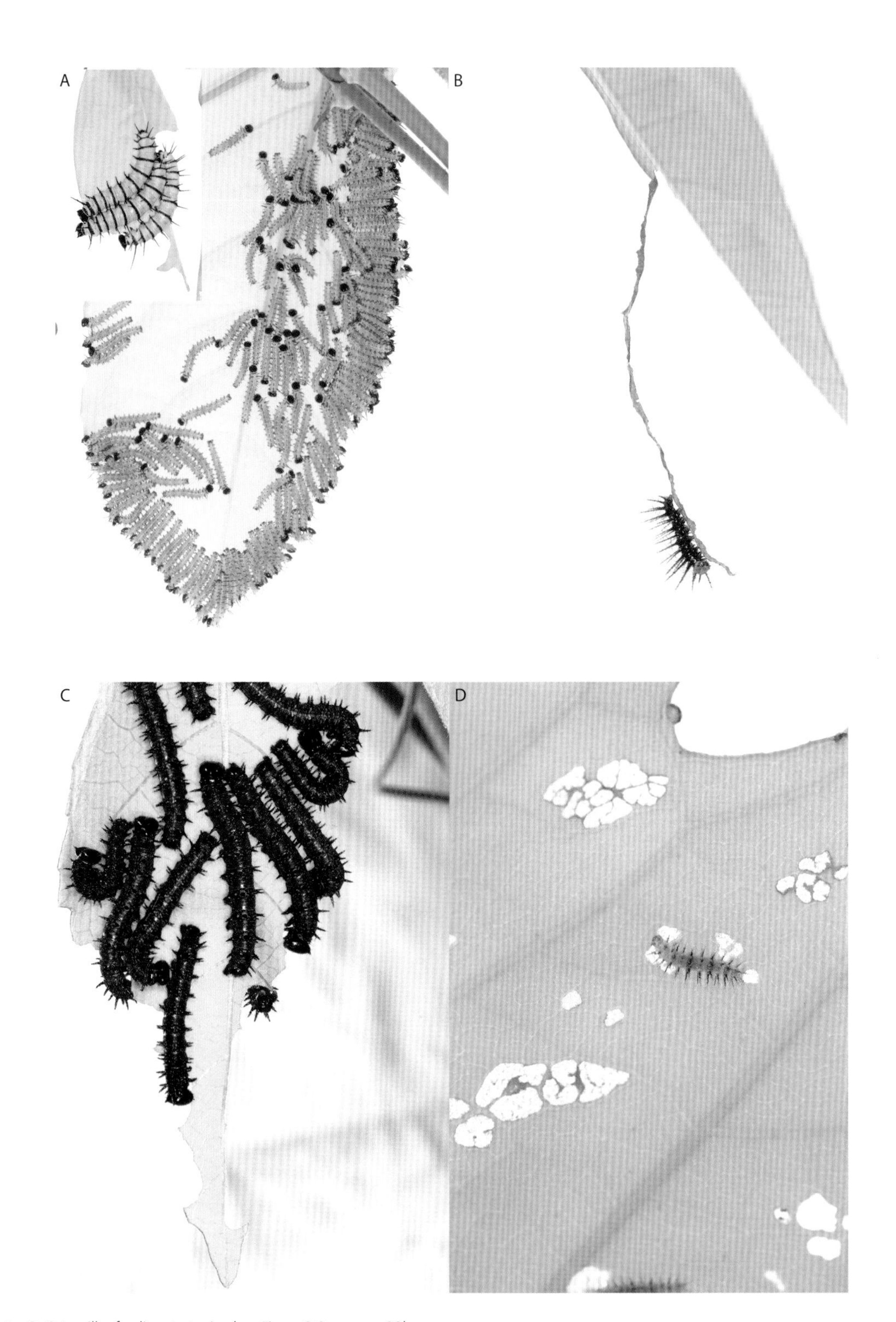

Plate 6 Caterpillar feeding strategies (see Figure 3.2 on page 33)

(A) *Heliconius doris* caterpillars feed in a coordinated manner in large groups. Third-instar caterpillars are shown in the main image and fifth-instar caterpillars are shown inset. (B) Early-instar caterpillars of *Dryas iulia* cut a narrow strip of leaf and sit at the end of the hanging thread. This is a strategy to avoid the attention of ants patrolling the leaves. (C) Caterpillars of *Dione juno* are a crop pest on commercial *Passiflora edulis*. They are highly gregarious but do not feed in such a coordinated manner as *H. doris*. (D) Early-instar caterpillars of *Eueides aliphera* scrape the underside of leaves, leaving characteristic 'windows' of translucent cuticle in the leaf.

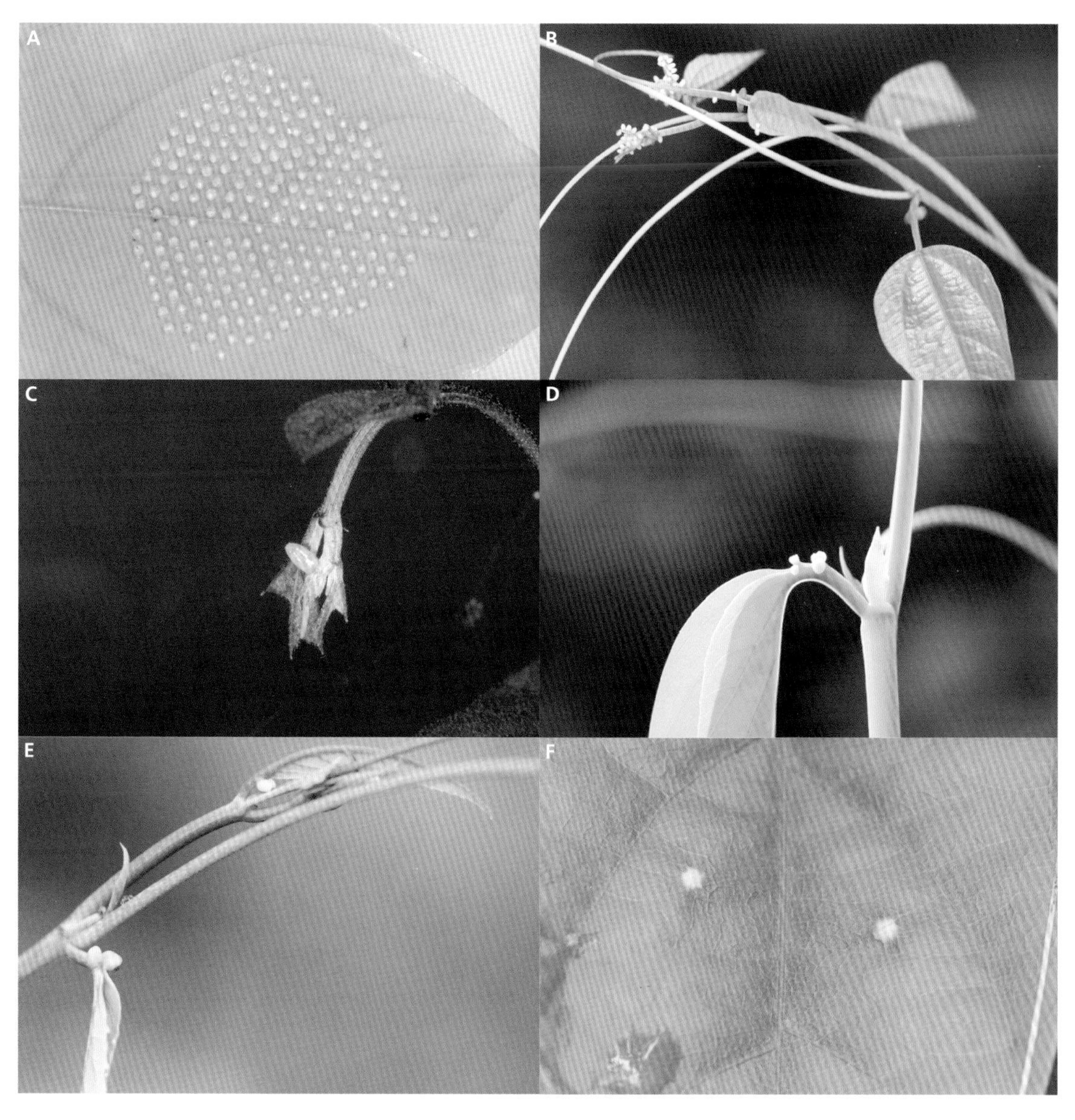

Plate 7 Eggs and egg mimics (see Figure 3.4 on page 40)

(A) Egg cluster of *H. doris* on *P. ambigua*; (B) eggs of *H. sara* on *P. auriculata*; (C) single egg of *H. erato* on *P. biflora*; glands on the petiole of (D) *P. quadrangularis* and (E) *Passiflora* spp. and on the leaf blade of (F) *P. auriculata* act as egg mimics to deter oviposition by *Heliconius* females.

Plate 8 Pollen-feeding behaviour (see Figure 4.2 on page 49)

(A) *H. melpomene aglaope* processing a pollen load. Note the large-grained *Psiguria* pollen on the proboscis. (B) The first flower on an inflorescence of *Psiguria warcsewiczii* is relatively large, and (C) subsequent flowers get smaller. (D) A single inflorescence of *Psiguria bignoniaceae* with over 250 flower scars, indicating almost a year of continuous flower production.

Plate 9 Roosting behaviour of *H. erato* (see Figure 5.3 on page 60)

One female is shown fanning a group of four individuals already roosting together. This photograph was taken in an insectary in Gamboa.

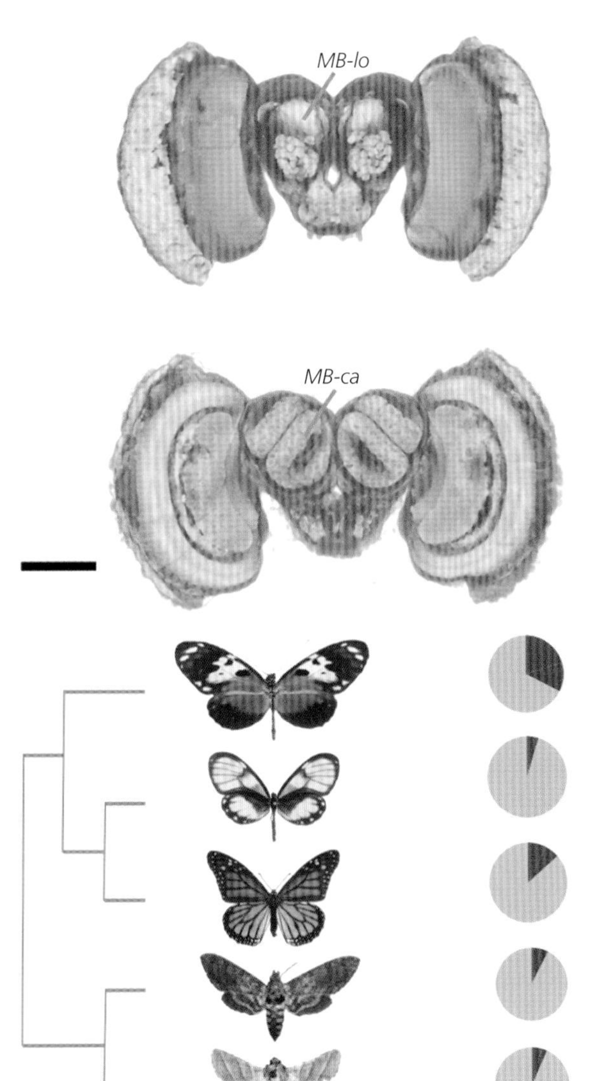

Plate 10 Expansion of mushroom bodies in *Heliconius* butterflies (see Figure 6.1 on page 67)

(A) 3D surface rendering of the brain of *H. hecale* (shown left, scale = 2.5 mm) viewed from the anterior (top) and posterior (bottom) indicating the mushroom body lobes (MB-lo) and calyx (MB-ca). Scale = 500 μm. (B) Pie charts show the proportion of the midbrain occupied by MB-calyx (dark red) and MB-lobe + peduncules (light red) in *H. hecale, Godyris zavaleta, Danaus plexippus, Manduca sexta*, and *Heliothis virescens* (top to bottom) (Merrill et al., 2015; Montgomery et al., 2016).

Plate 11 Mating behaviour and courtship sequence (see Figure 6.4 on page 74)

A courting pair of *H. ismenius* are shown at various stages in the courtship. (A) The male hovers over the female fanning his wings; (B) the male lands beside or even on top of the female; (C) the female can reject males by raising her abdomen and opening her pheromone glands; (D) the male attempts to mate by landing beside the female and bending his abdomen towards her; (E) during mating the male and female can stay attached for several hours. Note that panel E shows a different pair of individuals.

Plate 12 The behavioural sequence of pupal mating in *H. charithonia* (see Figure 6.6 on page 77)

Three males are shown contesting for a position on a female pupa (A), one in flight and two perched on the pupa. As the female begins to emerge, the male inserts his abdomen into the pupal case (B and C). In the inset (B), one male attempts to mate (left) while a second male is perched on the pupa (right). Finally, as the female ecloses from the pupa, she is already mating (D). Photos courtesy of Andrei Sourakov.

a) Mark-release-recapture (Benson 1972)

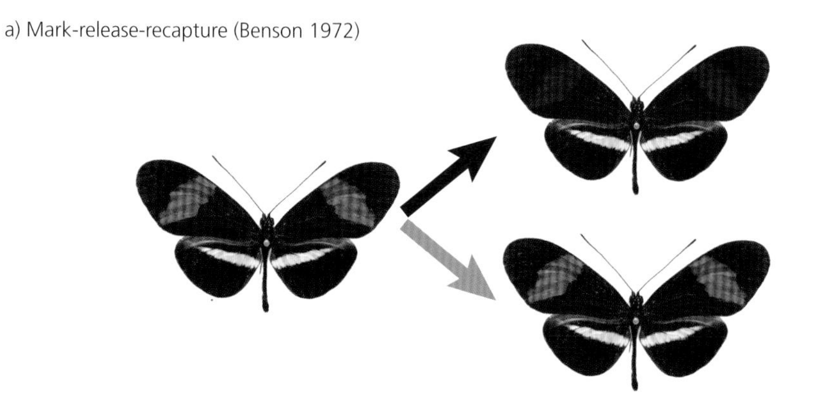

b) Mark-move-recapture across a hybrid zone (Mallet et al. 1989a)

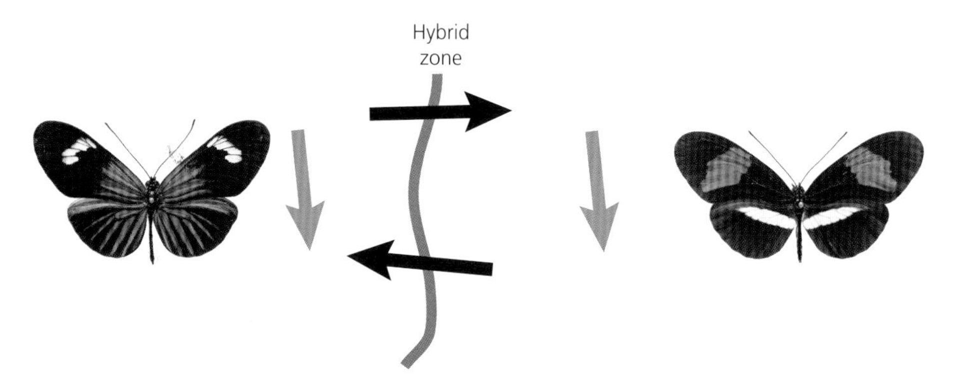

c) Mark-move-recapture between polymorphic populations (Kapan 2001)

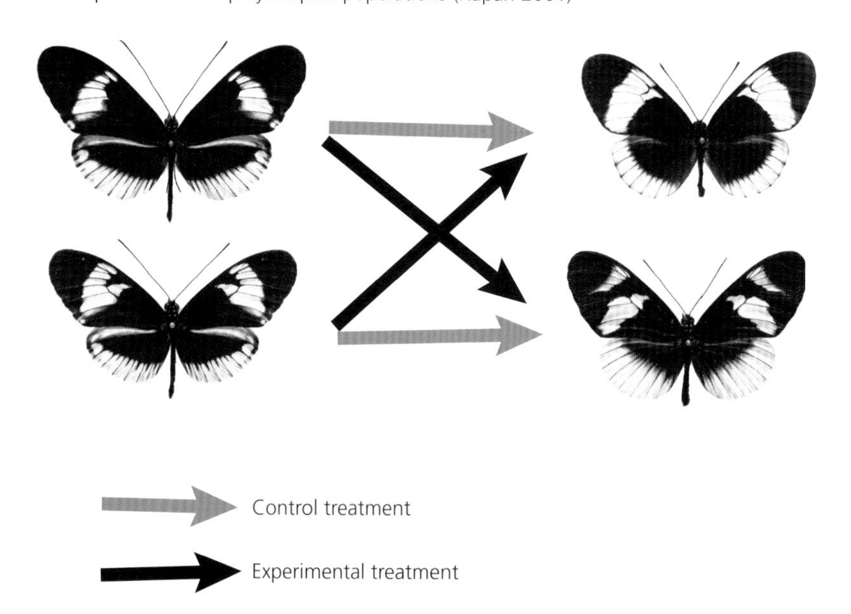

Plate 13 Summary of field tests of mimicry selection in *Heliconius* (see Figure 7.1 on page 92)

(A) Benson blacked out the forewing band of *Heliconius erato* in Costa Rica, with control butterflies similarly manipulated without a colour change; (B) Mallet moved butterflies across a wing pattern hybrid zone in Peru, with control butterflies being moved a similar distance but within the same wing pattern range; (C) Kapan moved polymorphic *Heliconius cydno* adults between populations with different morph frequencies, demonstrating that the locally abundant morph showed higher fitness. See text for full details and references.

Plate 14 Sympatric mimicry rings (see Figure 7.2 on page 94)

Coexisting mimetic species found in central Panama with upper (ventral; above) and underside (dorsal; below) wing patterns. Top row: *H. hecale melicerta, H. ismenius boulleti, H. sara magdalena*; second row: *H. cydno chioneus, H. sapho sapho, H. doris obscurus*; third row: *H. erato demophoon, H. melpomene rosina, H. hecalesia longarena*.

Plate 15 Distribution of geographic races of *Heliconius melpomene* (see Figure 7.3a on page 97)

Three forms lack images: *madeira* is similar to *thelxiope; intersectus* is the Isla de Marajó form with a distal yellow band otherwise similar to *thelxiopeia; euryades* is a postman form similar to *melpomene.* Note that *H. m. martinae* is polymorphic for the hindwing yellow band, with some individuals looking very similar to *H. m. rosina.* On the map, some specimens from the Neukirchen collection labelled as *n. ssp.* × *vicina* hybrids from the central Amazon are here assigned to the race *vicina.* I have not examined these specimens, and the extent of the race *vicina* may not extend as far east as indicated. Butterflies are shown with dorsal (left) and ventral (right) images. Geographic distribution data were compiled by Neil Rosser and maps drawn using CartoDB.

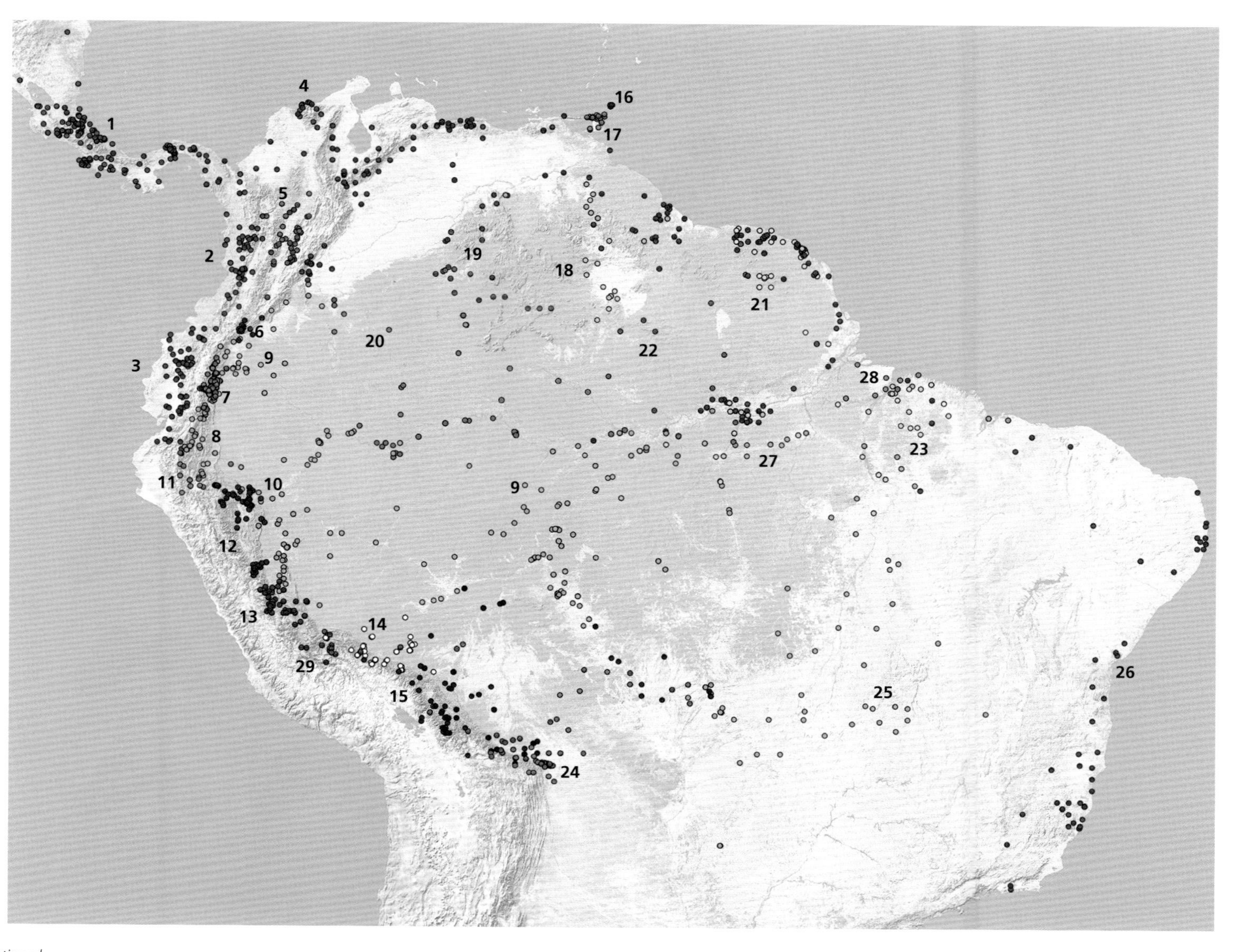

Plate 15 *Continued*

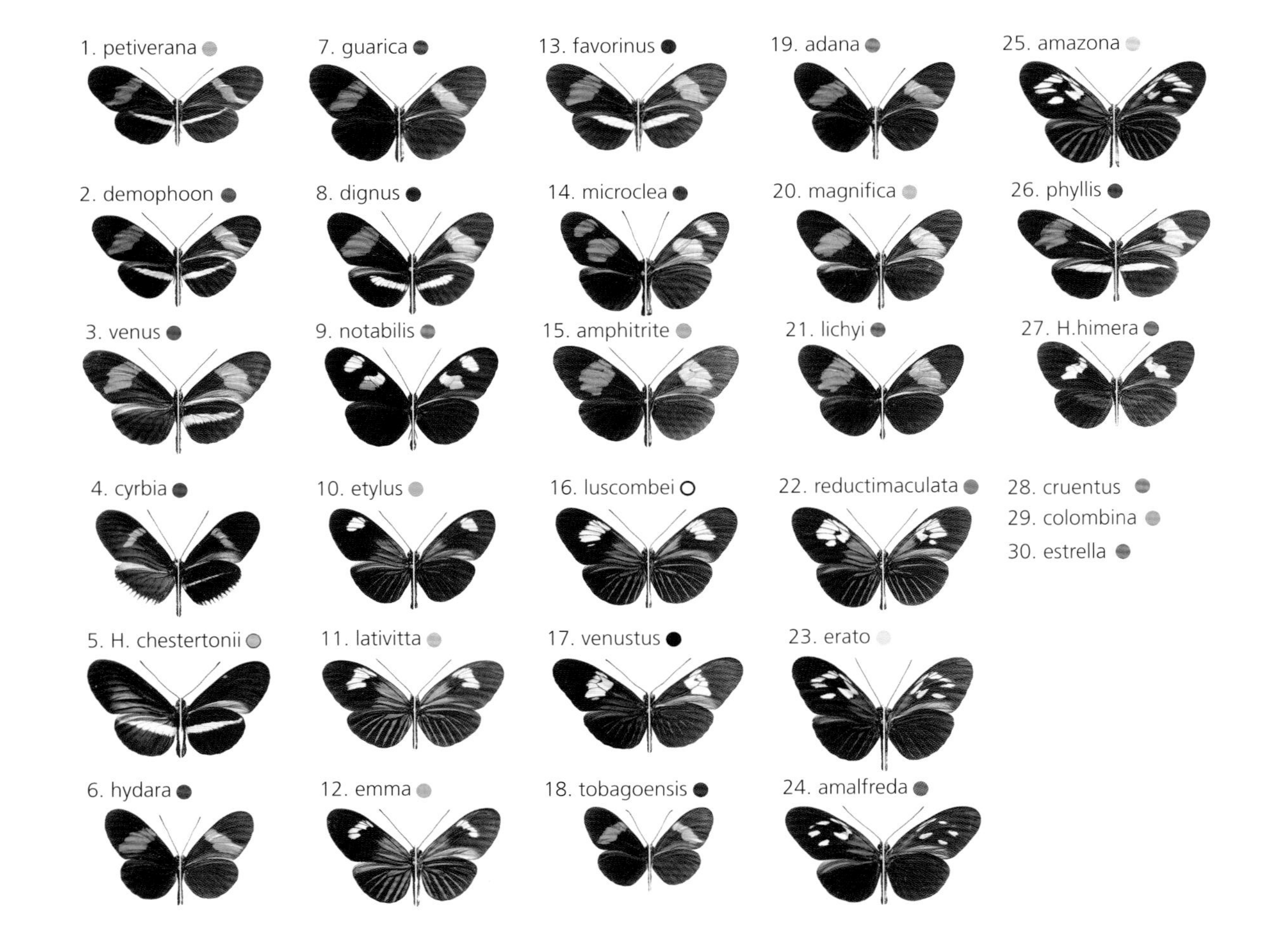

Plate 16 Distribution of geographic races of *Heliconius erato* (see Figure 7.3b on page 98)

Three forms lack images: *cruentus* is similar to and somewhat intermediate between *petiverana* and *demophoon* with a broad and slightly shorter hindwing yellow bar; *colombina* is similar to *demophoon; estrella* is the Isla de Marajó form with a distal yellow band otherwise similar to *erato*. Butterflies are shown with dorsal (left) and ventral (right) images. Geographic distribution data were compiled by Neil Rosser and maps drawn using CartoDB.

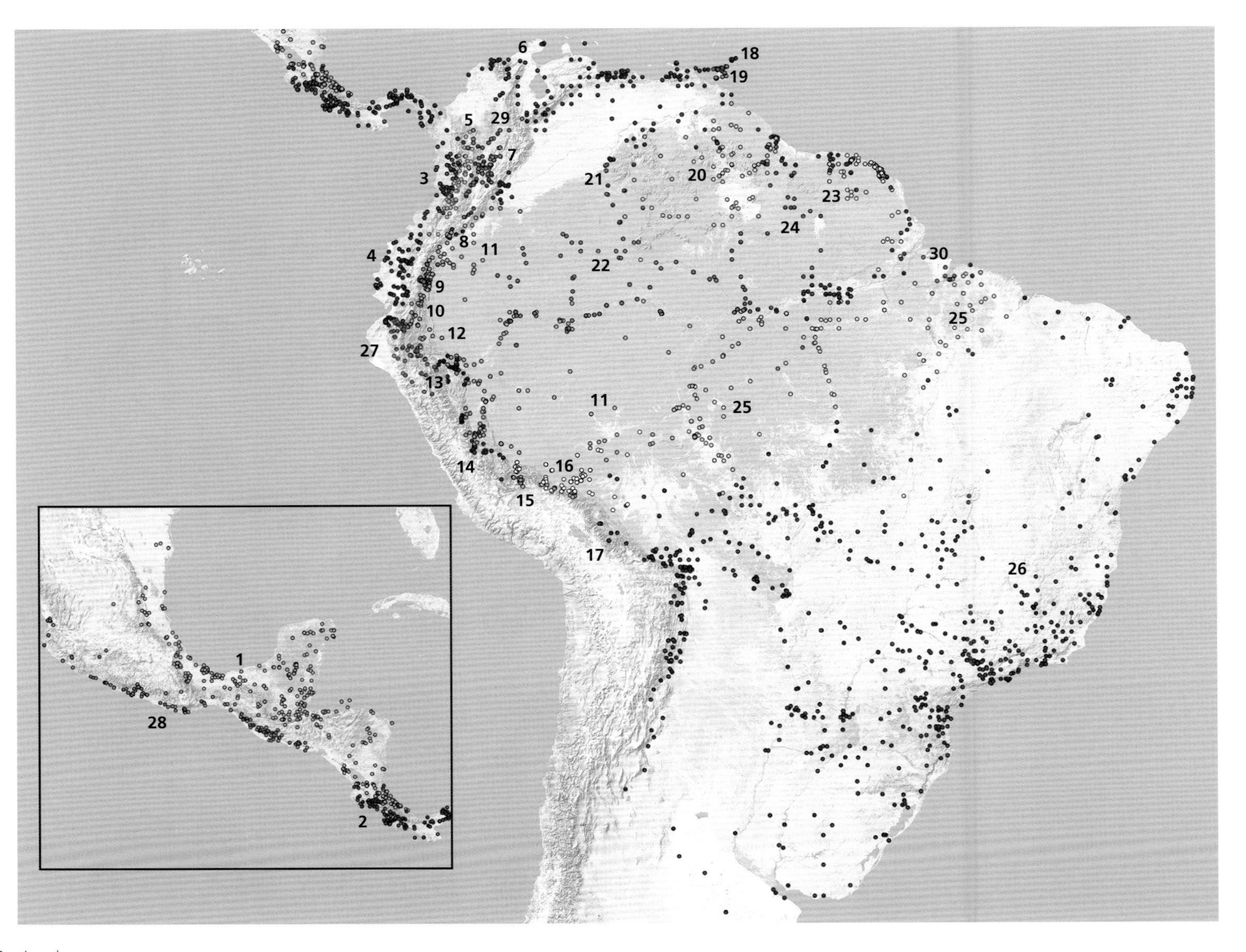

Plate 16 *Continued*

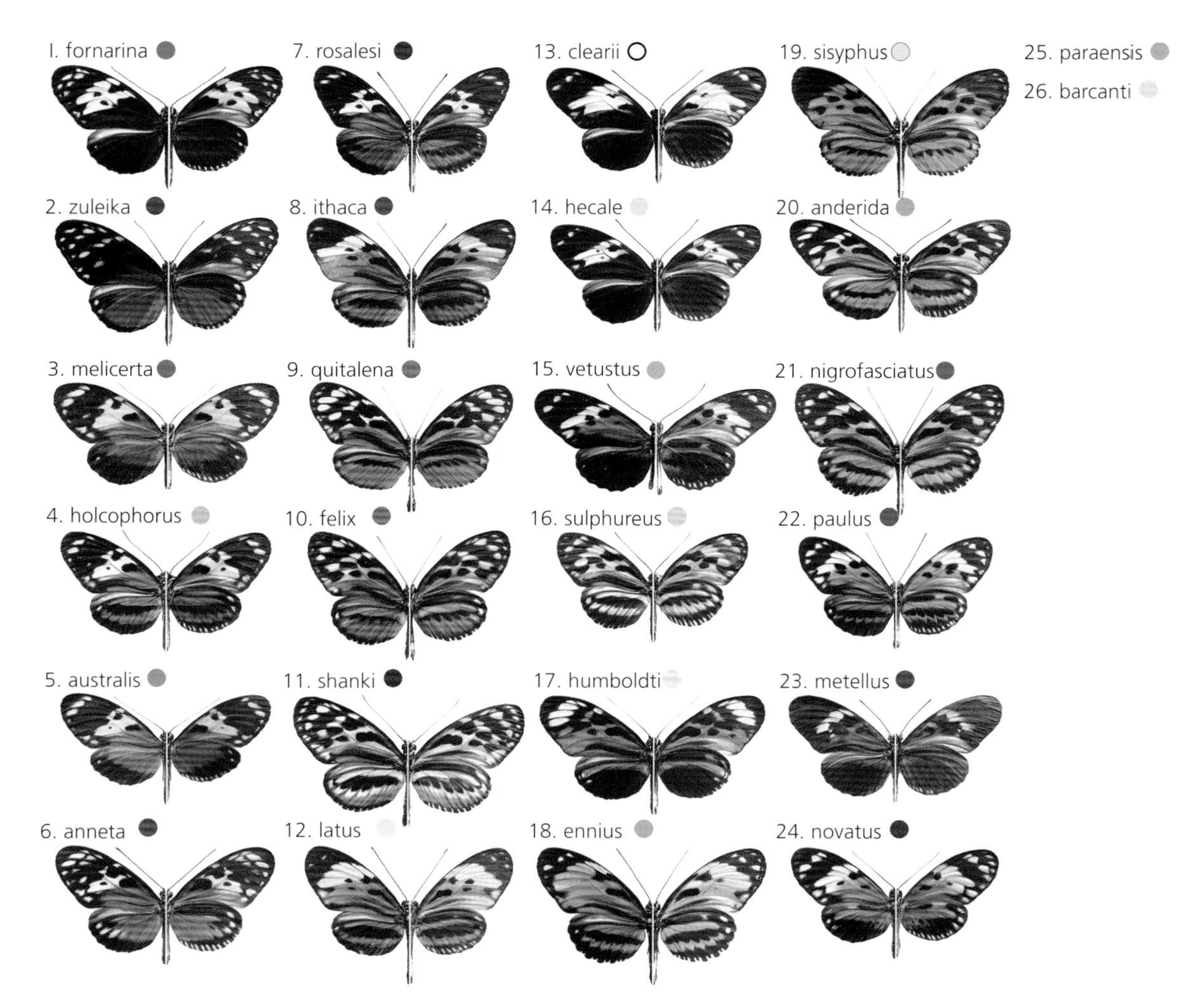

Plate 17 Distribution of geographic races of *Heliconius hecale* (see Figure 7.3c on page 99)

Three forms lack images: *paraensis* is similar to *paulus; barcanti* is similar to *clearii; zeus* (not mapped) was described from Bolivian specimens and is similar to *felix*; the distributions for these forms are combined. Note that *H. hecale ithaca* also has a form with a black hindwing (not shown) that is mimetic with *H. numata messene*. Butterflies are shown with dorsal (left) and ventral (right) images. Geographic distribution data were compiled by Neil Rosser and maps drawn using CartoDB.

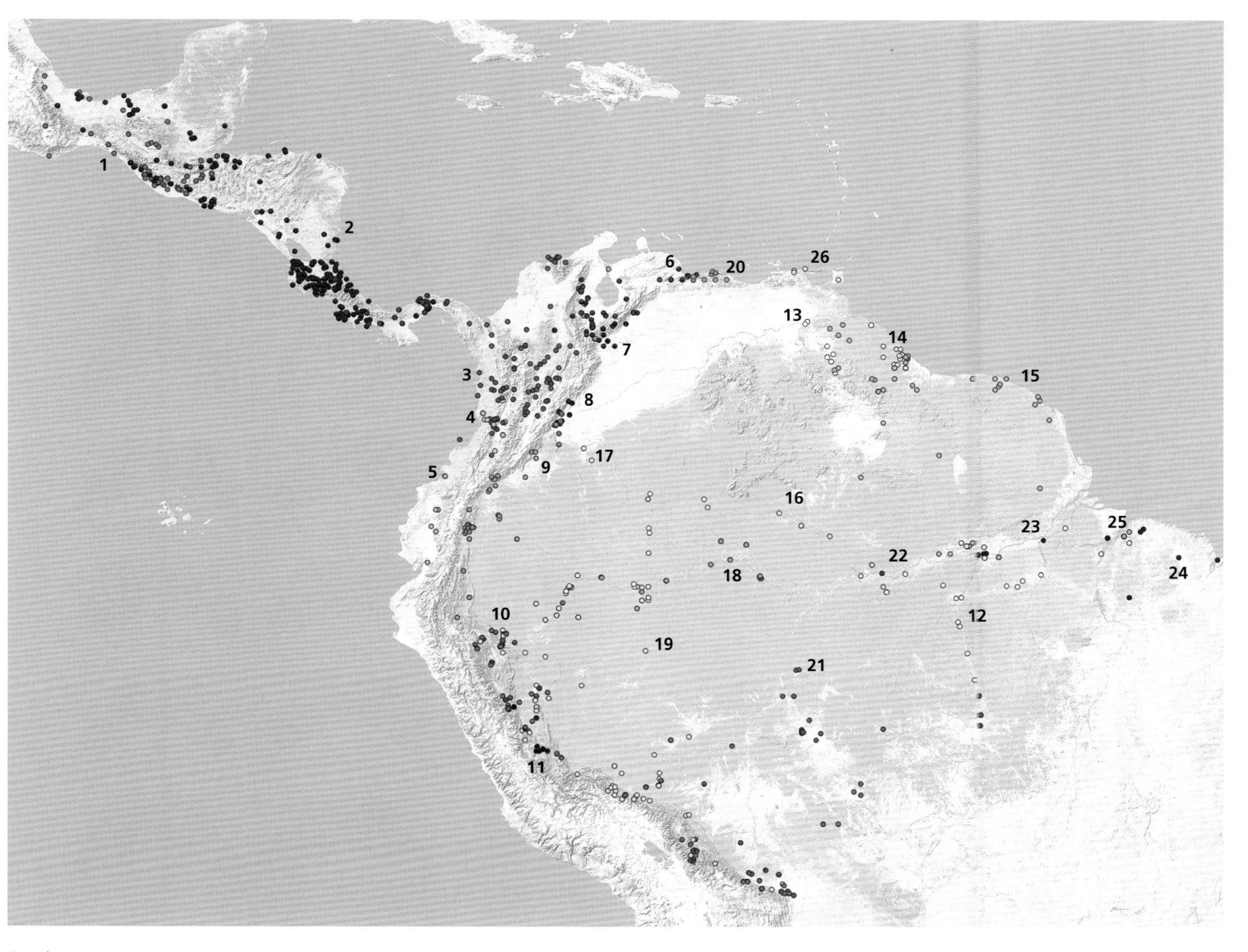

Plate 17 *Continued*

Plate 18 Distribution of forms of *Heliconius numata* (see Figure 7.3d on page 100)

The distinction between geographic races and within-species polymorphism breaks down in *H. numata*. As can be seen from the map, many of these taxa are sympatric forms controlled by alternate alleles at the *P* supergene, but for simplicity I consider all forms as subspecies. Note that some forms are combined for distribution mapping, as these are not distinguished in the available distribution data (*euphrasius, aurora*, and *elegans; lycraeus* and *timaeus; euphone* and *laura*). Forms not figured include *aulicus* (labelled as 27, pink circles on map) similar to *peeblesi* but with a more complete yellow forewing band; *geminatus* similar to *elegans* with more clearly defined yellow forewing bands; *isabellinus* similar to *elegans* but with a broad complete median forewing yellow band. Butterflies are shown with dorsal (left) and ventral (right) images. Geographic distribution data were compiled by Neil Rosser and maps drawn using CartoDB.

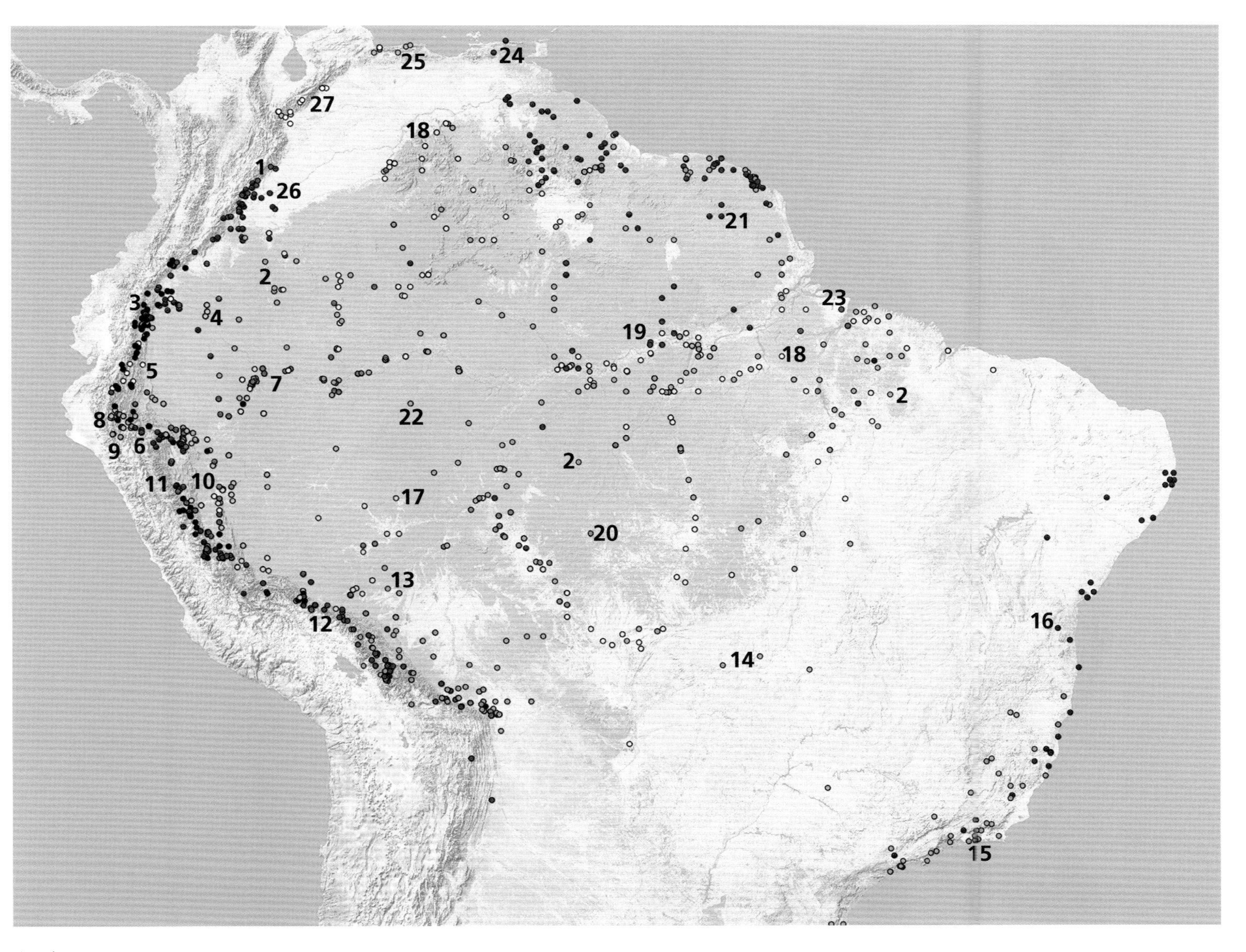

Plate 18 *Continued*

Plate 19 The Amazonian dennis-ray mimicry ring (see Figure 7.4 on page 101)

First row: *H. burneyi huebneri, H. aoede auca, H. xanthocles zamora*; second row: *H. timareta timareta* f. *timareta, H. doris doris, H. demeter ucayalensis;* third row: *H. melpomene malleti, H. egeria homogena, H. erato emma*; fourth row: *H. elevatus pseudocupidineus, Eueides heliconioides eanes, E. tales calathus*; bottom row: *Chetone phyleis*, a pericopine moth. These butterflies are from populations in both Ecuador and Peru and do not all occur in exactly the same locality.

Plate 20 Mimicry between *Melinaea* species and *Heliconius numata* forms (see Figure 7.6 on page 104)

First row: *Melinaea idae vespertina* and *H. ismenius metaphorus*; second row: *Melinaea satevis cydon* and *H. numata arcuella*; third row: *Melinaea marsaeus mothone* and *H. n. bicoloratus*; fourth row: *M. menophilus cocana* and *H. n. euphrasius*; fifth row: *M. menophilus menophilus* and *H. n. tarapotensis*. All photos by the author except *M. menophilus cocana*, courtesy of Keith Willmott.

Plate 21 Examples of sexual dimorphism in *Heliconius* (see Figure 7.7 on page 107)

Not many species of *Heliconius* show sexual dimorphism. Top two rows: Several silvaniform species such as *H. hecale quitalena* show dimorphism in the brightness of their wing patterns. Middle two rows: *H. demeter demeter* shows much more accurate mimicry in females as compared to males. Bottom two rows: *H. nattereri* is the most sexually dimorphic species in the genus *Heliconius*. Dorsal (above) and ventral (below) images are shown.

Plate 22 Phenotypes from a hybrid zone in eastern Ecuador (see Figure 8.1 on page 114)

There are three parental races that contribute variation to the hybrid zone, pictured here along the top row: *H. melpomene plesseni*, *H. m. malleti*, and *H. m. ecuadorensis*. Three major loci control the wing patterns: the red locus, *optix*, controls red/orange pattern elements; the shape locus, *WntA*, controls the shape of the forewing band (two spots or one); the yellow locus, *cortex*, produces the yellow forewing band. The form on the left column second from top is the putative 'stable hybrid' discussed on Page 181. The butterfly hybrids are all from the Neukirchen collection. Scale bar is 1 cm.

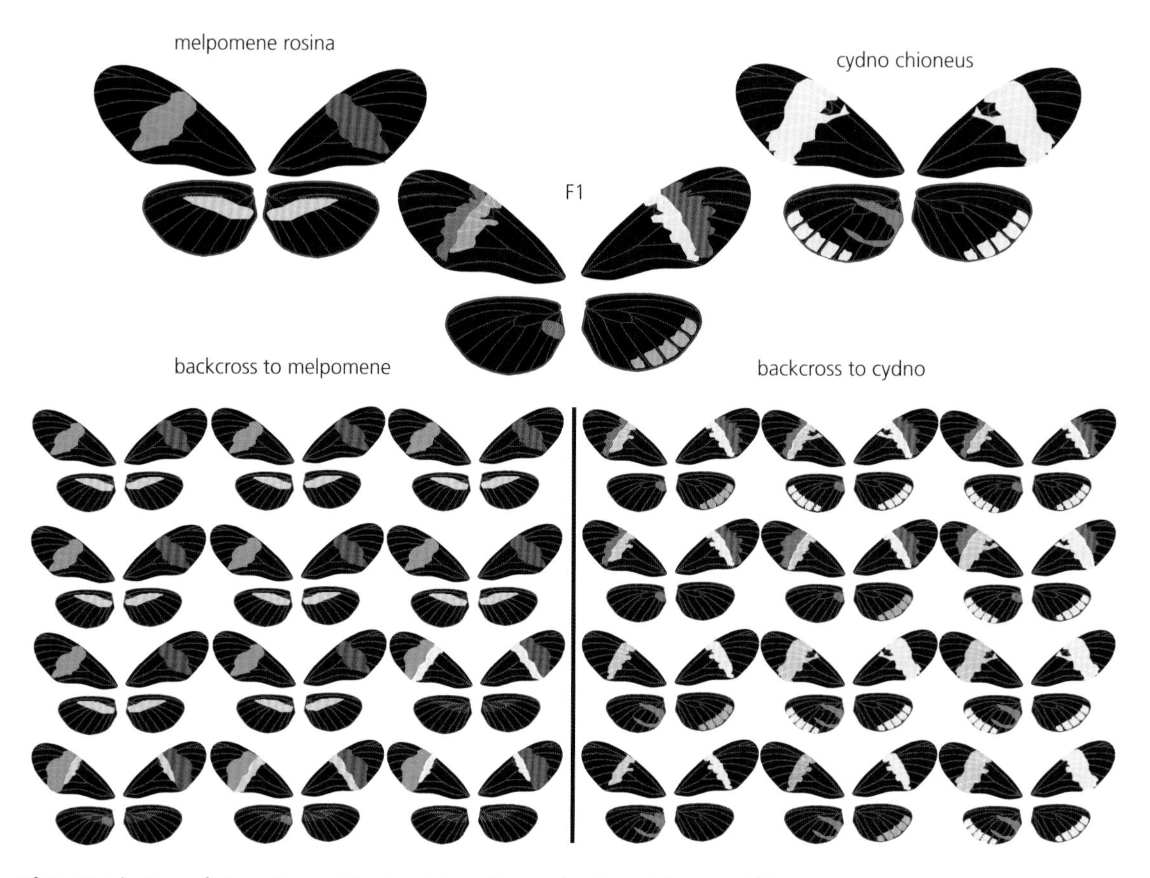

Plate 23 Inheritance of wing pattern variation in an interspecies cross (see Figure 8.2 on page 117)

Hybrid phenotypes for a cross between *H. melpomene rosina* and *H. cydno chioneus* from Panama. The backcross families are generated by crossing an F_1 male back to either a female *H. melpomene rosina* (left) or a male *H. c. chioneus* (right). The main pattern loci with alleles that segregate in this cross are the red locus *optix*, controlling the red forewing band; the yellow locus *cortex*, with alleles that control the yellow/white forewing band, yellow hindwing bar, and white margin; the shape locus *WntA*, which controls the hourglass-shaped forewing band element; and the colour locus *K*, which alters the colour of the forewing band from yellow to white. Drawn from crosses described in Naisbit et al. (2003).

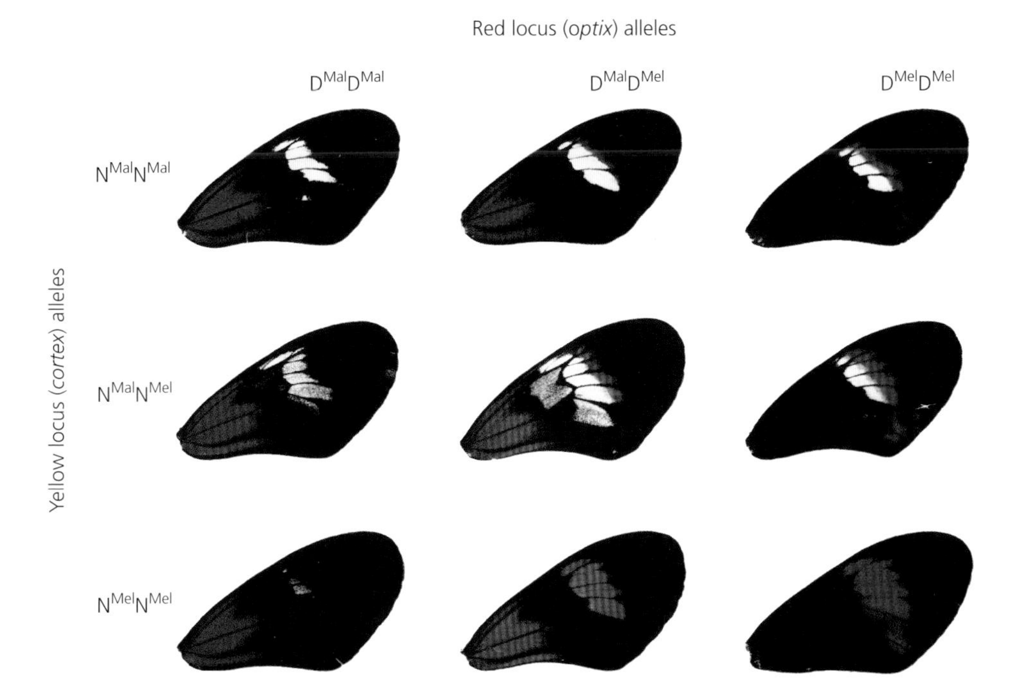

Plate 24 Epistatic interaction between the yellow and red loci (see Figure 8.3 on page 119)

In crosses between the red forewing banded postman races and the yellow forewing banded dennis-ray races of *H. melpomene* and relatives, there is an epistatic interaction between alleles at the yellow (*N* refers to the effect of this locus on the forewing band) and red (*D*) wing patterning loci. For example, the wings shown are from a cross between *H. m. malleti* (top left) and *H. m. melpomene* (bottom right). The full red band in the forewing is produced only in the absence of yellow alleles at the yellow locus. In hybrid genotypes, the presence of yellow band alleles at the yellow locus pushes the red band distally, to produce the characteristic double yellow and red band. This is similar to the phenotype of the putative hybrid species, *H. heurippa*.

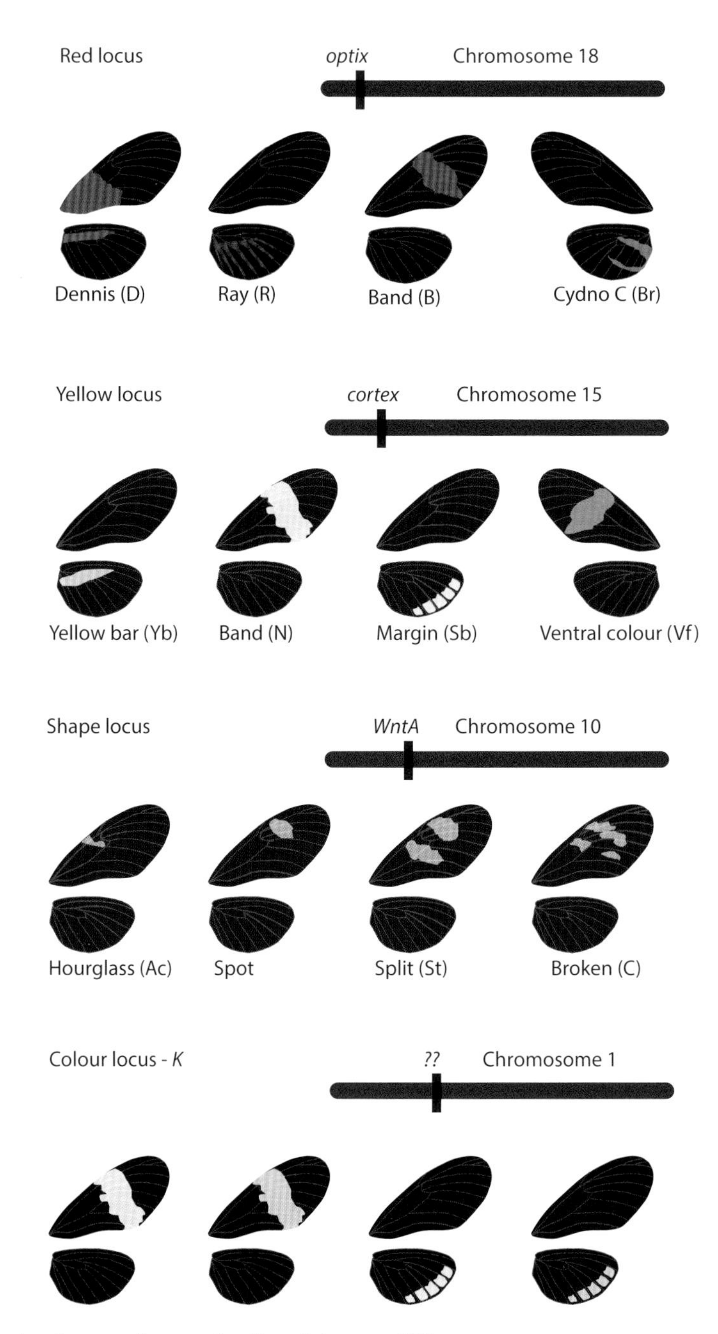

Plate 25 Summary of major wing patterning genes (see Figure 8.4 on page 122)

Major effects of four loci on wing patterns in the *H. melpomene* clade. Dorsal wing surfaces are depicted, except for wings in reverse orientation that show two ventral phenotypes. Allelic variants at the red and yellow loci, *optix* and *cortex,* control the presence/absence of the pattern elements depicted, except where the yellow locus alters the colour of the ventral red band from red/orange to pink (known as *Vf*). In contrast, the shape locus *WntA* has alternate alleles controlling different forewing band shape phenotypes, so rather than the presence/absence of elements, this locus controls the shape of elements whose presence is controlled by the other loci. The colour locus, *K,* alters patterns from white to yellow.

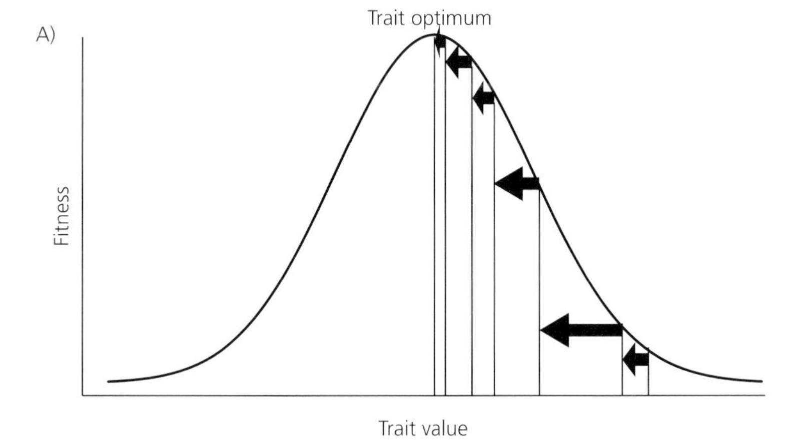

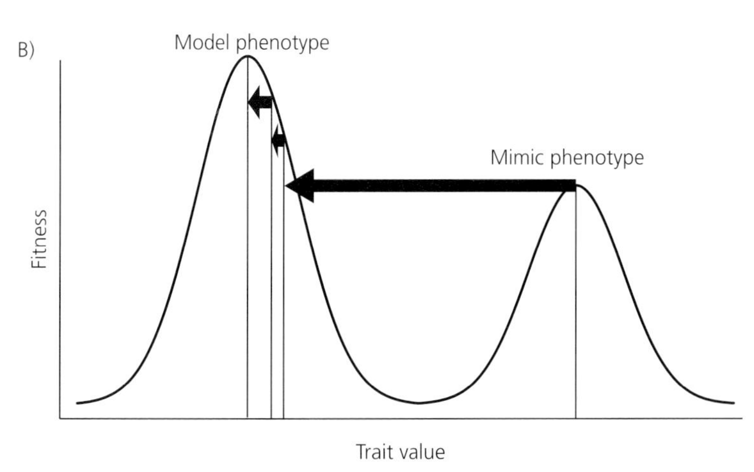

Plate 26 Adaptive landscapes in warning colour (see Figure 8.5 on page 127)

The curve represents the fitness of an individual at a range of phenotypic values. (A) An adaptive walk towards a single fitness peak. Arrows show possible mutations fixed as a population evolves towards the fitness peak. Fixation of large effect mutations is more likely early in the adaptive walk, with an exponential distribution of mutational effects expected overall. (B) Warning colour and mimicry are expected to show an especially rugged adaptive landscape, shown here with distinct peaks of fitness for the model and the mimic species. In order for the mimic to evolve similarity to the model wing pattern, a large mutational step may be required to produce sufficient initial similarity between the species.

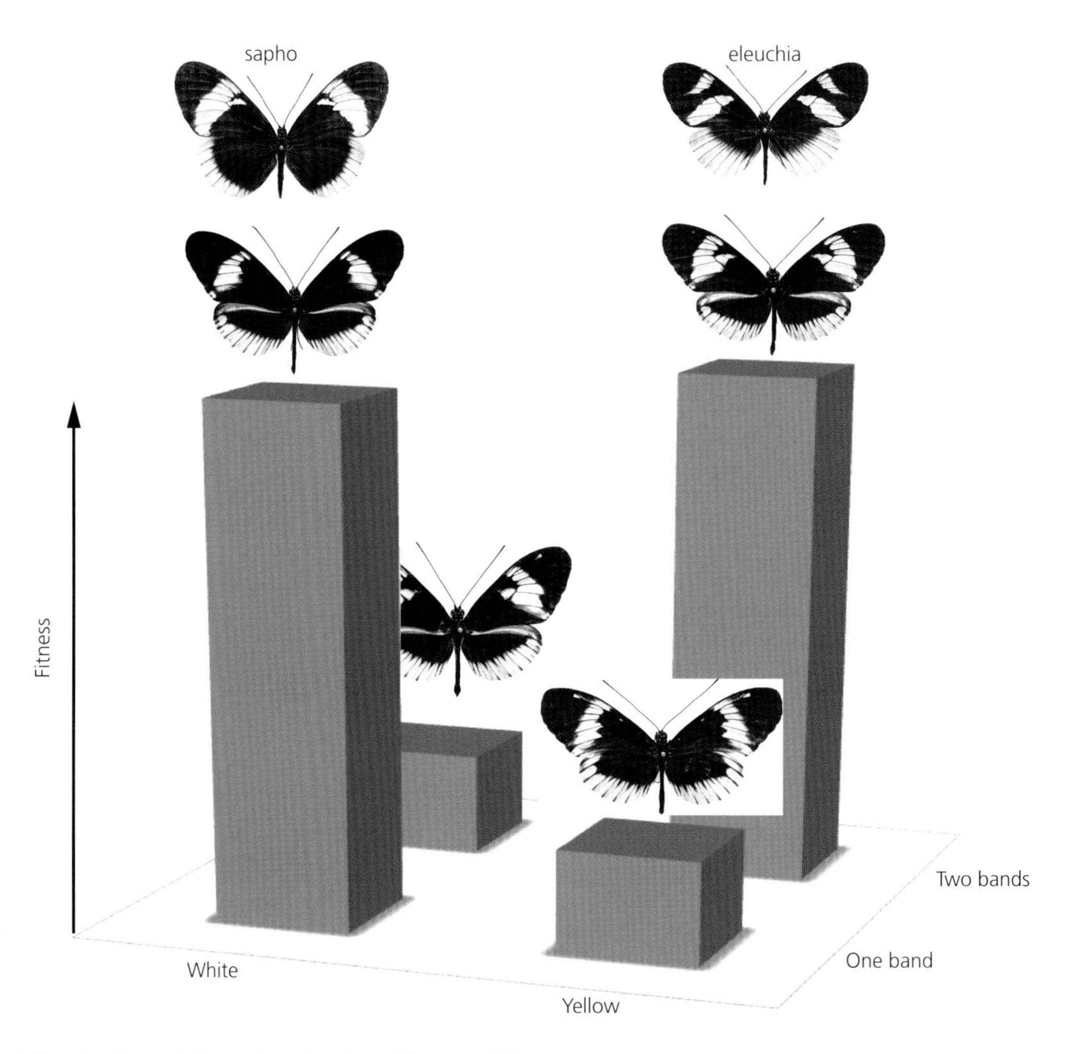

Plate 27 A two-locus adaptive landscape (see Figure 8.6 on page 130)

In western Ecuador, *H. cydno alithea* is polymorphic (four forms are shown below) and mimics *H. sapho* and *H. eleuchia* (top). Variation at the shape locus, *WntA*, acts to facilitate mimicry: although *H. cydno* never has two distinct bands, the *WntA* pattern element gives the appearance of a broken band that mimics *H. eleuchia*. The colour locus, *K*, controls the switch between yellow and white. Since there are two loci involved in this mimicry polymorphism, some of the phenotypes produced are less effective mimics. Recombination between the unlinked *K* and *WntA* loci produces these less fit combinations every generation. Note that fitness values are hypothetical—the joint fitness effects of the two loci have not been estimated in the wild.

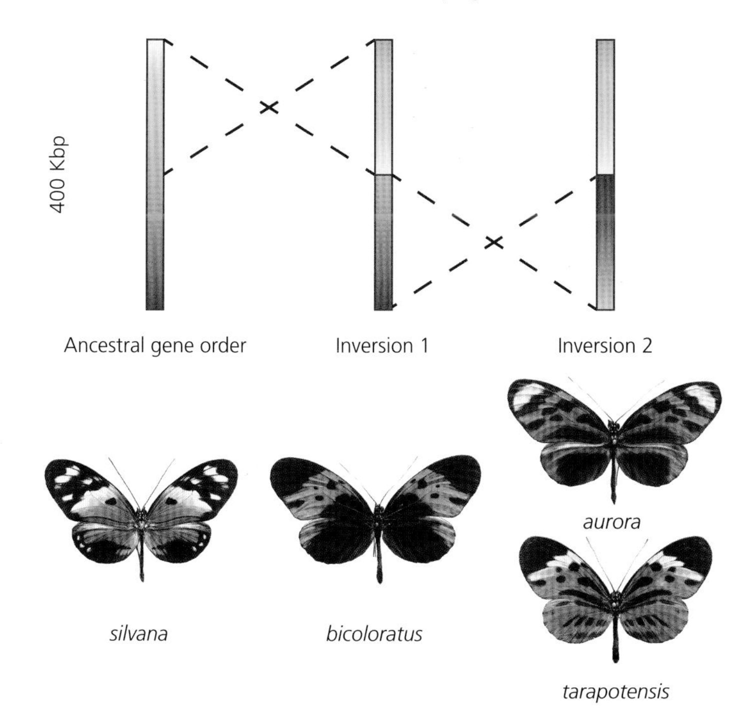

Plate 28 Structural variation associated with the *Heliconius numata* supergene (see Figure 8.7 on page 132)

At least two genetic inversions are associated with the *H. numata* supergene. The ancestral gene order, which matches that in *H. melpomene* and *H. erato*, is shown on the left and is associated with ancestral phenotypes such as *H. n. silvana*. Two sequentially derived inversions are associated with more dominant alleles and are shown middle and right. Redrawn from Joron et al. (2011).

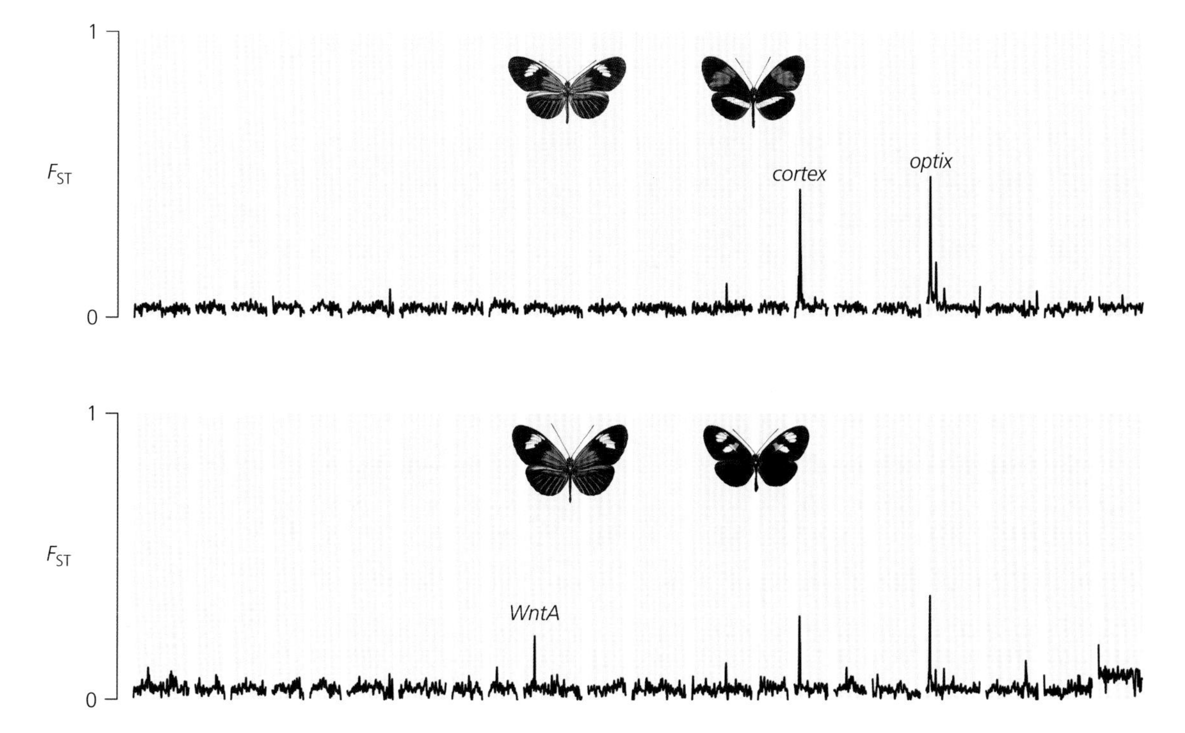

Plate 29 Genetic islands of differentiation between wing pattern races (see Figure 8.8 on page 135)

Genetic divergence is plotted across the whole genome between two pairs of races, *H. m. aglaope* and *H. m. amaryllis* (top) and *H. m. malleti* and *H. m. plesseni* (below). Each chromosome is shown by the grey shading. The highest divergence is seen at the wing patterning loci *WntA*, *cortex*, and *optix*, and corresponds to known differences identified from crossing experiments. Image drawn by Simon Martin.

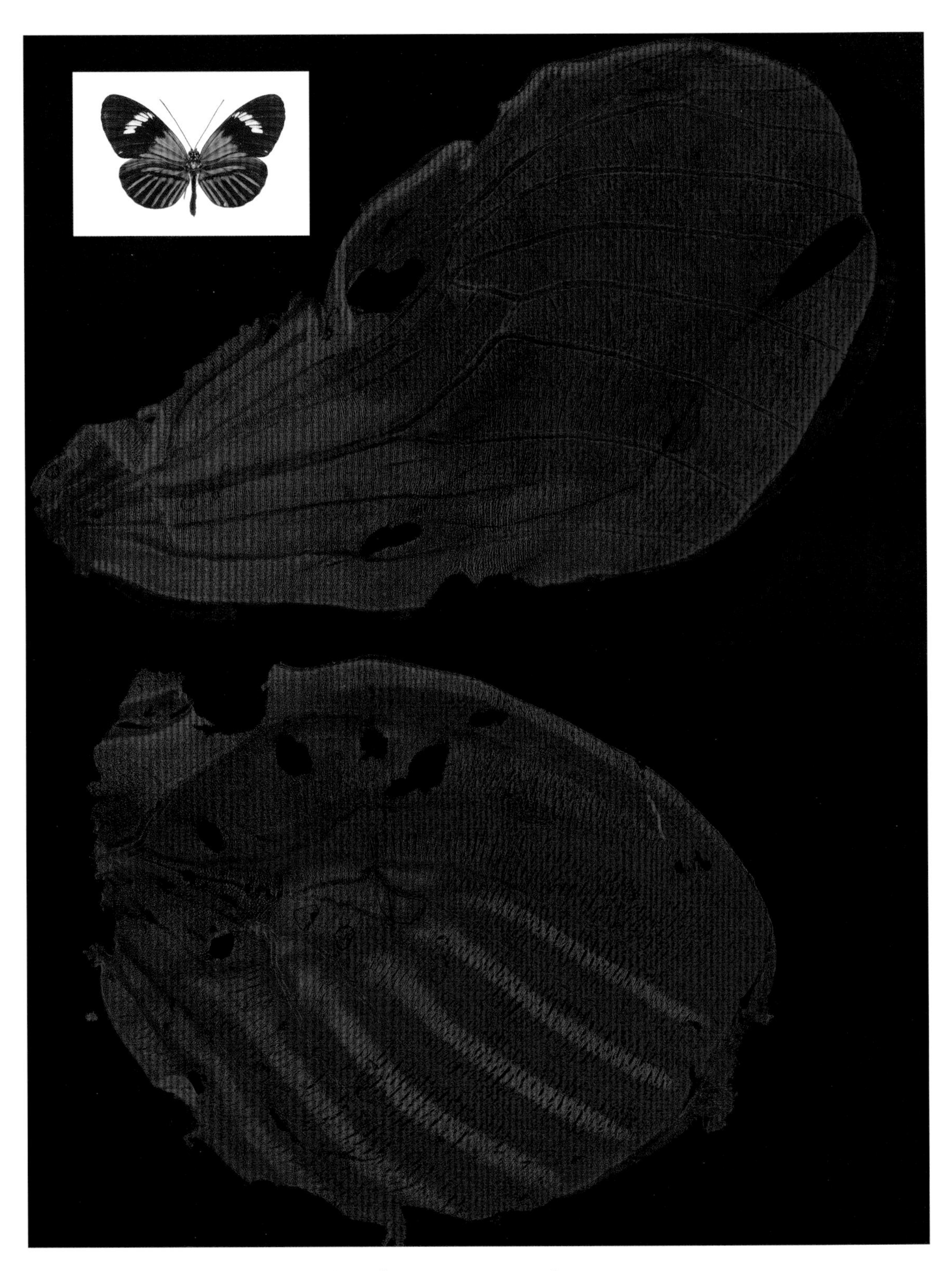

Plate 30 Antibody stain of Optix protein in pupal wings (see Figure 9.6 on page 152)

The Optix protein is localized during pupal wing development in regions that show perfect concordance with red patches in the adult wing. This is an individual of *H. elevatus pseudocupidineus*. This image was generated by Richard Wallbank.

The next line of evidence came from population genetic studies. Many thousands of generations of potential recombination in wild populations, compared to just a single generation of recombination in laboratory crosses, means that the former are potentially a more powerful method for identification of narrow regions under selection. Genetic studies of the closely related hybridizing races of *H. melpomene* and *H. erato* have shown that adjacent races differ little across the genome apart from narrow peaks of differentiation at the wing patterning loci described earlier (Figure 8.8).

Such studies were initially carried out with widely spaced genetic markers (Baxter et al., 2010; Counterman et al., 2010), but later repeated with complete sequencing of focal wing pattern regions (Nadeau et al., 2012), and later still with whole-genome sequences (Kronforst et al., 2013; Martin et al., 2013; Supple et al., 2013). These studies largely confirm the suspicions of earlier workers, that wing pattern races differ little except at a few loci of major effect (Turner, Johnson, & Eanes, 1979; Mallet, 1989) (Figure 8.8). They also provide strong population genetic confirmation of the broad patterns identified in the QTL studies described earlier. Overwhelmingly, large differences in the genome between races occur at loci already identified as being of major phenotypic effect in crossing experiments. Identification of specific genes causing pattern differences relied on combining these population genetic signatures with gene expression analyses, but I will leave the latter for Chapter 9 (Reed et al., 2011; Martin et al., 2012, Nadeau et al., 2016).

It has become common to conduct *Fst* outlier studies in order to detect regions showing high divergence relative to the genetic background (Beaumont & Balding, 2004; Bazin, Dawson, & Beaumont, 2010; Narum & Hess, 2011). Regions of high differentiation are hypothesized to represent regions involved in local adaptation, where selection has

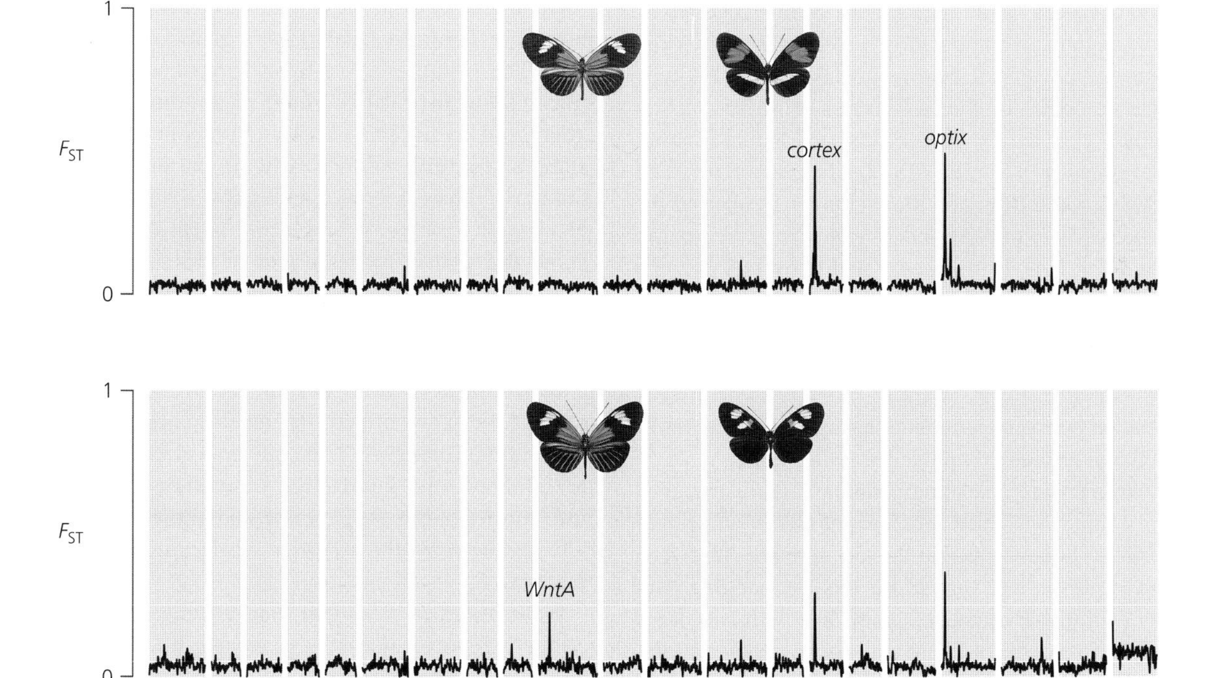

Figure 8.8 Genetic islands of differentiation between wing pattern races (see Plate 29)

Genetic divergence is plotted across the whole genome between two pairs of races, *H. m. aglaope* and *H. m. amaryllis* (top) and *H. m. malleti* and *H. m. plesseni* (below). Each chromosome is shown by the grey shading. The highest divergence is seen at the wing patterning loci *WntA*, *cortex*, and *optix*, and corresponds to known differences identified from crossing experiments. This image was generated by Simon Martin.

driven population divergence. In effect, we have conducted such a study in reverse: the regions underlying a known phenotype were first identified by laboratory mapping studies, and then shown to represent divergence outliers. In the face of widespread discussion about the status and causes of divergence outliers (Bierne et al., 2011; Cruickshank & Hahn, 2014), this work provides a confirmation that such *Fst* outliers can indeed be an effective method for detection of regions underlying local adaptation and divergence (Nadeau et al., 2012, 2014).

Most of the population genetic studies of *Heliconius* races to date have involved comparisons between 'pure' races with different wing patterns. However, a complementary approach is to use admixed populations in hybrid zones to conduct a genetic association study, sometimes termed 'admixture mapping' (see Buerkle & Lexer, 2008, for definitions of the related approaches of 'association mapping' and 'admixture mapping'). In hybrid zones, all possible wing pattern combinations occur within the same population, so we can look for regions of the genome associated with specific pattern elements. Many generations of ongoing recombination potentially offer even greater power to identify genotype-by-phenotype associations. Another advantage of an association study over the *Fst* outlier approach is that it correlates genetic variants with specific phenotypes. Led by Nicola Nadeau and Mayte Ruiz, we collected the species *H. melpomene* and *H. erato* across two parallel hybrid zones in Ecuador and Peru, to conduct a population genetic study of wing pattern variation (Nadeau et al., 2014). As well as all the known major loci, we were able to identify and map a genetic locus called *Ro* to linkage group 13, on the basis of an association between this phenotype and genetic variants on that linkage group (Nadeau et al., 2014).

In summary, isolating genomic regions involved in adaptation is now much easier than it was previously. The caveat is that the loci of major effect seen in *Heliconius* are very much the low-hanging fruit of evolutionary genetics: single loci with very large phenotypic effects. These may not be very representative of all kinds of adaptation, as many traits are more quantitative in nature. However, being a pragmatist, I have always believed in picking low-hanging fruit, and I believe we have learnt a lot about evolution from the identification of these genes.

8.20 Future directions

Heliconius wing patterns have become recognized as an example of a trait whose variation is dominated by a few major loci of large effect. However, we remain surprisingly ignorant of the distribution of effect sizes. There is considerable scope for better QTL analyses of wing pattern that include quantification of effect sizes. In particular, studies that combine large sample sizes with objective quantification of wing pattern variation should be used to quantify the distribution of effect sizes seen in families. This could be combined with association studies in wild populations, especially across hybrid zones, in order to characterize genetic variation controlling wing pattern in natural populations. A combination of QTL and association approaches with much larger sample sizes would lead to a more complete understanding of the number and effect size of loci controlling pattern adaptation.

It is well established that wing patterns are under strong selection. However, studies of natural selection have not yet attempted to measure the fitness effects of individual loci and genotypes in the wild. For example, across hybrid zones, we would like to know the relative fitness of common hybrid genotypes in order to understand both the epistatic and additive effects of wing pattern genes. Similarly, where species such as *H. melpomene* and *H. cydno* hybridize, it would be useful to understand the relative contributions of individual colour pattern loci towards selection against wing pattern hybrids. Ultimately, the 'effect size' of the various patterning loci should refer to their influence on fitness in different mimicry environments, not just the extent of their influence on appearance to a human observer. Thus, an important goal is to integrate our knowledge of pattern genetics into studies of fitness, which would allow us to really characterize the adaptive landscape of wing patterns.

Now that it is clear that mechanisms of dominance can evolve, at least in *H. numata*, it will be exciting to determine the molecular mechanism that underlies dominance evolution. How are patterns of dominance reversed in the *H. numata* supergene system? Does it result from specific gene products produced by the supergene locus?

We now have a fairly detailed picture of colour pattern genetics in the most phenotypically diverse *Heliconius* species: the *H. melpomene* and *H. erato* mimicry ring, *H. numata*, and, to a lesser extent, *H. hecale*. However, there is still a vast diversity of patterns in other species across the genus that remains poorly studied. It will be exciting to quantify the relative contributions of major and minor effect loci more widely across the genus and in related butterfly species. To what extent is the allelic variation and genetic architecture in the species that have been studied to date representative of the genus as a whole, and is it unusual compared to other butterflies?

CHAPTER 9

Development on the wing: how to make a wing pattern different

Chapter summary

It is an exciting time to be an evolutionary geneticist, because we are suddenly able to open what has previously been a black box—the genetic mechanism that links changes in DNA to changes in the appearance of an organism (Nadeau & Jiggins, 2010). Butterfly wings form as imaginal disks in the caterpillar, and are patterned by a gene regulatory network that increases in complexity through development, providing spatial information crucial for wing formation. Although relatively little is known of this network in *Heliconius*, evidence suggests that many patterns of gene expression are largely conserved between butterflies and fruit flies, and therefore provide spatial information about location in the wing to scale precursor cells. Genes such as *WntA* vary the information in this upstream patterning system in order to generate diversity. Slightly later in development, during the pupal stages, regulatory elements of genes such as *optix* interpret this spatial information and are expressed in patterns that pre-empt the final wing pattern. Every scale cell that is destined to be red expresses *optix*, which then turns on a battery of downstream genes that produce the correct scale structure and the red ommochrome pigments.

Some elements of *Heliconius* patterning are ancient, with genes such as *WntA* involved in pattern specification in other butterflies. Others are likely to be more derived, although the extent to which this is the case remains to be fully elucidated. There may be unique aspects to *Heliconius* wing development that have facilitated their rapid diversification.

Wing patterning genes in *Heliconius* are 'hotspots' for evolution: repeated evolutionary change has occurred at the same genetic locus. This is true for convergent evolution of similar patterns involving parallel genetic changes, but also for divergent evolution in which a great diversity of patterns can be controlled by complex alleles at a single gene. Evolutionary change might become focused on a one or a few genetic loci for several reasons. First, it seems likely that pre-existing complex *cis*-regulatory loci are more likely to acquire novel functions in wing patterning, as they already interact with the relevant transcription factors. Second, the shape of wing regulatory networks likely constrains evolutionary change to one or a few loci.

A sense of awe at the great diversity of *Heliconius* patterns is what inspired many of us biologists to study these insects. Much of this diversity has arisen in a short timescale, at least in evolutionary terms, of just a few million years (Kozak et al., 2015). This great diversity among very closely related forms therefore offers a window into how genetic changes can create new patterns, and inspires a number of questions: How are the precise patterns on a wing encoded in a strand of DNA? Why do *Heliconius* have such diverse patterns compared to their close relatives? Is mimicry made easier by shared developmental pathways? Why do so few genes control all of that diversity? Answers to these questions will require an understanding of the developmental processes that underlie the production of a butterfly

The Ecology and Evolution of Heliconius Butterflies. Chris D. Jiggins, Oxford University Press (2017).

DOI 10.1093/acprof:oso/9780199566570.001.0001

wing, and how those processes have been changed by evolution, a field known as evolutionary developmental biology, or evo-devo.

The diversity we see in *Heliconius* patterns lies somewhere between recent microevolutionary changes occurring over a few generations—such as the famous case of dark peppered moths that spread through the UK during the Industrial Revolution—and the macroevolutionary patterns of diversification seen, for example, in the diverse body plans of the arthropods. They have arisen recently enough that it is feasible to identify the exact DNA changes that produce different patterns. However, the diversity is old enough to represent complex novel phenotypes that have undergone a history of repeated natural selection, so they also offer an opportunity to understand how major changes in appearance are fine-tuned by natural selection. For me, at least, as we begin to understand the mechanisms that underlie this diversity, the sense of awe at their beauty and diversity can only increase.

9.1 *Heliconius* butterflies as a means of understanding the evolution of development

Before delving into the details of how genes control patterns in *Heliconius*, it is worth briefly considering how developmental genetics is studied in other organisms. The best-studied insect by far is the fruit fly, *Drosophila melanogaster*. Our first insights into the genes underlying development in the fly came from laboratory mutants with unusual phenotypes that arose by chance. Some of the most dramatic are the 'homeotic' mutants such as Antennapedia, which has legs where it should have antennae. Subsequently, large-scale mutant screens were carried out using artificial mutagens (Nüsslein-Volhard & Wieschaus, 1980). This approach to development is similar to figuring out how a Formula 1 car works by asking a blindfolded, hammer-wielding mechanic to smash components one by one, and observing the results. Once a mutant has been localized to a particular gene, its function can be further investigated with a variety of genetic tools. Of course, these tools are far more advanced now; one of the primary approaches is genetic manipulation: rather than waiting for genes to break at random, scientists now manipulate them experimentally, turning them on or off in novel contexts.

In contrast, virtually all we know about *Heliconius* has come from the study of naturally occurring variants—the diverse alleles seen in species such as *H. melpomene* are the butterfly equivalent of laboratory mutants in the fly. We can identify the genes underlying these variants by similar methods to those used to find mutant genes in flies. However, the conclusions that come from studying naturally occurring variants are likely to be different for two major reasons. First, the variants seen in natural populations have already been through the filter of natural selection—in our racing car analogy, they are the winners at the end of the Formula 1 season. They are the result of those few genetic changes that actually improved the performance of the car, rather than disrupting it. Unlike the approach used in *Drosophila*, we will not identify all the components of the engine in this way, but rather just those few that can be tweaked to improve performance. Second, the different variants at a single locus that we see in a species such as *H. melpomene* are not the result of a single change in the DNA, but rather the outcome of many generations of evolutionary change. So multiple changes have been made to our winning car over the season. The contrast between the pool of potential variation available (as exemplified by fly mutants) and the select few that are actually favoured by natural selection (as seen in *Heliconius*) will be a theme of this chapter.

9.2 The origins of novelty

A major challenge in evolutionary biology is to understand how novel phenotypes can arise without disruption of existing gene function. In general, organisms are well adapted to their current environment and subject to stabilizing selection. This is especially the case during development, which involves complex genetic interactions that need to be precisely coordinated to produce a functioning organism. The great majority of possible mutations that arise and alter developmental processes tend to be detrimental. This is especially true because a relatively small set of proteins are responsible for coordinating a huge variety of developmental processes.

For example, there are just six major signalling pathways that control much of the development of an animal, which are known as Hedgehog (Hh), Bone Morphogenetic Proteins, Wnt (Wingless/Int1), Steroid Hormone Receptor, Notch, and Receptor Tyrosine Kinase (Hayward, Kalmar, & Arias, 2008). Changing the action of these pathways might occasionally lead to useful new patterns or structures, but is far more likely to damage existing developmental processes.

The problem therefore arises because many, perhaps most, genes have several different jobs to perform. Any mutation in such a gene will often have multiple effects on different aspects of the phenotype. For example, in the fruit fly, mutations in the gene *yellow* were originally recognized because they have a more pale yellow appearance and lack dark melanin pigmentation. However, *yellow* mutant flies are also not very good at courtship behaviour and fail to get matings (Bastock, 1956). In fact, the *yellow* gene functions both in the central nervous system and in controlling pigmentation patterns, so mutations in *yellow* have effects on both pigmentation and behaviour.

This phenomenon is called *pleiotropy*—a single mutation affects several aspects of the phenotype. Evolutionary biologists tend to use pleiotropy in a slightly more specialized sense, to mean the side effects of a mutation with an evolutionarily relevant phenotype. For example, a new mutation might improve the fitness of adult butterflies by conferring greater mimetic resemblance to another species, but if changing that gene also disrupts early larval development, it is unlikely to be favoured by natural selection. The problems in larval development are 'pleiotropic' effects of a mutation that influences mimicry.

A number of mechanisms can overcome the problem of pleiotropy (Figure 9.1). One is gene duplication. By making a new copy of a gene, gene duplication results in one gene copy that can retain existing functions, while the duplicate gene copy can adopt new roles (in a process known as neo-functionalization), or sub-divide multiple existing functions (sub-functionalization). Duplication therefore decouples the different roles that a gene has to perform, allowing those to evolve in different ways (Ohno, 1970).

A good example in *Heliconius* is the evolution of a new UV-sensitive opsin protein that has a novel spectral sensitivity. Gene duplication and divergence allow a broader visual sensitivity through evolution of a new sensitivity in one copy while retaining the ancestral sensitivity in the other copy (Chapter 6 and Briscoe et al., 2010). The same process likely underlies the evolution of chemosensory systems (Briscoe et al., 2013).

Gene duplication can also play an important role in development. For example, gene families such as the Wnt signalling molecules and the Hox transcription factors have arisen through duplication. There are many related proteins that make up the Wnt 'family', which all arose from a common ancestor by gene duplication and divergence. However, gene duplication, which is common in chemosensory gene families, is probably not the main mode of evolution for signalling pathways and morphological development.

Developmental processes more commonly evolve through changes in gene regulation rather than gene duplication or protein-coding evolution (Carroll, Grenier, & Weatherbee, 2004; Gompel et al., 2005; Wray, 2007; Wittkopp & Kalay, 2012). That is, the sequence and structure of the protein remains conserved, but evolution alters where and when the proteins are expressed. This is known as *cis*-regulatory evolution, because it typically occurs through changes in the switches that turn genes on or off, which are located on the same DNA strand as the gene itself (i.e. in *cis*). The process is facilitated by the modular nature of these switches, or *cis*-regulatory elements that control gene expression. Enhancers and repressors control genes on the same DNA strand, often nearby but occasionally millions of base pairs away from the gene they control. They act by binding transcription factors (DNA-binding proteins) that act in combination to turn a gene on or off.

Thus, in order to control its target gene, a particular regulatory module will contain a DNA sequence that binds a specific set of transcription factors unique to a particular time and place during development. The action of such a module may therefore be quite specific—it acts as a switch to turn the gene on in one specific developmental context. This modular nature of gene regulation allows novel

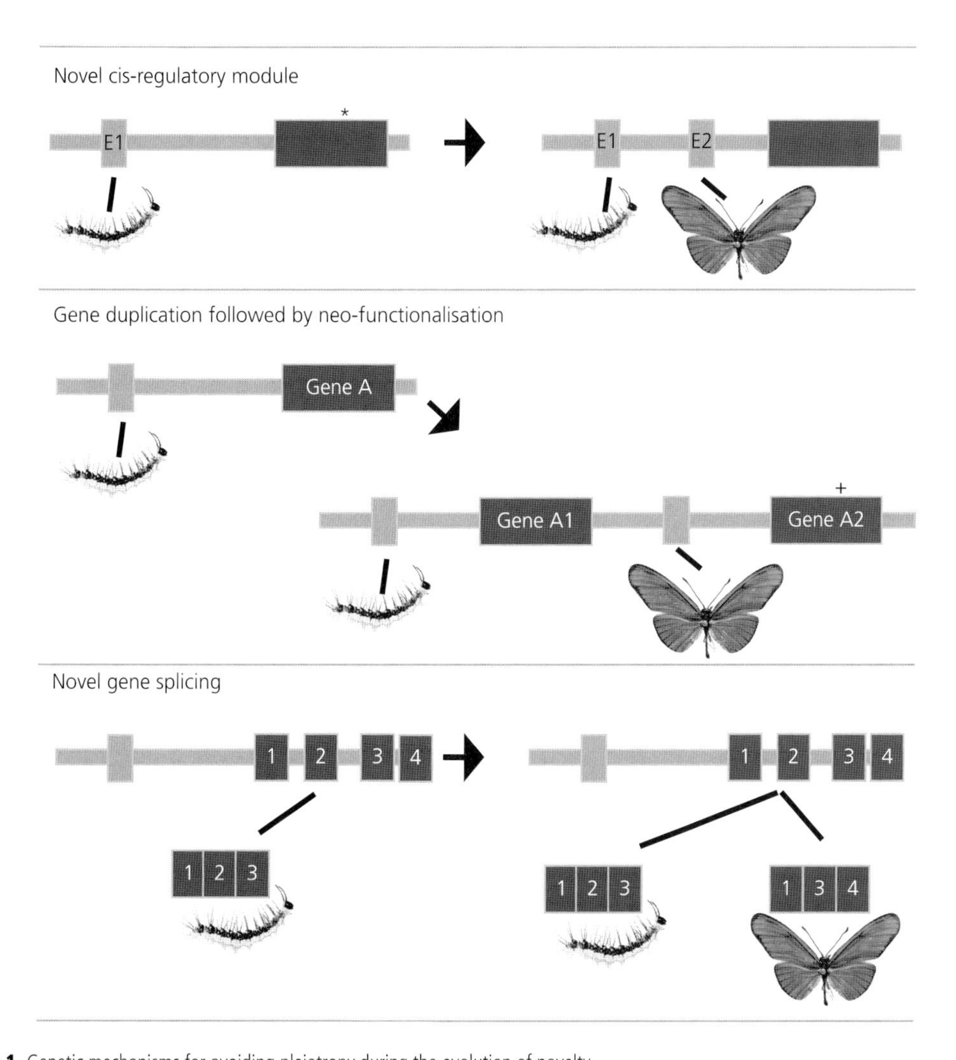

Figure 9.1 Genetic mechanisms for avoiding pleiotropy during the evolution of novelty

Genes can adopt new functions and avoid deleterious effects on their existing role through a variety of mechanisms. Gene duplication provides a new copy of the gene which can diverge in function. Modularity in regulatory control of genes also means that new functions can be adopted without disrupting the current role of the gene.

expression domains to arise for genes without disrupting existing functions. Put simply, genes can evolve a new switch that turns them on in a new context, without changing the switches that control existing functions. Similarly, evolution can tinker with an existing switch to alter one function of a gene, leaving other functions unchanged.

The *cis*-regulatory hypothesis for the evolution of novel morphologies was inspired by the widespread conservation of proteins such as signalling molecules and transcription factors, together with their frequent redeployment in different roles across the diversity of life. However, although proposed as a universal mechanism for developmental evolution, until recently there were relatively few good examples of evolutionary changes controlled in this way (Stern & Orgogozo, 2009). *Heliconius* wing patterns have proven to be an excellent example of evolutionary diversification through *cis*-regulatory evolution.

We are extending the problem of evolutionary novelty. In Chapter 8, I discussed how evolving new mimicry patterns can be difficult in the face of stabilizing selection driven by predator memory.

Here we can see that developmental processes also act to stabilize existing patterns. Novel mimicry mutations face a double challenge—they need to work developmentally and avoid breaking existing developmental processes, but they also need to work ecologically, by providing a fitness increase against the existing force of stabilizing mimicry selection. Before discussing evolutionary novelty in *Heliconius* specifically, I will outline the development of insect wings, and butterfly patterns more specifically. This will provide some background to the formation of wing patterns as a developmental process. A more complete description of wing development can be found in Nijhout's excellent book on the topic (Nijhout, 1991).

9.3 The time course of wing development

All butterfly wing patterns are constructed, like pixels on a computer screen, of a fine mosaic of differently coloured scales. So wing pattern formation is a problem of assigning a single colour state to each scale in a coordinated manner across the wing. During larval imaginal disk development, a complex pattern of spatial information is established across the wing tissue, resulting from a hierarchical process of increasingly complex gene expression patterns. The interacting genes involved in establishing this information are known as a gene regulatory network. Most of what we know about the process has been determined in the fly, but, where they have been investigated, these expression patterns are conserved in the Lepidoptera. In the pupal wing, development diverges between differently patterned forms, with differential development of scale precursor cells, resulting in the differently coloured patches of scales seen across the wing. The key question is, therefore, how can a conserved set of spatial information across the wing early in development be translated into the highly diverse colour patterns seen on the adult wing?

The more general problem of pattern formation—which is common to the development of all organisms and structures—is to go from an undifferentiated, homogenous system to a complex, organized structure. In the wing, the process is simplified by the constraints of the structure itself. There is no evidence for any extensive cell movement, so the formation of the wing pattern occurs in a largely fixed and two-dimensional monolayer of cells. Patterns across the wing must therefore be determined by local cell-cell communication, which can occur by direct contact between neighbouring cells or via propagation of signalling molecules over longer distances.

There is a clear and predictable time course of wing development. Like many adult structures in holometabolous insects (insects that undergo a pupal stage), the wings begin in the caterpillar as bundled sheets of tissue known as imaginal disks. The disks originate as an enlargement of epidermal cells in the larva, which fold inwards to produce the wing disk. During larval development, the disk folds over on itself to produce an inner double layer of cells, that will produce the upper (dorsal) and lower (ventral) wing surface, and an outer layer of cells, which is known as the peripodal membrane (Nijhout, 1991).

The two cell layers that will produce the wing soon become fused together across their whole area, but patterning occurs in each cell layer largely independently, to produce the upperside and underside wing patterns. The two layers are separated by an extracellular matrix known as the basal lamina. There are four imaginal disks corresponding to the four wings, which are located just above the second and third pair of true legs in the thoracic segments. By the fifth larval instar, the venation patterns provide the only visible structure to the wing, but there is considerable spatial information already laid out in terms of gene expression patterns. We have a much clearer understanding of what happens in the fruit fly, so I will switch to briefly describing early patterning events in the fly wing.

9.4 Early wing development in the fly

There is a rich landscape of spatial information across the developing wing in the fruit fly, in the form of localized expression of different genes, which is being established right from the first instar of larval development. The *engrailed* gene is expressed only in the rear (posterior) compartment of the wing from the first instar; *cubitus interruptus* in the front (anterior) compartment. *engrailed* activates another diffusible factor called *Hedgehog*. The Hedgehog protein binds with the receptor Patched, which activates *decapentaplegic* in a narrow stripe of cells along the boundary between the two compartments. Decapentaplegic and the wonderfully named Glass-bottom boat are diffusible factors that establish a gradient moving out from this central stripe in the wing. This gradient therefore provides spatial information across the whole wing that in turn acts on many further factors, some of which produce boundaries that define the location of the wing veins.

The lower (dorsal) and upper (ventral) wing surfaces are defined by expression of *apterous* and *vestigial*, respectively, in the second instar. At the boundary between them, a narrow strip of *wingless* expression forms along the wing margin. This regulates cell proliferation that extends the wing, and in turn activates *distal-less* that defines the distal portion of the wing. Two further factors, *teashirt* and *homothorax*, are initially expressed across the wing but are suppressed by *distal-less*, and therefore become localized in proximal regions near the body.

The fate of *teashirt* and *homothorax* indicates an important principle: turning off expression (by repressors) can be just important as turning on expression (by enhancers). Hence, there is establishment of a proximal-distal axis along the length of the wing, with *homothorax* and *teashirt* expressed near the body (proximal), and *wingless* and *distal-less* expressed at the wing tip (distal). Hindwings are distinguished from forewings by expression of *ultrabithorax* in the hindwing. In summary, this complex gene regulatory network of interacting genes establishes spatial information such that every region of each wing has a unique 'signature' defined by the particular combination of patterning factors expressed there.

9.5 Back to butterflies

Many of these genes and their expression patterns are conserved between flies and butterflies. It has been shown that *ultrabithorax, engrailed, cubitus interruptus*, and *apterous* are all expressed in butterflies in patterns very similar to their known role in flies (Carroll et al., 1994; Galant et al., 1998; Keys et al., 1999; Weatherbee et al., 1999; Brunetti et al., 2001). It seems likely, therefore, that early events in the development of the wing have been conserved throughout the evolution of the holometabolous insects, over some 300 million years of evolution (Nel et al., 2013).

The first visible manifestation of this patterning information are the wing veins. During the last larval instar, butterfly wing disks are invaded by tracheae (breathing tubes) forming the primary tracheal system (Nijhout, 1991). Rather oddly, these tracheae are not exactly the same as those found in the adult wing—instead, a second set of tracheae invades the wing during the mid-pupal stage (known as the secondary tracheae), which will form the wing veins in the adult wing. Most of the patterning of the adult wing vein system is evident in the primary tracheal system, but there are some differences.

The outer ring of tracheae in the late larval wing disk (known as the border lacuna) lies some distance from the outer edge of the disk itself. This outer ring of tracheae marks the future edge of the butterfly wing. Cells outside this region die by programmed cell death (apoptosis), such that the final shape of the butterfly wing is carved out; ornate tails on swallowtail butterfly wings, for example, are formed in this way. The genes *wingless* and *cut* (a transcription factor) have been shown to mark the region to be cut away by this molecular 'cookie cutter' (Macdonald, Martin, & Reed, 2010). Growing extra tissue and then killing it off seems an odd mechanism for shaping a wing, but is actually quite common in the development of complex organs, including our hands, where fingers are cut out of a larger piece of tissue (Gilbert, 2013). It seems likely that the evolution of this mechanism for wing development permitted the remarkable diversification of wing shape seen in the butterflies and moths.

In the prepupa and during pupation, the wing disks increase rapidly in size. Immediately after pupation, they are considerably enlarged but form an extremely delicate structure. Over the first 1–3 days of the pupal stage, this structure thickens through further cell divisions and cell growth, becoming much more robust. The scale cells become arranged in regular rows during the first 1–2 days of pupation. The organization of the rows is controlled by the Notch signalling pathway. Around 16 hours after pupation in *H. erato*, the Notch receptor molecule is upregulated in a grid across the wing, interspersed with regularly spaced cells showing a lower level of Notch expression. In other insects, Notch is used in similar ways to generate regularly repeating patterns through a process of lateral inhibition between adjacent cells that differentiate to express either Notch or its ligand, Delta, both of which are bound to the membrane. Lateral inhibition is a wonderful example of how a simple set of developmental rules can act to establish a complex pattern during development.

Once the grid is established, low-Notch cells in turn upregulate the *achaete-scute* transcription factor that induces scale cell development (Reed, 2004). This parallels similar processes that produce regularly spaced sensory bristles in *Drosophila*, and indeed provides evidence that lepidopteran wing scales are homologous to sensory bristles (Galant et al., 1998).

Subsequent scale cell development involves several cell divisions. The presumptive scale cell divides once perpendicular to the horizontal plane of the wing. The daughter cell towards the inside of the wing degenerates, while the other cell divides again at a 45° angle to the horizontal plane of the wing to produce a proximal daughter cell that will become the scale cell, and a more distal cell that will become the socket cell that anchors the base of the scale (Nijhout, 1991). The scale-building cell then becomes polyploid and very large, before sending out a projection that enlarges and flattens to form the scale. Extension of filaments of the protein polymer F-actin within the cell play a key role in scale cell elongation (Dinwiddie et al., 2014). Slightly later during pupation, F-actin also acts in formation of the longitudinal ridges that give wing scales their unique structures (Dinwiddie et al., 2014).

9.6 Different scale pigments and ultrastructures produce wing patterns

Each wing scale has a single colour. In *Heliconius*, the colours are mainly derived from chemical pigments. Thus, the red, orange, and brown colours are ommochrome pigments, xanthommatin, and its reduced form dihydro-xanthommatin. The difference between red and orange lies in the oxidation state of xanthommatin, with bright red being the reduced form and orange/brown the oxidized form (Brown, 1981; Nijhout, 1991). This difference is partly environmental, as the red colours fade in bright sunlight to orange, which can be clearly seen in older specimens. However, there is also a genetic component, as some patterns are already orange at eclosion, while others are bright red. A genetic locus (Or) has been described that controls this switch (Sheppard et al., 1985), but nothing is known about the gene (or genes) responsible; it has been speculated that it may represent a pigment-binding protein that stabilizes red pigment.

The yellow colour is 3-hydroxykynurenine, which is a biochemical precursor of the red pigments (Brown, 1967). However, during wing development, the pigmentation of yellow and red patches is very different. The red pigment is synthesized in situ in the wing scales as they develop. In contrast, rather than being synthesized in situ, the yellow pigment is taken up from the haemolymph into the scales just before eclosion (Reed, McMillan, & Nagy, 2008). This was demonstrated by injection of radio-labelled 3-OHK pigment into late-stage *H. himera* pupae, where it was taken up in a localized manner into the yellow forewing band region (Reed et al., 2008). Later it was suggested, based on gene expression patterns, that some localized synthesis of 3-OHK may also occur in yellow wing regions, although this has not been confirmed biochemically (Hines et al., 2012). Black patches, which also fade over time from black to a dark brownish colour, are pigmented with melanin (Nijhout, 1991).

Not all *Heliconius* colours result from pigments. The white colours represent an absence of pigment, and are a result of a 'structural' colour produced solely from scale ultrastructure (Gilbert et al., 1988). Similarly, iridescent green and blue colours in species such as *H. cydno*, *H. sara*, and some races of *H. erato* and *H. melpomene* are also structural.

The different pigment colours are associated with particular scale ultrastructures (Gilbert et al., 1988; Janssen, Monteiro, & Brakefield, 2001; Aymone et al., 2013). Three types of wing scales correspond to yellow/white (Type I), black (Type II), and red/orange (Type III) colours (Figure 9.2). The scale types differ in the spacing and frequency of the ridges and cross-ribs in the scale structure. It seems likely that the scale structures evolved to influence the appearance of the *Heliconius* patterns, and likely enhance brightness and hue of the coloured patches. Similar phenomena are seen in flowers, where cell shape enhances the appearance of pigment colours in the petal (Noda et al., 1994).

Figure 9.2 Scale structures associated with differently coloured scales

(A) Type I scales associated with yellow pigments; (B) Type II scales associated with black pigment; (C) Type III scales associated with red pigments. Images courtesy of Ana Aymone (Aymone, Valente, & de Araújo, 2013). Scale bars indicate 2 μm.

In some hybrids, there is a slightly altered scale structure without a change in pigment, illustrating the effect of scale structure alone on appearance. Thus, heterozygotes for the hindwing yellow bar in both *H. melpomene* and *H. erato* show an altered reflectance compared to surrounding melanic regions, which results from melanic scales with an altered Type II ultrastructure (Gilbert et al., 1988).

The association of scale structure with colour is much more tightly regulated in *Heliconius* than in *Bicyclus*. In the latter, experimental manipulation led to a variable relationship between structure and colour, whereas in *Heliconius*, wing damage can lead to patches of different colours (e.g. switching black to red), but the association of colour with scale structure is strictly maintained (Janssen et al., 2001). This suggests a tighter developmental link between pigmentation and scale development in *Heliconius*. Later work in *H. erato* showed that in the silvery-brown regions where the wings overlap, there is variation between individual scales, with ultrastructures corresponding to all three types (Aymone et al., 2013). This suggests that tight control of scale ultrastructure development may be restricted to the coloured wing regions. It would clearly be interesting to investigate colour-structure associations in outgroups to *Heliconius* and other butterflies more broadly to investigate the origins of the close regulation of colour and structure.

Therefore, *Heliconius* wing patterns are not just a result of altered pigmentation. In other organisms, the evolution of novel pigmentation patterns can result from changes in pigment synthesis enzymes. Most obviously, albinism represents a simple failure to produce melanin pigment in the usual quantity. Such changes can be understood from a purely biochemical perspective—break the chemical pathway needed to make the pigment, and the result is a change in the colour of the animal. *Heliconius* patterns are instead controlled by a patterning system that alters both scale structure and pigment in a coordinated manner. They cannot be understood solely through consideration of pigment biochemistry, but instead require an understanding of development, patterning, and spatial information in the wing.

This kind of distinction has led some to argue for a difference between 'physiological' adaptation, resulting from changes in enzymes, and 'morphological' evolution, resulting from changes in developmental patterning (Carroll et al., 2004). Pigmentation straddles the boundaries of this dichotomy: many pigment patterns can be altered by changes in enzymatic processes producing the pigment (physiological), but localization of pigments to particular body regions is a problem of pattern generation (morphological) (Stern & Orgogozo, 2009).

9.7 Variation in pattern is associated with changes in pigment enzyme expression

Expression of pigmentation enzymes is highly coordinated across the wing. The enzymatic pathways that produce ommochrome and melanin pigments are fairly well understood in other insects, notably *Drosophila*, and are likely to be similar in *Heliconius* (Figure 9.3). Genes such as *cinnabar* in the ommochrome pathway (Reed et al., 2008; Ferguson & Jiggins, 2009), and *ebony* and *tan* in the melanin pathway (Ferguson et al., 2011), show expression patterns correlated with wing colours in ways that make sense given their known functions in the fly: put simply, *kynurenine formamidase (kf)*, *vermilion*, and *cinnabar* make red and yellow, *tan* makes black, and *ebony* suppresses black. Hence, *kf*, *ebony*, and *cinnabar* are upregulated in red patches, and *tan* is upregulated in black regions (Reed et al., 2008; Ferguson & Jiggins, 2009; Hines et al., 2012).

Somewhat surprisingly, neither *ebony* nor *tan* were found to be strongly expressed in yellow wing

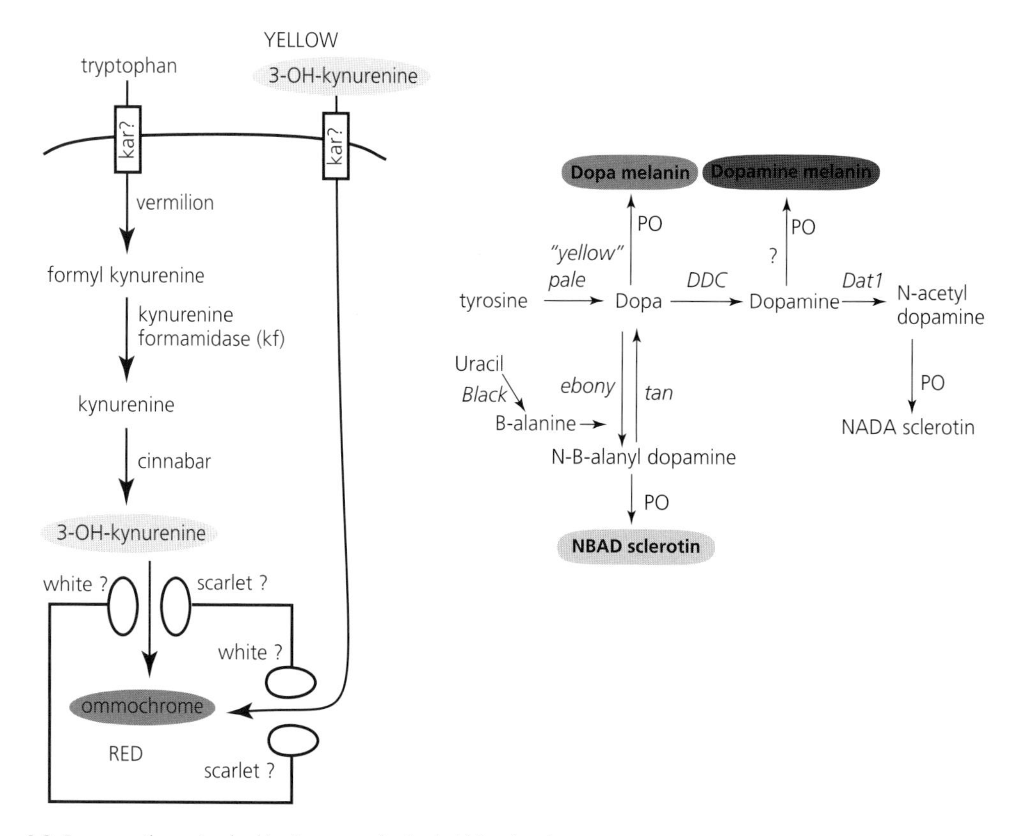

Figure 9.3 Enzyme pathways involved in pigment production in *Heliconius* wings

Left panel: Ommochrome pathway produces red (xanthommatin) and yellow (3-hydroxykynurenine) pigments. It is hypothesized that pigments are stored in a pigment granule and that uptake into the granule may involve ABC transporter molecules white and scarlet. Karmoisin is putatively involved in uptake of precursor molecules into the cell. Right panel: Melanin pathway produces black dopa melanin. PO indicates phenol oxidase and may act to cross-link precursor molecules to cuticle proteins. Dopa melanin and dopamine melanin produce the black colour on wings but are also a structural component of the insect cuticle. NBAD sclerotin is pale yellow whereas NADA sclerotin is translucent. Derived from Reed et al. (2008), Ferguson and Jiggins (2009), Ferguson, Maroja, and Jiggins (2011), Hines et al. (2012).

regions (Ferguson et al., 2011). An early study indicated that *vermilion* was upregulated in the forewing band in *H. erato*, but this was not the case in *H. melpomene*, nor was it supported in a later analysis of *H. erato* (Reed et al., 2008; Ferguson & Jiggins, 2009; Hines et al., 2012).

Other putative pigmentation genes also show localized expression across the wing, including members of the *yellow* gene family, the melanin gene *Dat1*, and putative ABC transporter genes that may be involved in ommochrome pigment uptake (Ferguson et al., 2010b; Hines et al., 2012). The expression patterns are surprisingly repeatable across species, with similar patterns found in the co-mimics *H. erato* and *H. melpomene*. In summary, there is strongly coordinated control of gene expression in production of the different scale types and colours.

In all of these studies, associations between colour and pigment enzyme expression are seen both between different wing regions in the same butterfly, and in the same wing region of differently patterned individuals. For example, wing pattern races within *H. melpomene* with a forewing red band show a region of *cinnabar* expression in the forewing band that is absent in mimetic races with a yellow forewing band. In Chapter 7, I discussed the genetic switches that regulate these changes in patterns. Now, we have seen that the expression of pigmentation enzymes also differs between these pattern forms, correlating with patches of colour on the adult wing.

There are thus two hypotheses for these observations. The pigmentation enzymes themselves may correspond to the loci that switch between different patterns; that is, differences in expression are controlled in *cis* by patterning loci. Alternatively, the differences in expression of pigmentation enzymes could be a downstream result of changes in the 'switch' loci; that is, differences in expression are controlled in *trans*.

It turns out that the latter is the case. None of the pigmentation enzymes are located at the regions of the genome that control variation in wing patterns (the red, yellow, shape and colour loci). It is therefore not genetic variation at these enzyme genes that directly controls variation in pattern. Rather, the pigmentation genes lie downstream of the genetic switches (Gilbert et al., 1988; Reed et al., 2008; Ferguson et al., 2011). Hence, genetic variation in the red locus, *optix*, regulates the expression of *cinnabar*, *ebony*, and *tan*, which then produce regions of red and black pigment.

In *Heliconius*, the timing of scale cell enlargement and maturation differs depending on the eventual colour pattern (Figure 9.4). Differences in the timing of development and gene expression are known as heterochrony, which can be an important mechanism for cellular differentiation.

The time course of scale development has been described in most detail in *H. erato*. At 60 hours after pupation the socket cells arise, at 72 hours the first scale elongation is evident, and by 96 hours the wing surface is covered with immature scales (Aymone et al., 2013). (Note that the exact timing is dependent on temperature and species.) At 120 hours—known as the pre-ommochrome stage—red and yellow/white-fated cells become mature earlier than the black-fated scales, and are evident as

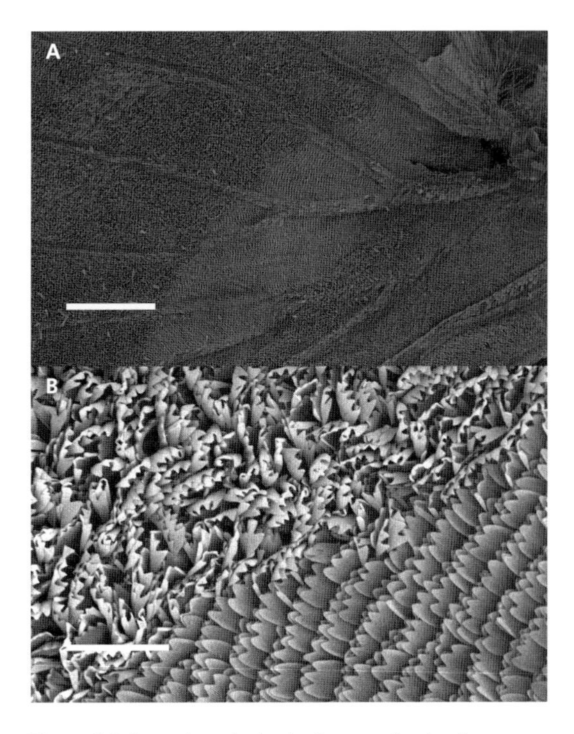

Figure 9.4 Heterochrony in the development of scale cells

The boundary between the hindwing yellow bar and the black wing region is shown in *H. erato* during late pupal development at two magnifications. The yellow scales have already matured and are rigid (lower right of each image), while the black scales are still soft and flexible (upper left of each image). Scale bars are 1 mm (A) and 100 μm (B). Photo credits to Ana Aymone (Aymone et al., 2013).

opaque regions in the wing (Ferguson & Jiggins, 2009; Aymone et al., 2013). The first pigments to develop are in red regions, which produce red colour evenly across the scales; this is termed the ommochrome-only stage. Subsequently, the melanin pigments appear first in the centre of the distal wing region, then spread in a wave towards the proximal and marginal wing regions. Finally, just before eclosion, the yellow wing regions turn from white to yellow as 3-OHK pigment is taken up from the haemolymph into the wing (Reed et al., 2008). In summary, different coloured patches on the wing are associated with differences in the timing of development, resulting in scales with different ultrastructure and ultimately pigment composition.

So far I have described the two ends of the development of a wing. At the beginning, a highly conserved set of patterning factors establish spatial information in the wing—in fact, they are so highly conserved that we can predict expression patterns in a butterfly from studying the wing of a fruit fly (which obviously looks completely different). At the end of the process, we have complex differences in spatial arrangement of differently structured and coloured scales that differ dramatically even between closely related populations within *Heliconius*. These differences are associated with changes in gene expression and the timing of scale development.

But what happens in the middle? How is conserved spatial information translated into the diversity of butterfly wing patterns?

We are still a long way from a complete understanding of this process, but major advances have come about from studying the loci identified in genetic studies of *Heliconius*. I will first describe two hypotheses about the development of *Heliconius* wing patterning that came before the modern era of molecular genetics. Then I will consider the molecular identity, mode of action, and origins of the major wing pattern loci described in the previous chapter.

9.8 Inferring the origins of patterns—the nymphalid ground plan

Across the nymphalid butterflies, it has been proposed that there is a 'ground plan' that underlies wing patterning (Figure 9.5). The idea was originally proposed by Schwanwitsch (1924) and subsequently Süffert (1927), later expanded by Nijhout (Nijhout & Wray, 1988; Nijhout, 1990, 1991). Many nymphalid patterns are composed of repeated elements often associated with wing venation, and there are clear homologies between such elements across different nymphalid species. Many of these elements form symmetry systems of repeated pattern elements. The simplest of these are parallel lines crossing the wing anterior–posterior, which are hypothesized to have formed around morphogen sources. The symmetry inherent in many of these elements is therefore thought to result from diffusion of a morphogen from a central source, producing symmetrical patterns around that source.

The hypothesis is supported by the involvement of Wnt morphogens in nymphalid wing patterning. The Wnt (or wingless) family of proteins are diffusible morphogens that act during development to signal between cells and establish patterns. For example, in wing discs dissected from fifth-instar larvae of *Euphydryas chalcedona*, expression of Wnt genes marks the development of the basal (*WntA*), discal (*Wnt1/Wnt6/Wnt10*), central (*WntA*), and external (*WntA*) symmetry systems (Figure 9.5) (Martin & Reed, 2014). These same genes are found to be associated with pattern elements in other species, including *Vanessa cardui* and *Agraulis vanillae*, demonstrating some level of homology between ground-plan elements across these species. The finding supports the nymphalid ground-plan hypothesis for homology across the nymphalid butterflies (Martin & Reed, 2014).

Another well-established element of the ground plan for which gene expression data supports widespread homology are the eyespots (ocelli) found around the margins of the wings in many butterflies. The origins and evolution of eyespots has been comprehensively reviewed by Antonia Monteiro (Monteiro, 2015). A remarkably delicate set of experiments has shown that eyespots result from morphogen diffusion gradients. Ablation of cells in the eyespot focus early in pupal development can remove eyespots completely, and translocation of just a few focal cells from one eyespot to another region of the wing can result in formation of a new eyespot (Nijhout, 1991; French & Brakefield, 1995). This confirmed the hypothesis

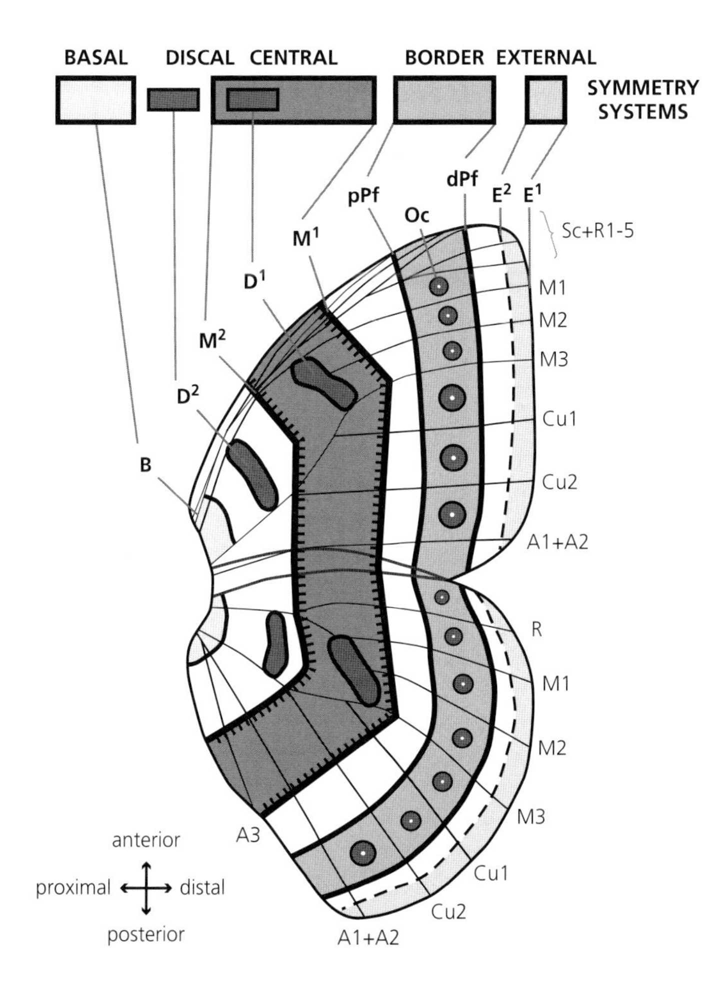

Figure 9.5 Nymphalid ground plan

An updated version of the Nymphalid ground plan (Martin & Reed, 2014) derived from the nomenclature of Schwanwitsch (1924). Elements are Basal (B), Discal (D^1 and D^2), Medial (M^1 and M^2), proximal and distal parafocal (pPf and dPf), Ocelli or eyespots (Oc), and External (E^1 and E^2). Wing veins are labelled to the right using standard nomenclature. Image courtesy of Arnaud Martin.

that eyespots result from outwards diffusion of a signal—perhaps a diffusible morphogen—from a central focus, which fits nicely with Nijhout's interpretation of the nymphalid ground plan (NGP). It remains unknown whether this apparently long-range signal actually acts across many cell diameters, or is the culmination of many short-range signalling interactions.

The NGP hypothesis therefore provides a comparative framework for studying wing patterns across the nymphalid butterflies. It recognizes similarities across the bewildering diversity of patterns we see in nature. However, there are two extreme interpretations that could be placed on the NGP.

Pattern elements across the butterflies could be truly homologous in the sense of sharing a common ancestor with the same pattern element. Alternatively, different lineages could evolve all their pattern elements independently, but rely on a shared set of regulatory information common to the development of all butterfly wings. Thus, because of the common conserved patterning of the wing, there are only limited permutations of pattern elements that can evolve easily, so we recognize NGP elements not because of true homology but rather because of this shared landscape of wing development.

It seems likely that the truth lies somewhere between these two extremes. For example, ancestral state reconstruction of eyespots across the nymphalids has shown that it is very likely that the common ancestor did not have all the eyespots in the NGP (Monteiro, 2015). At least for eyespots, therefore, the NGP does not represent an ancestral pattern.

Homology can also be investigated by studying genetic control of pattern elements. A number of genes, including *distal-less* and *spalt*, are expressed in the eyespot focus in late-stage larvae, and comparative studies across many species show that different combinations of genes are expressed in different butterfly lineages (Monteiro, 2015). Again, the differences in the genetic control of eyespots across lineages show that all eyespots are not created equal. The true extent to which NGP elements are reinvented by convergent evolution remains to be tested across the butterflies. This is not to deny the usefulness of the NGP, but rather to recognize the complexities inherent in its interpretation.

Nijhout extended the NGP hypothesis to *Heliconius*, and at this point the homologies are far less clear. Some of the outgroup heliconiines, notably *Agraulis vanillae* and *Dione* species, have a typical 'fritillary' pattern; homologies can be inferred between these patterns and ground-plan elements (Nijhout & Wray, 1988; Nijhout, 1991). Gene expression analyses show that Wnt genes associated with NGP elements are also expressed in association with pattern elements in *Agraulis vanillae*, consistent with the NGP hypothesis (Martin & Reed, 2014).

The majority of *Heliconius* patterns, however, with their blocks of red, orange, black, and yellow, are more difficult to fit into the NGP paradigm. Nijhout suggested that the small repeated pattern elements of the NGP have become enlarged to form the black and red patches on *Heliconius* wings. In this view, many of the contiguous areas of colour on the wing are actually fusions of multiple distinct pattern elements.

As with the NGP more broadly, Nijhout's work provides a framework for asking questions about the origins of *Heliconius* patterns. Have they arisen from elements that are homologous to existing elements in other nymphalid butterflies, as Nijhout proposed, or is there something unique in the development of *Heliconius* patterns that have freed them from the constraints of their ancestors and allowed the great diversification we see today? It seems likely that the bold blocks of colour in the *Heliconius* wing are an evolutionary innovation with only partial homology to the serially repeated elements of the ground plan.

9.9 An alternative model—Gilbert's windows and shutters

In contrast to Nijhout, Gilbert (2003) has emphasized the distinctness of *Heliconius* patterns and raised the question of 'why *Heliconius*, but not related genera, displays such a variety of patterns'. After extensive crosses, mainly between *Heliconius melpomene*, *H. cydno*, and *H. pachinus*, he devised a scheme of developmental 'windows' and 'shutters' to explain pattern variation. The phenotypes generated in a series of artificial 'hybrid swarms' allowed Gilbert to make inferences about the available patterning landscape, focusing mainly on the melpomene clade. Genetic variation from *H. melpomene* (presumably the red forewing band allele at the *optix* locus) is inferred to turn the forewing shutter red, 'revealing' a developmentally defined region on the wing.

A series of rare phenotypes produced by repeated backcrossing—that must result from unusual double recessive genotypes and epistatic interactions not normally seen in hybrids—allowed Gilbert to infer the position of developmental boundaries across the wing. The wing is seen as a window—regions outside the window are always black, but within the window, 'shutters' can turn particular regions from yellow/white to black or red.

The model has some commonalities with that of Nijhout. Most notably, black regions are seen as a composite result of different patterning elements—in this case, black areas can be both shutter and wall, with white and yellow as background (or window). However, the models differ in that Gilbert's windows and shutters are not vein-dependent—indeed, quite the opposite, as they cut across vein boundaries.

The windows-and-shutters model is actually quite consistent with modern developmental genetics. A common mechanism for patterning is to turn a gene on across a broad area and then refine expression by repressing it in particular regions. Thus, it would be very plausible for the window to represent an initial region of forewing band expression, which is then fine-tuned by turning off regions of expression (i.e. shutters) to produce the eventual pattern.

Unfortunately, this hypothesis has not been placed in the context of the genetic nomenclature of

John Turner, Phillip Sheppard, and others. In particular, both the positioning of the window itself and the shutters can vary between species and races, making it hard to make specific predictions about the action of the *optix, WntA*, and *cortex* alleles in relation to the shutters and windows. (It is said that John Turner, on visiting Austin, Texas, was amazed at the insectaries that Larry Gilbert had established, so it is unfortunate that the two of them never got together to test hypotheses regarding the genetic and developmental basis of wing patterns.)

I will describe what we know about the development of *Heliconius* patterns, before returning to the NGP and the windows/shutters models in the light of our current understanding. However, before that, a quick diversion into methods.

9.10 Methods for visualizing gene expression

I have already described some of the patterns of gene expression that play out across the developing wing, but so far have not explained how this information has been obtained. However, the methods have been so integral in advancing our understanding of *Heliconius* wing patterning from populations and crosses into the realm of developmental biology, that I will take a quick diversion to describe two of the key methods for detection of gene expression patterns.

The first is in situ hybridization of RNA probes, which specifically bind to the gene of interest and mark the location of mRNA molecules. The probes are labelled with a marker that can be visualized under a microscope, either by means of a chemical reaction or by fluorescence. Probes can be readily designed for any gene if the DNA sequence is known, which makes the technique readily applicable to different genes. However, the protocol is long and involved, and prone to artefactual signals. Especially during pupal stages, probes can produce a signal specific to particular wing regions that is an artefact of RNA molecules binding to differentiated scales across the wing.

A more robust technique is immunohistochemistry, in which fluorescent-tagged antibodies bind specifically to proteins and mark their presence or absence across the wing. The protocol is easier and more robust than in situ hybridization. Signals can be visualized readily using confocal microscopy at sufficient resolution to confirm precise cellular localization of the signal (e.g. transcription factor proteins are expected to be found primarily in the nucleus). Often antibodies developed for other insects also work in butterflies, but if this is not the case, generation of a specific antibody can be challenging, and is the major limiting step for application of this technique to new genes (Martin et al., 2014). Note that the two techniques are complementary, as they target different molecules (mRNA and protein).

Nonetheless, there are constraints on what is possible. Both techniques work well in late larval and early pupal wings, but at later pupal stages, the wing cuticle thickens and becomes impermeable to RNA probes and antibodies alike. Also, chitin is autofluorescent, and as it builds up, it increasingly interferes with the use of fluorescent probes to visualize RNA or protein. Studies of later pupal development have instead relied on dissection of wings, and quantification of mRNA molecules using PCR amplification for specific genes or transcriptomic analysis of all expressed genes (Ferguson & Jiggins, 2009; Hines et al., 2012). Fortunately, most of the critical formation of wing patterns seems to be complete by the time the wing tissue becomes inaccessible. Similarly, and perhaps more critically for understanding pattern formation, the prepupa and very early pupa are so delicate that wing dissection is challenging. These stages have not yet been studied, and thus there is a gap in our knowledge at a potentially critical stage in wing development.

9.11 Developmental evidence for the identity of patterning genes

As outlined in the previous chapter, genetic linkage maps and population genetic studies have been successful in localizing narrow regions of the genome that regulate major differences in patterning between races and species. Combining this population genetic evidence with developmental studies allowed identification of the exact genes involved. I will outline the evidence supporting the identity of the known wing patterning genes, and emphasize the evidence that they are primarily regulatory mutations, therefore supporting the *cis*-regulatory hypothesis for morphological evolution. This work is beginning to address the question of how changes in DNA sequence regulate the appearance of a butterfly wing.

9.12 The red gene optix

The first gene to be definitively identified was *optix*, a transcription factor at the red locus (Reed et al., 2011). A combination of population genetic association and expression data implicated *optix* as the most likely candidate gene within the red locus. At the time, *optix* itself showed the strongest signal of population differentiation, although it is now clear that a region of non-coding DNA some 60 kb to the 3′ end of *optix* is most closely associated with wing patterns in natural populations (Nadeau et al., 2012; Supple et al., 2013; Wallbank et al., 2016). There is also a remarkable correspondence between patterns of *optix* expression in developing pupal wings and red patches on adult wings. This was initially demonstrated using in situ hybridization and subsequently confirmed using immunohistochemistry (Figure 9.6) (Reed et al., 2011; Martin et al., 2014). Optix is now known to be localized in association with red wing patterns in *H. melpomene* from about 24 hours after pupation right through to day 7 when pigments are beginning to appear, and in *H. erato* from at least day 3 (Hines et al., 2012; Wallbank, personal communication). Importantly, races differing in expression pattern show no consistent differences in protein-coding sequence at *optix*. This implies that the difference between races is purely regulatory. The red locus therefore encodes precise instructions that turn on *optix* in specific regions of the wing, determining their fate as red-pigmented scales.

In the fruit fly, *optix* is involved in a wide range of developmental processes, including (but not limited to) eye development, and the same is likely true in *Heliconius*. The ancestral function of the Six gene family, to which *optix* belongs, is in specification of the forebrain, so a role in the eye may represent a retention of part of this ancestral function (Seo et al., 1999). It is therefore unsurprising that the coding sequence of the gene is highly conserved across the insects. The identification of *optix* fits with the paradigm of a conserved protein molecule with multiple

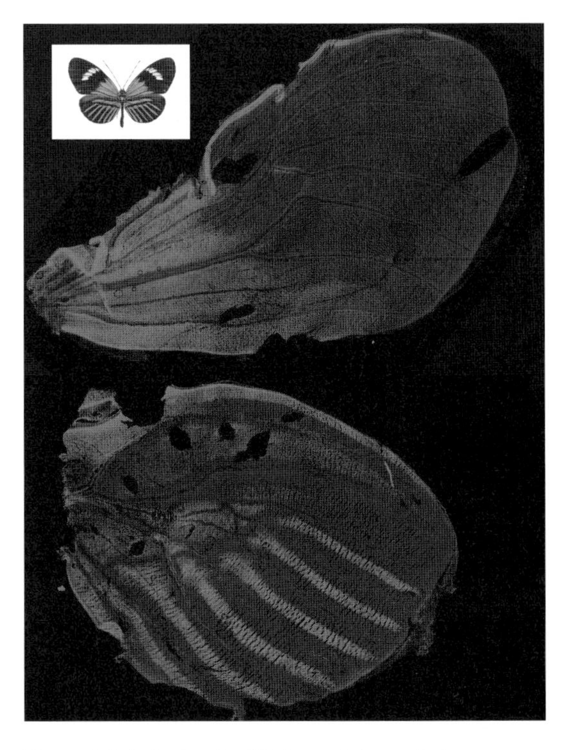

Figure 9.6 Antibody stain of Optix protein in pupal wings (see Plate 30)

The Optix protein is localized during pupal wing development in regions that show perfect concordance with red patches in the adult wing. This is an individual of *H. elevatus pseudocupidineus*. This image was generated by Richard Wallbank.

functions, that is deployed in a new developmental scenario by means of *cis*-regulatory evolution (see earlier).

A second gene at the red locus also shows patterns of expression associated with some but not all of the red wing pattern elements. A *kinesin* gene is expressed in red forewing bands, but not in red dennis or ray patterns. *H. melpomene* and *H. erato* both show expression of *kinesin* in red, but not yellow, forewing bands (Pardo-Diaz & Jiggins, 2014). The most strongly differentiated DNA sequence between races at the red locus lies almost midway between *optix* and *kinesin*, so this region might regulate expression of both genes. Although we still cannot rule out involvement of additional genes in pattern specification, for the rest of this chapter I will assume that *optix* is the sole causal factor of red patterns, bearing in mind that this has yet to be demonstrated experimentally.

9.13 Optix waiting in the wings

Expression of the *optix* gene is associated with red patterns in almost all *Heliconius* that have been tested to date (Martin et al., 2014). The only apparent exceptions are *H. hecale fornarina*, where red patterns show no evidence of *optix* expression, and the race *H. m. plesseni*, where the red/white forewing patch shows expression of *optix* in both red and white-fated scales. Although initially it seemed that *optix* expression was associated only with red patterns within *Heliconius*, later work suggested that similar associations are found in other Nymphalid butterflies as well (Arnaud Martin, personal communication). It therefore remains to be seen how widespread the colour-patterning function of *optix* is across the butterfly phylogeny. In addition, in all heliconiine species tested so far, *optix* is associated with the precursor cells for a set of spear-shaped 'wing coupling' scales that lie in the region of overlap between fore- and hindwing (Martin et al., 2014). These scales are thought to have a role in coupling the wings together during flight. The additional role in specification of wing scale morphology in the Lepidoptera may have predated its co-option into specification of wing pattern.

In *Agraulis vanillae* and *Dryas iulia, optix* is also associated with the specification of a small group of cells that go on to form a set of scales along three of the forewing veins (Cu1, Cu2, and 2A). These are only found in males, and appear to protect pheromone-producing scales found in the same wing region. The specification of these sex-specific scales is a derived role that has arisen in addition to the roles in red patterns and coupling scales (Martin et al., 2014).

Hence, *optix* has various roles in wing scale development including the control of wing colour patterns. Optix is expressed in the wings of *Drosophila*, so a role for the gene in wing development may be ancient among the insects. It is quite common for genes to be involved in various aspects of the development of a structure at slightly different times and places, and this offers some insight into how genes can adopt new functions. At some point, *optix* must have come under the control of switches that turn it on specifically in scale precursor cells. Because its expression is localized in the wing, it would have also involved interaction with transcription factors that define spatial information across the developing wing.

In addition, the role of *optix* in scale specification implies that the Optix protein evolved downstream regulatory links that allowed it to control scale morphology (presumably this involves regulating genes that control actin polymerization and therefore alter scale structure). In the case of red patterns, regulation also includes either direct or indirect regulation of pigmentation genes, including *cinnabar, ebony*, and *tan*. So once *optix* had evolved one of its functions in scale specification, it would have been pre-adapted to adopt new and related functions, eventually leading to the related roles in defining wing coupling scales, red pattern elements and, most recently in evolution, sex-specific scales in the *Agraulis/Dryas* group. An alternative suggestion, that *optix* was co-opted from an ancestral role in regulating ommochrome pigments in eye development (Monteiro, 2012), seems less likely given the multiple roles of *optix* in wing scale specification unrelated to pigmentation. Of course, this doesn't imply any forward planning on the part of evolution; rather, when selection favoured a new mode of pattern specification, *optix* was already waiting in the wings, ready to fulfil this function.

9.14 The origins of novelty at the *optix* locus

We can begin to understand more clearly this process whereby genes take on new functions by studying more recent evolutionary changes. In particular, we can ask, How do new red patterns arise at the *optix* locus?

Population genetic patterns indicate that dennis and ray patterns are derived from red bands in *H. melpomene* and its relatives. We used a large set of DNA sequence data to find exactly which sequences are most associated with these pattern elements. This is made possible because there are populations with one of these phenotypes but not the other. The Guiana races, *H. melpomene meriana* and *H. erato amalfreda*, both have dennis but not rays. In the *H. melpomene* clade, the polymorphic *H. timareta timareta* in Ecuador has rays but not dennis. And I was excited to catch a rare hybrid from Ecuador that had dennis but not rays. Analysis of the sequences of these populations and hybrids has identified small regions of a few thousand base pairs that are consistently associated with either dennis or ray patterns. They are adjacent to one another and

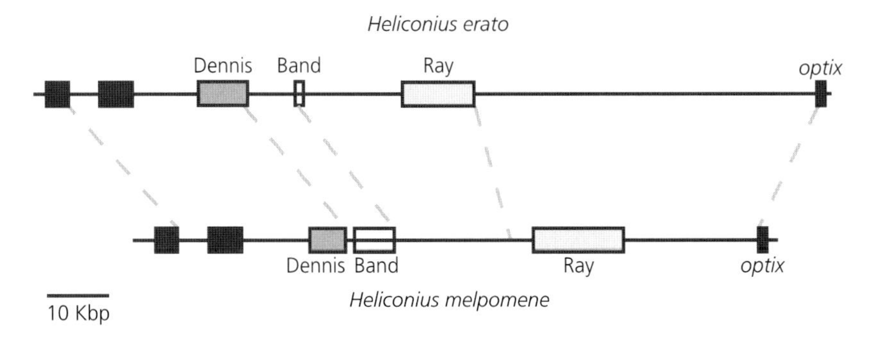

Figure 9.7 Relative location of dennis, ray, and band enhancers in *H. erato* and *H. melpomene*

The genome of the two species is depicted to scale in the region immediately downstream of the *optix* gene. The location of enhancers was mapped using natural recombinant phenotypes (such as *H. m. meriana*, which possesses dennis but not ray). The dennis enhancer has evolved independently in the two lineages in exactly the same position, while ray is almost orthologous but slightly displaced, and band is in the same region but differs in position relative to ray. This shows how enhancers for mimetic patterns can arise repeatedly from the same ancestral elements, or arise de novo in different positions. Image courtesy of Joe Hanly.

located between 60 and 110 kb from *optix* (Figure 9.7). These regions are presumed to act as modular enhancers, which turn on *optix* in dennis and ray regions of the developing wing, respectively.

How have these enhancers acquired the ability to turn on *optix* in such specific and localized wing regions? In the case of the dennis patch, Richard Wallbank and Carolina Pardo-Diaz in my group have identified a putative factor that may act upstream of *optix*. The gene *homothorax* is expressed in the basal region of the forewing. This expression pattern is conserved across all *Heliconius* that have been tested so far—it appears to be part of the conserved landscape of patterning the wing. Indeed, the gene is already known to have a similar function in *Drosophila*, where *homothorax* patterns the base of the wing imaginal disk, a region that will become the wing hinge.

In butterflies that are going to produce a red dennis patch on their wings, *optix* is also turned on in a region exactly coinciding with the cells expressing *homothorax*. All that is required to evolve a red patch on the base of the wing is a genetic change in the regulatory sequence of *optix*, such that the Homothorax protein starts binding to the DNA in that region of the genome. This could in principle be as simple as acquisition of Homothorax binding sites in the *cis*-regulatory module of *optix*—although in reality it is likely to be more complex than this. Because Homothorax binding sites are simple sequence motifs, it might have involved relatively few mutations.

Optix is already controlling all the molecular machinery needed to produce red scales. And the base of the wing is defined by Homothorax in all *Heliconius* butterflies. So all that is needed to produce a new patch of red on the wing is to bring the two together. Turning on *optix* at the right time point in development is likely to be sufficient to produce a new red patch. So a simple change in DNA sequence, involving acquisition of binding sites for an existing transcription factor, can lead to the evolution of a new red patch on the wing (Figure 9.8). Although we still

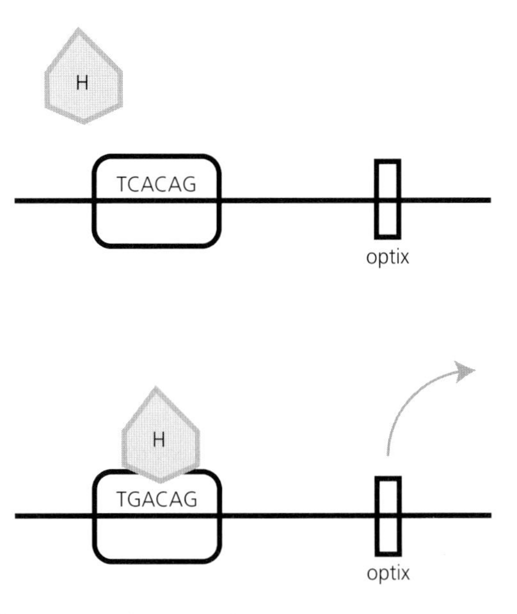

Figure 9.8 Novel interactions between genes can generate new patterns

H is an existing patterning factor that defines a particular wing region (such as Homothorax in the proximal dennis patch). Alteration of the DNA sequence in an enhancer region of *optix* can lead to binding of H, expression of *optix*, and a new red patch in the wing. A DNA sequence change can therefore alter patterns of gene regulation and lead to novel phenotypes.

lack direct evidence for Homothorax regulating *optix* expression, in principle the acquisition of a new link in the gene regulatory network provides a relatively straightforward mechanism by which an evolutionary novelty could arise.

9.15 The shape locus and Wnt signalling

The second major locus to be identified was the shape locus, which corresponds to the gene *WntA* (Martin et al., 2012). Again, the evidence came from a combination of genetic linkage mapping based on laboratory crosses, a population genetic signal of divergence between races, and differences in expression patterns of the *WntA* gene in developing wings (Martin et al., 2012; Papa et al., 2013; Gallant et al., 2014; Huber et al., 2015). In this case, the linkage map data was highly informative, especially in *H. erato*, where several large families narrowed down a region of the genome that included only the *WntA* gene and part of an adjacent chitin synthase gene. Unlike *optix*, *WntA* is expressed earlier in development in the final larval instar, where it also shows very diverse expression patterns associated with wing pattern phenotype (Figure 9.9).

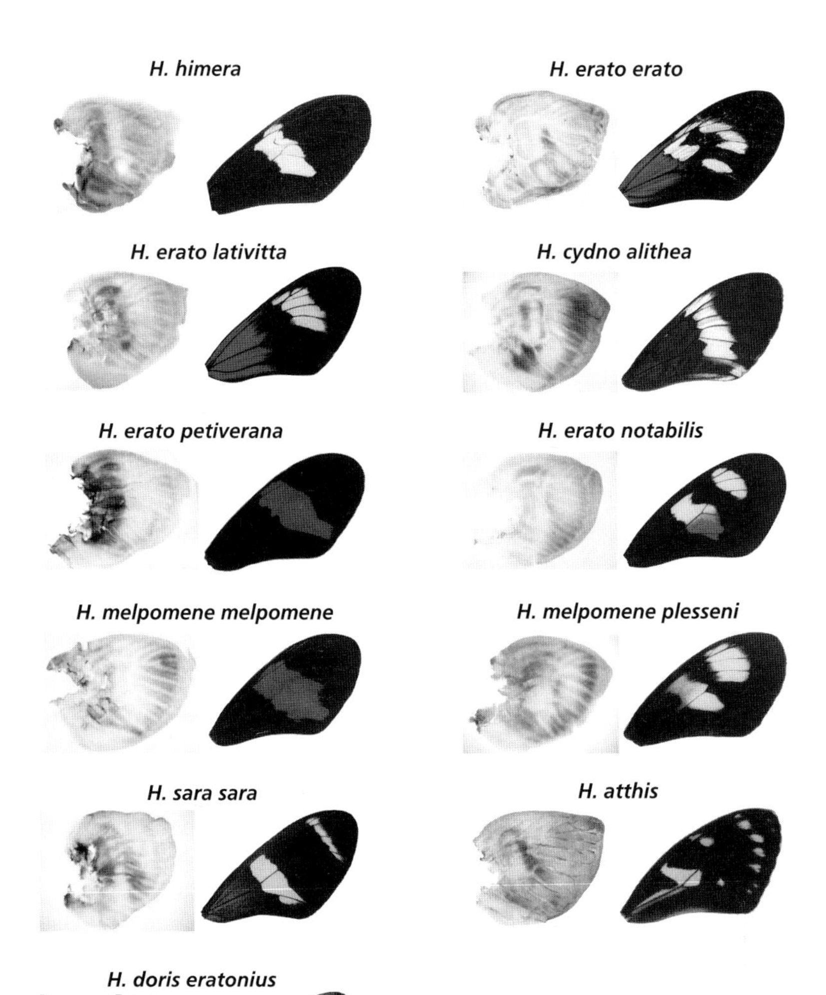

Figure 9.9 Expression patterns of *WntA* in larval wing discs of *Heliconius* species

WntA is expressed in larval wing discs in regions that are fated to become black in the adult butterfly. Note that not all black regions show *WntA* expression. Images courtesy of Arnaud Martin (Martin et al., 2012).

In general, *WntA* is expressed in association with black regions in the centre of the forewing, consistent with the known role for this locus in controlling forewing band shape in crosses. However, not all black wing regions show *WntA* expression, so there is not a one-to-one relationship between a wing colour and expression of *WntA*, as is the case for *optix* and red patches.

Additional evidence for the role of *WntA* comes from experiments that disrupt the normal functioning of the Wnt signalling pathway. Heparin is a molecule that binds Wnt family ligands and promotes their mobility through tissue. Injection of heparin into the wing tissue of developing *Heliconius* pupae leads to changes in adult wing pattern that are somewhat comparable to genetic effects of the shape locus (Martin et al., 2012). Notably, heparin leads to an increase in the extent of black patterning in the forewing, obscuring yellow and white pattern elements, a rough phenocopy of alleles at the shape locus that break up forewing band patterns.

The effects of heparin are not specific to WntA, and the wingless signalling pathway has long been known to be involved in wing patterning in butterflies, so these effects might represent a rather generic disruption of wingless patterning rather than a specific disruption of WntA action. Nonetheless, in combination with genetic mapping and expression studies, these experiments build a compelling case for a role of WntA in forewing pattern specification.

It has long been predicted that the evolution of patterns on animals might involve changes in diffusible molecules, or morphogens (Gilbert, 2013). However, to date there are surprisingly few examples of evolutionary changes regulated via morphogens. WntA is interesting, therefore, because it is a very different kind of molecule from Optix. Optix is a transcription factor that binds DNA in order to regulate expression of other genes. WntA is likely to act as a morphogen, a small diffusible glycoprotein that can transmit information between cells, albeit with the caveat that its mode of action remains to be demonstrated in *Heliconius*. Therefore, we anticipate that WntA is involved in establishing pattern across the wing, whereas *optix* is controlling the cellular response to that pattern. One comparable case in which a morphogen has been implicated in evolution of a new pattern concerns the origins of the many small wing spots found on the fly *Drosophila guttifera*. The multiple dark spots along its wing veins are marked by expression of the gene *wingless* (also known as *Wnt1*, a gene closely related to *WntA*) (Werner et al., 2010). Analysis of related species suggests that first a link evolved between *wingless* and pigmentation, such that existing patches of *wingless* expression became more strongly pigmented. Subsequently, novel areas of *wingless* expression formed in the wing to generate the characteristic spotting pattern in *Drosophila guttifera*. Changes in the expression of *wingless* are at least in part due to gain of *cis*-regulatory variation at the *wingless* locus (Koshikawa et al., 2015). As we have also seen for *optix*, this evolutionary tinkering with existing pathways—sequentially adding new links between existing genes that allow them to take on new functions—is typical of how evolution proceeds.

Similarly, by mapping evolutionary changes in natural populations, we know that the changes at WntA are controlled by genetic differences at the *WntA* locus itself (i.e. in *cis*). Alongside the work on *wingless* in *Drosophila*, this is therefore one of the first examples of an evolutionary change in patterning that has been shown to be caused by genetic changes at a putative morphogen gene itself. This highlights how studies of *Heliconius* can complement those in *Drosophila*. While *Heliconius* lack the remarkable genetic tools available in *Drosophila* to dissect how genes act during development, they nonetheless possess a much greater diversity of young allelic variation, which can offer insights into how and why particular genes are targeted by natural selection.

9.16 Patterning role for WntA is widespread across Nymphalid butterflies

WntA is associated with wing pattern across many butterflies, and it has been the target of selection in groups other than *Heliconius*. Allelic variation at *WntA* is associated with adaptive pattern variation in *Limenitis arthemis* (Gallant et al., 2014). *Limenitis arthemis arthemis* is the ancestral form of this species, which has a white transverse stripe—the white admiral butterfly that will be familiar to many

North American readers. *Limenitis arthemis astyanax* is a derived form that mimics the pipevine swallowtail, *Battus philenor*, in the southern USA. The white-band pattern difference between these species maps to the *WntA* locus. Surprisingly, unlike in *Heliconius*, there is no marked difference between the two *Limenitis* races in patterns of *WntA* expression, at least as detected by in situ hybridization in larval wing discs. While it is always hard to rule out the possibility that there is an expression difference at another time point, or a quantitative difference not detectable by the hybridization technique, this raises the possibility that there is a different mechanism of regulating pattern differences. Gallant et al. (2014) speculated that this might be due to differences in expression of isoforms of the gene due to differential splicing (see Figure 9.1).

Limenitis and *Heliconius* diverged about 65 million years ago, so the involvement of *WntA* in patterning implies a shared and relatively ancient role for this gene in nymphalid wing pattern specification. It seems likely that the *wingless* signalling pathway has a role in wing pattern establishment that is shared across all the Lepidoptera, but it remains unclear at which point the *WntA* gene specifically was co-opted into pattern specification.

9.17 The yellow locus

Although it was the first locus to be mapped using genetic markers (Joron et al., 2006), the yellow locus has proven to be the most intractable genetically. The genetic associations in wild populations are much more diffuse than at the *optix* locus, and there are no clear blocks of genetic sequence repeatedly associated with particular patterns as there are at *optix*. Nonetheless, there is a strong peak of association around a gene called *cortex*, which also shows differences in expression between wing patterning forms (Nadeau et al., 2016).

Intriguingly, this locus also controls adaptive variation in the peppered moth, *Biston betularia*, which has several different forms with varying degrees of melanization in the wing. The *typica* form is mottled white and is well camouflaged on lichen-covered tree trunks. In contrast, the *carbonaria* form is black all over and was better camouflaged on the darkened soot-covered tree trunks of Industrial Revolution Britain. The mutation responsible for this adaptive change is also located in *cortex*, implying that this locus is responsible for controlling wing pattern variation widely across the Lepidoptera (van't Hof et al., 2011). Work is needed to elucidate the mechanism by which this gene controls patterns, as *cortex* belongs to a family of cell cycle regulators and is not an obvious candidate for regulating patterning.

9.18 Revisiting the nymphalid ground plan and windows and shutters

The Nijhout nymphalid ground plan (NGP) and Gilbert's window-shutter model view *Heliconius* from two different perspectives. What have we learnt about these patterning models? Both models envisage the black wing regions as being fused components derived from different 'pattern' elements. The expression patterns of *WntA* do indeed suggest that black regions of the wing are composed of multiple regions defined in different ways during development. Specifically, *WntA* marks black regions in the centre of the wing that block out the forewing band and determine the shape of various yellow and white patterns, but its expression is not necessarily associated with more peripheral regions of black pigmentation (Martin & Reed, 2014). Intriguingly, this perhaps implies that the 'shutters' in the Gilbert model may be derived from the central symmetry system of the NGP, as marked by *WntA* expression.

One aspect of the NGP that seems outdated is the delineation of 'pattern' and 'background'. For example, Nijhout states, 'In studying pattern formation it is essential to distinguish between the pattern and the background upon which the pattern develops'(Nijhout & Wray, 1988, p. 347) and 'The majority of pattern elements consist of dark-coloured shapes on a lighter background' (Nijhout, 1990, p. 82). However, the expression and inheritance patterns of *optix* are hard to reconcile with the dichotomy between pattern and background. In *H. erato* crosses, *optix* expression turns on red in an otherwise yellow forewing band region. Under the NGP, in which yellow is background, this would imply that *optix* is controlling the colour of background elements.

This is also consistent with the expression of *WntA* in the black regions around these yellow/red patches. However, in the hindwing of both species and in the *H. melpomene* forewing, *optix* places red elements onto an otherwise black wing, such as the hindwing rays, which would be consistent with *optix* acting to colour black NGP pattern elements red.

Therefore in some cases *optix* controls red patches that correspond to 'background', while in other cases the same gene controls red patches that are 'pattern'. This reflects confusion in the NGP around whether red *Heliconius* elements are 'pattern' or 'background' (Nijhout & Wray, 1988). Furthermore, in crosses between Peruvian forms of *H. melpomene*, the hindwing yellow bar 'overprints' the hindwing portion of dennis, which is not consistent with yellow being background (Mallet, 1989). I suspect that once we have a full understanding of wing development, the distinction between pattern and background will become meaningless. Instead, there is a hierarchy of factors that interact by both activation and inhibition of downstream factors. The *optix* gene is a paintbrush that can be co-opted to colour wing regions already outlined by any part of this upstream patterning landscape.

Early formulation of the NGP also proposed that ground-plan elements are dependent on the positioning of wing veins as landmarks during development. For example, 'Development of pattern in each wing-cell (an area outlined by wing veins) is largely independent from that in other wing cells' (Nijhout, 1990, p. 83). Currently, there is mixed evidence for vein dependence of patterns in *Heliconius*. Clearly, the hindwing rays—stripes of orange scales lying between the veins—must be vein-dependent. However, the forewing bands cut across vein boundaries and instead appear to be dependent on a whole-wing patterning system. This is supported by a single 'veinless' mutant specimen that arose in the laboratory of Larry Gilbert, which had an intact forewing band pattern, demonstrating that *Heliconius* patterns can arise without the presence of veins (Reed & Gilbert, 2004), and a similar veinless mutant of *Papilio xuthus*, which had major elements of the pattern intact (Koch & Nijhout, 2002).

In addition, *WntA* expression domains appear to form a wing-wide patterning system. Thus, for example, the front-back white stripe of *Limenitis* probably involves regulation of *WntA* by whole-wing patterning information. Vein information can modulate the signal from this underlying whole-wing system to varying degrees. In some cases this modification may be very strong—producing entirely wing-cell-dependent repeated patterns such as eyespots (e.g. *Bicyclus anynana*) or the chevrons of the external symmetry system (*Euphydryas chalcedona*). This vein-dependent modification of a whole-wing patterning system is elegantly demonstrated by the veinless mutant *Papilio xuthus* (Koch & Nijhout, 2002). The patterns of WntA expression seen across the butterflies therefore represent regulatory integration of vein-dependent and vein-independent pattern information.

In summary, the NGP does not consist of vein-dependent repeated elements, but rather is part of the whole-wing patterning system, derived from a conserved set of genes that control the early stages of wing differentiation. The wing veins form part of this system, but we can perhaps think of both wing veins and wing pattern as being two outcomes of spatial patterning across the wing. All of this pattern information forms part of the gene regulatory network that establishes spatial information in the wing, and can then be interpreted by colour specification genes such as *optix*.

The process of wing patterning (as indeed for all developmental processes) is therefore a hierarchy of increasing complexity in a gene regulatory network, which moves from simple whole-wing spatial information established early in development, through to ever more complex spatial information later in development (Figure 9.10). The latter includes, but is not limited to, the wing venation system. Whole-wing spatial information may be established early, but in some cases, gene expression patterns are maintained through to late in development, and therefore this information is accessible to wing pattern evolution. Downstream wing patterning factors, such as *optix*, can tap into different levels of this hierarchy in an ad hoc manner and as required by the demands of evolution. So rather than wing patterning being seen as solely the outcome of a set of homologous vein-dependent elements, we should envisage a hierarchy of increasing complexity, with different levels of information available for patterning.

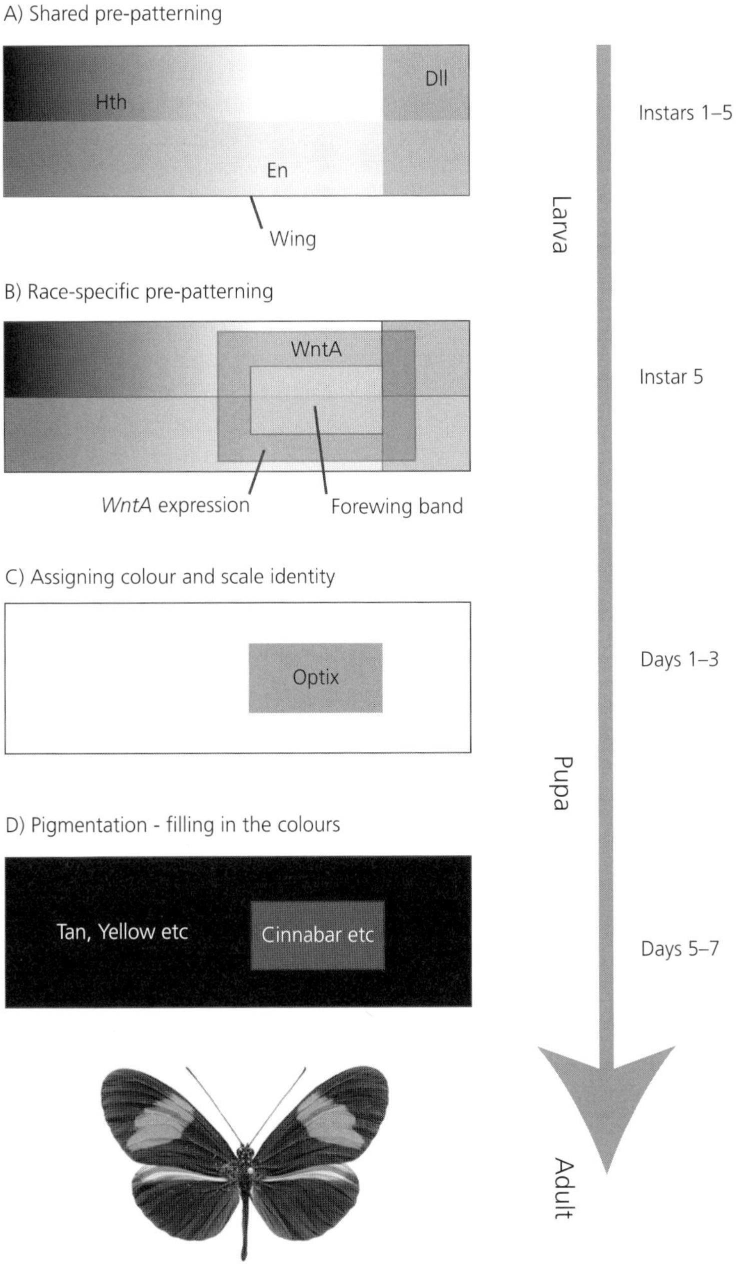

Figure 9.10 Summary of wing pattern development

(A) Early in larval development, many of the genes involved in early wing patterning are shared across the insects and therefore highly conserved in their expression domains. (B) Late larval stages start to differentiate in gene expression patterns between different wing pattern forms within *Heliconius*, notably at *WntA*. (C) During pupal stages, early patterning information is interpreted into different colours—such as *optix*, which defines presumptive red wing regions. (D) Finally, in the second half of pupation, the expression of patterning genes such as *optix* is converted into colours via localized expression of pigment synthesis enzymes including *cinnabar* (red) and *tan* (black).

9.19 Why hotspot genes?

Another related problem is the question of 'hotspot' genes. One of the surprising aspects of *Heliconius* wing pattern evolution is the repeated involvement of the same few genes in pattern evolution. This 'repeatability' in evolution is a general pattern emerging as mutations that control adaptive phenotypes are identified in many organisms. In laboratory strains of mice, one of the easiest traits to observe is the coloured patterns on the coat. In fact, in the laboratory, approximately 120 genes can influence coat colour (Mundy, 2005).

However, scientists have also extensively studied natural populations of mice to identify the genes that control coat colour in the wild. In Florida, for example, several populations of mice have adapted to live on sandy beaches. Here, mice with pale coats tends to blend in better with the background and are less likely to be eaten by predators. As such, pale mice have been favoured by natural selection in these beach habitats, leading to genetic differences in coat colour between beach and inland populations. On the Gulf Coast, coding sequence mutations in a gene called *MC1R* are partly responsible for this shift, while on the Atlantic coast the *agouti* gene is more important (Hoekstra et al., 2006).

The fact that two nearby populations have invented different solutions to the same problem might imply a lack of predictability to the genetic basis for adaptation, but in fact these two genes, *MC1R* and *agouti*, are repeatedly implicated in coat colour adaptation, not just in mice, but across the mammals and even in some birds (Mundy, 2005). Our own hair colour is a case in point—ginger hair is the result of molecular variation in *MC1R* (Valverde et al., 1995). So the question is, if so many genes can influence coat colour in the laboratory, why are just a handful of genes—the so-called hotspots for evolution—involved in cases of adaptation in natural populations (Stern & Orgogozo, 2009; Martin & Orgogozo, 2013)?

The phenomenon is not restricted to pigmentation patterns. Other well-studied examples include completely unrelated insect species that feed on chemically defended milkweed plants that have evolved the same amino acid changes to detoxify the toxins (Zhen et al., 2012). Similarly, thale cress plants repeatedly use the same two genes, *FRIGIDA* and *FLC*, to adapt their flowering time to local conditions (Johanson et al., 2000; Shindo et al., 2005), and among freshwater populations of sticklebacks, changes at a gene called *Pitx1* are always involved in the loss of pelvic spines (Chan et al., 2010).

Arnaud Martin and Virginie Orgogozo reviewed evidence from 1,008 alleles identified as controlling adaptive variation in animals, plants, and fungi (Martin & Orgogozo, 2013). These included cases of domestication over very short timescales, as well as natural adaptation over longer timescales. There were 111 genes that appeared more than once in the list—in other words, they have been the target of selection in two or more cases, with 35% of all the alleles in the dataset belonging to these genes. These genes could be from independent populations of the same species, different species, or even different kingdoms. This implies a highly non-random pattern in the distribution of genes targeted by natural selection.

Of course, there are a number of ascertainment biases that need to be taken into account in interpreting such data. It is often easier to identify a gene if someone else has already worked on it in another species—indeed, a whole class of studies involves the 'candidate gene' approach, which involves testing for the role of genes that have been identified in another species (Mundy, 2005). Furthermore, it is much easier to identify alleles of large phenotypic effect, so our understanding is inevitably biased towards cases in which single loci have major phenotypic effects (Rockman, 2012). Even taking into account these biases, however, most would agree that something unusual is going on here.

Convergent evolution provides one scenario for hotspot genes. It is quite common that different lineages will encounter similar selection pressures or challenges, and evolution often seems to come up with genetically similar solutions. To a surprising degree, it is now clear that when similar phenotypes evolve in different lineages, they often involve similar genetic changes. Mimicry is a special case of convergence, where the optimal phenotype is actually determined directly by the phenotype of the model species, rather than by some external, shared environmental challenge. As outlined earlier, in *Heliconius*, it is now clear that, to a fairly remarkable degree, convergent evolution involves utilization of a small handful of the same genetic loci (Table 8.1).

The clearest example is provided by the Amazonian rayed races of *Heliconius melpomene* and *H. erato*. Sequence data from the red locus shows that within each of these lineages, dennis-ray patterns have arisen independently, most likely from a red-banded postman ancestor. Other independently derived dennis-ray patterns occur, for example, in *Eueides tales* and *E. heliconioides*, but have not yet been investigated genetically. Therefore, genetic change at the *optix* locus underlies the independent evolution of similar phenotypes in at least two, and probably several more, lineages.

Genetic hotspots for wing pattern evolution represent more than just the same pattern arising multiple times, however. First, even for convergent mimetic patterns, the developmental mechanism is not necessarily identical—for example, the precise phenotype of the hindwing ray pattern is quite distinct in different lineages (Figure 7.4). More fundamentally, the hotspot genes are co-opted not just for convergence but for repeated evolution of a huge diversity of different patterns (Joron et al., 2006; Papa, Martin, & Reed, 2008b).

Thus, for example, the shape locus (*WntA*) is used to generate double bands in the co-mimics *H. m. plesseni* and *H. e. notabilis*, single bands in *H. cydno* and many races of *H. melpomene* and *H. erato*, spotted yellow forewing patterns in Amazonian races of the same species, and small yellow spots in *H. hecale* and *H. ismenius* that mimic ithomiine species. Similarly, it has been used to remove white bands when distantly related *Limenitis arthemis* mimics *Battus philen*or. Thus, a single gene underlies evolution of a huge variety of adaptive patterns that generate mimicry of a wide diversity of different species. In the terminology of Martin and Orgogozo (2013), this is truly both an 'inter-lineage' and 'intra-lineage' hotspot for evolution.

How surprising is this? In some cases of evolutionary hotspots, the same genes are targeted because only one gene carries out a particular task. An obvious example would be the repeated evolution of opsin genes in shaping the spectral sensitivity of visual systems. Altering opsin proteins is virtually the only means by which an organism can change its visual sensitivity to different wavelengths of light. However, for developmental processes such as patterning a wing, this is a less satisfactory answer. Hundreds of genes must be involved in the development of a wing, so why are only a few involved in pattern evolution? To test that this is indeed the case, we would like know the total number of genes in the genome that can possibly influence a trait, and then see whether a significantly reduced subset of those genes are involved in natural populations. Unfortunately, we cannot do this for *Heliconius* patterns, as we know virtually nothing about the kinds of mutations that influence wings. Probably the best we can do is the comparison between the 120 genes that influence coat colour in laboratory mice and the observation that just two genes, *agouti* and *MC1R*, repeatedly evolve in wild populations of mice with different coat colours (Mundy, 2005).

So why are there hotspots? There is something about hotspot genes such that the right mutations are more likely to occur in the first place, and if when they do, they are less likely to incur negative pleiotropic effects (Kopp, 2009; Stern & Orgogozo, 2009). A wide variety of possible reasons for hotspots have been proposed in the literature (Stern & Orgogozo, 2008, 2009; Gompel & Prud'homme, 2009; Kopp, 2009); in what follows, I will focus on those I think most relevant for understanding *Heliconius*.

9.20 Complexity of *cis*-regulatory elements

Once a particular genomic region has become a focus for complex *cis*-regulatory control, this may have a positive feedback effect and increase its chances of being targeted again. For example, a region of the genome with a high density of wing-specific *cis*-regulatory enhancers is more likely to be predisposed to adopt new wing-related functions. If binding sites are already present for stage- and tissue-specific transcription factors, it will be much easier for a new function to be adopted in that particular tissue.

In the case of *optix*, for example, expression needs to be driven in the wing scale precursor cells during early pupal development, to be specific to hindwings or forewings, and to show spatial localization across the wing. All of these likely require complex inputs to be interpreted correctly. The regulatory region of *optix* almost certainly interacts with Ultrabithorax (distinguishing hind- and forewing), Achaete-scute (defining the scale precursor cells), Engrailed (distinguishing the anterior/posterior compartment), and so on. Therefore, acquisition of a new interaction that produces a novel domain of *optix* expression does not require as many steps as the completely new evolution of a wing regulatory element at a different gene. This is a general phenomenon; multi-function developmental genes such as *optix* commonly have large and complex regulatory regions that are the outcome of many millions of years of repeated regulatory evolution targeted at the same locus (Ludwig et al., 2005; Hare et al., 2008).

The organization of enhancers along the DNA sequence seems to be conserved between *H. melpomene* and *H. erato*, which provides support for the idea that these regions are pre-adapted for adopting new functions (Figure 9.7). In both species, the order and approximate position of the *dennis, ray*, and *band* enhancers is broadly the same, even though these pattern elements have arisen independently in each species. It seems that within the *optix* regulatory locus, there are particular regions already predisposed to evolve particular kinds of red patterns, because of their existing landscape of interactions with transcription factors. Similar arguments have been made to explain repeated *cis*-regulatory evolution of the same enhancer element at the *yellow* gene in placing melanic spots onto the wings of fruit flies (Gompel & Prud'homme, 2009).

9.21 The hourglass shape of networks

In addition to the complexity of *cis*-regulatory control, particular positions in gene regulatory networks may be especially prone to evolutionary change. Perhaps mutations at some loci are just more likely to work than others. In order to work, of course, a mutation must produce a coherent outcome from development that is favoured by selection, but also avoid negative side effects (i.e. pleiotropy). For example, there is a charmingly named gene in the fruit fly called *shaven-baby* (*svb*). When it is mutated, the fly maggots lose many of their small hairs known as trichomes—hence *shaven-baby*. This gene is also the target of selection between species. A specialist fruit fly that lives only on the Seychelles, *Drosophila sechellia*, has bald maggots resulting from multiple mutations in the *cis*-regulatory region of the *svb* gene (Frankel et al., 2011; Stern & Frankel, 2013). The same gene is implicated in a parallel loss of trichomes in another species, *Drosophila ezoana* (some 40 million years divergent from *Drosophila sechellia*) (Stern & Frankel, 2013). Much like wing patterning genes in *Heliconius*, *svb* is a hotspot for mutations both within and between lineages.

Examination of the role of *svb* in making trichomes seems to explain why this particular gene is the focus of selection. In the epidermis of the developing maggot, expression of *svb* in a cell is enough to induce development of a trichome structure from a precursor cell. So *svb* integrates information from many upstream genes in order to 'decide' which cells will make trichomes. Like a computer circuit, the *cis*-regulatory region of *svb* combines inputs from many other genes in order to switch on *svb* in particular target cells. The circuit-board analogy holds for the output as well—cells either produce a complete trichome or none at all, so the information provided by *svb* is binary: trichome or no trichome. That information is then relayed to the large battery of downstream genes involved in making a trichome.

If one of the upstream patterning factors that provide the input to *svb* is mutated, it will likely have deleterious effects on many aspects of making a maggot that require spatial information (it is hard to be specific, because the upstream factors of *svb* are as yet unknown). Downstream, changing the genes that are involved in making the structure of the trichome will alter all trichomes, which is unlikely to be advantageous (*D. sechellia* still has trichomes—just fewer of them than its close relatives).

In summary, there is only one way to make new trichome patterns—by altering the spatial localization of *svb* through *cis*-regulatory evolution. Evolutionary change in trichomes becomes focused on the *svb* locus, such that even if each individual mutation has a small effect (and exhaustive experiments by David Stern and his group suggest that this is the case; Stern & Frankel, 2013), *svb* itself becomes a locus of major effect because over time many mutations accumulate at this locus. The network has an hourglass shape, with *svb*, known as the 'input-output' gene, lying at the pinch point in the network (Davidson & Erwin, 2006). In effect, *svb* acts as a unique switch to turn on the genetic module that produces trichomes.

I hypothesize that genes such as *optix* and *WntA* act in a similar way to *svb*, forming the input-output gene at the centre of the regulatory network for scale development. This remains speculative, as work is still underway to elucidate the regulatory network that controls patterns both upstream and downstream of genes such as *WntA*, but will become testable over the coming years.

Why have some regulatory networks evolved an hourglass shape? The answer may lie in the need to coordinate the development of repeated structures. Larval trichomes on the fly, and scales on a butterfly wing, have in common a repeated structure involving a complex developmental trajectory. Producing a coherent repeated structure requires the coordinated expression of many genes in the right sequence within each scale or trichome precursor cell. The output of a system for regulating the development of such a repeated structure therefore needs to be strongly digital. A cell should make a complete bristle, or not make one at all. A partial bristle would not be useful. Tight coordination of downstream events in development is therefore favoured in order to ensure a strongly canalized output. It seems plausible, therefore, to argue that highly repeated structures such as these might favour the evolution of an hourglass network shape in which a single gene takes on the role of switching the structure on or off.

Similarly, in *Heliconius* wing patterns, the development of scale structure and colour is strongly coordinated. It is interesting to note in this context that the coordination of colour and structure is stronger in *Heliconius* than in other butterflies (Janssen et al., 2001), and stronger in coloured regions of the wing than in regions that are less visible (Aymone et al., 2013). Similarly, there is strongly coordinated coexpression of several pigmentation enzymes associated with different coloured scales (Reed et al., 2008; Ferguson & Jiggins, 2009; Ferguson et al., 2011). The gene regulatory network for wing patterning is therefore highly modular, with regulatory modules representing multiple genes associated with particular scale colours. Perhaps during the evolutionary history of the *Heliconius* lineage, strong selection for visual signalling to predators might have favoured the evolution of tighter coordination of scale development in order to produce strongly contrasting coloured patches that more effectively signal distastefulness. Once such a network structure has evolved, evolutionary change in the wing patterns becomes focused on the single input-output gene at the centre of the network.

In one sense, therefore, an hourglass-shaped network constrains evolution, because it restricts change to a particular locus. I hypothesize that evolutionary change in red wing patterning is largely restricted to *cis*-regulatory change at *optix* for this reason (although, as *H. numata* demonstrates, such constraints are not universal). Conversely, however, this network structure might also facilitate diversification, as the expression of just a single gene needs to be changed to produce a new scale type. If a scale precursor cell required several inputs—perhaps one signal in order to make red, and another to make the Type III scale structure—then several genes would need to be changed to produce a new red patch on the wing. With the hourglass-shaped network, it seems that *optix* alone is both necessary and sufficient for the evolution of new red patches.

In summary, I have argued that both the shape of gene networks and the accumulation of complex regulatory enhancer networks in a genome region can contribute to biasing evolutionary change towards a particular locus. Much of this is speculative, given the rudimentary state of our current knowledge of the patterning network in *Heliconius*, but will increasingly be empirically tested over the coming years. Specifically, I predict that (1) the *optix* locus integrates information from many input signals in order to generate its spatially localized expression patterns; and (2) the Optix protein regulates many downstream targets. This may involve direct or indirect control of actins, pigmentation enzymes, and so on, but the important point is that there are many downstream targets. If *optix* were to regulate just a single downstream transcription factor that in turn upregulated red scale production, it would not support the hourglass hypothesis. Similarly, if the production of red scales went through several bottlenecks, with different genes up- and downstream of *optix* being both necessary and sufficient for red scale production, it would cast doubt on the uniqueness of *optix*.

9.22 Future directions

This is perhaps the chapter of this book that will become outdated most quickly. There is a huge amount of work that still needs to be done in order to understand both the upstream factors that determine the spatial information in the wing and regulate wing patterning genes, and the downstream targets of *WntA* and *optix*. Also, how do *cortex*, and perhaps other players at the yellow locus, act to regulate wing patterning? An understanding of the shape of these networks will shed further light on why *Heliconius* in particular have diversified so dramatically. It also remains unclear how the major patterning genes interact with one another. In crosses, there is strong epistasis between the *cortex* and *optix* loci in *H. melpomene*, but not *H. erato*—what is different between the species in how the loci interact during development?

Finally, and critically, we need direct experimental tests of the action of major patterning genes. Which mutations are critical for making a novel phenotype such as *dennis*, and how many mutations are involved in generating each new phenotype? Is *optix* expression alone necessary and sufficient to produce novel wing pattern pehnotyes? These questions will be answered once we have genome editing tools working in *Heliconius*, and can manipulate individual sequences to test specific hypotheses. Most exciting would be to move specific regulatory elements between different genetic backgrounds in order to test the specificity of their action in producing novel wing patterns. The ultimate goal is to set up an experimental system in which the influence of single mutational changes on patterning could be investigated.

CHAPTER 10

First steps: biogeography, hybridization, and the origins of novel patterns

Chapter summary

Warning colour patterns in *Heliconius* show great diversity as well as convergence due to mimicry. This chapter considers the various processes that might have generated this diversity. *Heliconius* have been proposed as an example of 'shifting balance', an evolutionary model involving a balance between drift and selection originally proposed by Sewell Wright. Although it does seem likely that alternative mimicry patterns can move around the species range (Phase III), there is only weak evidence for a role of genetic drift in driving evolutionary novelty in wing patterns. Alternative selectionist explanations for the origins of novel wing patterns cannot be ruled out. There is also little evidence for the 'refugia' model that proposes the origin of novel patterns in Pleistocene forest refugia. Patterns of genetic variation have been widely used to address the origins of wing patterning diversity, but in fact phylogeographic approaches tell us little about the wing patterns themselves. This suggests caution should be used in inferring the history of adaptive traits from 'neutral' molecular markers. Instead, sequencing of the genes that actually control wing pattern diversity has revealed a surprising and complex history. Within-species, disparate and disjunct populations with similar wing patterns show a common origin for such patterns, suggesting a complex history of movement of patterning alleles across the species ranges. Between species, convergent mimetic forms also commonly share alleles; this is most readily explained by adaptive introgression between species. Hence, closely related species evolve mimicry by sharing of alleles rather than independent convergence.

Mimicry is a theory of convergence and similarity. Mimicry selection should lead all species to look the same. And yet the most striking aspect of *Heliconius* is the great diversity of wing patterns. In Chapter 8, I have already discussed how multiple mimicry rings can coexist in a locality, and how a single species can maintain a stable polymorphism. However, a remaining puzzle is the origin of new patterns, and in particular the great geographic diversity seen across many species. How have *H. erato* and *H. melpomene* evolved such a huge amount of geographic variation?

This chapter considers one of the most frequent questions that is asked about *Heliconius* wing patterns: how and why do new patterns arise? Although it may be one of the most discussed questions in the literature, it is perhaps also the question with the least satisfactory resolution. Our understanding of race formation in *Heliconius* speaks to one of the 'big ideas' in evolutionary theory, with implications for understanding evolution more widely. And because race formation is likely to be the first step towards the evolution of new species, this chapter will lead us to the next chapter on speciation.

The Ecology and Evolution of Heliconius Butterflies. Chris D. Jiggins, Oxford University Press (2017).

In the second half of the chapter, I will describe what we have learnt about the origins of new patterns from sequencing the loci that control patterning on the wing. This will lead us to one of our most important recent discoveries, which is the ubiquitous importance of hybridization and introgression in wing pattern evolution. First I discuss two alternative explanations for the origins of novel patterns.

10.1 Two theories—shifting balance

The big idea is known as the shifting balance. The term was coined by Sewell Wright, one of the three mathematical biologists who developed the foundation for our modern understanding of evolution, now known as the modern synthesis (the other two being R. A. Fisher and J. B. S. Haldane). Wright differed from Haldane and Fisher in arguing that chance processes played a crucial role in allowing populations to explore new forms (Wright, 1931, 1982). He envisaged that once a species has evolved a particular set of characters that work well together, it would be hard for evolution to break up that combination—even though another alternative combination might work better overall. Essentially the intermediate combinations necessary to move to the new form, or adaptive peak, are less fit and represent a 'valley' in the adaptive landscape. In proposing this view of evolution, he clashed with Fisher, who emphasized the power of natural selection alone to drive evolution. Fisher saw random genetic drift solely as a hindering force, opposing selection and adaptation. The Fisherian view of natural selection predicts that adaptation proceeds most effectively in large populations. Shifting balance has been influential in promoting the opposite idea, that population sub-division and genetic drift could promote adaptation. The name is derived from the fact that the theory requires a balance between the evolutionary forces of drift and selection. The debate therefore addresses a fundamental question in evolution: is adaptation purely driven by the deterministic process of natural selection, or is there a major role for chance?

The shifting balance was proposed to involve three phases:

(1) Phase I: Random changes in allele frequency allow populations to drift away from their local state. Most of the time, this drift would be deleterious, but occasionally this might allow some individuals to edge closer to an alternative optimum phenotype. The population could move away from its current fitness peak and end up at the base of an alternative peak. Wright imagined that this alternative peak of fitness might represent a vacant ecological niche or a new ecological opportunity.
(2) Phase II: Natural selection would favour the new genotype in this new niche and lead it to increase in frequency, eventually to fixation. In other words, once the population has a phenotype that allows it to begin to exploit the new niche, natural selection will then act to perfect that phenotype and adapt the population better to its new ecological role.
(3) Phase III: The new genotype, which has increased fitness and is well adapted to the new niche, would send out migrants into nearby populations and gradually expand through the species range. The new form has higher fitness than surrounding populations so would tend to spread. Eventually, it might replace the original form, such that the whole species range has adapted into the new niche.

The shifting balance model of evolutionary change has been highly influential, but has been heavily criticized and, especially over the last 20 years, has largely fallen from favour (Coyne, Barton, & Turelli, 1997, 2000). The shifting balance is exactly that—a balance between the evolutionary forces of drift, selection, and migration. One of the major criticisms has been that for all stages to work, there needs to be a very particular kind of population structure. Small local and isolated populations may promote peak shifts in Phase I, but would make Phase III difficult, because it is hard to see how isolated populations can be vigorous enough to send out migrants and take over the species range. Conversely, large contiguous populations may facilitate the spread of new genotypes, but make it unlikely that genetic drift will be strong enough to produce a peak shift in Phase I (Coyne et al., 1997).

In addition, there is considerable empirical support for the Fisherian view of evolution, in which selection is universally more effective in large populations.

A large body of evidence from molecular evolution shows that selection is more effective in species with larger effective population sizes, and a similarly large body of evidence shows rapid evolution of quantitative characters under mass selection. Here I will discuss shifting balance and some of the criticisms in the context of *Heliconius* wing patterns.

Nonetheless, warning colour patterns pose a challenge to Fisherian mass selection. A large population of butterflies all sharing a common warning colour pattern is expected to show massive evolutionary inertia—predator memory will favour individuals that share the common warning pattern and select strongly against novelty. There is ample evidence for such selection in wild populations (Chapter 8). New patterns can evolve through mimicry of another existing species (Figure 8.5). However, mimicry cannot explain the origins of entirely novel patterns such as the many forms of *H. erato* and *H. melpomene* that are not seen in any other species. In the face of such inertia, therefore, how do novel patterns arise? An appeal to some form of non-deterministic chance process seems the only solution—as Turner and Mallet conclude, 'Maybe a deterministic explanation has eluded us; but we doubt it!' (Turner & Mallet, 1996, p. 843.

The challenge is to explain the evolution of completely novel patterns, which must have occurred with some frequency in order to produce the diversity we see today. Mallet has argued that the evolution of such novel patterns provides a convincing example of the shifting balance in action (Mallet & Singer, 1987; Mallet, 2010).

First, it is envisaged that populations somehow escape the stabilizing force of predator memory and become variable in wing colour pattern. This might happen through genetic drift alone—perhaps in a small locally isolated population. Alternatively, it might be combined with a local reduction in natural selection. Perhaps predators become locally extinct. I was struck by images of Hurricane Mitch, which hit Honduras in 1998 and ravaged a large area of Caribbean rainforest. After such an event, recolonizing butterflies might be briefly released from predation pressure and new wing patterns could become established. Alternatively, seasonal changes in prey availability might mean that at some times of year there are plenty of alternative prey, so predators 'play it safe' and avoid all *Heliconius*-like patterns, allowing prey populations to become more variable. As we saw in Chapter 8, we understand little about predator psychology and learning, which could lead to situations that favour establishment of novel forms. It has been suggested that predators might show neophobia, or avoidance of rare unusual prey, which could help in the early establishment of a novel pattern, although this does not seem to be consistent with what we know about selection in *Heliconius* hybrid zones (Sherratt, 2011).

We do occasionally observe populations that perhaps represent this process in action. These include polymorphic and non-mimetic populations, such as *H. timareta timareta* in Ecuador, and the high frequency of *H. melpomene* × *H. cydno* hybrids found in San Cristobal in Venezuela (Mavárez et al., 2006). These observations suggest that mimicry selection is occasionally relaxed enough to allow populations to drift in frequency.

However, invoking genetic drift may not be necessary to explain the origin of novel forms. Other forms of natural or sexual selection may kick in when mimicry selection is relaxed. For example, darker wing patterns may be better for thermoregulation at high altitudes. It has been suggested that some kinds of patterns might be better at signalling in particular light environments, which might favour new patterns (Benson, 1982; Endler, 1982). There is a broad association between postman patterns and open savannah habitats (in the llanos of Venezuela and the savannahs of Brazil), while the dennis-ray patterns dominate in the continuous forests of the central Amazon. Most strikingly, there is a narrow strip of the *H. melpomene melpomene*/*H. erato hydara* pattern along the coast of the Guiana shield and into the mouth of the Amazon. It seems likely that there is some ecological selection that maintains this race along the coastal habitat (Endler, 1982; Blum, 2008).

Richard Merrill has instead proposed that females might escape from harassment by ardent males through evolving new wing patterns (Merrill, personal communication). This would be a form of 'chase-away' sexual selection that could promote divergence and novelty, and perhaps provides a better explanation for the great diversity of patterns in the *H. melpomene*/*erato* mimicry ring.

In summary, a variety of processes could help new patterns get established. These may or may not involve a major role for genetic drift. The various selection-driven mechanisms might require a relaxation of predator selection, but could actually be hindered by drift and so would not require a reduction in local effective population size among the butterflies. This would therefore not represent shifting balance as envisaged by Sewell Wright.

These rare events are hard to observe and therefore to prove with any certainty what happened. Coyne at al. (1997) discuss *Heliconius* as a potential example of shifting balance, and it is primarily this ambiguity around Phase I that leads them to conclude, 'We believe that there is not yet enough evidence to strongly support any explanation' for the evolution of novel *Heliconius* patterns (Coyne et al., 1997, p. 659).

Phase II is more straightforward. As a novel pattern variant becomes more common, it will eventually reach a tipping point where the new pattern is recognized by local predators and becomes the locally favoured form. This is likely to be around 50% frequency, at which point the new pattern reaches a higher fitness than the original pattern. It is then straightforward for the new pattern to rise to fixation locally by natural selection driven by predator memory (indeed, Phase II of the shifting balance involves straightforward natural selection and is uncontroversial).

In Phase III, the new variant needs to spread out across the range of the species, taking over from earlier alternative patterns. This is perhaps where *Heliconius* really can make a contribution to the broader debate, because this part of the shifting balance has been considered controversial, but is happening in real time in *Heliconius*. In most of Panama, *H. erato* and *H. melpomene* both possess a broad yellow bar on the hindwing, but in eastern Panama, most individuals do not have the yellow bar. Indeed, this red-banded form stretches all the way from Colombia around the northern coast of South America into the mouth of the Amazon in Brazil.

When I first worked in the Canal zone in central Panama in 1998, it was extremely rare to find these Colombian butterflies without the yellow bar, but nowadays they are present in an appreciable frequency. In fact, the hybrid zone moved approximately 47 km in the 17 years between 1982 and 1999 (Mallet, 1986c; Blum, 2002), and more recent collections suggest that the movement continues in the same direction. Most of the collected butterflies were from *H. erato*, but the *H. melpomene* zone has also moved in parallel. This is perhaps the only case where we have directly observed evolutionary change in wing patterns in historical time in *Heliconius*.

What is rather extraordinary is that the rate and direction of movement was predicted by Mallet in 1986, based on dominance of the allele that removes the yellow bar (Mallet, 1986c). Theory predicts that dominant alleles will move at the expense of recessive ones, essentially because of the asymmetry whereby heterozygote genotypes teach predators to avoid the dominant phenotype. Nonetheless, it may not be dominance that is driving the movement—there has also been considerable deforestation in the Darien since 1982, and it is possible that the changing environment favours one pattern over another. Alternatively, perhaps the Colombian pattern is just generally better at teaching predators and is taking over, as envisaged by Wright in Phase III of the shifting balance. Whatever the reason behind the movement, it provides a clear case of recent rapid evolutionary change, involving replacement of one adaptive peak by another across a broadly continuous species range. A clear example of Phase III?

The cline in Panama is a change at a single locus, so hardly the kind of complex multilocus adaptation envisaged by Wright. Nonetheless, molecular evidence suggests that the Amazonian dennis-ray forms, which differ at 2–3 major loci from other races, have similarly spread out through the range of both *H. melpomene* and *H. erato* (Wallbank et al., 2016). Clines involving these races seem to be currently stable (Rosser et al., 2014), but their movement would be closer to the spread of multilocus genotypes as envisaged by Wright. Furthermore, far from being a rare event, the evidence suggests that movement of colour patterns around a continuous species range is a predominant mode of evolution in *Heliconius* (see later).

What are the implications for the shifting balance theory more generally? Given the extremely strong selection on colour pattern in *Heliconius*, and apparently large continuous population sizes, it seems unlikely that drift alone is sufficient to overcome selection, and generate new patterns. So the evidence from *Heliconius* is not exactly a ringing endorsement for shifting balance as a universal process in evolution, mainly because we have never directly observed Phase I and cannot rule out processes involving natural selection, rather than genetic drift in the evolution of novelty. Phase II is uncontroversial, and almost certainly does occur, especially in the perfection of mimetic patterns. Finally, *Heliconius* provide good evidence for the action of Phase III, and the spread of novel forms through a continuous geographic range.

10.2 Two theories—refugia theory

A second hypothesis for the origin of wing pattern races is that they arose in isolated forest refuges. Over the last 2.5 million years, a period known as the Quaternary, there have been global cycles of warming and cooling climate. During cooler and drier periods, when the glaciers reached their maximum extent in temperate regions, the wet tropical forests contracted. It has therefore been proposed that butterfly populations were restricted to forest islands, which promoted their differentiation into distinct races. During the 1970s, this hypothesis was developed to explain the Amazonian diversity of many organisms, including birds, by Haffer (1969); butterflies, by Keith Brown, Phillip Sheppard, and John Turner (1974; Brown, 1979); and plants, by Prance (1974).

Originally the motivation for development of refugia theory was the perceived requirement for geographic isolation during speciation (Haffer, 1969). The isolation of populations was a key stage in the theory of allopatric speciation, so refugia provided the necessary element of isolation that was thought necessary to explain tropical species diversity. However, John Turner suggested that, rather than genetic isolation, it was random extinctions among the butterfly fauna that provided the driving force for divergence (Turner and Mallet, 1996). If populations isolated in patches of forest undergo random extinction, this might lead to each refugium containing different combinations of mimetic butterflies, a process he termed 'biotic drift'. Natural selection for mimicry will then lead populations in different refugia to evolve different patterns. However, this theory can explain convergence and mimicry among existing patterns, but does not explain the origin of entirely novel patterns. Alternatively, refugia might lead to geographically restricted populations with reduced population sizes that could promote peak shifts, as in Phase I of the shifting balance.

However, the forest refugia hypothesis is problematic for a variety of reasons: (1) The pollen record is sparse in the central Amazon, but provides little evidence for isolated refugia—instead it seems more likely that the forest contracted, but central Amazonia remained forested during the Pleistocene, especially along the major rivers (Colinvaux et al., 1996; Bush, Gosling, & Colinvaux, 2011). (2) One line of evidence for refugia was the coincident distributions of butterfly subspecies, which are used to infer the location of refugia. These distributions are therefore implicitly considered to be independent data points, but of course if the different species are mimetic, they are not independent (Turner & Mallet, 1996; Dasmahapatra et al., 2010). Natural selection will act to ensure that they are coincident, so this provides little support for refugia. (3) Range distributions have never been tested against a null hypothesis—it isn't clear whether randomly throwing butterfly distributions onto a map of South America would lead to a similar degree of clustering between species as seen in the real data (Turner & Mallet, 1996). There are some clear patterns, such as clustering of boundary zones between races at large rivers or other major habitat discontinuities. Often one race will be found on only one side of a major river. However, such associations can be explained under various hypotheses (Turner & Mallet, 1996). Hybrid zones will tend to move into regions of low population density even across a continuous range (Barton & Gale, 1993). (4) Present-day refugial or island populations tend to contain the same few widespread species, and typically do not show marked colour pattern divergence (e.g. Caribbean *H. charithonia* and *H. erato*). There is therefore no evidence that isolation alone promotes diversification of wing pattern.

10.3 Patterns of species distributions

So can present-day distributions of species and subspecies diversity give any clues as to their origins? *H. melpomene* and *H. erato* certainly show some intriguing patterns in their racial distributions. Broadly, the species can be divided into five phenotypic classes. First, the centre of the range in both species is occupied by dennis-ray races, which are spread continuously across the Amazon basin. Within this area each species has many named forms, which differ primarily in the shape of the forewing yellow band, becoming more broken towards the east and the mouth of the Amazon. These races intermingle and do not show sharp boundaries, perhaps reflecting their relatively slight differences in appearance. Nonetheless, the Amazon River itself is a discontinuity, with racial boundaries commonly aligned along the river. This is also reflected in a peak of subspecies diversity along the river across the genus as a whole (Rosser et al., 2012).

The remaining classes form the so-called postman races, and are distributed all around the Amazon. They are named after the red and black uniforms of the Trinidad postal service. The archetypal postman pattern, the red and black *H. m. melpomene* and *H. e. hydara*, is distributed in a narrow strip along the coast, from the Darien all around northern South America as far as the mouth of the Amazon. The second class of postman races is similar in appearance but with a yellow hindwing bar; it shows perhaps the most remarkably disjunct distributions. Similar patterns are found in the savannah habitats of southern Brazil and in the north from Panama up to Mexico, as well as in isolated populations in eastern Colombia, eastern Peru, and Bolivia. The third postman class of phenotypes are the two-spotted forms found in Ecuador and Peru. These also show a complex leapfrog distribution, with isolated populations in Ecuador and Peru. These extraordinary disjunct distributions of very similar patterns require either multiple events of convergent evolution or a complex history of range expansions and contractions.

Finally, the fourth class of postman patterns is perhaps the simplest in terms of distribution. In the western Andes, both species have forms with blue iridescence and the hindwing yellow bar restricted to the underside of the wing. In Ecuador, both species show the white hindwing margin, a pattern that is strangely popular in other species in the same geographic region—notably *H. cydno* and *H. sara*. This illustrates a general phenomenon whereby coexisting species that are not mimetic nonetheless sometimes share pattern elements. Perhaps these shared partial patterns occasionally reinforce predator memory.

Overall, these patterns of racial distributions are intriguing but do not strongly differentiate alternative hypotheses for the origins of races—indeed, the alternative hypotheses described earlier have arisen from different researchers studying the same set of distribution patterns. One of the major lines of evidence for refugia theory in temperate zones has been genetic data, showing divergence between populations derived from different refugial populations. Such patterns are repeated again and again in different taxa, including plants, insects, and mammals. Are there similar patterns in tropical species?

10.4 Population structure and phylogeographic patterns

Heliconius populations have been studied using a wide variety of approaches and kinds of genetic data, ranging from protein variation (allozymes), mitochondrial DNA sequence data, small fragments of nuclear sequence data, anonymous AFLP markers, through to whole genomes (Turner et al., 1979; Brower, 1994b, 1996a; Flanagan et al., 2004; Quek et al., 2010; Martin et al., 2013; Nadeau et al., 2014). Indeed, perusal of these papers would provide a timeline of the history of population genetic techniques over the last 40 years. It is therefore a challenge to summarize these studies, but I will try to outline the main conclusions, focusing on the two co-mimic species *H. melpomene* and *H. erato*, which have been studied most intensively.

In broad outline, the species show similar patterns. Across wide biogeographic scales, there is considerable structure with a deep genetic difference between populations found on either side of the Andes. This vast mountain range reached its present height some 8–10 million years ago and represents an all but insurmountable barrier to a butterfly. The considerable genetic differences that have accumulated between populations on either side of the mountains indicate that both species are relatively ancient—the amount of difference between the eastern and western clades is similar to that between closely related species (e.g. *H. melpomene* and *H. cydno*, or *H. erato* and *H. himera*). This shows that both species have been present on both sides of the Andes for at least the last 1–2 million years.

However, within geographic regions there is little genetic structure between adjacent geographic races that have different wing patterns. This is true for allozymes (Turner et al., 1979), mitochondrial DNA (Brower, 1996a), nuclear gene sequences (Flanagan et al., 2004), and whole genomes (Martin et al., 2013; Nadeau et al., 2014). Wing pattern races really do differ little apart from a few major patterning loci. Even at AFLP markers—which seemed to show more geographic structure than other markers—clades of closely related butterflies from the same geographic area can include several different wing pattern forms (Quek et al., 2010). This pattern therefore contrasts strongly with genetic variation seen in north temperate species.

In organisms as diverse as hedgehogs, grasshoppers, and oak trees, phenotypically distinct populations identified as subspecies or closely related species have been found to be also quite strongly genetically differentiated (Hewitt, 1996, 2000). In other words, populations that look different are also genetically distinct. This can typically be traced back to the isolation of populations during repeated rounds of glaciation during the last 2 million years, which pushed organisms southwards into glacial refugia such as the Balkans and Iberian Peninsula.

Overall, then, the genetic data in *Heliconius* do not look very supportive of refugia theory. If geographic races had been isolated for long periods, and had only recently come back together, we would expect wing pattern hybrid zones to show strong genetic breaks across much of the genome, and this is not the case. We cannot completely rule out the explanation that populations might have been restricted to refugia for short periods of time, such that there was little impact on genetic diversity, or perhaps refugial populations have come together and been in contact for long enough that they have become extensively mixed. However, if the genetic effects of refugia are so transient, it is hard to see how such fleeting periods of isolation can have led to such a huge diversity of wing patterns. Sampling remains incomplete, and focused on a few accessible localities in Ecuador, Peru, and south-eastern Brazil with little in between, so it is possible that we might learn more from more comprehensive sampling across the species ranges. However, the data so far indicate that both species have been geographically widespread for a considerable period of time, with no strong evidence for genetic differentiation between wing pattern forms.

10.5 Coevolution or not?

The repeated convergence between *H. melpomene* and *H. erato* raises the question of whether one species came first. The two extreme alternatives are that (a) *H. melpomene* and *H. erato* coevolved together more or less simultaneously into the different races (Sheppard et al., 1985; Hoyal Cuthill & Charleston, 2012), and (b) that one species colonized a radiation already established in the other species, as originally suggested by Eltringham in 1916 (Eltringham, 1916; Mallet, 1999).

There are clues to this problem in phenotypic patterns of mimicry between the species. Although the species show detailed and repeated convergence in wing pattern—often extending to minute details, such as the colours of spots on the body (white in the west, yellow in the east)—there is a broad sense that *H. melpomene* is more impressionistic in its mimicry. For example, the border of the red forewing band is always more fuzzy in *H. melpomene*, a useful character to distinguish the species. Furthermore, the shape of the yellow hindwing bar differs between different regions in *H. erato* but is rather similar in *H. melpomene*. This perhaps suggests that the latter has adopted one common solution to mimicking slightly different phenotypes in *H. erato*. There is therefore a subjective sense in which *H. melpomene* seems to be mimicking the diversity seen in *H. erato*.

10.6 What does the order of branching in a phylogeographic tree tell us?

Many authors have addressed the question of co-evolution using genetic data. First, Brower (1996a) showed that French Guiana populations branched off first from *H. melpomene*, but western and Central American populations branched first from *H. erato*. From this he inferred that the two species had different histories. More recently, Quek et al. (2010) reached a similar conclusion but from a different branching order—western *H. erato* and east Brazilian *H. melpomene* being inferred as ancestral, albeit with weak branch support for nodes in the AFLP tree. Both suggested that it was most probable that *H. erato* diversified first; then *H. melpomene* spread out and evolved mimicry of the *H. erato* patterns. We subsequently inferred trees from genomic SNP data, which are very well supported and place western populations basal to both species, suggesting a more concordant history, although Brazilian samples were lacking (Nadeau et al., 2014). More recently, sequencing of Mexican butterflies led to the discovery of a deep mtDNA break in *H. erato* in Costa Rica, which was used to infer a North/Central American origin for that species (Hill, Gilbert, & Kronforst, 2013). Finally, Hoyal Cuthill and Charleston compared the branching order of populations in the two species and concluded that they were concordant with one another and provided significant evidence for codiversification—completely the opposite conclusion to most previous authors (Hoyal Cuthill & Charleston, 2012).

It is not that any of these inferences are entirely false—it seems likely that east Brazil is the clade that branches first in *H. melpomene*, but that western Ecuador and Central America diverged first in *H. erato*. Furthermore, as highlighted by Hoyal Cuthill and Charleston, there are similarities in the patterns of divergence in the two species. However, differences in sampling, outgroup selection, and tree reconstruction have meant that the branching order has jumped around between studies. This uncertainty highlights the extent to which the root of a tree depends on the particular samples chosen and tree construction methods, and should induce considerable caution in making strong conclusions about codiversification from patterns of branching within a species.

One issue is the inconsistency of branching order between studies, but another is the more fundamental question of whether biogeographical origins can be inferred from intraspecific branching patterns (Hey & Machado, 2003; Wakeley, 2003; Nielsen & Beaumont, 2009). Some of the methods widely employed for phylogeographic inference from phylogenetic trees have been strongly criticized (Panchal & Beaumont, 2007; Petit, 2008). If, hypothetically, *H. erato* had been continuously distributed across South and Central America for the last 10 million years—much longer than the coalescent age of the genetic variation segregating within the species—then the deepest surviving mitochondrial lineage would have to branch off somewhere in that range, but this might not tell us anything about the origin of the species. Continuously distributed species will tend to accumulate geographically distributed branching patterns as lineages become extinct at random (Irwin, 2002, 2012).

The placement of the root in an intraspecies tree is therefore highly influenced by chance events, but also by relative population sizes (Nielsen & Beaumont, 2009). A larger population will harbour more variation and is most likely to contain the deepest coalescent times. The root is therefore most likely to be in the region with the largest population size, irrespective of the original source population. In summary, branching order has been widely used to infer that *H. melpomene* and *H. erato* have very different population histories, but this should be treated with some caution.

10.7 Timing and effective population size

Nonetheless, there are some major differences in the biology of the two species that are reflected in the genetic data. In most places, you are much more likely to catch a *Heliconius erato* than a *Heliconius melpomene*. This is always frustrating when *H. melpomene* is the target, since it is not usually possible to tell which is which until you *have* caught them. On my first visit to French Guiana with Russ Naisbit, we struggled to find any *H. melpomene*, until the local collector Bernard Hermiers told me to look on the *Lantana* bush just behind the abandoned building that we had already visited. Sure enough, they were abundant. *H. melpomene* is rarer and more localized, but can be locally common, whereas *H. erato* is much more widespread and abundant. This greater abundance and more continuous population structure is reflected in greater genetic variation in *H. erato*, with a deeper genealogy and weaker population sub-division. In contrast, *H. melpomene* has less variation between individuals within a population, but more differences between populations.

Coalescent modelling of nuclear sequence data suggests that the effective population size of *H. erato* is about five times that of *H. melpomene*, and began to expand perhaps 2–3 million years ago, whereas *H. melpomene* is younger, dating to 1–2 million years (Flanagan et al., 2004). Genetic variation therefore reflects differences in the biology of the two species more than it does a shared history of isolation in refugia. Similar evidence comes from other taxa across the Amazon. In birds, ithomiine and heliconiine butterflies, divergence times vary enormously across different pairs of species or subspecies (Whinnett et al., 2005; Dasmahapatra et al., 2010; Smith et al., 2014), suggesting that individual taxa have very different histories. Overall, there is little evidence for concordant biogeographic histories that might suggest a history of shared separation into refugia. Instead, the data are more consistent with the idea that *H. erato* spread out first, and was subsequently followed by *H. melpomene*.

Biogeographic evidence also suggests that diversification of races can be very rapid. For example, the Ilha de Marajó in the mouth of the Amazon is thought to have been inundated as recently as 5,000 years ago, but now has phenotypically distinct subspecies in the genera *Heliconius* and *Eueides*, as well as in ithomiine genera *Napeogenes* and *Tithorea* (Brown, 1979; Mallet, 1993). Presumably these forms arose as the island was colonized from adjacent Amazonian populations rather than due to geographic isolation, as the island is not especially isolated.

10.8 Does phylogeography tell us anything about colour pattern evolution?

One of the initially puzzling outcomes of genetic studies was that patterns of genetic variation show remarkably little correspondence to wing pattern variation. In particular, divergent genetic clades found on either side of the Andes include populations with exactly the same wing patterns. For example, the race *H. e. hydara* has a range that spans the east-west mtDNA break, while similar postman patterns are found in disjunct areas of southern Brazil and the eastern Andes. Conversely, butterflies with dramatically different wing patterns found in the same geographic region can be genetically very similar to one another. The authors of many studies claim that wing patterns found in different clades must have arisen independently—after all, the genetic data seem to unequivocally show that these populations, similar in phenotype but divergent in genetic variation, have a distinct evolutionary origin (Brower, 1994b, 1996a; Flanagan et al., 2004; Quek et al., 2010).

Nevertheless, a shifting balance view of the world provides an alternative explanation, which turns out to be correct. Similar patterns in disjunct populations typically have a common evolutionary origin, and therefore different wing patterns moved around the species range, with little effect on genetic variation in the rest of the genome (Hines et al., 2011). Patterns of neutral variation across the genome tell us little about the history of the wing patterns.

It is an inevitable outcome of population genetic processes that different regions of the genome, even when sampled from the same individuals, have slightly different genealogical histories. This is especially true for genes controlling adaptive traits under strong selection, such as wing patterns that have an evolutionary history remarkably different to that of the rest of the genome. It is common in modern population genetics to compare patterns of variation in neutral markers to those putatively under natural selection. This forms the basis for tests such as the McDonald-Kreitman test, which compares patterns of divergence and variability in protein-coding DNA changes with those at synonymous sites in order to test for adaptive protein evolution.

Such an approach has been much rarer in the fields of phylogenetics and phylogeography. However, John Turner anticipated these patterns in 1979 in a paper entitled 'Contrasted Modes of Evolution in the Same Genome' (Turner et al., 1979). Turner and his colleagues used variation in proteins (known as allozyme markers) to show that most of the genome is genetically very similar across the ranges of several *Heliconius* species, showing only slight geographic variation, contrasting with the dramatic shifts in wing pattern alleles. Similar patterns of variation were seen in other species that had little wing pattern variation, such as *H. sara*. Studying the evolutionary history of wing patterns from the DNA sequences of those genes enables us to address questions of general evolutionary interest, such as: how old are wing pattern elements relative to the species? Which patterns are ancestral and which are derived? How often have particular mimetic patterns originated in *Heliconius*?

10.9 Phylogeography of wing patterns

Where patterns look similar, it turns out that they mostly share a common genetic origin, irrespective of geographic distribution. Across both the *H. melpomene* and *H. erato* radiation, patterns of variation at wing patterning loci are organized by the appearance of the butterflies rather than by geography (Figure 10.1). Similar-looking butterflies found in different geographic regions are most similar to one another in these regions, even though they may be quite unrelated in trees built from other regions of the genome. This is most evident at the red locus. Variation sampled from across the range of either species shows a very mixed signal across the red locus genomic interval. Most of the sequence looks most similar among individuals from the same geographic area, but moving into the core region between *optix* and *kinesin*, the signal switches, and individuals are most similar to others with the same wing pattern.

This is especially remarkable considering the vast geographic distances involved. Pairs of butterflies collected either side of hybrid zones in Peru, Ecuador, and French Guiana show genetic similarity related to their geographic localities across their entire genomes, except for this narrow region of the genome where they look more similar according to their appearance. For example, the dennis-ray butterflies from different localities are more similar to one another than to their red-banded sisters that live in the same locality (Hines et al., 2012; Nadeau et al., 2012; Supple et al., 2013). At least for the red locus, this clearly disproves the hypothesis that similar patterns found in disjunct geographic regions evolved independently.

Furthermore, in both species, the dennis-ray patterns found across the Amazon basin have evolved recently from red-banded ancestors. This has occurred independently in the two species, with the dennis-ray forms found in a derived position in a phylogeny of the red locus within both *H. erato* and *H. melpomene*. There is genetic evidence for recent expansion of alleles associated with dennis-ray phenotypes in both species (Hines et al., 2011). This implies that ancestral patterns are found around the periphery of the range, among disjunct populations in southern Brazil, along the eastern Andes, and in Central America, while more recent patterns are found in the core of the range. Mimicry rings in the peripheral populations are dominated by *H. melpomene* and *H. erato*. In contrast, the centre of the range is characterized by a much greater diversity of co-mimic species, with a wide variety of dennis-ray *Heliconius* and *Eueides* species as well as pierid butterflies and arctiid moths (Figure 7.4). Thus, the origin of the dennis-ray pattern in *H. melpomene* and *H. erato* led to the spread of this novel pattern that pushed ancestral patterns to the edge of the range. This happened independently and apparently in parallel in the two species.

The spread of these patterns, which are defined by allelic changes at 2–3 major loci, likely represents a clear case of the spread of multilocus genotypes through the species range, as envisaged under Phase III of the shifting balance. So, although it remains unclear whether these changes happened at the same time in the two species, there are nonetheless some fairly striking similarities between the two species in their patterns of diversification. The 'coevolution of mimicry' hypothesis remains alive if not exactly proven yet.

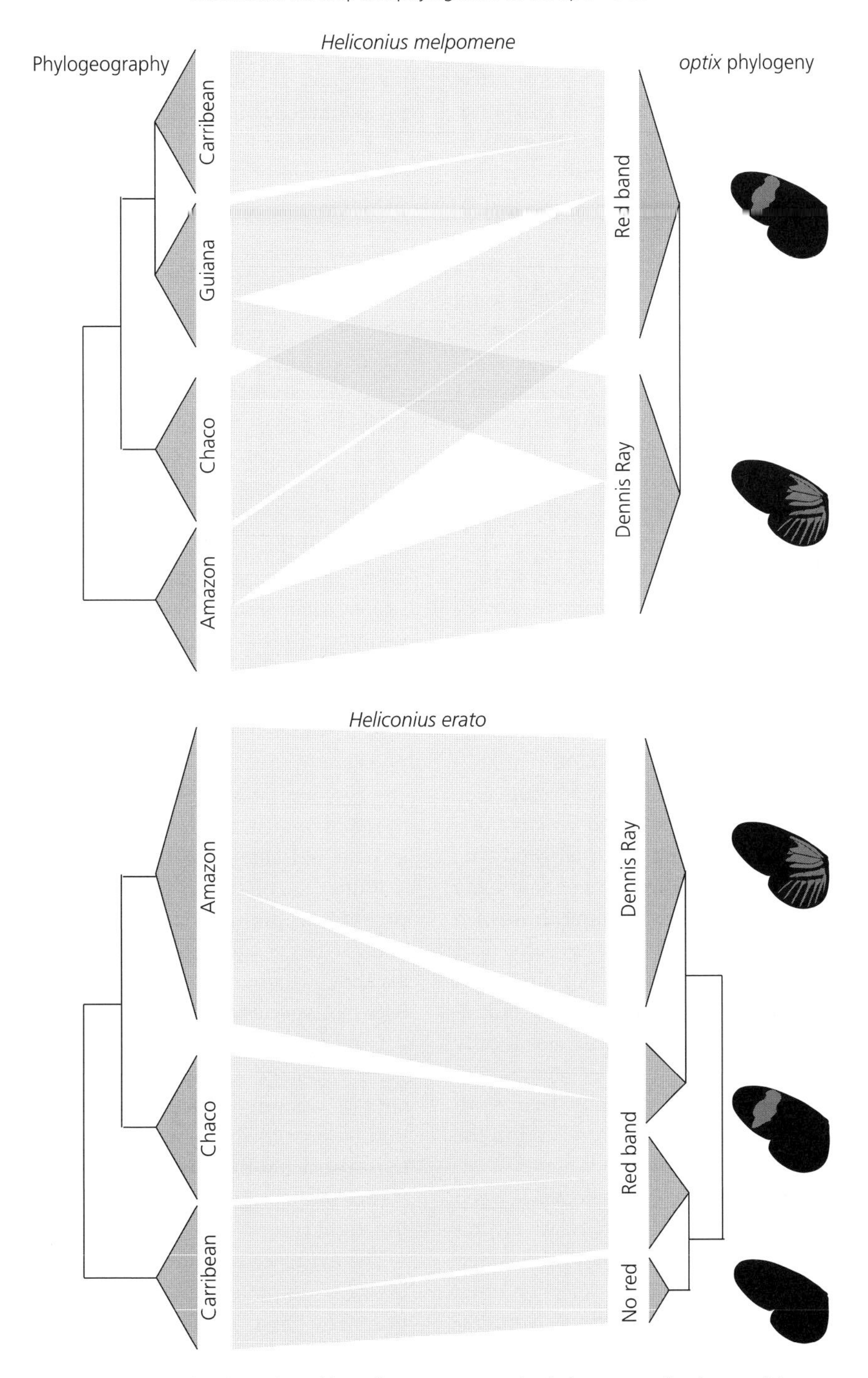

Figure 10.1 The contrasting intraspecific relationships of butterflies at wing patterning loci as compared to the rest of the genome

Relationships across the *H. melpomene* and *H. erato* populations at the genomic background and at the *optix* locus. Populations group by geography at most of the genome, but group by red patterns at regions close to the *optix* locus. Simplified from figures in Hines et al. (2011).

10.10 Signatures of selection on wing pattern loci

Where recent and rapid selection acts on a novel mutation, it acts to remove variation at surrounding sites. As the novel variant spreads through the population, it drags with it linked sites that are not under selection themselves, but just happened to be associated with the new mutation when it arose. After such an event, we expect to see a region of the genome with reduced variation and strong genetic associations between linked sites (linkage disequilibrium); such events are known as 'hard sweeps'. Good examples of such selective sweeps include genes involved in the domestication of maize (Palaisa et al., 2004), or the recent evolution of resistance to a male-killing parasite in *Hypolimnas* butterflies (Hornett et al., 2014). However, the signature of a selective sweep is expected to be weak or absent if selection acts on a variant that is already present in a population at an appreciable frequency (rather than a new mutation), a process known as a 'soft sweep' (Barrett & Schluter, 2008).

In *Heliconius*, there is not much evidence for hard sweeps at wing patterning loci (Baxter et al., 2010; Counterman et al., 2010). Genetic variability is somewhat reduced around patterning loci, and there is some evidence for higher linkage disequilibrium at wing patterning loci (Nadeau et al., 2012; Supple et al., 2013). However, genome-wide searches for selective sweeps using algorithms such as 'SweepFinder' do not identify wing patterning loci (Martin et al., 2016). Indeed, wing pattern loci may have a lower effective population size and reduced genetic variability because they are locally adapted and restricted to particular geographic regions—thus the effective population size is reduced at these regions for reasons unrelated to the original spread of the alleles involved.

The lack of a strong signal of a selective sweep may suggest that pattern variants are relatively ancient, such that any genetic signature of the sweep has been lost. However, in addition, selection often acts on variants that are acquired through migration or introgression, rather than a single novel mutation. As we shall see in the next section, wing pattern alleles are often considerably older than the populations in which they are found. Therefore selection may leave a signature more similar to soft rather than hard sweeps. Sweeps are commonly seen as an important signal of selection in the genome, and are often used to identify loci involved in adaptation, but the *Heliconius* phenotypes serve as a reminder that even loci known to be subject to strong selection may not necessarily show such patterns.

10.11 Introgression and hybridization

One of the major themes of the last two chapters has been to explain how evolution finds novel genetic variants that lead to new adaptations without disrupting existing functions. This is a problem at the ecological level, because of stabilizing selection by predators and their bitter memories (Chapter 8). But it is also a problem at the level of development because of the constraint of existing gene function on genetic change (Chapter 9). What if there were a process that could provide exactly the right genetic variants, already tested by selection and already adapted to a novel niche? This seems like an evolutionary holy grail, a sort of impossible Darwinian dream. Indeed, previous suggestions that mutations could arise that adapted organisms to a new environment in a single step have been derided as 'hopeful monsters' or, even better, 'evolution by jerks' (Goldschmidt, 1945; Turner, 1981).

Rather surprisingly, there *is* a process that can provide a population with well-adapted variants of large effect: hybridization. Genes can be exchanged between populations by migration, or between species by rare interspecies mating, and it is increasingly realized that even rare events of gene exchange can provide useful genetic variants for adaptive change. In *Heliconius*, hybridization is most likely between sympatric species that may also act as the model for mimicry, so provide a source of the 'right kind' of genetic variation in just the 'right place'.

The evidence for introgression comes from sharing of wing pattern alleles across species boundaries. The first evidence came from *Heliconius heurippa*, a species that has long been hypothesized to have a hybrid origin (Brown & Mielke, 1972; Gilbert, 2003). The hypothesis was largely derived from the distinctive red and yellow forewing band in *H. heurippa*, a pattern otherwise only known from hybrid butterflies. A region of sequence from a *kinesin* gene, located near *optix*, showed sequence similarity to *H. melpomene melpomene*, supporting the hypothesis that the red band on the forewing of *H. heurippa* had been acquired from *H. melpomene* (Salazar et al., 2010) (see also the discussion of hybrid speciation in Chapter 11, section 11.22).

Whereas *H. heurippa* has long been suspected to have hybrid origins, another species found down the eastern slopes of the Andes, *H. timareta*, has managed to remain hidden from biologists until very recently. Populations of *H. timareta* are so similar to the local races of *H. melpomene* in appearance that they have only been recognized with the help of genetic analysis (Brower, 1996b; Giraldo et al., 2008; Mérot et al., 2013; Nadeau et al., 2014). Moving down the eastern Andes from eastern Colombia, there is a striking leapfrog distribution of wing patterns: first a yellow band form (*H. t. linaresi*); then a dennis-ray (*H. t. florencia*); then a postman (*H. t. tristero*); then a dennis-ray (*ssp. nov.*); then a polymorphic population (*H. t. timareta*); then a dennis-ray (*H. t. timoratus*); and finally another postman (*H. t. thelxinoe*). (Sorry, if you are confused by that, but there is no simpler way to describe it!) Thus, in virtually all cases where *H. timareta* coexists with an *H. melpomene* postman or dennis-ray form, the two species are almost perfect mimics, except for the polymorphic *H. t. timareta*. Importantly, all of the major red pattern elements seen in the *H. timareta* forms are genetically homologous to similar phenotypes found in *H. melpomene* (Pardo-Diaz et al., 2012; The Heliconius Genome Consortium, 2012).

Perhaps the easiest way to visualize this shared variation is by constructing phylogenies across the red locus (*optix*) genome region. At the ends of the region, as in most of the genome, sequence variation clusters according to species. However, moving towards the regions that control wing pattern, genetic variation begins to cluster populations by appearance rather than species (Figure 10.2) (Pardo-Diaz et al., 2012; The Heliconius Genome Consortium, 2012). In other words, postman *H. melpomene* and *H. timareta* are more similar to one another than they are to rayed *H. melpomene* and *H. timareta*. The yellow-banded forms of *H. timareta* fall in with *H. cydno*, races of which have neither red bands nor dennis-ray red patterns. So, across the entire melpomene clade, currently consisting of five species, all races cluster at the red locus by appearance rather than by species or geography.

However, it gets even more surprising. The sister clade of silvaniform species mostly have blotchy yellow, orange, and black 'tiger' patterns that mimic ithomiine species (Figure 2.5). The silvaniforms shared a common ancestor with *H. melpomene* about 4 million years ago. However, there are two species, *H. elevatus* and *H. besckei*, that are the exception and join the *H. melpomene* mimicry ring. The former has a dennis-ray pattern; the latter a postman pattern. Until John Turner clearly separated the two species based on genitalia and subtle wing pattern characters (Turner, 1966b), *H. elevatus* was commonly confused taxonomically with *H. melpomene*. We now know that it is closely related to *H. pardalinus* and a member of the silvaniform clade (Kozak et al., 2015). However, despite the similarity in wing pattern, it was a complete surprise to discover that *H. elevatus* also clusters with dennis-ray *H. melpomene* forms at the red locus (The Heliconius Genome Consortium, 2012). There is similar evidence that *H. besckei* shares alleles with red-banded *H. melpomene* (Zhang et al., 2016).

In summary, therefore, across the entire melpomene and silvaniform clade, representing some 3–5 million years of evolution, there has been a single genetic origin for both dennis-ray and red-banded phenotypes. These variants are now found distributed widely across many apparently independent populations and species.

Although the red locus is the best documented case, allele sharing has also occurred at other major patterning loci. There is evidence for shared alleles at the yellow locus between *H. melpomene* and *H. timareta* where these species share a yellow hindwing bar near Tarapoto, Peru, and a yellow forewing band in Florencia, Colombia (The Heliconius Genome Consortium, 2012). However, the genetic signal is less clear at the yellow (*cortex*) locus. There has been no extensive analysis of the shape (*WntA*) locus to date, but many species vary in parallel across the Amazon in their forewing band shape, suggesting another possible arena for mimicry driven by allele sharing. The full extent to which patterns are shared between species

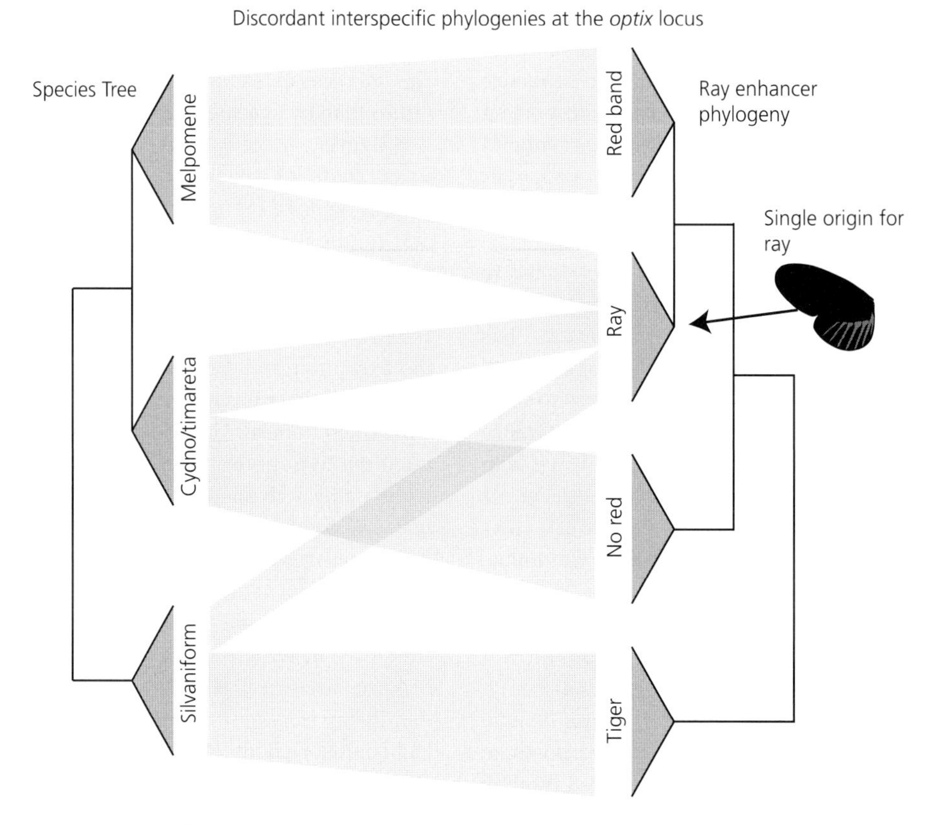

Figure 10.2 Contrasting interspecific phylogeny of the red locus as compared to the rest of the genome

The ray and dennis regions show a strongly divergent history as compared to the species tree. Most notably, the dennis-ray phenotype occurs in the species *H. melpomene*, *H. timareta*, and *H. elevatus*, but represents a single evolutionary origin at the ray locus. The phylogeny for the ray region is shown, but dennis is similar, except that dennis alleles originated within the silvaniform clade, whereas ray originated within *H. melpomene*. Redrawn from Wallbank et al. (2016).

remains to be documented, but it seems likely that it is pervasive.

How do distinct species come to share the same genetic variants? There are at least three possible explanations (Brower, 2013). First, it could be that the variation has arisen independently in each species, driven by selection for mimicry. If the same DNA changes were selected to produce mimetic wing patterns, we would see a pattern of apparently shared variation, which in fact had an independent origin in each species. We can rule out this explanation, because there are such a large number of shared nucleotide changes, including insertions and deletions, across a genomic region of up to 100 kb—the chances of all of these changes arising independently in several different species is vanishingly small. In addition, the extent of the shared region varies between individuals, populations, and species: there is a gradual decline in shared variation as one moves along the genome away from the *optix* regulatory locus. This is exactly what is expected from a gradual process of mixing between genomes through genetic recombination, but is hard to explain if the variants arose independently in each lineage.

The second explanation is that the distinct wing pattern forms have a single origin that predates the divergence and speciation of all the species involved, and so represent a shared polymorphism carried through speciation events. Under this scenario, the wing patterns would have been polymorphic in ancestral species before the divergence of *H. melpomene* and *H. timareta*. This may be plausible, but it is not parsimonious. Most present-day populations with such patterns are not polymorphic, but instead have monomorphic geographic races with one pattern form or another. For patterns found in multiple geographic races to be shared between species, speciation events would have to simultaneously capture wing pattern variants found across different populations. This would require some form of movement of speciation genes through the landscape to capture colour pattern alleles in different geographic regions, which seems unlikely. In addition, dating of wing pattern alleles indicates that their divergence is considerably more recent than the speciation history of the species involved, which is not consistent with an origin through ancient shared variation (Wallbank et al., 2016).

The third and most likely explanation is that patterns have been exchanged between species by occasional hybridization. Wild-collected hybrid specimens (Mallet et al., 2007) and genomic analyses (Martin et al., 2013) indicate that ongoing hybridization regularly occurs between the *melpomene* and *cydno/timareta* clades. If neutral genetic variation is exchanged on a regular basis, it seems likely that occasionally adaptive traits might also be exchanged. Alleles found at the red locus in these species are extremely similar genetically, indicating recent sharing (Wallbank et al., 2016). Furthermore, there are genetic signatures consistent with shared alleles having been exchanged through hybridization, such as enhanced linkage disequilibrium at the wing patterning loci (Pardo-Diaz et al., 2012). Sharing of alleles is the most parsimonious explanation, and in some cases would have been driven by selection for mimicry, providing an adaptive explanation for the spread of alleles between species.

10.12 Hybridization and recombination as a mechanism to generate evolutionary novelty

So far I have mainly discussed introgression as a mechanism that facilitates mimicry in *Heliconius*. In this context, acquisition of alleles from another species is obviously beneficial, because it facilitates resemblance between species in sympatry. However, Gilbert has also proposed that much of the evolutionary novelty in *Heliconius* might have been generated through hybridization (Gilbert, 2003), through what he termed the 'genetic toolkit'. At the beginning of the chapter I discussed the origins of novel patterns in terms of their establishment in populations; here I return to the problem of their genetic origins, which I first discussed in Chapters 7 and 8.

There is now considerable evidence in support of the idea that novel patterns can arise through shuffling of existing variants into new combinations. Best studied is the distinctive red and yellow forewing band of *H. heurippa*, which is not mimetic (although it has been suggested that there is some resemblance to *H. hecale ithaca*). The *H. heurippa* pattern arises from combining red band alleles at the red locus, derived from *H. melpomene*, with yellow band alleles at the yellow locus, from *H. cydno/timareta* (Mavárez et al., 2006). *H. heurippa* therefore represents a novel genotype produced by combining alleles at two different loci.

Novelty can also be generated by recombination between elements at the same locus. The dennis and ray patterns are usually found together in Amazonian populations, but can be separated in rare hybrid phenotypes that have one red patch but not the other (see Chapter 8). Furthermore, there are also established forms that separate the two elements. The Guyanese races *H. m. meriana* and *H. e. amalfreda* have a dennis patch without rays, while *H. timareta f. contigua* in Ecuador has ray but not dennis. Dennis and ray were originally considered two distinct loci, but the evidence for their separation as separate elements was speculative. However, genetic analysis has confirmed that these are two distinct loci, with recombination between tightly linked elements producing the patterns that have one element but not the other (Wallbank et al., 2016).

Thus, distinct combinations of dennis and ray patterns have arisen through recombination (Figure 10.3). The race *H. m. meriana* has a haplotype similar to other dennis-ray butterflies across the red locus, apart from a block of about 30 kb. In this short stretch of DNA, *H. m. meriana* looks genetically more similar to postman butterflies. Thus two recombination events have substituted in a block of sequence that removes the ray phenotype. Similarly, *H. timareta timareta f. contigua* has a recombination event that removes the dennis allele but retains the ray phenotype. In short, the regulatory modules at this locus are a bit like LEGO bricks—they can be mixed and matched to produce different combinations of phenotypes.

When we construct trees from these two regions of DNA, the dennis and ray regions of the red locus, they show very different patterns. The dennis alleles from *H. melpomene, H. timareta*, and *H. elevatus* are all most closely related to the silvaniform clade. In contrast, the ray allele is genetically extremely similar to other alleles from *H. melpomene*. The characteristic dennis-ray phenotype that dominates Amazonian mimicry rings therefore arose through recombination between haplotypes from different lineages. It seems that an early introgression event brought dennis from the silvaniform clade into *H. melpomene*; the ray element moved back into *H. elevatus* much more recently. The implication is that novel phenotypes can be generated by shuffling different combinations of genetic elements between lineages (Figure 10.3).

Surprising results can sometimes make sense in retrospect. In this case, the silvaniform butterflies often have orange patterns on the base of the forewing, so perhaps an origin for the dennis patch in this lineage should not be a complete surprise. One race of *H. hecale* in particular, *H. h. metellus* (Plate 17), has a phenotype extremely similar to the *H. melpomene* dennis patch. It seems likely that the dennis phenotype of *H. melpomene, H. timareta*, and *H. elevatus* arose from a similar-looking pattern in a silvaniform butterfly. What seemed to be an entirely novel pattern actually arose from tinkering with something that was already there, albeit in a completely different lineage. This ability for a 'toolkit' of modular patterning switches to be shuffled between populations and species must have helped *Heliconius* radiate rapidly. The evolution of new regulatory modules is likely to be slow. But once the modules have arisen, they can be reused by different species and in different combinations.

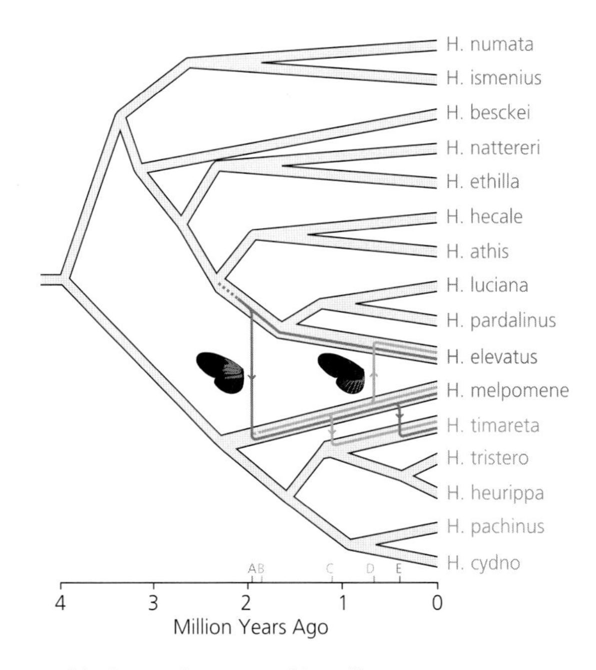

Figure 10.3 Introgression and origins of the dennis and ray regions of the red locus

Construction of dated phylogenies for the dennis and ray regions allows reconstruction of a putative history of introgression between *H. melpomene, H. timareta*, and *H. elevatus*. The first dennis-ray phenotype likely arose in this group when dennis introgressed from the silvaniforms into *H. melpomene*. Ray subsequently introgressed back into *H. elevatus*, and both dennis and ray introgressed into *H. timareta* within the last million years. Redrawn from Wallbank et al. (2016).

10.13 Hybrid zones demonstrate potential routes to evolutionary novelty

In Ecuador, Patricio Salazar has found what seems to be an early stage in the establishment of a hybrid pattern. Around the town of Puyo, in the eastern foothills of the Andes, there is a hybrid zone between the highland two-spotted forms, *H. m. plesseni* and *H. e. notabilis*, and the lowland dennis-ray forms, *H. m. ecuadorensis* and *H. e. etylus*. Around 1,000 m in altitude, there is a large plateau, thought to have formed when an ancient lake high in the Andes collapsed and a vast chunk of mountainside slumped down towards the Amazon. The hybrid zone lies across this forested plateau. Local collectors have long recognized that there is an abundant form in this hybrid zone, which generally looks like the lowland dennis-ray butterflies, but has a two-spotted yellow forewing band (Plate 22). Salazar (2012) carried out crosses in which he demonstrated that in *H. erato*, this phenotype results from combining a lowland allele at the red locus (*optix*) with the upland allele at the shape locus (*WntA*). The hybrid is therefore a homozygote at both loci—in other words, a potentially stable form that can reproduce itself.

The abundance of this particular hybrid reflects the fact that the clines at the two loci are displaced relative to each other. The cline at the red locus occurs 17 km farther down the mountain relative to the cline at the shape locus. This means that in the middle of the hybrid zone there is a region where most of the butterflies have the lowland allele at the red locus but the upland allele at the shape locus, producing the two-spotted phenotype (Figure 10.4).

The reason for displacement of the clines remains unclear. One possible explanation is non-additive fitnesses of hybrid genotypes. For example, if the two-spotted hybrid were to have higher fitness than other phenotypes, perhaps because it is recognized

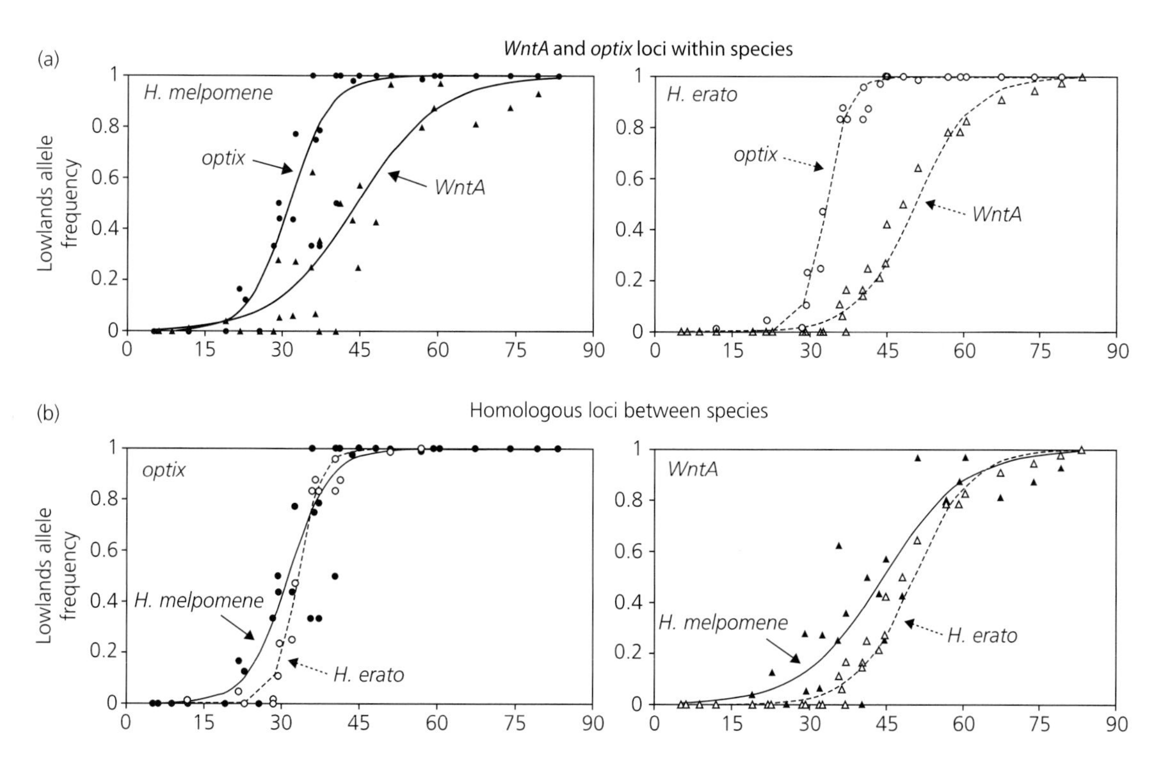

Figure 10.4 Discordant clines in a hybrid zone in Ecuador

The centre of this hybrid zone is characterized by an abundant hybrid form with two forewing spots. This results from displacement of the clines at two alleles: an upland shape locus (*WntA*) allele and a lowland red locus (*optix*) allele. Note that the clines are more similar at homologous loci (lower panel) than at unlinked loci within species (upper panel) (Salazar, 2012). See Plate 22 for phenotypes found in this hybrid zone.

as similar to parental forms more than other hybrids, this could act to push apart clines at the two major loci. Whatever the mechanism, this case provides another example in which alleles at major loci can be combined into different combinations to generate novelty.

Intriguingly, there is a more established mimetic pair near the hybrid zone that may result from a similar process. The races *H. m. ecuadorensis* and *H. e. etylus* each have a small yellow spot in the forewing homologous in position with the more distal spot of the two-spotted *H. m. plesseni* and *H. e. notabilis*. It has previously been suggested that forewing shape alleles from *H. m. plesseni* were combined with the genetic background of a rayed form such as *H. m. malleti* to produce the pattern. The hypothesis could now be tested with genetic data relatively easily.

Other patterns in the genus may have a hybrid origin, as well. For example, Gilbert (2003) recreated aspects of the double band phenotype of *H. pachinus* by hybridization of *H. melpomene* and *H. cydno*, and there is genetic evidence to suggest that the yellow band in Colombian *H. cydno* has similarly arisen through introgression from *H. melpomene* (Camilo Salazar and Juan Enciso, personal communication). Genetic analysis of the red locus indicates that the broad red hindwing band phenotype shared by *H. hecalesia* and *H. clysonymus* is also a result of introgression (Kozak, 2015).

10.14 Establishment of novel hybrid patterns

Novel patterns that arise through introgression will face the same challenges to establishment as any novel aposematic pattern. In particular, they must reach a high enough frequency through relaxed selection or genetic drift in order to become established as a novel phenotype. Brower (2013) has argued that the difficulty of establishing a new pattern argues against adaptive introgression in the case of *H. heurippa*, but in fact introgression could make establishment more probable. The new variant could be introduced at a higher frequency by hybridization than if it arose through de novo mutation. A wing pattern mutation is unlikely to occur twice in the same population, but a hybridization event could be repeated.

10.15 Wider importance of introgression

Some have contested the idea of hybridization and introgression of wing patterns, but have failed to offer a plausible alternative interpretation of the *Heliconius* data (Brower, 2013). As similar patterns emerge in other taxa, it seems likely that this is a far more common mode of evolution during adaptive radiations than has been appreciated. For example, in stickleback fish, the same alleles for reduced armour plating are repeatedly recruited during adaptation to freshwater environments (Colosimo et al., 2005). These alleles move around the species range by continued gene flow from freshwater into marine habitats, where they are found at low frequency. In Darwin's finches, alleles for beak shape are shared across many species in the radiation, and reused as species adapt to different feeding strategies (Lamichhaney et al., 2015a). In perhaps the greatest vertebrate adaptive radiation, there is evidence from cichlid fish in African lakes for ancient alleles being shared among different species across the radiation (Seehausen, 2004; Loh et al., 2013). Even in humans, there is evidence for genomic regions from Neanderthals that provide useful variation, for example, in adaptation to high altitudes (Racimo et al., 2015).

All these examples demonstrate the reuse of ancient variants into novel evolutionary contexts. The process therefore effectively decouples the origins of genetic variation from the selective event in which they are favoured. This perhaps offers a solution to one of the puzzles of regulatory morphological evolution. On the one hand, detailed studies of regulatory modules suggest that they have been assembled by the accumulation of many mutations of small effect, presumably over relatively long evolutionary timescales (Rebeiz et al., 2009; Frankel et al., 2011). On the other hand, rapid evolutionary radiations seem to involve novel morphologies arising repeatedly and on very short timescales, for example, within *H. melpomene* and *H. erato*, or across the African lake cichlid radiations. One of the challenges has therefore been to understand how so much diversity can arise so rapidly.

Hybridization, introgression, and shuffling of genetic modules by recombination offer a means to produce convergence and repeated evolution of similar structures, but also to generate new morphological combinations that can facilitate adaptation to new environments. It seems likely that this is a more common mechanism for evolutionary diversification than has been appreciated until recently.

10.16 Future directions

One of the major outstanding questions is the extent to which drift versus natural selection drives the evolution of novel wing patterns. Although the question may be difficult to address directly, there is plenty of scope to investigate the plausibility of alternative scenarios. For example, comparing predation pressures across different seasons and habitats could give insights into the possibilities for temporarily relaxed selection on wing patterns. Further investigation of natural selection on wing patterns in polymorphic populations, such as areas where interspecific hybrids are especially abundant, might provide insights into how selection acts in the early stages of establishment of novel wing pattern forms. The possibility of chase-away sexual selection can be investigated by studies of male harassment on different and especially novel wing pattern forms. Studies of the efficacy of signals in differing light environments could provide insight into the extent to which wing patterns are locally adapted to send efficient signals to both predators and mates. Finally, genetic data from wing patterning loci could provide additional evidence for the action of natural selection when novel wing pattern variants become established.

Although there are intriguing parallels between the diversification patterns of *H. melpomene* and *H. erato*, the long-running argument about the extent to which they diversified together is far from settled. The increasing availability of genomic sequence data offers an opportunity to make much better inferences of the historical demography of these species. This should lead to far more robust conclusions regarding the degree to which they have codiversified, and the extent to which populations were reduced in size and fragmented during the Quaternary. These studies could be extended to the *H. numata, H. hecale*, and ithomiine mimicry rings to similarly establish their timing and patterns of diversification. More broadly, *Heliconius* could be an excellent test case for studying historical demography across tropical ecosystems using genomic data. This has much wider implications for understanding past and future population changes in response to the changes in the tropical environment.

However, in addition to the historical demography of the genome, addressing the codiversification question will require additional data from the major wing patterning loci. This will offer the opportunity to address the timing of diversification and the frequency of introgression across the whole radiation. Are the spectacular examples of pattern sharing seen in the melpomene clade unique, or is it a general pattern seen across multiple species? How often do patterns arise through de novo mutation compared to introgression and shuffling of existing variation? Such data may also allow reconstruction of ancestral wing patterns, which could allow us to trace the true history of mimicry evolution. Using data from the whole genome, it will also be possible to ask whether introgression is a phenomenon primarily limited to wing patterns, or has it played a broader role in the evolution of other adaptations?

CHAPTER 11

Completing the process: adaptive radiation and speciation

Chapter summary

Speciation is the process that generates biodiversity; it has been the focus of a large body of *Heliconius* research. *Heliconius* have undergone a recent radiation and show evidence for faster speciation than closely related lineages. The increased rate of diversification is associated with a number of evolutionary innovations that may have played a role in promoting speciation, including pollen feeding, a reduction in chromosome number, and diversification of mimicry patterns.

A survey of reproductive isolation across multiple species pairs demonstrates that speciation is a multidimensional process in *Heliconius*. Closely related species are isolated by a variety of barriers, including strong assortative mating, ecological selection against hybrid colour patterns, and genetic incompatibility in the form of hybrid sterility. There is evidence for linkage between genes controlling some of these traits, notably colour pattern, mate preference, and hybrid sterility. Such genetic associations are thought to facilitate speciation with gene flow. Similarly, wing colour pattern acts as a multiple-effect trait, which simultaneously generates post-mating ecological isolation through selection against hybrids, and pre-mating isolation due to colour-mediated mate choice.

Heliconius provide several examples whereby hybridization has contributed to speciation. Hybrid colour patterns can lead to mimicry or novel phenotypes and simultaneously generate reproductive isolation between the hybrid lineage and its parent. Crosses have shown that hybrids can have novel derived mate preference phenotypes as well as wing patterns, which could also facilitate speciation.

11.1 What is an adaptive radiation, and is Heliconius an example?

Adaptive radiations are some of the most spectacular evolutionary phenomena on earth—they are groups of species that have undergone both rapid adaptation through natural selection and species diversification (Heard & Hauser, 1995). Well-known examples include Hawaiian honeycreepers, cichlid fish in African lakes, and Caribbean *Anolis* lizards. A reviewer of one of our papers questioned whether *Heliconius* should be included among these iconic examples as an adaptive radiation. I was initially taken aback—well, of course they should! However, this led me to reflect on what we mean by the term and the ways in which *Heliconius* fit the definition. There is much argument about exactly what constitutes an adaptive radiation, but here I will follow the definition of Richard Glor (2010), who outlines three major characteristics: (a) multiplication of species and common descent, (b) adaptation through natural selection, and (c) extraordinary diversification. Does *Heliconius* fit the bill?

First, the genus *Heliconius*, and perhaps also its sister *Eueides*, represent a group of species that are

The Ecology and Evolution of Heliconius Butterflies. Chris D. Jiggins, Oxford University Press (2017).

DOI 10.1093/acprof:oso/9780199566570.001.0001

diverse (60 spp in total) and share a common ancestor. They share derived traits that characterize the group—pollen feeding, large brain size, and increased longevity in the case of *Heliconius;* warning colour and extensive mimicry if we count *Heliconius* and *Eueides* together. They have therefore jointly colonized a particular niche space in the neotropics and radiated into numerous species. Second, they show a wide diversity of adaptations among species, which include diverse strategies for host-plant exploitation, mating tactics, adult feeding strategies, and, most obviously, a huge diversity of warning colour patterns. Almost all species have some geographic diversity in their wing patterns, but it reaches its peak among those species that you have read about the most in this book—*H. melpomene, H erato, H. numata*, and *H. hecale. Heliconius* are therefore unusual among the classic adaptive radiations in that there is such rampant within-species diversification as well as between-species adaptive diversity. This might suggest that the radiation is still very much ongoing. Importantly, as we shall see in this chapter, there is good evidence that divergent adaptive traits also contribute to reproductive isolation and speciation, so there is a direct link between adaptation and speciation. The concept of adaptive radiation therefore combines the two key processes of Darwinian evolution: adaptation and speciation. In this chapter, I will argue that speciation is a common outcome of divergent adaptation—the two processes are inextricably linked. So *Heliconius* clearly fit the first two criteria.

Finally, is the diversification 'extraordinary'? Extraordinary is difficult to define, but generations of field biologists have certainly found the diversity of *Heliconius* wing patterns spectacular. There are other groups of butterflies that show diversity in wing pattern, both within and between species, but the combination of pattern disparity and diversity in *Heliconius* seems greater than in any other group, although this has never been formally quantified.

We can ask a more specific question about species diversity: is there evidence for a significant increase in diversification associated with the *Heliconius* lineage, either an increase in net diversification over time, or an increase in species diversity relative to other groups? Phylogenetic trees can be used to address both questions.

The most recent phylogenetic hypothesis for the Heliconiini is well supported and has nearly complete species coverage (Kozak et al., 2015): across the Heliconiini, the phylogeny shows a steady increase in diversification rates throughout the last 11 million years (Figure 11.1). Furthermore, the increase is driven almost entirely by speciation in the genera *Heliconius* and to a lesser extent *Eueides*. The other genera in the group, such as *Dryas, Dryadula, Podotricha*, and *Agraulis*, show almost no diversification during this period and are very species poor. The genus *Philaethria* has also diversified over the same period, but shows little diversity in wing pattern. Instead, *Philaethria* shows considerable variation in chromosome number, perhaps suggesting an independent, non-adaptive radiation (Suomalainen & Brown, 1984).

Overall, the phylogeny therefore supports the idea that *Heliconius* and *Eueides* have significantly accelerated in species diversification. Although species diversity does not rival radiations such as the cichlid fish, the really impressive aspect of *Heliconius* is the intraspecific diversity in wing pattern, which is not captured in species-level analyses. I'll leave the reader to decide whether this is 'extraordinary'.

Another common characteristic of adaptive radiation is that an initial rapid burst of diversification is followed by a slowing down as niches are filled. Some have considered this 'early-burst' model as the key signature of adaptive radiation (Moen & Morlon, 2014). An early-burst pattern is expected, for example, when a lineage colonizes an empty island and then diversifies into a limited set of available niches. *Heliconius* show no evidence for slowing down in diversification rates (Kozak et al., 2015). Perhaps this is because the radiation is still underway, so the burst is still happening. Inevitably, diversification rates must slow at some point, so perhaps over longer timescales diversification in *Heliconius* would come to resemble the early-burst model.

11.2 Timing of diversification

When did *Heliconius* diversify? Any phylogenetic hypothesis based on DNA data alone has no calibration in time. The tree gives the pattern of branching between taxa, and the branch lengths, or the

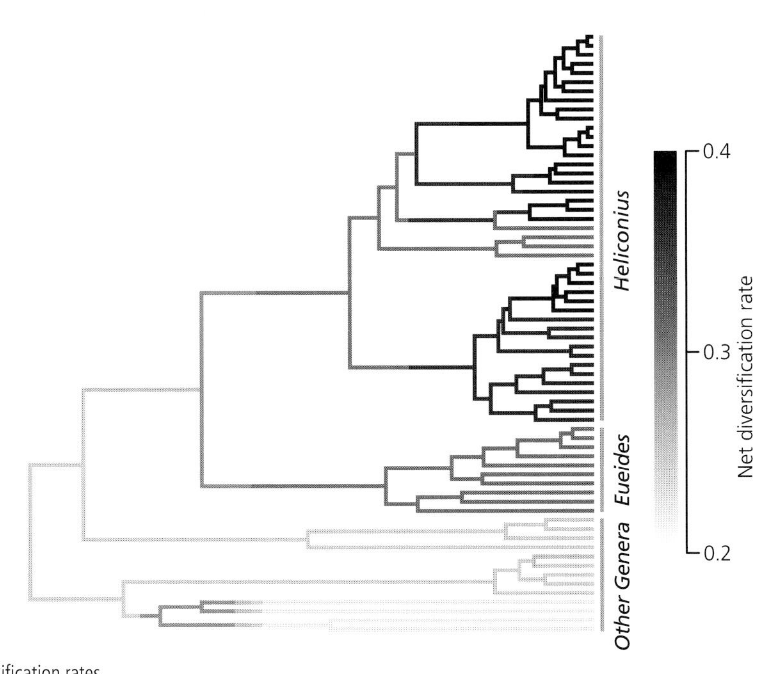

Figure 11.1 Diversification rates

Net diversification (speciation-extinction) rate changes through the phylogeny; it is higher in *Heliconius* compared to outgroup taxa (Kozak et al., 2015).

ages of branches relative to one another. In order to work out the timing of the radiation, we need to use fossils or some other form of external date calibration. Dating the diversification of butterflies more broadly is challenging, because there are so few butterfly fossils. Niklas Wahlberg has led the efforts to produce dated trees for the butterflies (Wahlberg et al., 2009); for example, using seven fossil butterflies to date the radiation of the nymphalids. These fossils can only give minimum ages for branching points in the tree. For example, a 34 million-year-old fossil that is inferred to be related to the genus *Hypanartia* can give a minimum age for the split of *Hypanartia* from its closest relatives (Wahlberg et al., 2009). Nonetheless, these dates have been questioned (Garzón-Orduña et al., 2015) and may alter in the light of new discoveries.

Because there are no fossils within the Heliconiini, dating the radiation depends on extrapolating from dates inferred in the rest of the Nymphalidae. This gives an origin for the Heliconiini around 25 million years ago, and divergence of *Heliconius* and *Eueides* around 18 million years ago. In most extant species, however, diversity arose within the last 5 million years in *Heliconius*, and slightly earlier in *Eueides*.

Because the rise in rate of speciation has been gradual, it isn't clear that diversification has been correlated with any specific external environmental factor. The Andes underwent a second phase of major uplift around 12 million years ago, which would have altered the topology and likely generated a great diversity of habitats. This may have played a role in promoting speciation, as most species' diversity has arisen since then, but such an association is speculative. It is, however, clear that much of the species diversity is older than the Pleistocene, which began around 2.5 million years ago, so cannot be explained by the more recent cycles of global cooling. Similar patterns are seen more broadly in other neotropical taxa (Rull, 2011).

11.3 What triggered the radiation?

One of the concepts that commonly underlies discussion of adaptive radiation is the idea of key innovations (Heard & Hauser, 1995). These are particular adaptations that trigger diversification, perhaps by allowing colonization of a previously unexploited resource. In plant-feeding insects, such events are often associated with the evolution of a novel host range. However, as we saw in Chapter 3, this is not the case in *Heliconius*, as *Passiflora* feeding evolved in the ancestors to the Heliconiini. What then was the trigger for *Heliconius* radiation?

I will say at the outset that much of what follows is speculative, because the particular combination of traits seen in *Heliconius* is unique; it is therefore difficult to disentangle which were important in promoting diversification. Rather than a single innovation, there are a suite of traits including pollen feeding and associated shifts in life history and behaviour that are associated with the radiation (Gilbert, 1972, 1991). Pollen feeding made available much higher quality resources for adults, leading to greater longevity and the evolution of traplining behaviour. This likely allowed a shift to feeding on new shoots, which are more nutritious and have a greater concentration of cyanogens that can be used for chemical defence, but often represent a rare and ephemeral resource. At some point *Heliconius* also evolved the ability to synthesize their own cyanogenic compounds, leading to greater toxicity and independence from host-plant compounds (Engler-Chaouat & Gilbert, 2007).

These changes would have made the butterflies more toxic, and increasingly reliant on aposematism. Warning colour is associated with a change in flight physiology, such that butterflies can be active in cooler conditions, earlier in the day, and in overcast weather, allowing them to compete for floral resources usually dominated by hummingbirds and bees (Gilbert, 1991). Aposematism also leads to increased ecological opportunities to forage and exploit resources unhindered by fear of predation. This can help with pollen feeding, which is far more time consuming than feeding on nectar, and therefore likely to incur a predation risk. As we shall see later in the chapter, diversification in wing pattern has promoted speciation, so increased reliance on aposematism can be directly linked to increased speciation rates (Jiggins, 2008).

The combination of high quality adult food and freedom from predation must have favoured the evolution of increased investment in brain size and behavioural sophistication. This is presumably what led to traplining behaviour and long-term memory, and opened niches for feeding on rare *Passiflora* shoots and *Psiguria* flowers. These novel niche dimensions might have helped promote coexistence of species and perhaps reduced extinction rates of new species. However, pollen feeding arose in the genus *Heliconius* and is not shared with *Eueides*, and the increase in speciation rate began in the lineage leading to *Eueides*. The ecological benefits of pollen feeding are therefore certainly not the only possible explanation for *Heliconius* diversification.

Another factor to consider is the developmental potential to produce phenotypic diversity. There are likely to be differences in the way that patterns develop in *Heliconius* as compared to its relatives, although we are in the early stages of understanding what such differences are (Martin et al., 2014). So innovations in wing development may also have promoted diversification of patterns. Within the radiation, once a large diversity of regulatory modules controlling different wing patches had evolved, the exchange and mixing of these genetic modules through hybridization and introgression would have further promoted diversification (Wallbank et al., 2016).

Chromosome structure changed dramatically on the lineage leading to *Heliconius*, with ten chromosomal fusions on the branch separating *Eueides* (N = 31) and *Heliconius* (N = 21) (Brown et al., 1992; Kozak et al., 2015). John Davey has suggested that this dramatic change in chromosome structure might have played a role in promoting diversification, perhaps by reducing recombination rates across the fused chromosomes (Davey et al., 2016).

Finally, we also cannot rule out the null hypothesis that stochastic differences in speciation and extinction rates could explain the difference in diversification rate between *Heliconius, Eueides*, and their relatives. Since many of the candidate key innovation traits are unique characteristics of *Heliconius*—most notably pollen feeding—there are no phylogenetically independent contrasts that could permit a statistical test for association with diversification rate. In summary, there are a suite of characteristics that are unusual or unique in *Heliconius*, but it remains unclear the extent to which they have contributed to speciation. In contrast, speciation is ongoing right now, and we can study the processes that cause speciation in current populations. This is likely to be a more fruitful approach to understanding diversification.

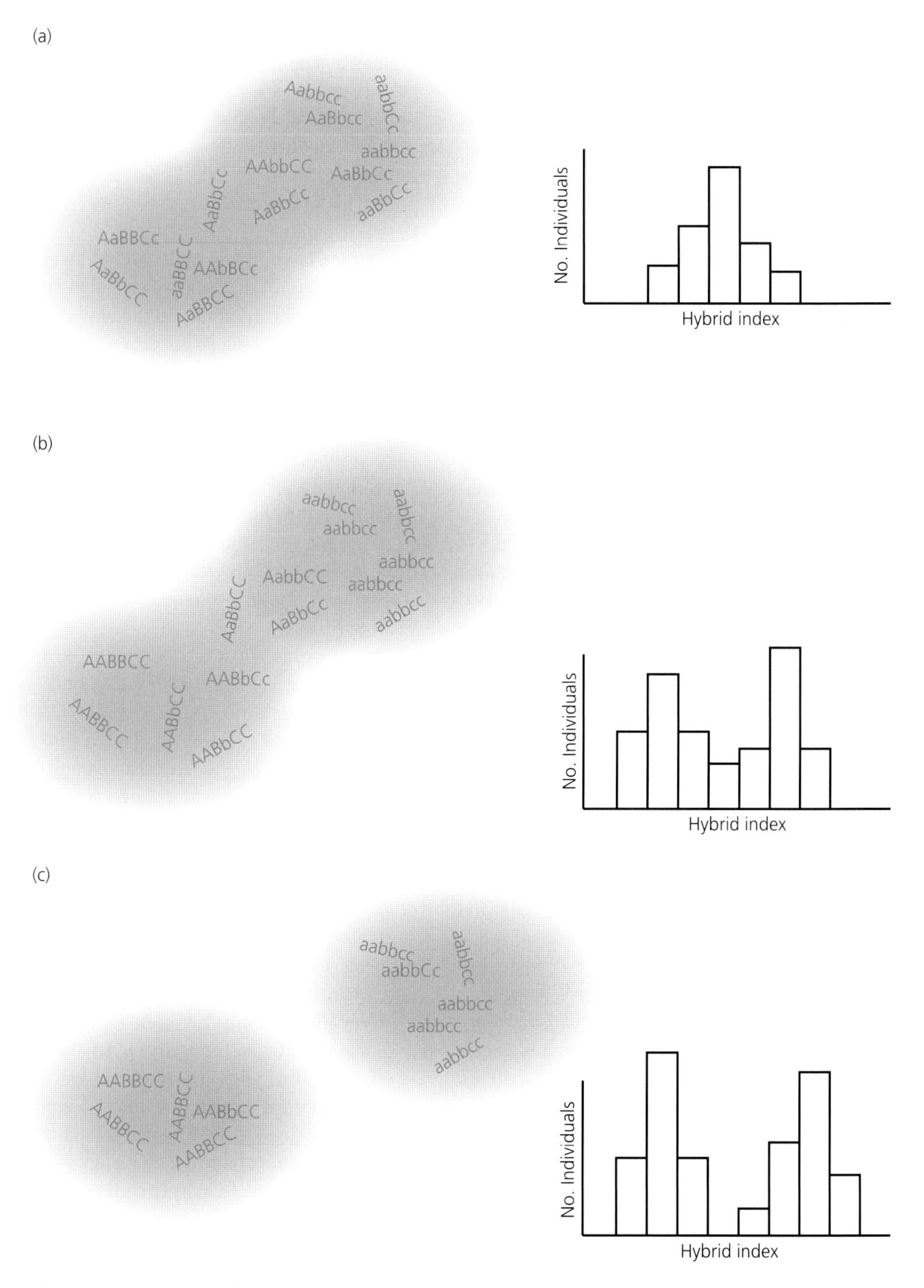

Figure 11.2 Speciation continuum and bimodal hybrid zones

There is a continuum of stages of divergence that can be used to study speciation, which can perhaps be best observed in hybrid zones (Jiggins & Mallet, 2000). Hypothetical clouds of genotypes are shown, plotted both as clouds in a two-dimensional genotype space, and as a one-dimensional hybrid index. (A) Randomly hybridizing wing pattern races that mix fully when in contact; (B) through forms that produce a significant proportion of hybrids but remain distinct; (C) reproductively isolated species where hybrids occur but at very low frequency.

11.4 The species continuum—parapatric races

Speciation is typically a process that takes place over thousands or even millions of years, so we rarely get to see the entire process. However, by studying taxa at many different stages of isolation, we can infer a great deal about the process (Figure 11.2). In *Heliconius*, there are a multitude of different levels of divergence and isolation.

At the earliest putative stage of divergence, many colour pattern races differ in little but their wing patterns, and are separated by hybrid zones. These vary from broad regions of mixing that stretch across hundreds of kilometres—such as the various races that meet in the Amazon basin—to narrow clines of 10 km in width between forms that differ more dramatically in appearance. The hybrid zones provide an opportunity to study the extent of reproductive isolation between taxa that are in an early stage of speciation. Indeed, we have argued that a continuum of different degrees of reproductive isolation is evident in hybrid zones (Jiggins & Mallet, 2000). The earliest stages of differentiation in *Heliconius* therefore seem to be parapatric: geographically separate but with abutting ranges (Mallet, 1993).

Another plausible intermediate step on the road to speciation are polymorphic populations such as *H. cydno alithea* and *H. numata*. In the case of *H. cydno alithea*, there is weak assortative mating between yellow and white forms that represents incipient reproductive isolation (Chamberlain et al., 2009). In contrast, *H. numata* morphs show disassortative mating, which is likely to prevent divergence towards speciation (Mathieu Chouteau, personal communication). Overall, however, polymorphism is rare in *Heliconius* as compared to the great diversity of geographic races, and it therefore seems more likely that the early stages of speciation are parapatric or allopatric rather than sympatric (Mallet, 1993).

11.5 The species continuum—incipient parapatric species

There are also hybrid zones in which divergent populations meet but do not mix freely with one another (Figure 11.2b). The best studied are two species closely related to *H. erato*, both found in dry Andean valleys: *H. chestertonii* in the Cauca Valley in Colombia and *H. himera* in the dry forests of southern Ecuador and the Marañon Valley in Peru. In both cases, where these forms meet *H. erato*, the hybrids form a minority of the population compared to parental genotypes. This deficit of hybrids indicates that reproductive isolation is stronger than in most *Heliconius* hybrid zones.

Unfortunately, these are not especially easy places to work. The Colombian hybrid zone lies in a region of occasional conflict between the FARC and the Colombian military, in the Western Cordillera of the Andes. Despite the dangers, Mauricio Linares and his students Carlos Arias and Astrid Muñoz have sampled this zone and shown that hybrids between *H. chestertonii* and *H. e. venus* are found at a frequency of around 25% (Arias et al., 2008). The Ecuadorean zone is safer, but lies in a series of heavily deforested and remote valleys in the far south of the country. I worked with Owen McMillan for my PhD research in this area; we showed that where *H. himera* and *H. e. cyrbia* meet, hybrids form around 10% of the population (Jiggins, McMillan, et al., 1996). We called these 'bimodal' hybrid zones, because hybrids are outnumbered by their parents (Jiggins & Mallet, 2000). A sufficient degree of reproductive isolation has accumulated to allow the parental forms to coexist with only occasional hybridization.

Political difficulties in Colombia are also a problem for another set of interesting taxa. The melpomene group includes several geographic forms in the early stages of speciation. *H. cydno, H. timareta, H. pachinus*, and *H. heurippa* form a clade sister to *H. melpomene*. The key area for understanding the relationships of these taxa lies in eastern Colombia where, moving from north to south, *H. cydno* is first replaced by *H. heurippa* and then by a series of forms of *H. timareta* that have only recently been discovered.

These areas are poorly studied and difficult to work in. While trying to sample in a possible contact zone between *H. heurippa* and *H. cydno*, Mauricio Linares was kidnapped by the FARC guerrilla and held for three days near a town called Yopal. His subsequent attempts to return to the area have not been successful. The fourth species in the group, *H. pachinus*, is another geographic replacement of *H. cydno* found in Costa Rica.

The status of these taxa therefore remains unresolved: sequence data clearly indicate that *H. cydno*, which is primarily found in the western Andes, and *H. timareta*, found in the east, are genetically distinct; but *H. heurippa* and *H. pachinus* are arguably geographic forms of *H. timareta* and *H. cydno*, respectively. In summary. there are a number of parapatric but partially reproductively isolated 'species' of *Heliconius* that can be seen as an intermediate stage of speciation.

11.6 The species continuum—sympatric and parapatric species

Showing greater genetic divergence, there are also pairs of distinct parapatric species that rarely hybridize. These include *H. charithonia* and *H. peruvianus* (Jiggins & Davies, 1998), *H. demeter* and *H. eratosignis* (Dasmahapatra et al., 2010), *H. sapho* and *H. hewitsoni*, and *H. congener* and *H. eleuchia*.

Finally, at similar or greater genetic distances, are distinct but closely related species that coexist in sympatry. The best studied are *H. melpomene* and *H. cydno/timareta* that coexist broadly through the Andes and Central America. Another example is *H. elevatus* and *H. pardalinus*, which are extremely similar genetically and sympatric across the Amazon. *H. sara* and *H. leucadia* are an unusual case of two sympatric sister species with almost identical wing patterns that coexist throughout the Amazon. In some of these cases, hybrids are known to occur, such that there is still the potential for gene flow—the best studied being *H. melpomene* and *H. cydno/timareta*, which I discuss in more detail later.

11.7 Using the speciation continuum to study speciation

This continuum of forms at different levels of divergence and reproductive isolation has become known as the speciation continuum. We can use this continuum of different stages to learn about the process of speciation. By studying divergent forms alive today, we can plot a potential route to the evolution of new species. In *Heliconius*, this runs from wing pattern races that meet and hybridize freely, through parapatric species that meet and hybridize at a lower frequency, to sister species that can coexist and hybridize much more rarely. The observation of these states in the present day demonstrates that they are stable and plausible intermediates on the road to new species. The argument is similar to one used to explain the evolution of complex structures, such as the eye. The existence of light-sensing structures of varying sensitivity and complexity in different animals supports the idea that the complex eye evolved through a series of viable intermediate steps, but it also helps us reconstruct what those steps might have looked like (Nilsson, 2009).

Of course, the idea that these cases form a speciation continuum is a hypothesis. For the latter stages, we cannot know for sure whether they have indeed passed through earlier stages similar to those observed today. *H. melpomene* and *H. cydno* may or may not have been through a stage of parapatry similar to *H. himera* and *H. erato*. Furthermore, there is no inevitability that earlier stages will go on to evolve into good species. Incipient species such as *H. chestertonii* seem to be stable and might continue to form hybrids with *H. erato* into the foreseeable future, or they may collapse back into a freely hybridizing swarm and 'de-speciate', as has been seen in other organisms (Seehausen, 2006). Nonetheless, the existence of plausible intermediate steps can help build up a possible historical scenario for the formation of species, much like it has for evolution of the eye. Furthermore, by studying examples of closely related species that still hybridize, we can identify processes that contribute to speciation before it becomes irreversible.

11.8 Hybridization is common, but declines with genetic distance

The speciation continuum represents a gradual decline in hybridization rates (Figure 11.3). In the zone of contact between species such as *H. himera* and *H. erato*, hybrids can reach high frequency (10%), but they are restricted to a narrow geographic zone of overlap. In contrast, between broadly sympatric species such as *H. melpomene* and *H. cydno*, hybrids are extremely rare on an individual basis. I have never caught a definite *H. melpomene* × *H. cydno* hybrid in the wild, even though these are the commonest interspecies hybrids known from museum

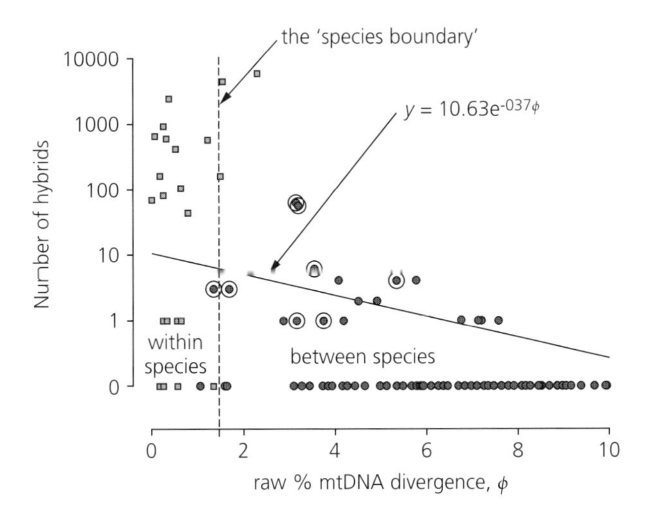

Figure 11.3 Decline in hybridization with genetic distance

The numbers of natural hybrids known between pairs of species plotted on a logarithmic scale against the genetic distance between species (average uncorrected DNA divergence). Haloes around points indicate that backcrosses are known from wild specimens. A least-squares exponential fit of just the species data is shown. However, data are also shown for intraspecific hybridization for comparison, with smaller square points representing the equivalent numbers of intraspecific hybrids in world collections (not used in curve fitting). These data were estimated from the number of intraspecific hybrids in the W. Neukirchen collection (Mallet et al., 2007).

specimens. In total, 68 wild-caught hybrid specimens are known; Mallet has estimated that hybrids occur with a frequency of around 0.05% in natural populations of sympatric *H. melpomene* and *H. cydno* (Mallet et al., 2007). However, this figure is difficult to verify, as collectors like oddities, so museum collections do not represent a random sample of wild populations—hybrids may be rarer than this in most wild populations.

This is not always the case, however. At a single site, San Cristobal in Venezuela, in October and December 2003, Jesús Mavárez collected 103 butterflies, including 60 *H. cydno*, 36 *H. melpomene*, and 7 hybrids: 6 backcrosses to *H. cydno* and 1 backcross to *H. melpomene* (; Mavárez et al., 2006; Mallet et al., 2007). At an overall frequency of 7%, the hybrids were two orders of magnitude greater than the frequency estimated from museum specimens. Genetic analysis and the wing pattern phenotypes indicate that these were at least second- or third-generation backcrosses, suggesting that they must have resulted from multiple generations of hybridization (Mavárez, personal communication). This intriguing observation shows that even between species that rarely hybridize, hybrids can be locally common.

In a wide-ranging survey of museum collections worldwide, Mallet has collected data on the existence of hybrids between all Heliconiini species (Figure 11.3) (Mallet et al., 2007). In total, he identified 161 specimens that are likely interspecific hybrids, all from the genera *Heliconius* and *Eueides*. The majority are from *H. melpomene* × *H. cydno* and *H. himera* × *H. erato*, but in total 16 species are involved in at least one hybridization event (note that *H. chestertonii* was not considered a separate species to *H. erato*, so these hybrids were not included here). This indicates that over 25% of heliconiine species are involved in hybridization with at least one other species (26–29%, depending on species definitions). This is a higher proportion than found in similar surveys of other groups, including birds, European mammals, and European butterflies (9%, 6%, 11%, respectively), although similar proportions are found in American warblers or birds of paradise (Mallet, 2005). In reality, the precise numbers are not comparable, because genetic distances between species vary considerably across groups. However, one conclusion is clear: in all these groups, entities traditionally considered as distinct species can and do hybridize with one another at appreciable frequencies.

The identity of these hybrid specimens—and therefore the occurrence of interspecies hybridization—has been aggressively challenged by Andrew Brower (Brower, 2010, 2013): 'In sum, so many of the Mallet et al. records are dubious, at least for *H. cydno* × *H. melpomene* "hybrids", that this dataset must be discounted' (Brower, 2013, p. 4). He has argued that they are essentially all fraudulent specimens either generated in insectary hybridizations or incorrectly identified.

However, some specimens have been verified with molecular analysis, so are indisputable. Most remarkably, a male *H. melpomene* × *H. ethilla* hybrid collected near Moyobamba in northern Peru has been verified as a likely F_1 using molecular analysis, demonstrating natural hybridization between the silvaniform and melpomene clades that have been separated for around 4 million years (Dasmahapatra et al., 2007). An *H. melpomene* × *H. cydno* hybrid was collected by Keith Brown in Colombia in 1971 (Brown & Mielke, 1972); Mavárez has collected hybrids in Venezuela; and Linares, Kronforst, and Gilbert have all collected interspecies hybrids included in these analyses (Mallet et al., 2007). Furthermore, Mallet has spoken to many of the amateur collectors who collected the hybrid specimens found in private collections to confirm their identity.

In short, even if many of the museum hybrids are fake and the collectors are lying—which seems unlikely—there is at least some good evidence for recent hybridization supported by molecular analysis and from well-documented sources. A further line of evidence in support of the reality of these hybrids is that collection localities of hybrids correspond to areas where putative parental species have the correct combinations of patterning genes that can give rise to the hybrid patterns. Thus, although it is possible that a few of the hybrid specimens may be problematic, the broad picture of hybridization in *Heliconius* is well supported.

11.9 What is a species?

Closely related *Heliconius* taxa therefore exhibit wide variation in their levels of genetic divergence, their frequency of hybridization, and their degree of geographic overlap. Where should we draw the line and consider the break for a species?

One possibility would be to exclude all possibility of hybridization—in effect, the strictest possible application of the biological species concept. This would effectively lump all of the melpomene and silvaniform group into one species (currently 15 species). It would go against the stable taxonomy of the last 40 years, and would lose many ecologically, phenotypically, and genetically distinct taxa into a single species. The 15 current species in this group are distinct ecological entities, and we need some way to communicate about their biology effectively, so I believe that such an approach would be unhelpful.

Alternatively, we could adopt a looser biological species concept and allow *some* hybridization. In the hybrid database, there is a big jump in the number of museum hybrids between *H. ethilla* × *H. besckei* with 6, and *H. himera* × *H. erato* with 57—perhaps that is the place to draw a line in the sand. However, that would leave *H. melpomene* and *H. cydno* as conspecifics, which makes no sense biologically—they are phenotypically and ecologically distinct taxa that coexist across a broad area. Under any natural history definition, they would be distinct species.

The frequency of hybrids in sympatry would be a better criterion, which might leave *H. himera*/*H. erato* as the same species (10% hybrids) but would separate *H. melpomene* and *H. cydno* (~0.1% hybrids). However, the biological species concept is problematic, because if we define species on the grounds of interbreeding but then allow some hybridization, it is unclear how much is too much.

In reality, what biologists do is not to measure interbreeding itself, but rather study the outcome of interbreeding—i.e. genetic differences between groups of individuals. Bird species can be readily distinguished using a field guide without the need to wait for the breeding season and determine whether individuals can interbreed or not. It would therefore make sense to use a species definition based on the observation of measurable differences between groups of individuals.

Inspired by work on *Heliconius*, Mallet (1995) has proposed the 'genotypic cluster definition', which does just that. Where populations coexist, they are defined as distinct species if they form two distinct multilocus genotypic clusters. This pattern-based definition separates the definition and identification of species from the processes that keep them apart.

We can recognize species based on genetic and phenotypic patterns, and then study the multitudinous processes that allow them to coexist. Furthermore, the definition recognizes that species can persist even in the face of persistent interbreeding and gene exchange. I believe that this definition is more consistent with the reality of how biologists recognize species.

In *Heliconius*, I will apply this genotypic cluster definition of a species. This generally means that most phenotypically distinct geographic races are considered as subspecies, on the grounds that hybridization occurs freely where they come into contact. In contrast, where geographic forms come into contact and overlapping populations are dominated by parental genotypes, they are considered distinct species. *Heliconius erato* and *Heliconius himera*, for example, meet in a narrow hybrid zone in southern Ecuador, and hybrids form about 10% of the population. Since most individuals collected in these regions can be assigned to one or the other cluster, we consider them distinct species (Jiggins, McMillan, King, et al., 1997). The genotypic cluster definition does therefore tend to recognize distinct species at quite an early stage of speciation, when interbreeding is still ongoing but at a reduced rate. There are of course a wealth of other possible species definitions, but I don't intend to discuss them here.

11.10 Relationships between species and the genomic evidence for hybridization

The existence of hybrids between species does not necessarily imply that species can exchange genes. If hybrids were always sterile or simply unsuccessful in finding mates or surviving, they might represent a genetic and evolutionary dead end.

Do species pairs that produce hybrids in the wild also show evidence for genetic exchange? The challenge is to distinguish between the various causes of discordant histories across the genome and identify the signal of hybridization (Figure 11.4). We can now tackle this question using genomics, by obtaining sequence data from multiple individuals sampled from wild populations (Martin et al., 2013).

We have taken advantage of the geographic distribution of *H. melpomene* and *H. cydno* to estimate the extent to which they share genes where they live together. *H. cydno* is found only in the Andes and Central America, so we can use *H. melpomene* individuals sampled from eastern South America to provide a control sample (Bull et al., 2006; Martin et al., 2013). These individuals live thousands of kilometres from the current range of *H. cydno*, so have not had any recent contact with *H. cydno*. (This strategy has other benefits, as the delights of fieldwork in French Guiana include dropping into the local cafe for pan du chocolat and coffee for breakfast, before heading out into the nearby rainforest in a rented Renault Clio field vehicle; the few roads are so good that a four-wheel drive is barely necessary.)

Simon Martin analysed genomes sampled from French Guiana, Panama, and Peru, and the results were a surprise. Around 40% of the genome was more similar between *H. melpomene* and *H. cydno* in Panama than it was between populations of *H. melpomene* from across its range (Figure 11.5). In other words, for a remarkably large proportion of the genome, the two species were more similar to each other where they occur together, than to other populations of the same species.

Before we can attribute this similarity to hybridization and gene flow, however, we need to rule out alternative explanations (Figure 11.4). Some shared genetic variation is expected between species because of their shared history. Any single population contains genetic variation, and when such a population is split in two, the new populations will continue to share alleles for some period of time. The process of sorting ancestral variation depends on population size, with larger populations continuing to share variation longer.

However, we can control for this to a large degree with our sampling design. If shared variation were simply due to shared ancestral variants, we would expect a similar amount of sharing between *H. cydno* and any *H. melpomene* population. This is not the case, because only about 2% of the genome shows clustering of *H. cydno* with *H. melpomene* from French Guiana, compared to 40% with Panama. And this seems to be true of any sympatric pair of populations, as a similar pattern is seen between *H. timareta* and *H. melpomene* in Peru. These populations also share much more variation with each other in sympatry than with other populations of either species (Martin et al., 2013).

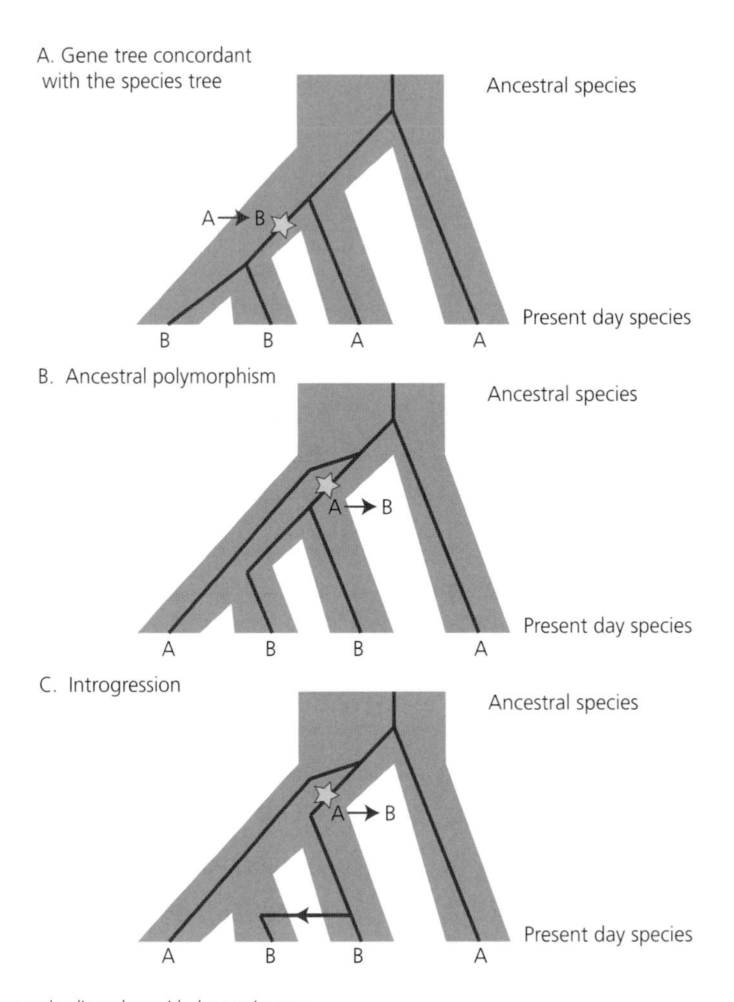

Figure 11.4 Gene trees may be discordant with the species tree

A hypothetical four-species tree, showing a gene tree that follows the species tree (A), and two possible discordant trees due to either (B) shared ancestral genetic variation or (C) recent hybridization and introgression. The pattern of single nucleotide variation across the four taxa is shown below each tree, demonstrating that similar patterns of allele sharing can arise under either shared ancestral variation or recent gene flow. The ABBA-BABA method for detecting gene flow counts the number of sites that show an ABBA pattern compared to BABA (not shown). A similar number of each class are expected due to shared ancestral variation, but an excess of one class, such as ABBA sites, indicates gene flow (Green et al., 2010).

The sampling design is similar to studies of hybridization between modern humans and Neanderthals. Modern humans spread out of Africa and came into contact with their Neanderthal relatives in Europe. Archaeological evidence suggests that the two overlapped for perhaps as little as 3–5,000 years, about 40,000 years ago (Higham et al., 2014). Genome sequences acquired from modern humans in Europe and Africa, in addition to ancient Neanderthal bones, allows for a similar test for allele sharing as we have conducted in the butterflies (Green et al., 2010). Modern humans from Africa provide a 'control' population that has never been in contact with Neanderthals, much like our *H. melpomene* population from French Guiana. There is more funding available for human archaeology than *Heliconius* palaeontology, so we have a much clearer idea of the history of contact in the case of the humans, but the genetic comparison is similar.

Around 1.15–1.38% of our genomes are derived from Neanderthal ancestry (Sankararaman et al., 2014). I am especially proud of my own hybrid percentage, which, at 2.8%, seems to put me at the upper end of Neanderthal-ness (this was estimated

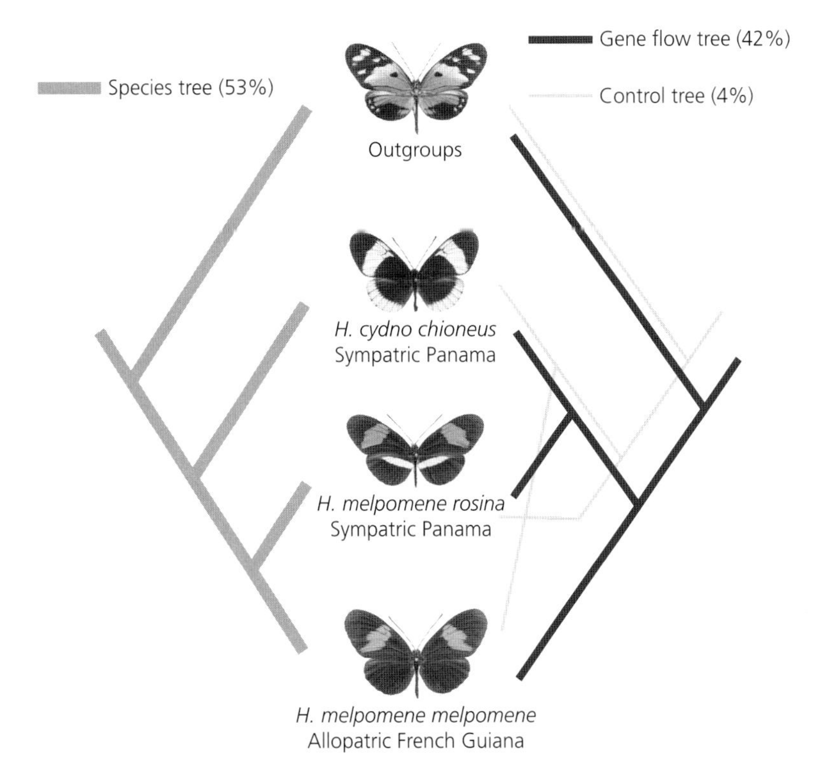

Figure 11.5 Discordant gene trees provide evidence for persistent gene flow between species

Four-taxon trees were calculated in 100 kb windows across the genome. There were 2,848 trees in total. The majority agreed with the expected branching pattern of the species (53%), but a large proportion (42%) showed a pattern that grouped populations in sympatry (gene flow tree), compared to only 4% that showed the reverse (control tree) (Martin et al., 2013).

by the genotyping service 23andMe so isn't actually very comparable). However, even that pales to insignificance when compared to *Heliconius melpomene*, in which some 40% of the genome is shared with its close relative. Although many have argued that gene flow between species is common, I think the extent to which this is actually true is really a huge surprise and shakes our current understanding of species.

What are the implications of all this gene mixing for the relationships between species? In the case of *H. melpomene* and *H. cydno*, a phylogeny inferred from whole genome sequences reconstructs the two species as mutually monophyletic groups. All *H. melpomene* are closer relatives to one another than any is to *H. cydno*, and vice versa. However, there is no particular reason this should be the case, and a little more gene flow could mean that, on average, Panamanian individuals of the two species are closer to each other than to other populations of their same species.

Indeed, it seems that something similar is happening in *H. elevatus* and *H. pardalinus*, two species that are broadly sympatric across the Amazon but in which most of the genome is almost panmictic (Dasmahapatra and Mallet, personal communication; Kryvokhyzha, 2015). Only a few narrow regions of the genome group individuals from across the geographic range according to recognised species relationships. A similar situation has been described in *Anopheles* mosquitoes, where gene flow between species is so extensive that only a very small proportion of the genome (perhaps less than 10%) corresponds to the known morphological and ecological species boundaries (Neafsey et al., 2015).

In essence, species can in some cases therefore be defined by just a few regions of the genome. The genotypic clusters in this case are sets of genes that are found in strong association with one another across the genome and that define whether an individual is an *H. elevatus* or *H. pardalinus*. The rest of the genome can be mixed to varying degrees through hybridization.

11.11 Patterns of reproductive isolation in the *Heliconius* melpomene clade

After reading about the extent of hybridization and genetic mixing between species such as *H. melpomene* and *H. cydno*, you may be seriously doubting their status as species, or just wondering about my rather lax definition of species. Species just aren't meant to mix up their genes in this way. So now I will describe the very extensive barriers to gene flow between these species, which will hopefully convince you once again that these really are good species by any definition.

The mechanisms that reduce mixing between species—and therefore allow species to persist and separate— fall under the catch-all term 'reproductive isolation' but are also sometimes termed 'barriers to gene flow'. We have spent many years studying the populations of *H. melpomene* and *H. cydno* around Gamboa, in the Canal zone of Panama, so I will focus initially on these and summarize the barriers roughly in the order in which they occur in the life-cycle.

First, the two species are found in different habitats. We often work along a dirt road known as Pipeline Road, which follows the line of a disused oil pipeline installed parallel to the Panama Canal during the Second World War. It runs from the village of Gamboa into the rainforest of the Soberanía National Park. *H. melpomene* are found around Gamboa, and along the disturbed habitats beside the canal. *H. cydno* are seldom seen in these areas, but rather are found deeper into the forest, starting from around 3 km along Pipeline Road, but more commonly after about 7 km.

The two overlap, but during our collections in 1999, only about 25% of individuals were found in areas where the two species come into contact (Estrada & Jiggins, 2002). They also tend to differ in the pollen species found in their pollen loads, reflecting different habits in visiting flowers. Finally, the two species have different strategies for exploiting larval host plants, with *H. cydno* oliphagous, laying on numerous species of *Passiflora*, and *H. melpomene* essentially a specialist on *P. menispermifolia* (Table 3.1; Merrill et al., 2013). Even where they are seen flying together, females of the two species may be searching for host plants in slightly different places.

If the species do encounter each other, the strongest barriers to gene flow occur through their mating behaviour. Individuals of the two species simply do not like to mate with each other. Over the years, we have carried out many hybridization experiments in which we mate *H. cydno* and *H. melpomene* in the insectary. However, these are tiresome because it is hard to persuade them to mate. Perhaps this should be unsurprising given how rare hybrids are in the wild (see earlier). Wild males are much more vigorous that those raised in cages, so we always need to collect *H. melpomene* males from the field. We then hang female *H. cydno* pupae, raised with great care from eggs in the insectary, in cages with these wild *H. melpomene* males. Even in this extreme scenario, in which several males are enclosed in a confined space, with no choice and exposure to freshly emerged virgin females, matings are very rare. Occasionally a male will mate with a freshly emerged female before she has had a chance to harden her wings after eclosion. Mating in the reverse direction—between *H. cydno* males and *H. melpomene* females—is even more rare.

This scenario illustrates several important aspects of mate choice in *Heliconius*. In the first hour or two after emergence, females cannot reject males, so there is no scope for females to choose. However, matings are rare even during this window, demonstrating that males are choosy. However, females are also choosy, and reject males once their wings have hardened. The rare matings that do occur are always in the first hour or so after eclosion, because after that females strongly reject suitors of the wrong species. In summary, both males and females strongly prefer mates of their own species.

Controlled experiments confirm the low probability of interspecific mating between species (Jiggins, Naisbit, et al., 2001a; Naisbit, Jiggins, & Mallet, 2001). In fact, interspecies matings never occurred in such experiments, mainly because the sample sizes were not large enough to capture rare events.

We have carried out both 'no-choice' trials, in which a single virgin female and a male of different species are placed together in a small cage, and 'choice' trials, with a male and a female of each species (Jiggins, Naisbit, et al., 2001a; Naisbit et al., 2001). The cage experiments are useful, but ideally we would also like to estimate the frequency of mating in the wild. However, the rate of interspecies mating is so low that it is impractical—in many years of collecting around the Gamboa area in Panama and bringing females into the insectaries, we have never collected a female that produced hybrid offspring. This suggests that the rate of interspecies mating is vanishingly low in wild populations.

If the two species do successfully mate, the hybrid eggs and larvae hatch and survive without any measurable problems. Similarly—although this has never been quantified—adult hybrids survive, pollen feed, and live as long in insectaries as do their parents. Although hybrids survive well in the insectary, they have unusual wing patterns. We therefore expect that these might not do so well in the wild, because predators will not have had a chance to experience and avoid their wing patterns. Mimicry theory therefore predicts ecological selection against hybrids.

I had several failed attempts to measure this selection against hybrids in the wild, either tethering butterflies on strings (they didn't like that much), releasing marked butterflies that had been sterilized with superglue (too many just few away and were never recovered), or painting *H. cydno* to look like hybrids (this worked, but failed to demonstrate survival differences, perhaps due to small sample sizes). Richard Merrill finally managed to design an experiment that did show selection against hybrids, using 1,440 plasticine butterfly models with printed wings (Merrill et al., 2012). The printed colours were matched as closely as possible to real butterflies, using computer models of bird vision. Only 58 of these were attacked, but in total 30 of the attacks were on F_1 patterns, with only 10 attacks on *H. melpomene* and 18 on *H. cydno*. Several years earlier, Vanessa Bull had carried out a series of experiments using live birds caught around the Gamboa area and shown a similar pattern, with hybrid butterflies more likely to be attacked (Merrill et al., 2012).

Overall, these model experiments in the field combined with predator experiments show that selection does indeed act against F_1 hybrids. Of course, this is exactly what we expect for rare wing patterns, and is consistent with the many experiments supporting frequency-dependent selection on warning colour discussed in Chapter 7, but it confirms strong selection specifically against F_1 interspecies hybrids.

Hybrids also have difficulty obtaining mates. F_1 hybrids mate readily among themselves but generally fail to obtain matings with either *H. melpomene* or *H. cydno* (Naisbit et al., 2001). Given the rarity of F_1 hybrids in the wild, it is far more likely that hybrids would interact with their parental species than with one another. This 'unsexy hybrid' effect probably has different implications for each sex. Females only need to mate once, so hybrid females will probably find a mate eventually, but male hybrids are likely to suffer more strongly reduced fitness as a result of lower mating success.

If hybrids do manage to survive and mate, there is yet another problem to overcome. Female hybrids, at least in the crosses that we have achieved, are completely sterile. They can mate and sometimes lay eggs, but these invariably fail to hatch (Naisbit et al., 2002). Males, on the other hand, are fine, and can be mated, albeit reluctantly, to either species to produce viable offspring. This allows gene flow between the species. In the backcross offspring of these matings, the females again show varying degrees of sterility, depending in part on which Z chromosome they inherit (Naisbit et al., 2002). In summary, *H. melpomene* and *H. cydno* are, by virtually any definition of the term, distinct biological species, despite the abundant evidence for genetic exchange between them.

11.12 Quantification of reproductive isolation along the continuum—pre-mating isolation

I have used the Panama populations of *H. melpomene* and *H. cydno* to give an overview of different forms of reproductive isolation in *Heliconius*, partly because they are the best-studied species pair to date. However, there have been a series of studies of reproductive isolation of many different species pairs at varying degrees of divergence. We can now quantify in some detail levels of reproductive isolation for many pairs of species. Claire Mérot used a method for quantification of isolation developed by Sobel and Chen (2014), and collated data from across the melpomene clade (Table 11.1). Looking across many species pairs can help determine the order in which barriers evolve, and their relative importance at different stages of the process.

The first barriers to evolve between populations are caused by differences in wing pattern. Divergent wing pattern races can be seen as a candidate early stage in speciation, because predator selection on wing pattern produces a partial barrier to gene flow. As outlined in Chapter 7, individuals crossing a wing pattern hybrid zone are strongly

Table 11.1 Quantification of reproductive isolation in the melpomene/cydno clade

	Spatial	Mate choice	F1 hatch rate	Larval survival	Adult survival	F1 mating to Sp. 1	F1 mating to Sp. 2	F1 fertility
Within species								
CYxCW	?	0.26	–	–	0.19	–	0.26	–
CWxCY	?	0.31	–	–	0.19	–	0.26	–
MaxM	x	0.40	–	–	0.33	–	–	–
MxMa	x	0.00	–	–	0.33	–	–	–
MGxMP	x	1.00	0.00	–	x	x	x	0.00
MPxMG	x	0.48	0.00	–	x	x	x	0.43
Parapatric species								
CCxH	x	0.56	0.00	0.00	x	x	x	0.00
HxCC	x	0.98	0.00	0.00	x	x	x	0.00
CxP	x	0.91	–	–	–	0.00	0.94	–
PxC	x	1.00	–	–	–	0.00	0.94	–
Sympatric species								
CPxMP	0.73	1.00	0.00	?	0.35	0.20	0.52	0.31
MPxCP	0.73	1.00	0.00	?	0.35	?	?	?
TtxM*	0.77	0.86	0.00	0.00	–	0.46	0.00	0.32
MxTt*	0.77	0.85	0.00	0.00	–	0.87	0.00	0.08
HxMC	0.75	0.93	0.00	0.00	–	0.44	0.29	0.33
MCxH	0.75	0.90	0.00	0.00	–	0.75	0.20	0.00
TfxMm*	?	0.96	0.00	?	?	?	?	0.33
MmxTf*	?	0.96	0.00	?	?	?	?	0.19
CPxMG	x	0.78	0.00	?	x	x	x	0.49
MGxCP	x	1.00	0.00	?	x	x	x	0.33
CCxMC	?	0.82	0.00	0.00	?	?	?	0.33
MCxCC	?	0.88	0.00	0.00	?	?	?	0.22

Legend: Numbers represent the strength of particular barriers to gene flow estimated for each pair of species. x means allopatric or parapatric comparisons so ecological habitat segregation not applicable; – means that no data are available but likely to be 0; ? means missing data. Each pair of taxa is shown by two rows representing reciprocal crosses (female first). * indicates mimetic species pair.

Taxonomic abbreviations: CY: *H. cydno alithea* yellow form; CW: *H. cydno alithea* white form; Ma: *H. m. aglaope*; M: *H. m. amaryllis*; MG: *H. m. melpomene* French Guiana; MP: *H. m. rosina* Panama; CC: *H. c. cordula* Colombia; H: *H. heurippa*; C: *H. c. galanthus*, Costa Rica; P: *H. pachinus* Costa Rica; CP: *H. c. chioneus* Panama; Tt: *H. t. thelxinoe*; MC: *H. m. melpomene* Colombia; Tf: *H. t. florencia*; Mm: *H. m. malleti*. Data compiled by Claire Mérot.

selected against. This leads to selection against immigrants from divergent populations, a form of reproductive isolation that Patrick Nosil has termed 'immigrant inviability' (Rundle & Nosil, 2005). Genomic evidence shows that wing pattern races differ strongly at just a small number of genomic regions (Nadeau et al., 2012). Wing pattern races are an example of local adaptation, in which selection is reducing gene flow at just a few genetic loci. These forms therefore exhibit a form

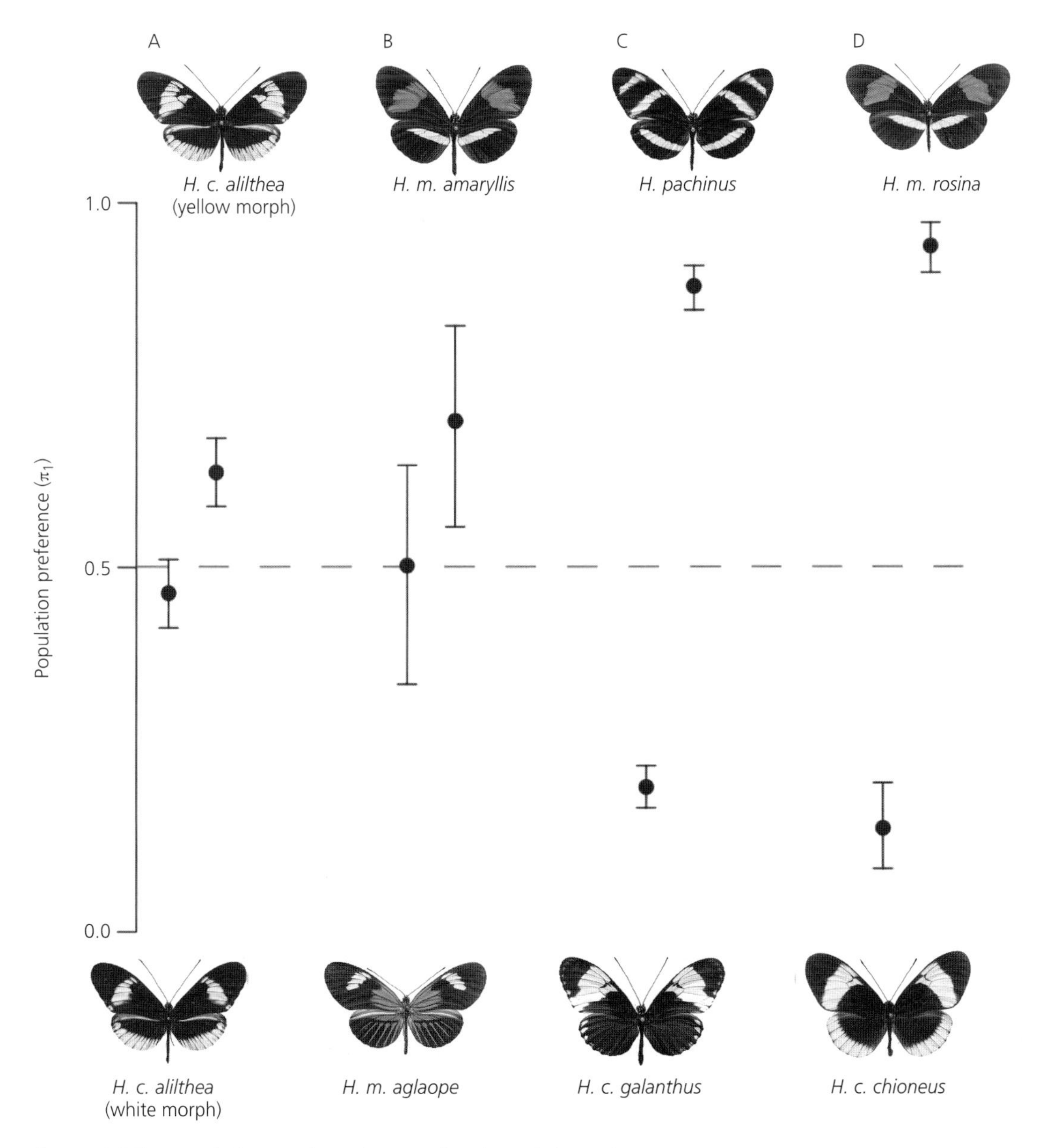

Figure 11.6 The strength of mate preference increases along the speciation continuum

Population-level mate preferences for (A) within-population comparison between yellow and white morphs of *Heliconius cydno alithea*; (B) warning-colour race comparison between *H. melpomene aglaope* and *H. melpomene amaryllis*; (C) parapatric species comparison between *H. cydno galanthus* and *H. pachinus*; (D) sympatric species comparison between *H. cydno chioneus* and *H. melpomene rosina*. 1 = complete preference for the red or yellow female type of the pair; 0 = complete preference for the white or rayed female type of the pair. Reproduced from Merrill et al. (2011a).

of reproductive isolation that is highly localized in its effect on the genome.

Acting in addition to divergence in wing pattern and mimicry is assortative mating based on colour cues (Figure 11.6). Even sympatric colour morphs within species, such as forms of *H. cydno alithea*, show weak preferences to mate with others that share their colour pattern (Chamberlain et al., 2009). Geographic races of *H. melpomene* show similar preferences (Merrill et al., 2011a); they are more likely to court artificial models with their own wing patterns (Jiggins, Estrada, & Rodrigues, 2004). So, even within species, there is evidence for pre-mating isolation due to colour pattern–based mating preferences, combined with predation selection on wing patterns, which can act as both pre-mating (immigrant inviability) and post-mating isolation (through selection against hybrid wing patterns).

Pre-mating barriers are much stronger and consistent in comparisons between both parapatric and sympatric species (mostly > 80%). Variability between individuals in preference is also lower at later stages along the continuum (Merrill et al., 2011a), suggesting that they are subject to stronger selection. Overall, the evidence therefore suggests that mate preferences are some of the first barriers to arise during speciation, and that they continue to increase in strength during the process of divergence (Merrill et al., 2011a; Mérot, 2015).

In contrast, the ability of F_1 hybrids to find themselves a mate is quite variable between different pairs of populations (Mérot, 2015). This perhaps suggests that the genetic basis for mate preferences varies between populations—the behaviour of hybrids depends on the genetic dominance of mating preferences and their patterns of inheritance in hybrid genomes. However, the reasons for variability in F_1 mating success are largely unknown.

11.13 Quantification of reproductive isolation along the continuum—post-mating isolation

As mentioned earlier, wing pattern differences between populations also cause ecological post-mating isolation. Rare hybrid phenotypes are selected against by predators. Such selection may be relatively weak in inter-racial hybrid zones, because hybrids can dominate hybrid zone populations, so are not especially rare. In such hybrid zones, selection mainly acts at the edges of the zone, where individuals migrate into a population dominated by one of the parental species. In addition, many wing pattern alleles show almost complete dominance, so in some hybrid zones, the F_1 hybrids appear very similar to one of the parental forms. However, where pre-mating isolation is strong and colour pattern divergence much greater—as commonly the case with interspecific hybridization—post-mating mimicry selection is also expected to be strong.

Incompatibilities that reduce the fertility of hybrids evolve later in the process. Comparisons of the most closely related races and species are generally entirely compatible with one another genetically. That is, hybrids are both viable and fertile when raised in the laboratory. However, most sympatric species show hybrid sterility limited to females. In *H. timareta* × *H. melpomene* crosses it is asymmetric, with female hybrids having an *H. timareta* father being fertile, but those with an *H. melpomene* father being sterile (Table 11.1). In contrast, *H. cydno* × *H. melpomene* crosses produce sterile females in both directions, as far as has been determined (Naisbit et al., 2002). The sterility of female, but not male, hybrids fits with a well-established pattern in speciation biology known as Haldane's rule, which states that if the sexes differ, it is the heterogametic sex that is sterile or inviable (heterogametic refers to the sex with divergent sex chromosomes, XY males in mammals, ZW females in butterflies).

One of the major puzzles to early evolutionists was the origin of hybrid sterility—after all, if there is one trait that should be ruthlessly eliminated by selection, it is surely sterility. Almost by definition, genes that cause sterility cannot be passed on to the next generation. The simplest genetic model for sterility involves a single gene, such that Population 1 is aa and Population 2 is AA, with sterile hybrids Aa—but this is difficult to evolve because the population needs to pass through a stage of mainly sterile Aa individuals to get from aa to AA.

There is, however, a simple solution to this problem. If at least two genes are involved, then sterility can be caused by an interaction between alleles at each locus. Population 1 evolves a new allele A, so its genotype is AAbb, while Population 2 evolves a new allele B, so its genotype is aaBB. Hybrids are AaBb. The interaction between A and B can cause sterility because these two alleles have never been found together in the same individual during the history of the two lineages (Figure 11.7).

The model was described by Dobzhansky (1936) in an elegant paper on hybrid sterility in *Drosophila*, and later elaborated by Müller (1942), so became known as the Dobzhansky-Müller model. Allen Orr (1996) noticed that William Bateson had outlined something very similar in 1909, so it is now more correctly known as the Bateson-Dobzhansky-Müller model (BDM).

At first sight, it seems quite surprising that one sex should be completely sterile while the other is completely fertile. Why such an extreme sex difference? However, such patterns are actually common and form one of the few rules in evolutionary biology; as noted earlier, named after J. B. S. Haldane. Indeed, the fact that it is almost always the heterogametic sex that first shows sterility or inviability in interspecific crosses is expected under the BDM model if the alleles causing incompatibility are mostly recessive (Coyne & Orr, 2004). In the F_1 generation, male butterflies will have one copy of the Z chromosome from each parent, so recessive alleles from each parent are

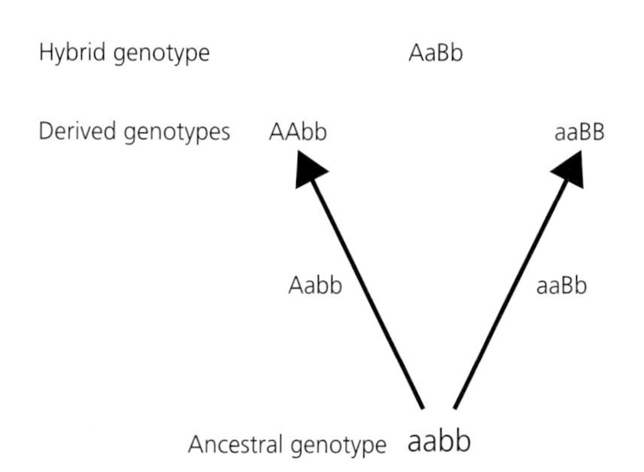

Figure 11.7 The two-locus model for the evolution of Bateson-Dobzhansky-Müller (BDM) incompatibilities

The ancestral genotype, aabb, diverges at different loci in the two lineages to produce daughter species AAbb and aaBB. Hybrids are AaBb, and are unfit because of interaction between the derived A and B alleles, which have never been in the same individual during the divergence of the two species.

masked, whereas the female has just a single Z, so recessive alleles are expressed. Hence, recessive alleles on the sex chromosome are expressed in the heterogametic sex but not the homogametic sex. This has been called 'dominance theory' and provides a general explanation for Haldane's rule in both male- and female-heterogametic taxa.

When fertile F_1 melpomene × cydno males are crossed to the parental species (a backcross), their offspring show variation in degree of fertility. In some cases, these families show a large effect of the Z chromosome, such that, for example, hybrids with a mostly *H. melpomene* genome but an *H. cydno* Z chromosome are more likely to be sterile than those with an *H. melpomene* Z chromosome (Jiggins et al., 2001b; Naisbit et al., 2002). This 'large-Z' effect is also seen in *Drosophila* and is consistent with the dominance theory explanation for Haldane's rule (Coyne & Orr, 2004). More generally, patterns of sterility across the melpomene clade suggest that sterility results from gradual accumulation of genetic factors that interact to cause sterility, and are largely recessive in nature. Nonetheless, the precise nature of the genes for sterility in *Heliconius* remains unknown.

In addition to dominance theory, there are many other explanations for the evolution of incompatibilities and Haldane's rule. Those that have some empirical support include faster-male theory, which suggests that sexual selection drives faster divergence and accumulation of alleles that cause sterility in males (Wu, Johnson, & Palopoli, 1996); faster-X theory, which implicates more rapid evolution on the sex chromosomes (see also Chapter 8) (Charlesworth et al., 1987); genomic conflict, whereby alleles involved in conflicts (e.g. meiotic drive) evolve rapidly and cause incompatibility between populations (Hurst & Pomiankowski, 1991); and, finally, interaction with microorganisms, such as endosymbiotic bacteria or even gut microbiota (Brucker & Bordenstein, 2012).

None of these ideas have direct support in *Heliconius*, but some may be important. We can rule out faster-male theory, because it is females that are sterile. Indeed, female-heterogametic taxa provide one of the main lines of support for the general importance of dominance theory, because these taxa separate the effects of maleness from those of sex chromosome inheritance. It seems likely that faster-X effects (or faster-Z, in our case) are important, although testing for the distribution of incompatibility factors on different chromosomes is challenging, and may not be feasible in *Heliconius*. Genomic conflict certainly could be driving the fixation of alleles in divergent populations, but this will probably not be testable until we can identify genes causing sterility in *Heliconius*. In mice, for example, identification of the genetic mechanism of hybrid sterility has clearly implicated conflict during meiosis as a cause of the evolution of incompatibility (Campbell, Good, & Nachman, 2013). We can largely rule out endosymbiotic bacteria, based on patterns of inheritance that are not compatible with factors inherited through the maternal line (Jiggins et al., 2001b), but a more general influence of the microbiota has never been tested (this could fairly easily be done by treating hybrids with antibiotics).

There is one case of hybrid incompatibility that does not show sex-specific effects. When *H. erato* and *H. chestertonii* are crossed, there is a partial reduction in the hatch rate of hybrid eggs in the F_1 generation that affects males and females equally (Muñoz et al., 2010). Furthermore, in backcrosses, expression of sterility depends on the direction of the original F_1 cross. These patterns might be compatible with the effect of an endosymbiont or other maternally inherited factor, although a survey of endosymbiont bacteria did not show a consistent difference between the two species—*H. chestertonii* is infected with *Wolbachia* bacteria, but they are at a low frequency, so don't seem to explain incompatibility with *H. erato* (Muñoz et al., 2011). The genetic basis for this incompatibility is therefore unknown and will require further study.

In summary, female-heterogametic taxa such as butterflies, in which females are ZW, provide an important test of theories to explain Haldane's rule. In general, although the sophistication of genetic analysis remains crude in butterflies compared to what is possible in *Drosophila*, the patterns in *Heliconius* have supported theories derived from flies, and in particular the importance of dominance theory as a general explanation for Haldane's rule.

Overall, post-mating isolation due to sterility is a barrier similar in strength to predator selection against hybrids. Furthermore, divergence in wing pattern and assortative mating arise much earlier along the speciation continuum. The speciation of *H. himera* and *H. erato* has occurred with no evidence for hybrid sterility or inviability, demonstrating that speciation can occur without such incompatibilities. Genetic incompatibilities therefore are not absolutely necessary for speciation, and play a role primarily in the latter stages.

11.14 The problem of speciation

Speciation is a multifactorial process involving a wealth of different processes that keep species apart. The challenge is to understand the forces that drive populations to diverge, and the mechanisms that allow them to do so without being blended into one genetic soup by hybridization. In other words, as populations diverge but continue to interbreed, what maintains them as different and prevents the production of intermediates with mixed-up combinations of the parental alleles? How are the multiple genetic differences required to adapt to different niches maintained in association with one another?

This problem was perhaps best outlined by Joe Felsenstein, in a classic paper entitled 'Skepticism towards Santa Rosalia, or Why Are There So Few Kinds of Animals?' (Felsenstein, 1981). The Santa Rosalia of the title refers to an earlier paper by Hutchinson (1959) in which he describes the concept of the ecological niche, with reference to his own work on water beetles near a shrine dedicated to Santa Rosalia in Sicily.

Felsenstein (1981) described a simple mathematical model involving three genetic loci. Two influence ecological traits and are involved in adaptation to two alternative niches (Locus A and B). The third is a mating locus, and causes assortative mating between individuals as they adapt to these niches (Locus C). Speciation results when most of the individuals in the model are well adapted to one niche and have a corresponding mating type—that is, most genotypes are either aa, bb, cc or AA, BB, CC. The difficulty is that recombination between individuals produces intermediate genotypes (aA, BB, cc; AA, bB, Cc, etc.). If the hybrids are abundant, then instead of two species, the population remains a single swarm of variable individuals. Felsenstein's contribution was to highlight that it is hybridization and subsequent recombination between loci that impedes speciation. Overcoming the antagonism between selection and recombination is key when divergent populations remain in contact.

As the speciation literature shows, many different mechanisms can help to maintain associations between ecological and mating traits and so permit speciation in the face of gene flow (Smadja & Butlin, 2011). In the following sections, I review these mechanisms as they apply to *Heliconius*.

11.15 Geography and the role of allopatry

A widely held view of speciation is that the most important process that permits build-up of associations between traits is geographic isolation (allopatry). If populations are completely isolated, the problem of hybridization and recombination goes away; genetic differences can accumulate gradually until populations become completely incompatible. In allopatry, differences can accumulate readily by genetic drift or by natural selection. There is therefore an inevitability about allopatric speciation: given enough time, populations will eventually accumulate incompatibilities and become distinct species.

However, complete allopatry does not seem an especially likely scenario for *Heliconius* speciation. As we saw in Chapter 10, there is little evidence to support the idea that there were rainforest refugia across the Amazon during periods of global cooling. Peak species diversity is found in the continuous forests of the upper Amazon, where there are no major barriers between populations, and analysis of branch lengths on the *Heliconius* tree also indicates that the shortest terminal branches (i.e. those branches separating the most recent speciation events) are found in the eastern Andes and upper Amazon (Rosser et al., 2012). The evidence therefore suggests that the part of the *Heliconius* range with the greatest continuity of habitat is also the region with the highest speciation rate (Rosser et al., 2012).

Another method of testing for the geography of speciation are age-range overlap plots, where the degree of overlap between species is plotted against their age inferred from a phylogeny. Such plots for *Heliconius* indicate a high degree of range overlap between closely related species, suggesting that speciation occurs predominantly in populations that are in geographic contact (Rosser et al., 2015). Analysis of more than 58,000 locality records showed that 32–40% of heliconiine sister species show complete (>0.95) range overlap and 50–65% have range overlap >0.5. The same methods show very different patterns in birds and mammals, where closely related taxa are far more commonly allopatric (Fitzpatrick & Turelli, 2006; Phillimore et al., 2008). It does therefore seem likely that the geographic context of speciation varies between taxonomic groups.

Nonetheless, geography certainly is important in *Heliconius* speciation. As described earlier, there is a continuum—from parapatric races through to fully reproductively isolated species—suggesting that parapatric races are the earliest stage of diversification. Although geographic races hybridize freely where they come into contact, these contact zones can be very narrow relative to the range of the species. For example, some of the steepest clines are around 10 km in width, between races that are distributed over hundreds of kilometres. So most individuals across the range never come into contact with one another. Geographic structure therefore helps to maintain associations between the alleles that produce differences in wing pattern and mating preferences early in divergence (Jiggins, Estrada, & Rodrigues, 2004).

There is still disagreement over the importance of allopatry in speciation. Perhaps the only certainty is that the geographic context of speciation will continue to generate controversy. As is commonly the case in science, the volume of debate is inversely correlated with the amount of evidence; in this case, the problem lies in tracking the history of isolation and contact between species deep into the past in order to know whether populations have been in contact. Scientists can interpret the same data differently depending on their expectations. Those expectations are in part driven by study organism; for example, herbivorous insect biologists are typically more open to sympatric speciation, whereas bird biologists tend to favour allopatry. The evidence from age-range correlation plots suggests that the differences may be based on real differences in speciation patterns between these groups.

Fortunately, there is some hope of resolving this impasse with actual data. Genomic analyses have the potential to estimate the degree to which populations have been exchanging genes during divergence. As outlined earlier, the genomic data from *Heliconius* suggest that speciation with gene flow has been common, at least within this group.

11.16 Speciation with gene flow

An increasing acceptance of the view that speciation often involves some degree of ongoing gene flow has meant that speciation biology has moved on from debates about geographic context. The term that perhaps best encapsulates this shifting focus is 'speciation with gene flow', which covers all speciation that does not involve complete allopatry. The term emphasizes the fact that even geographically isolated populations are likely to exchange migrants at least occasionally, and that most speciation events likely involve gene exchange at some period during the process of divergence.

Although gene flow is a force that can impede speciation, there are also processes that can drive speciation with gene flow that do not act on allopatric populations. For example, competition for resources can push populations apart into different niches. This is most commonly envisaged as colonization of an empty niche, and therefore avoidance of competition in the ancestral habitat. However, theory suggests that frequency-dependent selection can also cause speciation in a continuous resource distribution (Dieckmann & Doebeli, 1999). Finally, selection against the production of maladaptive hybrids through reinforcement—which necessarily occurs only when diverging populations are in contact—might make speciation with gene flow faster than speciation in allopatry.

11.17 Ecological speciation and magic traits

One common cause of speciation with gene flow is ecological divergence. Speciation driven primarily by ecological processes has been termed 'ecological speciation' (note that ecological speciation can also occur in allopatry, so ecological speciation and speciation with gene flow are overlapping but distinct categories). The maintenance of trait associations is made easier if traits under ecological selection are also directly involved in causing reproductive isolation. *Heliconius* are a good example, because divergence in wing pattern influences both mating and ecological adaptation due to mimicry.

H. melpomene and *H. cydno* in Panama fall into different mimicry rings, which mimic *H. erato* and *H. sapho*, respectively. Genetic evidence suggests that *H. erato* and *H sapho* are older than *H. melpomene* and *H. cydno*, so the latter pair probably diverged to mimic the former rather than the other way around. Thus, a shift in mimicry is associated with the speciation of *H. melpomene* and *H. cydno*. Many of the different elements of reproductive isolation seen between the species are tied to this ecological divergence.

First, mimicry rings are associated with different habitats, so the divergence in mimicry between *H. melpomene* and *H. cydno* is also associated with divergence in habitat. The black and white mimicry ring tends to be found in deeper forest, while the red and black ring is found in more open areas. The habitat difference is consistent across the range of the two species, although in the Andes there is also a pronounced altitudinal shift. *H. cydno* and its co-mimics (*H. sapho* and *H. eleuchia*) are found at higher elevations than *H. melpomene* and *H. erato*.

Second, wing pattern divergence also leads to disruptive selection against hybrids. Intermediate hybrid wing patterns are rare in the wild and selected against by predators, as described earlier (Merrill et al., 2012).

Finally, wing pattern plays an important role in assortative mating. Initially inspired by observations in the field, we were interested in the role of colour as a potential cue in mate selection. An old trick for catching *Heliconius* in the field is to frantically wave a red rag, which can attract high-flying males into reach of the net. When in the field near Gamboa, James Mallet instead tried waving a white rag, and successfully attracted black and white *H. cydno* males. This suggested that colour might be a species-specific cue, and inspired a series of experiments to look at the role of colour in mate attraction. In fact, males respond well to fluttering model butterflies, which can be either made from real wings or from printed paper wings. Male butterflies show strong preferences for fluttering models with their own wing pattern (Jiggins, Naisbit, et al., 2001a; Jiggins, Estrada, & Rodrigues, 2004; Kronforst et al., 2006a; Merrill et al., 2011b).

Thus wing pattern is a single trait under strong ecological selection and also contributes to assortative mating. Theoretical analyses of sympatric speciation have typically considered such traits to be rather unlikely. For example, Maynard-Smith (1966) briefly considered 'pleiotropism', whereby 'the gene . . . adapting individuals to different niches may [itself] cause assortative mating', and rejected it as 'very unlikely' (p. 643). Subsequently, Gavrilets coined the term 'magic trait' to describe the same phenomenon—and by implication also considered such traits to be unusual (Gavrilets, 2004; Servedio et al., 2011). The term magic trait is perhaps unhelpful, and the less catchy but descriptive 'multiple-effect trait' has been suggested instead (Smadja & Butlin, 2011). A variety of potential empirical examples of such traits have been described, providing a nice example of empirical evolutionary studies being inspired by theory (Servedio et al., 2011).

It is important to distinguish the different ways ecological traits can cause populations to become reproductively isolated. Perhaps the most general example of magic traits are herbivorous insects that mate on their host plants. In this case a switch to a new host immediately ensures that individuals mate with others that have made a similar switch. This has been termed an 'automatic magic trait', and because it is trivial to demonstrate how this process can cause speciation, it has not been widely addressed in theoretical models (although it is likely to be important in nature) (Servedio et al., 2011).

Most magic trait models instead deal with situations more similar to that in *Heliconius*, where a signal used in mating is also under divergent ecological selection. In this scenario, divergence in the so-called magic trait alone is not sufficient to cause assortative mating, but needs to be followed by the evolution of divergent mate preferences (Jiggins, Emelianov, & Mallet, 2006; Smadja & Butlin, 2011). In some cases, the reverse can be true—mate preferences may diverge as a side effect of ecological adaptation, and be followed by divergence of the mating signals (Seehausen et al., 2008). In summary, many ecological differences characterize sympatric *Heliconius* species, but there are reasons to suppose that wing patterning differences contributed disproportionately to the origins of reproductive isolation.

11.18 Was a colour shift needed for speciation?

The importance of mimicry in *Heliconius* speciation has been called into question by the discovery of a series of races of *H. timareta* that are sympatric and mimetic with *H. melpomene* in the eastern Andes (Jiggins, 2008). Their existence demonstrates that differences in mimicry are not necessary for coexistence of *H. melpomene* and *H. cydno/timareta* species. This doesn't negate earlier work on *H. cydno*—colour pattern *is* an important factor in the coexistence of *H. melpomene* with *H. cydno*.

It does, however, raise the question of whether the initial divergence of the *H. melpomene* and *H. cydno/timareta* lineages involved a shift in wing pattern, as originally suggested. After all, when two species share a trait, the obvious inference would be that it is due to common ancestry. However, genetic evidence suggests that mimicry in the *H. timareta* populations resulted from secondary convergence due to introgression (Pardo-Diaz et al., 2012). Other forms of reproductive isolation must have secondarily evolved to compensate for similarity in wing pattern between these species. In particular, assortative mating is extremely strong between *H. melpomene* and *H. timareta*, and depends on chemical cues (Giraldo et al., 2008; Mérot, 2015).

11.19 Linkage of trait and preference

If multiple traits are involved in speciation and controlled by different genes, then genetic linkage between such genes can also make speciation easier. In *Heliconius*, wing pattern is a cue that is used for mate choice, but populations also differ in their preferences for wing pattern. Richard Merrill produced backcross hybrids by mating a first-generation male hybrid back to *H. cydno*, and demonstrated that the

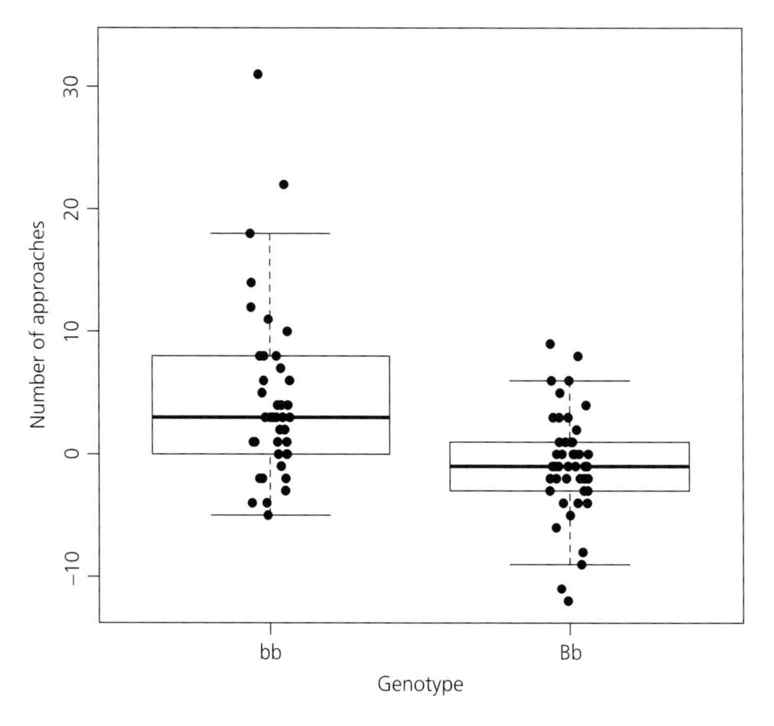

Figure 11.8 Association of mate preferences with inheritance of red wing patterns

The data show a difference in mating preferences between brothers from the same backcross family. Males that inherit a red band (shown as Bb genotypes) show a stronger preference for approaching red-banded mates as compared to their brothers with no red band (bb genotypes) (Merrill et al., 2011b).

mate preferences of those hybrids were associated with their wing patterns (Figure 11.8) (Merrill et al., 2011b). The hybrids with red wing patches preferred to court *H. melpomene*, while their brothers with all-white wings preferred *H. cydno* females. This locus explained 34% of the difference between the parental species preferences, suggesting that it is a major effect locus for mate preference. There is therefore a genetic association between mate preference genes and the red locus that controls red colour in the forewing.

Similar patterns are seen in another species pair, which shows genetic associations at a different locus. *Heliconius pachinus* and *H. cydno* differ in wing pattern and colour, in part controlled by the colour locus, *K*. These two geographic forms also show strong assortative mating. Marcus Kronforst presented males with mounted butterflies showing different combinations of colour and pattern. The yellow/white colour difference was a more important cue for mate choice than pattern (Kronforst et al., 2006a). He then looked at the colour preferences of 29 F_2 hybrids, and showed an association between segregation of colour and mate preference. In other words, hybrids that had yellow wings also preferred yellow mates.

Surprisingly, other traits also show similar patterns of linkage. In the same set of crosses mentioned earlier, Richard Merrill showed that hybrid sterility is associated with the shape locus, *WntA*. Female hybrids that have a mostly *H. cydno* genome—but an *H. melpomene* allele at the shape locus—are less likely to lay eggs than their sisters with the *H. cydno* allele. This is likely to be an interaction between the shape locus and other genes in the *H. cydno* genome that makes females partially sterile (Merrill et al., 2011b). Curiously, when the females do lay eggs, there is also an association of the shape locus with preference for different host plants (Merrill et al., 2013). There is an intriguing possibility that hybrid fertility and host-plant choice might have a common genetic basis, such that sterility results from a breakdown of host choice. Perhaps sterile hybrids just can't decide where to lay their eggs.

Finally, hybrid females are less likely to mate with *H. cydno* males if they have an *H. melpomene* allele at the yellow locus, *cortex* (Merrill et al., 2011b). Again, this association is expressed in the backcross to *H. cydno* families, and in this case explained 12% of the variance in mating outcome. The segregation of this locus had no detectable effect on male interest in the females—males courted the different yellow locus genotypes equally, so we interpret the effect as representing genetic variation in female choice. In summary, genetic associations are known between different 'speciation traits' and all four of the major wing patterning loci, *optix*, *cortex*, *WntA* and K.

All these genetic associations mean that hybrids are more likely to inherit combinations of alleles that work well together—those combinations found in their parents. As two incipient species diverge but continue to hybridize, these genetic associations make it easier for selection to maintain species differences. Fewer hybrids are produced with the 'wrong' combinations of alleles that suffer reduced fitness. As a result, selection has to eliminate fewer individuals in order to maintain species differences.

Kronforst suggested that preference might be a pleiotropic effect of the wing pattern gene (Kronforst et al., 2006a). Specifically, because ommochrome pigments are found both in the eye and the wing, the same gene might be responsible for controlling both traits. The data are inconclusive, as in both of the preference experiments the number of individuals tested was sufficient only to identify broad genomic regions. However, even if the same gene were involved, it seems unlikely that the same mutations would influence both wing pattern and mating. Matching wing patterns and mating preferences are more likely linked genetic changes rather than pleiotropic effects of the same mutation.

11.20 Evidence for reinforcement

Reinforcement involves selection directly favouring speciation. This contrasts with most speciation processes, which envisage divergence as a side effect of differences accumulated through drift or natural selection. Reinforcement arises where two populations are sufficiently divergent that there is some cost to mating with the other species. This was originally envisaged as the cost of producing unfit hybrids, in what has been termed the classic view of reinforcement (Servedio & Noor, 2003). This could be hybrid sterility, inviability, or ecological selection against intermediate genotypes.

Wing colour patterns of *H. melpomene* and *H. cydno* would be a good example of this kind of trait, where hybrids suffer from increased predation in the wild. Individuals that mate with the other species therefore end up producing hybrid offspring with reduced fitness. Selection will therefore favour individuals that choose to mate with their own type, and avoid mating with the other species. Although the process was originally envisaged as happening after a period of allopatry, similar processes can drive sympatric speciation. The process of reinforcement therefore occupies a special place in speciation biology as a process in which selection directly favours reproductive isolation.

The classic evidence for reinforcement comes from character displacement in sympatry. Where species overlap in only part of their range, reinforcement predicts stronger isolation in populations that are in contact with one another. That is, there is displacement of mating traits where species are in contact. For example, *H. melpomene* populations from French Guiana occur thousands of kilometres from the current range of *H. cydno*, and therefore have not experienced reinforcement selection.

Sure enough, in comparisons between populations of *H. melpomene* from Panama and those from French Guiana, the Panamanian populations are far more choosy and less likely to mate with *H. cydno* (Jiggins, Naisbit, et al., 2001a). Kronforst found similar patterns in Costa Rica on a much smaller geographic scale. The yellow *H. pachinus* replaces white *H. cydno* on the Pacific side of Costa Rica, while *H. melpomene* is found throughout. *H. melpomene* from the Caribbean side are more likely to court *H. pachinus*, and those on the Pacific side are more likely to court *H. cydno*. It seems that *H. melpomene* has evolved to avoid the species it is in most contact with. *H. cydno* and *H. pachinus* also come into occasional contact with one another in the central valleys of Costa Rica, so there is also potential for reinforcement selection between these species. In fact, *H. cydno* closer to the central valley region show stronger rejection of *H. pachinus* than those farther away (Kronforst, Young, & Gilbert, 2007).

One criticism of these studies is the lack of replication inherent in comparing just one allopatric and one sympatric population of each species. However, the cumulative evidence from multiple paired comparisons does suggest that there is a real and consistent pattern of stronger assortative mating where species are in contact.

Evidence for character displacement indicates that selection acts to increase reproductive isolation where species overlap. However, there are various ways by which this might come about. The classic view is that displacement is favoured by selection to avoid producing unfit hybrids. However, it has been pointed out that selection to avoid unfit hybrids acts only indirectly on mate preferences: the cost of hybridization is not incurred until the following generation. Such selection therefore relies on associations between alleles for mate preference and those for hybrid inviability (Servedio, 2001).

In contrast, forms of selection that act directly on the genes for mate preference may be more effective (Servedio, 2001). For example, increased isolation could be favoured by interspecific ecological competition. Flying in different habitats might avoid competition for host plants; avoiding courtship with the other species could be favoured to reduce time-wasting courtships of the wrong species that rarely result in mating.

Several of these selection pressures seem plausible in *Heliconius*. If females typically reject mates from the wrong species, selection would favour males that avoid wasting their time chasing such females. If females are often harassed by males of the other species, they might benefit from sending stronger signals of rejection that more clearly distinguish them from other species. Competition for host plants could also lead females to avoid areas where the other species are flying (although this probably doesn't explain the strong preferences seen in insectary cages described earlier).

Overall, it is challenging to separate the various reinforcement-like selection pressures. In *Drosophila*, reinforcement has been confirmed using experimental evolution (Higgie, Chenoweth, & Blows, 2000; Matute, 2010), and this might prove a more amenable system for distinguishing causes of character displacement. Nonetheless, if we consider reinforcement in the broad sense to include all these different selection pressures (Servedio & Noor, 2003), we can conclude that there is good evidence in *Heliconius* for reinforcement, and that interactions between species play a role in divergence. Reinforcement processes cannot occur between allopatric populations, so the widespread evidence for reinforcement provides further support for speciation with gene flow.

11.21 Plasticity and learning of preferences

Given what we know about the braininess of *Heliconius* (Chapter 6), it seems plausible that they might alter their mate preferences through learning. This could make speciation easier and faster, because a population that evolved a new wing pattern might then learn a preference for that pattern, leading to rapid speciation.

Nevertheless, despite all the evidence for learning in ecological contexts, there is no good evidence for learnt mating preferences. Experiments in which males were kept with virgin females of different wing patterns had no effect on their subsequent colour preferences (Jiggins, Estrada, & Rodrigues, 2004). We were also specifically interested in whether there might be self-matching mate preference—whether males might somehow see their own wing patterns and learn to prefer them. Merrill carried out an experiment in which hybrid male wings were blacked out upon emergence and then tested for preference. There was no effect on preference: red-banded males still preferred red females (Merrill et al., 2011b).

Earlier, Jocelyn Crane (1955) carried out a similar experiment in *H. erato*, blocking individuals from any conditioning to red by painting out the red markings after emergence, and going to great pains to ensure that the butterflies were not able to see any red—even to the extent of using tiny black felt blindfolds on their eyes while painting their wings! They were then isolated from other individuals before testing. This had no detectable influence on subsequent courtship responses to red (Crane, 1955).

In summary, all the experiments so far indicate that responses to colour by adult butterflies are innate and genetically determined. Nonetheless, I would not rule out future experiments detecting learnt influences on mating behaviour. Experiments directed at testing female mate-preference learning would be a good place to start.

Perhaps it is not surprising that mate preferences for colour are hardwired, given the monomorphism of colour patterns in most *Heliconius* populations. Learning is presumably costly and therefore favoured only where it has some benefit—such as in dealing with variability in space or through time (Cornwallis & Uller, 2010). For example, there is evidence for learning of mate preferences in *Bicyclus anynana,* which also show marked phenotypic plasticity in their wing patterns between dry and wet seasons (Westerman et al., 2012). Learning presumably allows individuals to tailor their preferences to the local population. Somewhat counter to this, in damsel flies there is evidence that learning contributes to isolation between sympatric species, akin to reinforcement (Svensson et al., 2010). There may also be costs to learning. For example, in Darwin's finches, individuals that learn the wrong song can end up mating with the wrong species (Grant, 1999). Overall, it remains to be seen whether there is a common role for learning and behavioural plasticity in speciation.

11.22 Hybrid speciation

One controversial idea in speciation biology is that hybridization could lead to the evolution of new species. Biologists have often seen hybridization as a negative force in speciation. After all, it is the maintenance of species differences in the face of hybridization and recombination that provides the major challenge to speciation with gene flow. However, as we saw in the last chapter, hybridization can also introduce useful variation for evolutionary change. Wing pattern alleles exchanged between species have led to adaptive evolution of mimicry. Thus it seems plausible that occasionally variants introduced through hybridization could also contribute to speciation. Hybrid speciation is therefore more than just hybridization and introgression. The hybridization event needs to directly contribute to reproductive isolation (Jiggins, Salazar, et al., 2008; Schumer, Rosenthal, & Andolfatto, 2014).

The best-studied case of hybrid speciation in *Heliconius* is *H. heurippa*. This species is found on the slopes of the eastern Andes near the town of Villavicencio. It has long been considered a possible hybrid on the grounds that its yellow and red forewing band phenotype is otherwise only found in hybrids (Brown & Mielke, 1972). The double band is the typical result of interaction between red band alleles at the red locus, *optix*, and yellow forewing alleles at the yellow locus (*cortex*, phenotypic effects of this locus on the forewing band are known as *N*). In fact, the *H. heurippa* colour pattern is quite similar to some *H. melpomene* × *H. cydno* hybrid specimens found in museums. *H. heurippa* is ecologically similar to the *H. cydno* and *H. timareta* lineage, which it replaces geographically. The species share an altitudinal range of 800–1,700 m, are similarly generalist in host-plant preferences, and tend to fly in more forested areas.

Led by Mauricio Linares and his student Camilo Salazar, there has been a long-term effort to investigate the hybrid speciation hypothesis in *H. heurippa*. One of the first steps was to demonstrate that the *H. heurippa* wing pattern could indeed be recreated by crossing the putative parents (Mavárez et al., 2006). First, *H. c. cordula* × *H. m. melpomene* F_1 hybrids were generated and then crossed back to *H. cydno*. Those with the red band were then mated to produce a stock of hybrids that were homozygous for the red band allele from *H. melpomene* on a largely *H. cydno* genetic background.

These looked very similar to wild *H. heurippa*, which prompted Carl Zimmer, a well-known science writer, to write, 'Darwin, meet Frankenstein'—implying that Linares had ghoulishly recreated the species in the lab (Zimmer, 2006). Of course, that wasn't really true, as the laboratory hybrids were outwardly similar to *H. heurippa* but genetically far more variable. In fact, they still exhibited a high degree of hybrid sterility, which made carrying out further experiments with this stock challenging. However, the crosses demonstrate how easily hybridization can create new pattern combinations that can potentially contribute to evolution.

As a postscript to these experiments, we discovered that *H. heurippa* is actually closer to *H. timareta* than *H. cydno* (Nadeau et al., 2013), which means that we probably used the wrong parent for the crosses. A form known as *H. timareta linaresi* also has a solid yellow forewing band and is a more likely candidate as the parent species for *H. heurippa*. Unpublished work by Sonia Rodriguez has shown that the *H. heurippa* pattern can also be recreated using crosses involving *H. timareta linaresi*. Of course, these crosses do not prove that this was the route taken by evolution for the formation of *H. heurippa*. However, the plausibility of this hypothesis is strongly supported by the fact that among all the rampant geographic variation in the parental species, the precise allelic variants that produce the *H. heurippa* pattern are those found immediately adjacent to the current range of the species.

Sequencing of the regulatory region of the red locus has shown unequivocally that *H. heurippa* alleles share a common ancestor with red-banded *H. melpomene* alleles (Salazar et al., 2010; Pardo-Diaz et al., 2012), and are genetically divergent from alleles at the same locus found in *H. cydno* or *H. timareta* populations without red forewing bands. This provides genetic confirmation of homology through shared ancestry between the *H. heurippa* red band with the similar phenotype found in *H. melpomene*, further supporting the hybrid origin hypothesis.

Has the novel *H. heurippa* pattern contributed to reproductive isolation? Given the widespread evidence for a role of wing patterns in mating, it is not a complete surprise that the novel wing pattern in *H. heurippa* is also involved in mate selection. *H. heurippa* males approach the wing patterns of either red *H. melpomene* or yellow *H. cydno* about half as frequently as their own pattern with both red and yellow bands (Mavárez et al., 2006). This shows that present-day populations of *H. heurippa* use the double band phenotype to find mates. It doesn't, however, directly address how these specific preferences arose.

Maria Clara Melo, another student in Linares's group, wanted to test whether the mate preferences of *H. heurippa* might have come about through hybridization. I was sceptical that this would be the case, but the experiments did indeed show that 'reconstructed' backcross hybrids show a similar mate preference for yellow and red bands as wild *H. heurippa*. This suggests that there is an additive basis to preference, such that a preference for yellow from *H. cydno/timareta* and a preference for red from *H. melpomene* can be combined so that hybrids prefer a yellow and red phenotype. The genetic linkage of mate preference with wing pattern should make it easier (Merrill et al., 2011b). This demonstrates that the derived mate preferences of the hybrid species could have arisen as a product of hybridization (Melo et al., 2009) (Figure 11.9).

One common line of evidence for hybrid speciation is a hybrid genome. Hybrid species are expected to show a mosaic of variation across the whole genome from two or more parents. This has sometimes been termed a 'mosaic' genome. However, there is no evidence that *H. heurippa* has a greater contribution from *H. melpomene* than any other member of its clade, apart from at the red wing pattern locus. As we have seen, all members of the *H. cydno/timareta* clade show considerable evidence for genetic mixing with *H. melpomene*, so the whole group shows plenty of shared variation. In *H. heurippa*, it seems likely that further evidence for introgression may become apparent from genome sequencing, but for the moment, the hybrid status of *H. heurippa* is based largely on evidence from wing patterns and mate preference. This led us to coin the term 'hybrid trait speciation' to describe cases in which evidence for a hybrid origin is limited to particular traits that contribute to speciation (Jiggins, Salazar, et al., 2008).

There has been discussion about how best to define hybrid speciation (Abbott et al., 2013; Schumer et al., 2014). Despite many claims of hybrid speciation, there are few examples in the literature in which a direct role for hybridization in causing reproductive isolation has been demonstrated (Schumer et al., 2014). *H. heurippa* has contributed to the debate by providing one of the most convincing animal examples in which hybridization and introgression have led to a new phenotype that directly causes reproductive isolation. There are also several excellent examples in plants, in which, for example, hybrid sunflower lineages have derived ecological tolerances, and in some cases these hybrid traits can be recreated in the laboratory by artificial hybridization (Rieseberg et al., 2003).

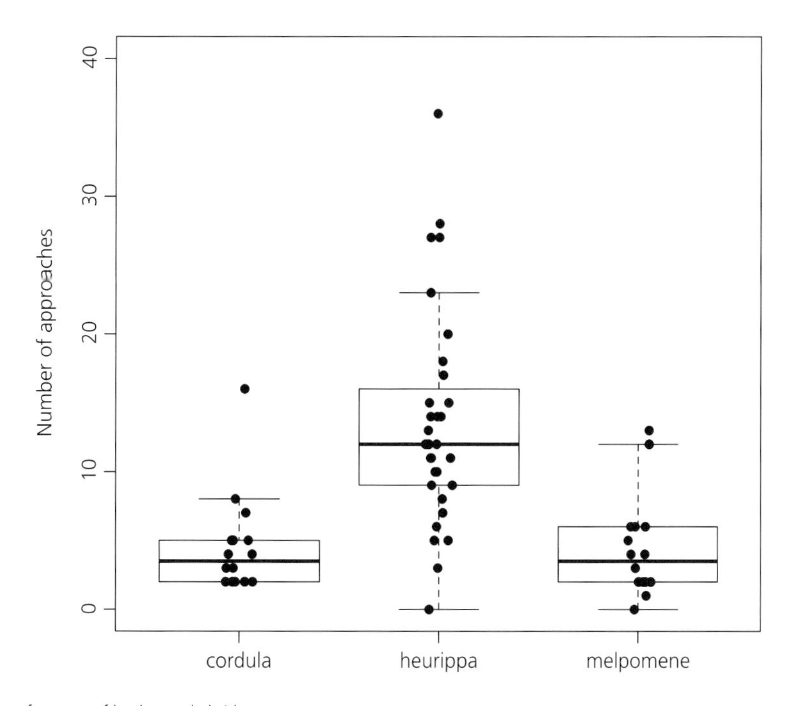

Figure 11.9 Mate preferences of backcross hybrids

The first generation of backcross hybrids between *H. melpomene melpomene* and *H. cydno cordula* actually prefer mates with hybrid-like *H. heurippa* wing patterns. This demonstrates a possible route for hybrid speciation, where hybridization can lead not only to a novel wing pattern but also a novel mating behaviour (Melo et al., 2009).

Many other putative examples of hybrid species in animals involve lineages with some degree of reproductive isolation from parental taxa, which have arisen with genetic contributions from at least two parental species (Mavárez & Linares, 2008). It seems likely that hybridization has contributed useful adaptive variation in many of these examples, but in most cases there is no strong evidence that hybridization has directly contributed to speciation. However, as genetic and genomic approaches for linking genotype to phenotype become more accessible, I predict that more clear examples of hybrid traits being directly involved in speciation will be uncovered.

The extensive empirical studies of *H. heurippa* prompted a student of Sergey Gavrilets to develop a theoretical model of hybrid speciation based on empirical parameters (Duenez-Guzman et al., 2009). This detailed model included predator selection, mate choice, habitat choice, and hybrid sterility. The results suggested that hybrid speciation could happen fairly rapidly (hundreds or thousands of generations), and therefore supports the plausibility of the hybrid origin explanation for the evolution of *H. heurippa*.

Nonetheless, the simulations never predicted stable coexistence of the hybrid and its parental species, suggesting that either the current coexistence of these species is only transient, or perhaps, more likely, that the model was lacking key parameters that would stabilize coexistence. The model perhaps highlights our more general ignorance of how multiple mimicry rings can arise and coexist (Chapters 8 and 10), as coexistence of mimicry rings is a more general problem that is not specific to hybrid speciation. In the case of *H. heurippa*, perhaps one key aspect that was missing from the model is geographic variation in the adaptive landscape of mimicry. The presence of *H. erato* should act to stabilize the *H. melpomene* wing pattern, whereas the absence of the *H. cydno* co-mimics found in the western Andes, *H. sapho* and *H. eleuchia*, might contribute to establishment of the novel *H. heurippa* phenotype.

Is *H. heurippa* a hybrid species? Perhaps the strongest argument against this claim is that *H. heurippa* may not actually be a distinct species at all. Originally we suspected that *H. heurippa* was derived from *H. cydno*, but later genomic analysis showed that it is actually closer to *H. timareta* (Nadeau et al., 2013). Subsequently we realized that there are previously unknown populations of *H. timareta* distributed along the eastern Andes with different wing patterns.

So although *H. heurippa* was previously thought to be an isolated population, genetically and geographically distant from its closest relatives, we now know that it lies at the junction of two lineages. *H. cydno* stretches from Mexico down into the northern Andes, reaching just into the eastern slopes in Colombia. In the east, *H. timareta* replaces *H. cydno* from Colombia down the Andes through Ecuador into Peru. In between lies the narrow range of *H. heurippa* along about 200 km of the eastern Andes.

H. heurippa could therefore be considered a race of *H. timareta* (actually, since the name *H. heurippa* is older and therefore takes precedence, the whole complex would have to be renamed *H. heurippa*). We have yet to discover a contact zone between *H. heurippa* and *H. timareta* (it is a difficult region of Colombia), but this would be needed in order to know for sure whether the two are reproductively isolated in the wild. There is pre-mating reproductive isolation in insectary experiments, but *H. heurippa* and *H. timareta* are genetically very similar to one another (Nadeau et al., 2013). In summary, *H. heurippa* is on the cusp of being a distinct species.

Whatever the status of *H. heurippa*, the extensive work on this butterfly and its close relatives has demonstrated that hybridization and introgression can directly cause reproductive isolation. As I have emphasized elsewhere in this chapter, speciation in *Heliconius* is a multifactorial process that requires divergence along many different axes, and it is probably unlikely that a single hybridization event would produce a completely new species in one step. Thus it seems likely that hybrid speciation is part of that process, rather than representing a fait accompli.

It may be that there are more complete examples of hybrid speciation in *Heliconius*. Perhaps the most promising is *H. elevatus*, which is genetically similar and broadly sympatric with its close relative *H. pardalinus*, but shares its wing pattern genes with the far more distantly related species *H. melpomene* (The Heliconius Genome Consortium, 2012). If it can be shown that *H. elevatus* is the recipient in this genetic exchange, and that this event triggered speciation, then this might provide an unequivocal case of hybrid speciation. In any case, hybridization clearly does contribute in a major way to both adaptation and speciation in *Heliconius*.

11.23 Future directions

The study of speciation is being revolutionized by genomic data, which provide new insights into speciation and divergence history. Such data, along with appropriate model-fitting, can be used to address the degree to which species have diverged with ongoing gene flow, and hopefully identify whether there have been periods of complete isolation during speciation. This is exciting, as we will finally be able to address the age-old, and still largely unresolved, debate about allopatry versus sympatry in speciation. Using genomic data for hybridizing species, we can also address the factors that restrict gene flow. These include the typical processes of divergent selection, as well as factors that came to the fore later, such as patterns of recombination and background selection across the genome.

There are also exciting prospects for identification of genes involved in speciation. We already know a great deal about the genetic basis for wing pattern in *Heliconius*, and in some cases these genes also act as 'speciation genes'. However, speciation also involves changes in mating preferences, host-plant use, hybrid sterility, and other ecological factors. Little is known of the genetic basis of these traits—identifying specific genes controlling species differences is likely to provide insights into their history and patterns of divergence. As we have seen from the wing patterning genes, identification of specific genes underlying adaptive traits can bring insights into their history of divergence, and in particular the dramatic role of adaptive introgression. Is it a widespread phenomenon, or are wing patterns unusual?

Even without identifying specific genes, studies of the genetic basis for trait differences both within and between species can address a variety of questions. We are interested in the extent to which genes controlling speciation traits are genetically linked, and their distribution across the genome. How do effect size distributions differ for the traits that underlie speciation—are other ecological traits apart from wing pattern more polygenic in nature? Comparison of the genetic architecture between species pairs at different levels of divergence, and in allopatric and sympatric comparisons, could be used to determine whether the trait linkage already observed is an ancestral characteristic that facilitates speciation, or a derived result of speciation with gene flow within the melpomene/cydno group.

It is clear that pre-mating isolation is perhaps the most important factor in speciation, but we would like to dissect the behavioural and sexual factors that diverge in order to lead to such isolation. What is the relative importance of visual versus pheromonal cues? How and why do pheromonal cues diverge during speciation—are they primarily driven by sexual or natural selection? Is there any role for learning of preferences and plasticity in brain architecture?

There is considerable scope for more ecological studies of speciation. Our understanding of habitat divergence and host-plant use is based on studies of just a handful of populations and species pairs. This understanding is bound to deepen if we study more species in their natural habitats. In particular, studies of habitat segregation in species-rich Amazonian habitats would be especially useful, where niche partitioning presumably needs to be more fine-grained to permit species coexistence. Outstanding questions include the following: How much does habitat segregation contribute to speciation? Does immigrant inviability contribute to speciation through selection against individuals that fly into the 'wrong' habitat? What selective forces drive the evolution of host-plant strategies?

This chapter began with a discussion of the increasing speciation rates in *Heliconius* compared to related genera, but we still have little idea of what caused the difference. Identification of factors correlated with increased diversification may require comparative analysis across a wider array of insect species. However, *Heliconius* are a good test case in which the results of such comparative analyses could be directly tested. Even within the heliconiines, there is potential for comparative analyses to investigate the roles of host-plant adaptation, colour pattern diversification, and mimicry in species formation and extinction. Ultimately, it is hoped that we can integrate broader-scale comparative analyses with in-depth studies of focal species to get a clear picture of how and why species are formed.

CHAPTER 12

Taxonomic list

Gerardo Lamas and Chris Jiggins

The following list represents the work of Gerardo Lamas and is based on his most recent list of taxonomic names for the neotropical butterflies (Lamas, 2004). A few alterations, additions, and corrections to that list have been made, notably with assistance from Mathieu Joron for the subspecies of *H. numata*. The generic names *Neruda* and *Laparus* are subsumed within *Heliconius* on the basis of recent molecular taxonomy (Kozak et al., 2015). The localities listed represent the country of origin (where known) of the type specimen(s), not necessarily the entire geographic range of the taxon. Synonyms are listed in italics. References are in the *Annotated Bibliography of the Neotropical Butterflies and Skippers* by Gerardo Lamas (2016), available at http://www.butterfliesofamerica.com/L/Biblio.htm.

Taxon	Locality
AGRAULIS Boisduval & Le Conte [1835]	
vanillae (Linnaeus, 1758) (*Papilio*)	
a) vanillae (Linnaeus, 1758) (*Papilio*)	America
passiflorae (Fabricius, 1793) (*Papilio*), repl. Name	America
b) forbesi Michener, 1942	Peru
c) galapagensis Holland, 1890	Galapagos
d) incarnata (Riley, 1926) (*Dione*)	Mexico (DG)
vanillae ab. *comstocki* (Gunder, 1925) (*Dione*)	USA (CA)
vanillae insularis tr.f. *fumosus* (Gunder, 1927) (*Dione*)	USA (CA)
vanillae f. *incarnata* tr.f. *margineapertus* (Gunder, 1928) (*Dione*)	USA (CA)
vanillae f. *incarnata* tr.f. *hewlettae* (Gunder, 1930) (*Dione*)	USA (CA)
e) insularis Maynard, 1889	Bahamas
f) lucina C. Felder & R. Felder, 1862	'Brazil (AM)'
vanillae catella (Stichel [1908]) (*Dione*), hyb.	Peru
g) maculosa (Stichel [1908]) (*Dione*)	Brazil (ES); Paraguay
argentea (Larrañaga, 1923) (*Papilio*)	Uruguay
vanillae f. *superargentata* (Giacomelli, 1925) (*Dione*)	Argentina
h) nigrior Michener, 1942	USA (Fl)
[**n. sp.**] Lamas, MS	Peru

The Ecology and Evolution of Heliconius Butterflies. Chris D. Jiggins, Oxford University Press (2017).
 DOI 10.1093/acprof:oso/9780199566570.001.0001

DIONE Hübner [1819]

glycera (C. Felder & R. Felder, 1861) (*Agraulis*)	Venezuela; Colombia
glycera (Ménétriés, 1857) (*Agraulis*), nom. nud.	Venezuela
glycera gnophota Stichel, 1907	Colombia
juno (Cramer, 1779) (*Papilio*)	
a) juno (Cramer, 1779) (*Papilio*)	Surinam
b) andicola (H. W. Bates, 1864) (*Agraulis*)	Ecuador
c) huascuma (Reakirt, 1866) (*Agraulis*)	Mexico
huascama Auctt., missp.	
d) miraculosa Hering, 1926	Peru
e) suffumata K. S. Brown & Mielke, 1972 (ICZN, Art. 45.6.2)	Paraguay
juno ab. *suffumata* Hayward, 1931	Paraguay
moneta Hübner [1825]	
a) moneta Hübner [1825]	?
b) butleri Stichel [1908]	Colombia
c) poeyii Butler, 1873	Mexico [(OA)]

PODOTRICHA Michener, 1942

judith (Guérin-Méneville [1844]) (*Cethosia*)	
a) judith (Guérin-Méneville [1844]) (*Cethosia*)	Colombia
euchroia (Doubleday [1847]) (*Colaenis*)	Colombia
b) caucana (Riley, 1926) (*Colaenis*)	Colombia
c) mellosa (Stichel, 1906) (*Colaenis*)	Ecuador
d) straminea (Riley, 1926) (*Colaenis*)	Ecuador
telesiphe (Hewitson, 1867) (*Colaenis*)	
a) telesiphe (Hewitson, 1867) (*Colaenis*)	Ecuador
tithraustes var. *diaphana* (Niepelt, 1915) (*Colaenis*)	Peru
b) tithraustes (Salvin, 1871) (*Colaenis*)	Ecuador

DRYADULA Michener, 1942

phaetusa (Linnaeus, 1758) (*Papilio*)	'Indiis'
f. *stupenda* (Stichel [1908]) (*Colaenis*)	Panama
f. *lutulenta* (Stichel [1908]) (*Colaenis*)	Paraguay
f. *deleta* (Stichel [1908]) (*Colaenis*)	Paraguay

DRYAS Hübner [1807]

Alcionea Rafinesque, 1815

Colaenis Hübner [1819]

iulia (Fabricius, 1775) (*Papilio*)	
a) iulia (Fabricius, 1775) (*Papilio*)	America
iulia juncta W. P. Comstock, 1944	Puerto Rico
b) alcionea (Cramer, 1779) (*Papilio*)	Surinam

iulia titio (Stichel [1908]) (*Colaenis*)	Bolivia
c) carteri (Riley, 1926) (*Colaenis*)	Bahamas
d) delila (Fabricius, 1775) (*Papilio*)	America
cillene (Cramer, 1779) (*Papilio*)	'Surinam'
e) dominicana (A. Hall, 1917) (*Colaenis*)	Dominica
f) framptoni (Riley, 1926) (*Colaenis*)	St. Vincent
g) fucatus (Boddaert, 1783) (*Papilio*)	Dominican Republic
julia [sic] *hispaniola* (A. Hall, 1925) (*Colaenis*)	[Dominican Republic]
h) lucia (Riley, 1926) (*Colaenis*)	St. Lucia
i) martinica Enrico & Pinchon, 1969	Martinique
j) moderata (Riley, 1926) (*Colaenis*) (ICZN, Art. 45.5.1)	Honduras
iulia delila f. *moderata* (Stichel [1908]) (*Colaenis*)	Honduras
k) nudeola (M. Bates, 1934) (*Colaenis*) (ICZN, Art. 45.5.1)	Cuba
iulia cillene f. *nudeola* (Stichel [1908]) (*Colaenis*)	Cuba
l) warneri (A. Hall, 1936) (*Colaenis*)	St. Kitts
m) zoe L. D. Miller & Steinhauser, 1992	Cayman Islands

PHILAETHRIA Billberg, 1820
Metamandana Stichel [1908]

andrei Brévignon, 2002	
a) andrei Brévignon, 2002	French Guiana
b) orinocensis Constantino & Salazar, 2010	Venezuela
browni Constantino & Salazar, 2010	Venezuela
constantinoi Salazar, 1991	Colombia
diatonica (Fruhstorfer, 1912) (*Metamandana*)	Honduras
dido (Linnaeus, 1763) (*Papilio*)	
a) dido (Linnaeus, 1763) (*Papilio*)	South America
b) chocoensis Constantino, 1999	Colombia
c) panamensis Constantino & Salazar, 2010	Panama
neildi Constantino & Salazar, 2010	
a) neildi Constantino & Salazar, 2010	Venezuela
b) tachiraensis Constantino & Salazar, 2010	Venezuela
c) winhardi Constantino & Salazar, 2010	Colombia
ostara (Röber, 1906) (*Metamorpha*)	
a) ostara (Röber, 1906) (*Metamorpha*)	Colombia
b) araguensis Constantino & Salazar, 2010	Venezuela
c) meridensis Constantino & Salazar, 2010	Venezuela
romeroi Constantino & Salazar, 2010	Venezuela
wernickei (Röber, 1906) (*Metamorpha*)	Brazil (RS/SC)
pygmalion (Fruhstorfer, 1912) (*Metamandana*)	Brazil (PA)
pygmalion metaensis Constantino & Salazar, 2010	Colombia

Taxon	Locality
EUEIDES Hübner, 1816	
Mechanitis [Illiger], 1807, suppr. (ICZN, Op. 232)	
Semelia Doubleday [1845]	
Evides Agassiz, 1846, emend.	
Semelia Erichson [1849], preocc. (Doubleday [1845])	
Semelia Boisduval, 1870, preocc. (Doubleday [1845])	
aliphera (Godart, 1819) (*Cethosia*)	
a) aliphera (Godart, 1819) (*Cethosia*)	Brazil
b) cyllenella Seitz, 1912	?
alipheraab. cillenula Seitz, 1913, missp.	
cyllenula Seitz, 1924, emend.	
c) gracilis Stichel, 1903	Honduras; Costa Rica
emsleyi K. S. Brown, 1976	
a) emsleyi K. S. Brown, 1976	Colombia
b) esmeraldensis Cast, 2013	Ecuador
heliconioides C. Felder & R. Felder, 1861	
a) heliconioides C. Felder & R. Felder, 1861	'Ecuador'
eanes ab. *eanides* Stichel, 1903	Peru
eanes ab. *farragosa* Stichel, 1903, hyb.	Peru
eanes ab. *aides* Stichel, 1903	Peru
eanes ab. *riffarthi* Stichel, 1903, hyb.	Peru; Bolivia
eanes ab. *felderi* Stichel, 1903	'Ecuador'
eanes ab. *pluto* Stichel, 1903	Peru
eanes heliconioides f. *felicitatis* H. Holzinger & R. Holzinger, 1993	[Peru]
b) eanes Hewitson, 1861	Peru
c) koenigi H. Holzinger & R. Holzinger, 1993	Peru
eanes koeningi H. Holzinger & R. Holzinger, 1993, missp.	
isabella (Stoll, 1781) (*Papilio*)	
a) isabella (Stoll, 1781) (*Papilio*)	Surinam
b) cleobaea Geyer, 1832	Cuba
anaxa Ménétriés, 1857	'Nicaragua'; Cuba
c) dianasa (Hübner [1806]) (*Nerëis* [sic])	Brazil (BA)
dianasa ab. *decolorata* Stichel, 1903	Brazil (BA)
d) dissoluta Stichel, 1903	Peru
isabella hippolinus ab. *margaritifera* Stichel, 1903	Peru
isabella pellucida ab. *vegetissima* Stichel, 1903	Ecuador
isabella hübneri [sic] var. *olga* Neustetter, 1916	Peru
e) ecuadorensis Strand, 1912	Ecuador
f) eva (Fabricius, 1793) (*Papilio*)	'Surinam'
anaxa Ménétriés, 1855, nom. nud.	Nicaragua
zorcaon Reakirt, 1866	Mexico (Ver.); Honduras; Guatemala
cleobaea zorcaon ab. *adusta* Stichel, 1903	Panama

g) hippolinus Butler, 1873	Peru
ab. *personata* Stichel, 1903	Peru
ab. *brunnea* Stichel, 1903	Peru
imitans Seitz, 1912	?
h) huebneri Ménétriés, 1857	'Brazil'
dynastes C. Felder & R. Felder, 1861	Venezuela; Colombia
pellucida Srnka, 1885	Colombia
isabella seitzi Stichel, 1903	Colombia
isabella arquata Stichel, 1903	Colombia
isabella hübneri [sic] ab. *spoliata* Stichel, 1903	Colombia
isabella isabella f. *perimacula* Boullet & Le Cerf, 1910	Colombia
i) melphis (Godart, 1819) (*Heliconia*)	Antilles
cleobaea var. *monochroma* Boullet & Le Cerf, 1910	Haiti
j) nigricornis R. G. Maza, 1982	Mexico (GR)
k) [n. ssp.] K. S. Brown, MS	Brazil (CE)
lampeto H. W. Bates, 1862	
a) lampeto H. W. Bates, 1862	Brazil (AM)
lampeto romani Bryk, 1953	Brazil (AM)
b) acacetes Hewitson, 1869	Ecuador
lampeto ab. *fuliginosa* Stichel, 1903	Ecuador
lampeto ab. *carbo* Stichel, 1903	Ecuador
lampeto ab. *amoena* Stichel, 1903	Ecuador; 'Bolivia'
lampeto carbo Seitz, 1913 (ICZN, Art. 45.6.2)	Ecuador
ambatensis Oberthür, 1916	Ecuador
c) apicalis Röber, 1927	Ecuador
lampeto fuliginosus f. *pallida* Riffarth, 1907	Ecuador
d) brownsbergensis Gernaat & Beckles, 2010	Surinam
e) concisa Lamas, 1985	Peru
f) nigrofulva Kaye, 1906	Guyana
lampeto copiosus Stichel, 1906	Guyana
g) [n. ssp.] K. S. Brown, MS	Brazil (AM)
libitina Staudinger, 1885	
a) libitina Staudinger, 1885	French Guiana
b) malleti Lamas, 1998	Peru
lineata Salvin & Godman, 1868	Guatemala
lybia (Fabricius, 1775) (*Papilio*)	
a) lybia (Fabricius, 1775) (*Papilio*)	'India'
hypsipyle (Cramer, 1777) (*Papilio*)	Surinam
b) lybioides Staudinger, 1876	Panama
c) olympia (Fabricius, 1793) (*Papilio*)	America
leucomma H. W. Bates, 1866	Panama
d) orinocensis K. S. Brown & F. Fernández, 1985	Venezuela

e) otelloi K. S. Brown & F. Fernández, 1985	Venezuela
f) salcedoi K. S. Brown & F. Fernández, 1985	Venezuela
pavana Ménétriés, 1857	Brazil
thyana C. Felder & R. Felder, 1860	[Brazil]
f. *ferruginea* Raymundo, 1931	Brazil (RJ)
procula Doubleday [1847]	
a) procula Doubleday [1847]	Venezuela
b) asidia Schaus, 1920	Guatemala
c) browni H. Holzinger & R. Holzinger, 1974	Venezuela
d) edias Hewitson, 1861	Colombia
edias luminosus Stichel, 1903	Venezuela
e) eurysaces Hewitson, 1864	Ecuador
f) kuenowii Dewitz, 1877	Colombia
edias umbratilis Röber, 1927	Colombia
g) vulgiformis Butler & H. Druce, 1872	Costa Rica
tales (Cramer, 1775) (*Papilio*)	
a) tales (Cramer, 1775) (*Papilio*)	Surinam
thalestris (Godart, 1819) (*Heliconia*), repl. name	Surinam
b) barcellinus J. Zikán, 1937	Brazil (AM)
eanes f. *lucretius* J. Zikán, 1937	Brazil (AM)
c) calathus Stichel, 1909	Ecuador
d) cognata Weymer, 1890	Colombia
e) franciscus K. S. Brown & H. Holzinger, 1973	Venezuela
f) guyanensis Lacomme, 2011	French Guiana
g) michaeli J. Zikán, 1937	Peru
h) pseudeanes Boullet & Le Cerf, 1910	Venezuela
i) pythagoras W. F. Kirby [1899]	Brazil (PA)
thales [sic] Hübner [1810] (not Cramer, 1775)	
thales [sic] *heraldicus* Stichel, 1903	Brazil (PA)
tales f. *zernyi* Neustetter, 1928	Brazil (PA)
j) surdus Stichel, 1903	Brazil (PA); Guyana
thales [sic] *surdus* ab. *aquilifer* Stichel, 1903	Brazil (PA)
thales [sic] f. *reducta* Neustetter, 1931	Brazil (PA)
k) tabernula Lamas, 1985	Peru
l) xenophanes C. Felder & R. Felder, 1865	Colombia
crystalina A. Hall, 1921	Colombia
m) [n. ssp.] K. S. Brown, MS	Brazil (PA)
vibilia (Godart, 1819) (*Cethosia*)	
a) vibilia (Godart, 1819) (*Cethosia*)	Brazil
mereaui (Hübner, 1819) (*Colaenis*), nom. nud.	
mereaui (Hübner, 1823) (*Colaenis*)	'Brazil'
ab. *pallens* Stichel, 1903	Brazil (ES)

b) louisi Brévignon, 2006 — French Guiana
vibilia romeroi H. Holzinger & R. Holzinger, 1994, unavail. (ICZN, Art. 15.1) — Venezuela
d) unifasciatus Butler, 1873 — Brazil (AM)
vibilia var. *arcita* Stichel, 1903, nom. nud.
e) vialis Stichel, 1903 — Costa Rica
f) vicinalis Stichel, 1903 — Ecuador
g) [n. ssp.] K. S. Brown, MS — Colombia

HELICONIUS Kluk, 1780
Heliconius Latreille, 1804, preocc. (Kluk, 1780)
Migonitis Hübner, 1816, preocc. (Rafinesque, 1815)
Sunias Hübner, 1816
Ajantis Hübner, 1816
Apostraphia Hübner, 1816
Sicyonia Hübner, 1816, suppr. (ICZN, Op. 382)
Heliconia Latreille, 1818, emend.
Laparus Billberg, 1820
Crenis Hübner, 1821
Phlogris Hübner [1825]
Podalirius Gistel, 1848, preocc. (Latreille, 1802), repl. name
Blanchardia Buchecker [1880], preocc. (Castelnau, 1875)
Neruda J. R. G. Turner, 1976

antiochus (Linnaeus, 1767) (*Papilio*)
a) antiochus (Linnaeus, 1767) (*Papilio*) — 'Indiis'
zobeide Butler, 1869 — Brazil (PA); Peru
ab. *divisus* Staudinger, 1897 — Brazil (PA)
ab. *alba* Riffarth, 1900 — Peru; Surinam
antiochus alba ab. *trimaculata* R. Krüger, 1933 — Brazil (PA)
b) aranea (Fabricius, 1793) (*Papilio*) — ?
c) araneides Staudinger, 1897 — Venezuela
d) salvinii Dewitz, 1877 — [Venezuela]
e) [n. ssp.] Lamas, MS — Peru

aoede (Hübner, 1816) (*Migonitis*)
a) aoede (Hübner, 1816) (*Migonitis*) — Brazil
aoede (Hübner, 1808) (*Nerëis* [sic]), nom. nud.
b) aliciae (Neukirchen, 2000) (*Neruda*) — Ecuador
c) astydamia Erichson [1849] (*Heliconia*) — Guyana
d) auca (Neukirchen, 1997) (*Neruda*) — Ecuador
e) ayacuchensis Neukirchen, 1992 (*Heliconius* (*Neruda*)) — Venezuela
f) bartletti H. Druce, 1876 — Peru

Taxon	Locality
vedius Staudinger, 1885	Peru
aoede f. *metella* Neustetter, 1932	Colombia
g) centurius Neukirchen, 1994 (*Heliconius* (*Neruda*))	French Guiana
h) cupidineus Stichel, 1906	Peru
bartletti var. *reducta* Oberthür, 1920 (*Heliconia*)	Peru
i) emmelina Oberthür, 1902 (*Heliconia*)	Guyana
j) eurycleia K. S. Brown, 1973	Brazil (MT)
k) faleria Fruhstorfer, 1910	Brazil (MT)
l) lucretius Weymer, 1891	Brazil (AM)
aoede aoede f. *postalbimacula* Bryk, 1953	Brazil (AM)
m) manu Lamas, 1976	Peru
n) philipi K. S. Brown, 1976	Bolivia
o) [n. ssp.] K. S. Brown, MS	Brazil
astraea Staudinger, 1897	
a) astraea Staudinger, 1897	Brazil (AM)
b) rondonia K. S. Brown, 1973	Brazil (RO)
atthis Doubleday, 1847 (*Heliconia*)	Ecuador
bourcieri Becker, 1851 (*Heliconia*), nom. nud.	
f. *gerstneriana* Descimon & Mast, 1971	Ecuador
besckei Ménétriés, 1857 (*Heliconia*)	Brazil (RJ/MG)
besckii Ménétriés, 1857 (*Heliconia*), missp.	
epiphyllis Riffarth, 1901, nom. nud.	
besckei principalis Röber, 1921	Brazil (SC/SP)
f. *intermixta* Neustetter, 1931	Brazil (RJ)
burneyi (Hübner [1831]) (*Migonitis*)	
a) burneyi (Hübner [1831]) (*Migonitis*)	Brazil
var. *jeanneae* Boullet & Le Cerf, 1909	French Guiana
f. *oberthüri* [sic] Boullet & Le Cerf, 1909, preocc. (not Riffarth, 1903)	['Peru']
f. *roseni* R. Krüger, 1928	'Surinam'
b) ada Neustetter, 1925	Brazil (RO)
c) anjae Neukirchen, 1995	Brazil (AM)
d) boliviensis Neukirchen, 1995	Bolivia
burneyi jeanneae f. *reducta* Boullet & Le Cerf, 1909	Bolivia
e) catharinae Staudinger, 1885	[Brazil (PA)]
burneyi serpensis Kaye, 1919, hyb.	Brazil (AM)
burneyi f. *catharinae* ab. *vitiosa* R. Krüger, 1933	Brazil (PA)
burneyi f. *miraculosa* Kotzsch, 1936	Brazil (PA)
f) huebneri Staudinger, 1897	[Peru]
g) jamesi Neukirchen, 1995	Peru
h) koenigi Neukirchen, 1995	Peru
i) lindigii C. Felder & R. Felder, 1865	Colombia

j) mirtarosa Orellana [2007]	Venezuela
k) skinneri K. S. Brown & F. Fernández, 1985	Venezuela
charithonia (Linnaeus, 1767) (*Papilio*)	
a) charithonia (Linnaeus, 1767) (*Papilio*)	America
charitonia (Linnaeus, 1767) (*Papilio*), missp.	
b) antiquus Lamas, 1988, repl. Name	St. Kitts; Antigua
charithonia punctata A. Hall, 1936, preocc. (not Neustetter, 1907)	St. Kitts; Antigua
c) bassleri W. P. Comstock & F. M. Brown, 1950	Colombia
d) churchi W. P. Comstock & F. M. Brown, 1950	Haiti
e) ramsdeni W. P. Comstock & F. M. Brown, 1950	Cuba
f) simulator Röber, 1921	Jamaica
g) tuckeri W. P. Comstock & F. M. Brown, 1950	USA (FL)
charitonia [sic] *tuckerorum* Lamas, 1988, emend.	
h) vazquezae W. P. Comstock & F. M. Brown, 1950	Mexico (CP)
chestertonii Hewitson, 1872 (*Heliconia*)	Colombia
damysus Hopffer, 1874	'Bolivia'
hydara ab. *nocturna* Riffarth, 1900, hyb.	'Venezuela'
clysonymus Latreille [1817]	
a) clysonymus Latreille [1817]	[Colombia]
clysonimus [sic] *fischeri* Fassl, 1912, hyb.	Colombia
clysonimus [sic] *flavopunctatus* Fassl, 1912, hyb.	Colombia
micrus Seitz, 1912	[Venezuela]
clysonymus apicalis Joicey & Kaye, 1917, nom. nud.	
clysonymus apicalis ab. *semirubra* Joicey & Kaye, 1917, hyb.	Colombia
clysonymus perbellus Stichel, 1923	Colombia
hygianus fischeri f. *leonis* H. Holzinger & R. Holzinger, 1970, hyb.	Colombia
hygianus fischeri f. *leoncitonis* H. Holzinger & R. Holzinger, 1970, hyb.	Colombia
b) hygiana Hewitson, 1867 (*Heliconia*)	Ecuador
hygiana ab. *albescens* Kaye, 1916	Ecuador
c) montanus Salvin, 1871	Costa Rica
d) tabaconas K. S. Brown, 1976	Peru
congener Weymer, 1890	
a) congener Weymer, 1890	Ecuador
apseudes var. *paranapurae* Staudinger, 1897	Peru
b) aquilionaris K. S. Brown, 1976	Colombia
c) ocannensis Stichel, 1905, repl. Name	Colombia
ocannus Staudinger, 1885, preocc. (not Buchecker [1880])	Colombia
cydno Doubleday, 1847 (*Heliconia*)	
a) cydno Doubleday, 1847 (*Heliconia*)	Colombia
ab. *mediocydno* Neustetter, 1907	Colombia

f. *punctata* Neustetter, 1907	Colombia
b) alithea Hewitson, 1869 (*Heliconia*)	Ecuador
cydno ab. *haenschi* Riffarth, 1900	Ecuador
cydno broncus Stichel, 1906, hyb.	'Peru?'
f. *egregia* Riffarth, 1907	Ecuador
alithea (?var.) ab. *minor* Neustetter, 1907	Ecuador
alithea f. *neustetteri* Riffarth, 1908	Ecuador
c) barinasensis Masters, 1973	Venezuela
d) chioneus H. W. Bates, 1864	Panama
e) cordula Neustetter, 1913	Venezuela
cydno ab. *larseni* Niepelt, 1916	Colombia
cydno flaveola Joicey & Kaye, 1917, hyb.	Venezuela
cydno perijaensis Masters, 1973	Venezuela
f) cydnides Staudinger, 1885	Colombia
cydno var. *epicydnides* Staudinger, 1897	Colombia
cydno var. *subcydnides* Staudinger, 1897	Colombia
interrupta Riffarth, 1901, nom. nud.	
f. *werneri* Neustetter, 1928	Colombia
f. *azteka* Neustetter, 1928	Colombia
f. *semicydnides* H. Holzinger & R. Holzinger, 1968	Colombia
f. *tenebrosa* H. Holzinger & R. Holzinger, 1968	Colombia
g) gadouae K. S. Brown & F. Fernández, 1985	Venezuela
hahneli Staudinger, 1885, hyb.	Venezuela
h) galanthus H. W. Bates, 1864	Guatemala
diotrephes Hewitson, 1869 (*Heliconia*)	Nicaragua
cydno ab. *stübeli* [sic] Riffarth, 1900	Costa Rica
piera Riffarth, 1901, nom. nud.	
galanthus f. *exornata* Riffarth, 1907, hyb.	Costa Rica
galanthus ab. *subrufescens* Schaus, 1913, hyb.	Costa Rica
i) hermogenes Hewitson [1858] (*Heliconia*)	[Colombia]
temerinda Hewitson, 1873 (*Heliconia*)	Colombia
hermogenes ab. *lutescens* Kaye, 1916	Colombia
f. *anthelea* Neustetter, 1925	Colombia
cydno dolores Apolinar, 1926	Colombia
j) lisethae Neukirchen, 1995	Colombia
k) wanningeri Neukirchen, 1991	Colombia
l) weymeri Staudinger, 1897	Colombia
weymeri ab. *gustavi* Staudinger, 1897	Colombia
weymeri submarginatus Fassl, 1912	Colombia
submarginalis Seitz, 1912, missp.	
weymeri ab. *sulphureomaculata* Fassl, 1914	Colombia

cydno gustavi ab. *flavomaculata* Seitz, 1916	Colombia
cydno hermogenes ab. *leucosticta* Apolinar, 1926	Colombia
cydno hermogenes ab. *xanthosticta* Apolinar, 1926	Colombia
cydno gerstneri H. Holzinger & R. Holzinger, 1968, hyb.	Colombia
cydno gerstneri f. *pseudoweymeri* H. Holzinger & R. Holzinger, 1968, hyb.	Colombia
cydno gerstneri f. *denhezi* H. Holzinger & R. Holzinger, 1968, hyb.	Colombia
cydno gerstneri f. *flavissima* H. Holzinger & R. Holzinger, 1968, hyb.	Colombia
n) zelinde Butler, 1869	W coast of America
f. *inca* Neustetter, 1928	?
cydno cydnides f. *flavidior* Neustetter, 1928, hyb.	Colombia
cydno cydnides f. *confluens* Neustetter, 1928, hyb.	Colombia
cydno cydnides f. *albidior* Neustetter, 1928, hyb.	Colombia
demeter Staudinger, 1897	
a) demeter Staudinger, 1897	Peru
f. *similis* Neustetter, 1931	Brazil (AM)
demeter angeli Neukirchen, 1997	Peru
b) beebei J. R. G. Turner, 1966	Guyana
c) bouqueti Nöldner, 1901	French Guiana
xanthoceras Oberthür, 1902 (*Heliconia*)	French Guiana
eueidina Oberthür, 1916 (*Heliconia*)	[French Guiana]
egeriformis (Joicey & Kaye, 1917) (*Eueides*)	French Guiana
automatia Oberthür, 1925 (*Heliconia*)	French Guiana
eratoformis Neustetter, 1931	French Guiana
d) karinae Neukirchen, 1990	Brazil (PA)
e) neildi Neukirchen, 1997	Ecuador
g) terrasanta K. S. Brown & Benson, 1975	Brazil (PA)
h) titan Neukirchen, 1995	Brazil (AM)
i) turneri K. S. Brown & Benson, 1975	Brazil (AM)
j) zikani K. S. Brown & Benson, 1975	Brazil (AM)
k) [n. ssp.] Lamas, MS	Peru
doris (Linnaeus, 1771) (*Papilio*)	
a) doris (Linnaeus, 1771) (*Papilio*)	Surinam
quirina (Cramer, 1775) (*Papilio*)	Surinam
amathusia (Cramer, 1777) (*Papilio*)	Surinam
delila (Hübner [1813]) (*Nerëis* [sic])	Surinam
crenis (Hübner, 1816) (*Migonitis*), repl. Name	Surinam
brylle (Hübner, 1821) (*Crenis*), repl. Name	Surinam
dorimena Doubleday, 1847 (*Heliconia*), nom. nud.	Bolivia
mars Staudinger, 1885	Peru
erato ab. *metharmina* Staudinger, 1897	[Brazil (AM)]
erato ab. *tecta* Riffarth, 1900	Peru

doris-caerulea [sic] Riffarth, 1901, nom. nud.
doris rubra Stichel, 1906, nom. nud.
doris nigra Stichel, 1906, nom. nud.
doris caeruleatus Stichel, 1906 — Peru
doris delila f. *albina* Boullet & Le Cerf, 1909 — [Peru]
var. *le moulti* [sic] Boullet & Le Cerf, 1910 — French Guiana
ab. *gibbsi* Kaye, 1919 — Guyana
doris dialis Stichel, 1923 — Colombia
f. *albescens* Neustetter, 1926 — French Guiana

b) dives Oberthür, 1920 (*Heliconia*) — Colombia
doris alberato Avinoff 1926 — Venezuela
doris azurea Avinoff, 1926 — Venezuela
doris viridis f. *viridana* Stichel, 1906 — Colombia
doris eratonia f. *suavior* Kaye, 1914 — Venezuela
doris metharmina f. *fascinator* Kaye, 1914 — Venezuela

c) obscurus Weymer, 1891 — Colombia
erato aristomache Riffarth, 1901 — Ecuador

d) viridis Staudinger, 1885 — Panama
erato var. *eratonius* Staudinger, 1897 — Panama
erato var. *transiens* Staudinger, 1897 — Honduras
erato luminosus Riffarth, 1901 — Panama

e) [n. ssp.] Neukirchen, MS — Trinidad

egeria (Cramer, 1775) (*Papilio*)

a) egeria (Cramer, 1775) (*Papilio*) — Surinam
isaea (Hübner, 1816) (*Migonitis*), repl. Name — Surinam
ergatis Godart, 1819 (*Heliconia*), repl. Name — Surinam
ab. *clearista* Oberthür, 1923 (*Heliconia*) — French Guiana

b) egerides Staudinger, 1897 — Brazil (AM)
egeria christiani Neukirchen, 1997 — Brazil (PA)

c) homogena Bryk, 1953 — Brazil (AM)
egeria-hyas [sic] f. *asterope* J. Zikán, 1937 — Brazil (AM)

d) hyas Weymer, 1883 — [Brazil (AM)]
egeria mariasibyllae Neukirchen, 1991 — Brazil (PA)

e) keithbrowni Neukirchen, 1992 — Venezuela

f) [n. ssp.] Lamas, MS — Peru

eleuchia Hewitson [1854] (*Heliconia*)

a) eleuchia Hewitson [1854] (*Heliconia*) — Colombia

b) eleusinus Staudinger, 1885 — Colombia
sapho primularis ab. *deflava* Joicey & Kaye, 1917, hyb. — Ecuador
ceres Oberthür, 1920 (*Heliconia*), hyb. — Colombia
primularis ceres Stichel, 1923, preocc. (not Oberthür, 1920), hyb. — Colombia

c) primularis Butler, 1869 — Ecuador

elevatus Nöldner, 1901	
a) elevatus Nöldner, 1901	'Amazonas'
f. *griseoviridis* Neustetter, 1938	Peru
elevatus schmidt-mummi [sic] Takahashi, 1977	Brazil (AM)
elevatus willmotti Neukirchen, 1997	Ecuador
b) bari Oberthür, 1902 (*Heliconia*)	French Guiana
c) lapis Lamas, 1976	Peru
d) perchlora Joicey & Kaye, 1917	Bolivia
e) pseudocupidineus Neustetter, 1931	Peru
elevatus f. *nigromacula* Neustetter, 1932	Peru
elevatus f. *nöldneri* [sic] Neustetter, 1938	Peru
f) roraima J. R. G. Turner, 1966	Guyana
g) schmassmanni Joicey & Talbot, 1925	Brazil (MT)
melpomene thelxiope f. *aquilina* Neustetter, 1925	Brazil ([RO])
h) taracuanus Bryk, 1953	Brazil (AM)
i) tumatumari Kaye, 1906	Guyana
elevatus sonjae Neukirchen, 1997	Brazil (PA)
j) zoelleri Neukirchen, 1990	Venezuela
erato (Linnaeus, 1758) (*Papilio*)	
a) erato (Linnaeus, 1758) (*Papilio*)	Surinam
vesta (Cramer, 1777) (*Papilio*), hyb.	Surinam
erythrea (Cramer, 1777) (*Papilio*), hyb.	Surinam
andremona (Cramer, 1780) (*Papilio*), hyb.	Surinam
udalrica (Cramer, 1780) (*Papilio*), hyb.	Surinam
ulrica (Hübner, 1816) (*Migonitis*), missp.	
cynisca Godart, 1819 (*Heliconia*)	French Guiana
vesta tellus Oberthür, 1902 (*Heliconia*), hyb.	French Guiana
erato oberthürii [sic] Riffarth, 1903, hyb.	Surinam
f. *fuliginosa* Riffarth, 1907	French Guiana; Surinam
ab. *hemicycla* Joicey & Kaye, 1917, hyb.	French Guiana
ab. *protea* Joicey & Kaye, 1917, hyb.	French Guiana
ab. *constricta* Joicey & Kaye, 1917, hyb.	French Guiana
f. *albida* Joicey & Kaye, 1919	French Guiana
ab. *telloides* Joicey & Talbot, 1925, hyb.	French Guiana
f. *roseoflava* Neustetter, 1926, hyb.	French Guiana
f. *fumata* Neustetter, 1926, hyb.	French Guiana
f. *latiflava* Neustetter, 1931, hyb.	French Guiana
f. *nigrobasalis* Neustetter, 1931, hyb.	French Guiana
b) adana J. R. G. Turner, 1967 (ICZN, Art. 45.6.2)	[Trinidad]
hydara ab. *adana* Seitz, 1913	[Trinidad]
hydarus hydarus f. *vitellina* Stichel, 1919	Trinidad

Taxon	Locality
c) amalfreda Riffarth, 1901	Brazil (PA)
elimaea Erichson [1849] (*Heliconia*), hyb.	Guyana
cybele ab. *cybelina* Staudinger, 1897	Brazil (AM)
vesta ab. *leda* Staudinger, 1897, hyb.	French Guiana
phyllis ab. *amalfreda* Riffarth, 1900	Brazil (AM)
erato cybellinus f. *helena* Riffarth, 1907, hyb.	Brazil (PA); Surinam
erato erato ab. *cybelellus* Joicey & Kaye, 1917, hyb.	Brazil (AM)
melpomene f. *pyritosa* J. Zikán, 1937	Brazil ([PA])
melpomene-pyritosa [sic] var. *fumigata* J. Zikán, 1937	Brazil ([PA])
erato andremona f. *juanita* Neustetter, 1938, hyb.	Brazil (Rr)
f. *heydei* Stammeshaus, 1982	Surinam
d) amazona Staudinger, 1897	Brazil (PA)
vesta. Hübner [1807] (*Nerëis* [sic]) (not Cramer, 1777)	
philadelphus W. F. Kirby [1899]	Brazil
androdaixa Seitz, 1912, hyb.	[Brazil (PA)]
e) amphitrite Riffarth, 1901	Peru
erato estrella f. *simplex* Riffarth, 1906, hyb.	Peru
erato phyllis f. *sperata* Riffarth, 1907, hyb.	Peru; 'Bolivia'
melpomene hyperplea Dyar, 1913	Peru
ab. *unipuncta* Joicey & Kaye, 1917	Peru
f) colombina Staudinger, 1897	Colombia
hydara var. *antigona* Riffarth, 1900	Colombia
g) cruentus Lamas, 1998	Mexico (CH)
erato punctata Beutelspacher, 1992, preocc. (not Neustetter, 1907)	[Mexico (JA)]
h) cyrbia Godart, 1819 (*Heliconia*)	America
cyrbia ab. *diformata* Riffarth, 1900, hyb.	Ecuador
cyrbia cyrbia f. *bella* Riffarth, 1907	Ecuador
i) demophoon Ménétriés, 1855	Nicaragua
chiriquensis Riffarth, 1900, nom. nud.	
j) dignus Stichel, 1923	Colombia
dignus f. *discerpta* Stichel, 1923, hyb.	Colombia
erato estrella f. *problemata* Neustetter, 1928, hyb.	Colombia
estrella f. *meliorina* Neustetter, 1928, hyb.	Colombia
estrella f. *glaucina* Neustetter, 1928, hyb.	Colombia
dignus f. *elvira* Niepelt, 1928, hyb.	Colombia
k) emma Riffarth, 1901	Peru; 'Ecuador'
augusta Riffarth, 1901, nom. nud.	
erato estrella f. *palmata* Stichel, 1906, hyb.	Peru
erato estrella f. *agnata* Stichel, 1906, hyb.	Peru
l) estrella H. W. Bates, 1862	Brazil (PA)
m) etylus Salvin, 1871	Ecuador
ab. *insignis* Kaye, 1916, hyb.	Ecuador

n) favorinus Hopffer, 1874	'Bolivia'
amaryllis var.? *pseudamaryllis* Staudinger, 1897	Peru
erato eratophylla Joicey & Kaye, 1917, hyb.	Peru
favorinus pseudoanacreon Neustetter, 1932, hyb.	Peru
o) guarica Reakirt, 1868	Colombia
guayana Herrich-Schäffer, 1865, nom. nud.	?
euryas Boisduval, 1870 (*Heliconia*)	[Venezuela]
p) hydara Hewitson, 1867 (*Heliconia*)	Colombia
callycopis (Cramer, 1777) (*Papilio*), hyb.	Surinam
carolina (Herbst, 1790) (*Papilio*), artefact	?
coralii Butler, 1877, hyb.	Brazil (AM)
erato f. *palantia* Möschler, 1883, hyb.	Surinam
cyrbia var. *juno* Riffarth, 1900, hyb.	?
petiverana ab. *tristis* Riffarth, 1900, hyb.	Panama
phyllis var. *viculata* Riffarth, 1900, hyb.	Surinam
phyllis ab. *dryope* Riffarth, 1900, hyb.	Surinam
phyllis ab. *callista* Riffarth, 1900, hyb.	Surinam
palantes Riffarth, 1901, nom. nud.	
erato callycopis f. *belticopis* Joicey & Kaye, 1917, hyb.	Surinam
erato rubrizona Joicey & Kaye, 1917, hyb.	Brazil (PA)
q) lativitta Butler, 1877	Brazil (AM); Ecuador
vesta var. *vestalis* Staudinger, 1885	[Brazil (AM)]
erato estrella f. *ochracea* Riffarth, 1907, hyb.	Ecuador
erato estrella f. *feyeri* Niepelt, 1908, hyb.	Ecuador
erato estrella f. *anactorina* Stichel, 1923, hyb.	Colombia
erato estrella f. *sanguinella* Stichel, 1923, hyb.	Colombia
erato estrella f. *perplexa* Stichel, 1923, hyb.	Colombia
microclea beata f. *zoraida* Neustetter, 1925, hyb.	Ecuador
erato estrella f. *aurivillii* Bryk, 1953, hyb.	Brazil (AM)
r) lichyi K. S. Brown & F. Fernández, 1985	Venezuela
s) luscombei Lamas, 1976	Peru
t) magnifica Riffarth, 1900	Guyana
u) microclea Kaye, 1907	Peru
f. *microfluens* Descimon & Mast, 1971, hyb.	Peru
v) notabilis Salvin & Godman, 1868	Ecuador
erato estrella f. *beata* Riffarth, 1907, hyb.	Ecuador
erato simplex f. *rosacea* Riffarth, 1907, hyb.	Ecuador
erato estrella f. *ilia* Niepelt, 1908, hyb.	Ecuador
erato estrella f. *rothschildi* Niepelt, 1909, hyb.	Ecuador
microclea notabilis radiata Oberthür, 1916 (*Heliconia*), hyb.	Ecuador
notabilis rubrescens Röber, 1921, hyb.	Ecuador
microclea notabilis f. *flavopunctatus* Neustetter, 1925, hyb.	Ecuador

w) petiverana Doubleday, 1847 (*Heliconia*)	Mexico; Honduras
mexicana Boisduval, 1870 (*Heliconia*)	Mexico ([VZ]); Honduras
petiverea Riffarth, 1901, nom. nud.	
x) phyllis (Fabricius, 1775) (*Papilio*)	Brazil
roxane (Cramer, 1775) (*Papilio*)	'Surinam'
phyllidis Grose-Smith & W. F. Kirby, 1892	Bolivia
anacreon Grose-Smith & W. F. Kirby, 1892, hyb.	Bolivia
amatus Staudinger, 1897, hyb.	Bolivia
phyllis ab. *artifex* Stichel, 1899	Paraguay; Brazil (ES)
f. *diffluens* Riffarth, 1907	'Peru'; Bolivia
erato anacreon f. *anaitis* Riffarth, 1907, hyb.	Bolivia
f. *athene* Neustetter, 1909	Bolivia
phyllis ab. *miletus* d'Almeida, 1928	Brazil (RJ)
phyllis ab. *cohaerens* Hayward, 1931	Argentina
phyllis ab. *alicia* Schweizer & Kay, 1941	Uruguay
y) reductimacula Bryk 1953	Brazil (AM)
f. *venusta* Bryk, 1953	Brazil (AM)
z) tobagoensis Barcant, 1982	Tobago
aa) venus Staudinger, 1882	Colombia
molina Grose-Smith, 1898, hyb.	Colombia
erato f. *extrema* Kaye, 1919, nom. nud.	Colombia
erato extrema Kaye, 1919, hyb.	Colombia
ab) venustus Salvin, 1871	Bolivia
anactorie Doubleday, 1847 (*Heliconia*), hyb.	Bolivia
anactorie ab. *sanguineus* Staudinger, 1894, hyb.	Bolivia
phyllis ab. *ottonis* Riffarth, 1900, hyb.	Bolivia
locris Riffarth, 1901, nom. nud.	
erato diva Stichel, 1906, hyb.	Bolivia
phyllis ab. *confluens* Seitz, 1913, hyb.	Bolivia
phyllis f. *anacreonides* Neustetter, 1925, hyb.	Bolivia
phyllis f. *krügeri* [sic] Neustetter, 1925, hyb.	Bolivia
phyllis f. *flavomixta* Neustetter, 1925, hyb.	Bolivia
phyllis f. *leonora* R. Krüger, 1927, hyb.	Bolivia
phyllis anacreon f. *clelia* Neustetter, 1927, hyb.	Bolivia
phyllis anactorie f. *henrici* R. Krüger, 1929, hyb.	Bolivia
phyllis venustus f. *pseudoleda* Neustetter, 1932, hyb.	Bolivia
xanthocles f. *denieri* Hayward, 1939	Bolivia
eratosignis (Joicey & Talbot, 1925) (*Eueides*)	
a) eratosignis (Joicey & Talbot, 1925) (*Eueides*)	Brazil (MT)
b) tambopata Lamas, 1985	Peru
c) ucayalensis H. Holzinger & R. Holzinger, 1975	Peru

d) ulysses K. S. Brown & Benson, 1975	Bolivia
ethilla Godart, 1819 (*Heliconia*)	
a) ethilla Godart, 1819 (*Heliconia*)	'Antilles'
metalilis var. *flavidus* Weymer, 1894	Venezuela
ethilla metalilis f. *depuncta* Boullet & Le Cerf, 1909	Trinidad
b) adela Neustetter, 1912	Peru
c) aerotome C. Felder & R. Felder, 1862	'Brazil (AM)'
d) cephallenia C. Felder & R. Felder, 1865	Surinam
e) chapadensis K. S. Brown, 1973	Brazil (MT)
f) claudia Godman & Salvin, 1881	Panama
assimilis Röber, 1921	Panama
g) eucoma (Hübner [1831]) (*Eueides*)	Brazil
eucoma (Hübner, 1816) (*Eueides*), nom. nud.	
h) flavofasciatus Weymer, 1894	Brazil (PA)
i) flavomaculatus Weymer, 1894	Brazil (Pe)
j) hyalina Neustetter, 1928	Brazil (Rr)
k) jaruensis K. S. Brown, 1976	Brazil (RO)
l) latona Neukirchen, 1998	Colombia
m) mentor Weymer, 1883	Colombia
orchamus Weymer, 1912	Colombia
n) metalilis Butler, 1873	Venezuela
o) michaelianius Lamas, 1988, repl. name	Peru
aurotome [sic] f. *clarus* Michael, 1926, preocc. (not Fabricius, 1793)	Peru
p) narcaea Godart, 1819 (*Heliconia*)	'Antilles'
eucrate (Hübner [1823]) (*Mechanites* [sic])	?
satis Weymer, 1875	Brazil
eucrate var. *infuscata* Staudinger, 1885	Brazil
narcaea ab. *brunnescens* Neustetter, 1907	Brazil (RJ)
connexa Seitz, 1912	Brazil ([RJ])
q) nebulosa Kaye, 1916	Peru
r) neukircheni Lamas, 1998	Peru
s) numismaticus Weymer, 1894	Brazil (PA)
t) penthesilea Neukirchen, 1994	Brazil (AM)
u) polychrous C. Felder & R. Felder, 1865	Brazil ([SP])
physcoa Seitz, 1912	[Brazil (SP)]
v) semiflavidus Weymer, 1894	Colombia
eucoma var. *daguanus* Staudinger, 1897	Colombia
eucoma var. *juntana* Riffarth, 1900	Colombia
w) thielei Riffarth, 1900	French Guiana
numatus isabellinus ab. *fusca* Boullet & Le Cerf, 1909	French Guiana
x) tyndarus Weymer, 1897	Bolivia

y) yuruani K. S. Brown & F. Fernández, 1985 — Venezuela

godmani Staudinger, 1882 — Colombia

hecale (Fabricius, 1776) (*Papilio*)

a) hecale (Fabricius, 1776) (*Papilio*) — Surinam

pasithoë (Cramer, 1775) (*Papilio*), preocc. (not Linnaeus, 1767) — Surinam;

pasithoë [sic] var. *fulvescens* Lathy, 1906 — Guyana

b) anderida Hewitson [1853] (*Heliconia*) — [Venezuela]

zagora Riffarth, 1901, nom. nud.

rebeli Neustetter, 1907 — 'Colombia'

anderida estebana Kaye, 1914 — Venezuela

c) annetta Riffarth, 1900 — Colombia

d) australis K. S. Brown, 1976 — Ecuador

e) barcanti K. S. Brown, 1976 — Venezuela

f) clearei A. Hall, 1930 — Guyana

g) ennius Weymer, 1891 — Brazil (AM)

h) felix Weymer, 1894 — Peru

felix var. *concors* Weymer, 1894 — Peru

versicolor Weymer, 1894 — 'Brazil (AM)'

quitalena felix f. *umbrina* Neustetter, 1931, hyb. — Peru

i) fornarina Hewitson, 1854 (*Heliconia*) — ['S. Amer.']

anderida fornarina f. *bouvieri* Boullet & Le Cerf, 1909 — Guatemala

anderida f. *styx* Niepelt, 1921 — Guatemala

j) holcophorus Staudinger, 1897 — Colombia

metaphorus var. *semiphorus* Staudinger, 1897 — Colombia

eucherius Weymer, 1906 — Colombia

k) humboldti Neustetter, 1928 — Peru

humboldti f. *alexander* Neustetter, 1928 — Peru

l) ithaca C. Felder & R. Felder, 1862 — Colombia

vittatus Butler, 1873 — Colombia

marius Weymer, 1890 — Colombia

ithaka [sic] f. *cajetani* Neustetter, 1909 — [Colombia]

ithaca f. *hero* Weymer, 1912 — Colombia

aristiona indecisa Joicey & Kaye, 1917 — [Colombia]

ithaka ithaka [sic] f. *sulphureofasciata* Neustetter, 1925 — Colombia

ithaka [sic] *vittatus* f. *nigroapicalis* Neustetter, 1925 — Colombia

m) latus Riffarth, 1900 — Brazil (PA)

gradatus xinguensis Neustetter, 1925 — Brazil (PA)

n) melicerta H. W. Bates, 1866 — Panama; Colombia

clara (Fabricius, 1793) (*Papilio*), preocc. (not Cramer, 1775) — 'Surinam'

albucilla H. W. Bates, 1866 — Panama

etholea Riffarth, 1901, nom. nud.

anderida melicerta f. *zygia* Riffarth, 1907	Colombia
anderida melicerta f. *muzoënsis* [sic] Neustetter, 1909	Colombia
o) metellus Weymer, 1894	Brazil (PA)
vetustus boyi Röber, 1923	Brazil (PA)
p) nigrofasciatus Weymer, 1894	Brazil (AM)
q) novatus H. W. Bates, 1867	Brazil (PA/MA)
schulzi Riffarth, 1899	Brazil (PA)
schultzi Seitz, 1924, emend.	
r) paraensis Riffarth, 1900	Brazil (PA)
s) paulus Neukirchen, 1998	Brazil (PA)
t) quitalena Hewitson [1853] (*Heliconia*)	Ecuador
u) rosalesi K. S. Brown & F. Fernández, 1976	Venezuela
v) shanki Lamas & K. S. Brown, 1976	Peru
w) sisyphus Salvin, 1871	Peru
jonas Weymer, 1894	Brazil (AM)
x) sulphureus Weymer, 1894	Brazil (AM)
quitalenus denticulatus Riffarth, 1907	Brazil (AM)
? *sulphureus subsulphureus* Röber, 1919, nom. dub.	?
y) vetustus Butler, 1873	Guyana
clarissa Riffarth, 1901, nom. nud.	
hecale naxos Neukirchen, 1998	Brazil (PA)
hecale fraternitas Brévignon, 2006	French Guiana
z) zeus Neukirchen, 1995	Bolivia
felix var. *concors* Weymer, 1894, in part	
aa) zuleika Hewitson, 1854 (*Heliconia*)	Nicaragua
jucundus H. W. Bates, 1864	Panama
xanthicus H. W. Bates, 1864	Panama
chrysantis Godman & Salvin, 1881	Nicaragua
discomaculatus Weymer, 1891, hyb.	Honduras
zuleika ab. *albipunctata* Riffarth, 1900	Panama
zuleika ab. *dentata* Neustetter, 1907	Panama
hecalesia Hewitson [1854] (*Heliconia*)	
a) hecalesia Hewitson [1854] (*Heliconia*)	Colombia
b) eximius Stichel, 1923	Colombia
c) formosus H. W. Bates, 1866	Panama
d) gynaesia Hewitson, 1875 (*Heliconia*)	?
e) longarena Hewitson, 1875 (*Heliconia*)	Colombia
hecalesia ernestus K. S. Brown & Benson, 1975, hyb.	Colombia
f) octavia H. W. Bates, 1866	Guatemala
g) romeroi K. S. Brown & F. Fernández, 1985	Venezuela
hecuba Hewitson [1858] (*Heliconia*)	

a) hecuba Hewitson [1858] (*Heliconia*) Colombia

hecuba intermedius Riffarth, 1907, hyb. Colombia

b) cassandra C. Felder & R. Felder, 1862 Colombia

c) choarina Hewitson, 1872 (*Heliconia*) Ecuador

dismorphia (Buchecker [1880]) (*Blanchardia*), nom. nud. 'Colombia'

f. *cacica* Neustetter, 1928 Ecuador

hecuba bonplandi Neukirchen, 1991 Ecuador

d) creusa H. Holzinger & R. Holzinger, 1989 Colombia

crispus var. *crespinus* E. Krüger, 1925, hyb. Colombia

e) crispus Staudinger, 1885 Colombia

hecuba salazari Neukirchen, 1993 Colombia

hecuba walteri Salazar, 1998 Colombia

f) flava K. S. Brown, 1979 (ICZN, Art. 45.5.1) Ecuador

hecuba choarinus f. *flava* Neustetter, 1928 Ecuador

g) lamasi Neukirchen, 1991 Ecuador

h) tolima Fassl, 1912 Colombia

hecuba tolima Seitz, 1913, preocc. (Fassl, 1912) Colombia

hermathena Hewitson [1854] (*Heliconia*)

a) hermathena Hewitson [1854] (*Heliconia*) [Brazil]

b) duckei K. S. Brown & Benson, 1977 Brazil (PA)

hermathena f. *rubropunctata* D'Abrera, 1984, nom. nud. Brazil

c) renatae K. S. Brown & F. Fernández, 1985 Venezuela

d) sabinae Neukirchen, 1992 Brazil (AM)

e) sheppardi K. S. Brown & Benson, 1977 Brazil (AM)

hermathena sheppardorum Lamas, 1988, emend.

f) vereatta Stichel, 1912 Brazil (PA)

hermathena f. *hydarina* Stichel, 1912 Brazil (PA)

heurippa Hewitson [1854] (*Heliconia*) Colombia

hewitsoni Hewitson, 1875 (*Heliconia*) Panama

hewitsoni Staudinger, 1876, preocc. (not Hewitson, 1875) Panama

hierax Hewitson, 1869 (*Heliconia*)

a) hierax Hewitson, 1869 (*Heliconia*) Ecuador

f. *semibrunea* Niepelt, 1923 Ecuador

b) leveri Attal, 1999 Ecuador

himera Hewitson, 1867 (*Heliconia*) Ecuador

hortense Guérin-Méneville [1844] (*Heliconia*) Mexico

ismenius Latreille [1817]

a) ismenius Latreille [1817] [S. Amer.]

fritscheï [sic] Möschler, 1872 (*Heliconia*) 'Peru or Colombia'

ocanna Buchecker [1880] Colombia

faunus Staudinger, 1885 Colombia

faunus var. *antioquensis* Staudinger, 1885	Colombia
var. (ab.?) *hermanni* Riffarth, 1899	[Colombia]
distincta Riffarth, 1901, nom. nud.	
f. *immoderata* Stichel, 1906	Colombia
ab. *albofasciatus* Neustetter, 1907	Colombia
f. *defasciatus* Neustetter, 1909	Colombia
ismenius abadiae Apolinar, 1926	Colombia
b) boulleti Neustetter, 1928	'French Guiana'
sylvana sylvana [sic] f. *sticheli* Boullet & Le Cerf, 1909	'French Guiana'
c) clarescens Butler, 1875	Panama
d) fasciatus Godman & Salvin, 1877	Panama
e) metaphorus Weymer, 1883	Ecuador
catilina Riffarth, 1901, nom. nud.	
ismenius faunus f. *hoppi* Neustetter, 1928, hyb.	Colombia
f) occidentalis Neustetter, 1928	Colombia
g) telchinia Doubleday, 1847 (*Heliconia*)	[Honduras]
h) tilletti K. S. Brown & F. Fernández, 1976	Venezuela
lalitae Brévignon, 1996	French Guiana
leucadia H. W. Bates, 1862	
a) leucadia H. W. Bates, 1862	Brazil (AM)
b) andromeda Neukirchen, 1996	Ecuador
c) birgitae Neukirchen, 1996	Peru
d) pseudorhea Staudinger, 1897	Brazil (AM)
luciana Lichy, 1960	
a) luciana Lichy, 1970	Venezuela
b) watunna Lichy, 1960	Venezuela
melpomene (Linnaeus, 1758) (*Papilio*)	
a) melpomene (Linnaeus, 1758) (*Papilio*)	America
lucia (Stoll, 1781) (*Papilio*), hyb.	Surinam
var. *tyche* H. W. Bates, 1862, hyb.	Brazil (AM)
var. *hippolyte* H. W. Bates, 1862, hyb.	Brazil (AM/PA)
rufolimbatus Butler, 1873, hyb.	Brazil (PA)
mutabilis Butler, 1877, hyb.	Brazil (AM)
funebris Möschler, 1877, hyb.	Surinam
amor Staudinger, 1885, hyb.	Brazil (AM)
timareta ab.? *erebia* Riffarth, 1900, hyb.	French Guiana
ab. *atrosecta* Riffarth, 1900	Brazil (PA)
ab. *lucinda* Riffarth, 1900, hyb.	Surinam
ab. *melpomenides* Riffarth 1900, hyb.	Surinam
ab. *karschi* Riffarth, 1900, hyb.	Brazil (PA)
var. *melanippe* Riffarth, 1900, hyb.	Surinam

ab. *diana* Riffarth, 1900, hyb.	Surinam?
gaea Riffarth, 1901, nom. nud.	
jussa Riffarth, 1901, nom. nud.	
justina Riffarth, 1901, nom. nud.	
ab. *collis* Joicey & Kaye, 1917, hyb.	French Guiana
ab. *primus* Joicey & Kaye, 1917, hyb.	French Guiana
ab. *melpina* Joicey & Kaye, 1917, hyb.	French Guiana
melpomene cybele ab. *dianides* Joicey & Kaye, 1917, hyb.	French Guiana
melpomene cybele ab. *elegantula* Joicey & Kaye, 1917, hyb.	French Guiana
melpomene cybele ab. *maris* Joicey & Kaye, 1917, hyb.	French Guiana
melpomene eltringhami Joicey & Kaye, 1917, hyb.	French Guiana
ab. *compacta* Joicey & Kaye, 1919, hyb.	French Guiana
ab. *faivrei* Joicey & Kaye, 1919, hyb.	French Guiana
f. *flavorubra* Neustetter, 1926	French Guiana
f. *lydia* Neustetter, 1926, hyb.	French Guiana
f. *luteipicta* Neustetter, 1926, hyb.	French Guiana; Guyana
f. *aurelia* Neustetter, 1926, hyb.	French Guiana
f. *bang-haasi* [sic] Neustetter, 1926, hyb.	French Guiana
f. *rubroflammea* Neustetter, 1927, hyb.	French Guiana
f. *laurentina* Neustetter, 1927, hyb.	French Guiana
f. *rufolinea* Neustetter, 1928, hyb.	French Guiana
f. *trimacula* Neustetter, 1931, hyb.	French Guiana
melpomene melpomene alabanda Lever, Chazal & Lacomme, 2008, hyb.	French Guiana
b) aglaope C. Felder & R. Felder, 1862	'Brazil (AM)'
melpomene ab. *mirabilis* Riffarth, 1900, hyb.	Peru
melpomene riffarthi Stichel, 1906, hyb.	Peru
melpomene riffarthi f. *rubescens* Stichel, 1906, hyb.	Peru
f. *cognata* Riffarth, 1907	Peru
aglaope f. *flavotenuiata* Neustetter, 1931	Peru
aglaope f. *magnimacula* Neustetter, 1932, hyb.	Peru
aglaope f. *daira* Neustetter, 1932, hyb.	Peru
c) amandus Grose-Smith & W. F. Kirby, 1892	Bolivia
penelope ab. *pelopeia* Staudinger, 1894, hyb.	Bolivia
penelope ab. *penelamanda* Staudinger, 1894, hyb.	Bolivia
aphrodyte Staudinger, 1897, hyb.	Bolivia
penelope ab. *penelopeia* Staudinger, 1897, emend.	
phyllis f. *flammea* Niepelt, 1925, hyb.	Bolivia
penelope f. *biedermanni* Niepelt, 1926, hyb.	Bolivia
penelope f. *amandoides* Neustetter, 1926, hyb.	Bolivia
penelope f. *carnea* Neustetter, 1926, hyb.	Bolivia
d) amaryllis C. Felder & R. Felder, 1862	'Brazil (AM)'

amaryllis ab. *pseudo-penelamanda* [sic] Michael, 1912, hyb.	Peru
melpomene melpophylla Joicey & Kaye, 1917, hyb.	Peru?
e) anduzei K. S. Brown & F. Fernández, 1985	Venezuela
f) bellula J. R. G. Turner, 1971 (ICZN, Art. 45.5.1)	Colombia
amaryllis amaryllis f. *bellula* Stichel, 1923	Colombia
amaryllis amaryllis f. *permira* Stichel, 1923, hyb.	Colombia
amaryllis amaryllis f. *degener* Stichel, 1923, hyb.	Colombia
amaryllis amaryllis anacreontica Stichel, 1923, hyb.	Colombia
amaryllis amaryllis f. *perrara* Stichel, 1923, hyb.	Colombia
melpomene aglaope f. *parva* Neustetter, 1928, hyb.	Colombia
melpomene mocoa Brower, 1996	Colombia
g) burchelli Poulton, 1910	Brazil (Go)
f. *curvifascia* Talbot, 1928, hyb.	Brazil (MT)
h) cythera Hewitson, 1869 (*Heliconia*)	Ecuador
vulcanus ab. *modesta* Riffarth, 1900, hyb.	Ecuador
hypna Riffarth, 1901, nom. nud.	
vulcanus sticheli Riffarth, 1907, hyb.	Ecuador
vulcanus tenuistriga Kaye, 1920	Ecuador
i) ecuadorensis Emsley, 1964 (ICZN, Art. 45.5.1)	Ecuador
melpomene aglaope f. *ecuadorensis* Neustetter, 1909	Ecuador
equatorensis Seitz, 1912, missp.	
melpomene equadoriensis Seitz, 1913, missp.	
j) euryades Riffarth, 1900	Peru
amaryllis euryades f. *rubrica* Stichel, 1919, hyb.	Peru
k) flagrans Stichel, 1919	Trinidad
l) intersectus Neustetter, 1928	Brazil ([PA])
m) madeira Riley, 1919	Brazil (AM)
n) malleti Lamas, 1988	Ecuador
melpomene aglaope f. *iris* Riffarth, 1907, hyb.	Ecuador
melpomene aglaope f. *isolda* Niepelt, 1908, hyb.	Ecuador
melpomene aglaope f. *adonides* Niepelt, 1908, hyb.	Ecuador
melpomene aglaope f. *rubripicta* Niepelt, 1908, hyb.	Ecuador
melpomene aglaope f. *gisela* Niepelt, 1908, hyb.	Ecuador
melpomene aglaope f. *anna* Neustetter, 1909, hyb.	Ecuador
melpomene aglaope f. *dione* Neustetter, 1909, hyb.	Ecuador
melpomene aglaope f. *gratiosa* Niepelt, 1909, hyb.	Ecuador
amaryllis amaryllis f. *rufata* Stichel, 1923, hyb.	Colombia
amaryllis amaryllis f. *aglaspis* Stichel, 1923, hyb.	Colombia
melpomene aglaope f. *aurofasciata* Neustetter, 1928, hyb.	Colombia
melpomene aglaope f. *paula* Neustetter, 1928, hyb.	Colombia
aglaope f. *paulina* Niepelt, 1928, hyb.	Colombia

aglaope f. *carminata* Niepelt, 1928, hyb.	Colombia
o) martinae Cast & Le Crom, 2012	Colombia
melpomene martinae f. *flavomartina* Cast & Le Crom, 2012	Colombia
melpomene martinae f. *mixta* Cast & Le Crom, 2012	Colombia
p) meriana J. R. G. Turner, 1967, repl. name	Surinam
cybele (Cramer, 1777) (*Papilio*), preocc. (not Fabricius, 1775)	Surinam
deinia Möschler, 1877, hyb.	[Surinam]
faustina Staudinger, 1885, hyb.	French Guiana
melpomene ab. *eulalia* Riffarth, 1900, hyb.	Surinam
melpomene funebris f. *obscurata* Riffarth, 1907	Surinam
fascinatrix Seitz, 1912, hyb.	French Guiana
melpomene cybele ab. *faustalia* Joicey & Kaye, 1917, hyb.	French Guiana
melpomene cybele ab. *negroida* Joicey & Kaye, 1917, hyb.	French Guiana
melpomene cybele ab. *cybeleia* Joicey & Kaye, 1919, hyb.	French Guiana
melpomene f. *lavinia* Neustetter, 1926, hyb.	French Guiana
f. *alba* Stammeshaus, 1982	Surinam
melpomene meriana novalis Lever, Chazal & Lacomme, 2008, hyb.	French Guiana
melpomene meriana penthesilea Lever, Chazal & Lacomme, 2008, hyb.	French Guiana
melpomene meriana hyperion Lever, Chazal & Lacomme, 2008, hyb.	French Guiana
q) michellae Neukirchen, 1997	Peru
r) nanna Stichel, 1899	Brazil (ES/MG)
bidentatus Riffarth, 1901, nom. nud.	
mayi d'Almeida, 1928	Brazil (RJ)
s) penelope Staudinger, 1894	Bolivia
penelope ab. *pluto* Staudinger, 1897, hyb.	Bolivia
melpomene ab. *margarita* Riffarth, 1900, hyb.	Bolivia
f. *praxedis* Neustetter, 1925, hyb.	Bolivia
penelope f. *noctis* Neustetter, 1926, hyb.	Bolivia
penelope f. *excellens* Neustetter, 1926, hyb.	Bolivia
penelope f. *aida* Neustetter, 1926, hyb.	Bolivia
penelope f. *amneris* Neustetter, 1926, hyb.	Bolivia
penelope f. *rufofascia* Neustetter, 1926, hyb.	Bolivia
melpomene burchelli f. *obscurifascia* Talbot, 1928, hyb.	Brazil (MT)
penelope f. *flavodiscalis* Neustetter, 1931, hyb.	Bolivia
t) plesseni Riffarth, 1907	Ecuador
unimaculata Hewitson, 1869 (*Heliconia*), hyb.	Ecuador
radiatus Riffarth, 1901, nom. nud.	
batesi plesseni f. *rubicunda* Niepelt, 1907	Ecuador
batesi plesseni f. *pura* Niepelt, 1907	Ecuador
melpomene aglaope f. *niepelti* Riffarth, 1907, hyb.	Ecuador
batesi plesseni f. *adonis* Riffarth, 1907, hyb.	Ecuador

batesi plesseni f. *corona* Niepelt, 1908	Ecuador
batesi plesseni f. *diadema* Niepelt, 1908, hyb.	Ecuador
melpomene aglaope f. *fraterna* Niepelt, 1909	Ecuador
microclea virgo Oberthür, 1916 (*Heliconia*)	Ecuador
xenoclea plesseni f. *dido* Neustetter, 1925, hyb.	Ecuador
xenoclea plesseni f. *clytie* Neustetter, 1927, hyb.	Ecuador
xenoclea plesseni f. *mimetica* Neustetter, 1928, hyb.	Ecuador
u) pyrforus Kaye, 1907	Guyana
vulcanus f. *immarginata* Neustetter, 1938	Brazil (Rr)
v) rosina Boisduval, 1870 (*Heliconia*)	Costa Rica; 'Mexico'
w) schunkei Lamas, 1976	Peru
melpomene aglaope ab. *rubra* Stichel, 1906	Peru
aglaope var. *incarnata* Stichel, 1906, nom. nud.	
x) tessa Barcant, 1982	Tobago
y) thelxiope (Hübner [1806]) (*Nerëis* [sic])	[Brazil]
z) thelxiopeia Staudinger, 1897	French Guiana
thelxiope var. *aglaopeia* Staudinger, 1897	French Guiana
melpomene var. *augusta* Riffarth, 1900, hyb.	Surinam
judith Riffarth, 1901, nom. nud.	
milesia Riffarth, 1901, nom. nud.	
melpomene thelxiope ab. *punctarius* Joicey & Kaye, 1917, hyb.	French Guiana
melpomene thelxiope ab. *lucindella* Joicey & Kaye, 1917, hyb.	French Guiana
melpomene thelxiope ab. *majestica* Joicey & Kaye, 1917, hyb.	French Guiana
melpomene thelxiope ab. *stygianus* Joicey & Kaye, 1917, hyb.	French Guiana
melpomene thelxiope ab. *negroidens* Joicey & Kaye, 1917, hyb.	French Guiana
melpomene f. *athalia* Neustetter, 1927, hyb.	French Guiana
melpomene f. *penelopides* Neustetter, 1927, hyb.	French Guiana
melpomene f. *nigrointerrupta* Neustetter, 1931, hyb.	French Guiana
melpomene thelxiopeia diotima Lever, Chazal & Lacomme, 2008, hyb.	French Guiana
aa) vicina Ménétriés, 1857	Brazil
ab) vulcanus Butler, 1865	'Guyana'; Panama
amaryllis rosina f. *fumosa* Neustetter, 1925, hyb.	Colombia
ac) xenoclea Hewitson [1853] (*Heliconia*)	?
batesi Riffarth, 1900	'Ecuador'; Peru
xenoclea var. *superba* Lathy, 1906, hyb.	Peru
xenoclea var. *confluens* Lathy, 1906, hyb.	Peru
xenoclea ab. *zio* Röber, 1919, hyb.	Peru
f. *integra* Descimon & Mast, 1971, hyb.	Peru
metharme Erichson [1849] (*Heliconia*)	
a) metharme Erichson [1849] (*Heliconia*)	Guyana
thetis Boisduval, 1870 (*Heliconia*)	'Nicaragua'

anaclia Riffarth, 1901, nom. nud.	
b) makiritare (K. S. Brown & F. Fernández, 1985) (*Neruda*)	Venezuela
c) perseis Stichel, 1923	Colombia
metis (Moreira & C. Mielke, 2010) (*Neruda*)	Brazil (MA)
nattereri C. Felder & R. Felder, 1865	Brazil ([BA])
fruhstorferi Riffarth, 1899	Brazil (ES)
numata (Cramer, 1780) (*Papilio*)	
a) numata (Cramer, 1780) (*Papilio*)	Surinam
pione (Hübner, 1816) (*Eueides*), repl. name	Surinam
var. *melanops* Weymer, 1894	French Guiana
var. *guiensis* Riffarth, 1900	Guyana
numata isabellinus f. *intermedia* Boullet & Le Cerf, 1909	French Guiana
f. *melanopors* Joicey & Kaye, 1917	French Guiana
b) arcuella H. Druce, 1874	Peru
seraphion Weymer, 1894, hyb.	Peru
numatus praelautus Stichel, 1906	Peru
aristiona aurora f. *deflavata* Neustetter, 1932, hyb.	Peru
c) aristiona Hewitson [1853] (*Heliconia*)	'Colombia'
aristiona Doubleday, 1847 (*Heliconia*), nom. nud.	Bolivia
aristiona var. *peruana* Hopffer, 1879	Peru
aristiona var. *splendidus* Weymer, 1894, hyb.	Bolivia
d) aulicus Weymer, 1883	[Venezuela]
e) aurora H. W. Bates, 1862	Brazil (AM)
gordius Weymer, 1894, hyb.	Brazil (AM)
aristiona aurora f. *michaeli* Neustetter, 1931	Peru
aristiona idalion f. *excelsa* Neustetter, 1932, hyb.	Colombia
f) bicoloratus Butler, 1873	Peru
bicoloratus var. *phalaris* Weymer, 1894, hyb.	'Brazil (AM)'
g) elegans Weymer, 1894	Peru
floridus Weymer, 1894, hyb.	Bolivia; Peru
h) ethra (Hübner [1831]) (*Eueides*)	Brazil
dryalus Hopffer, 1869, repl. name	Brazil
ethra ab. *brasiliensis* Neustetter, 1907	Brazil (ES)
ethra ?var. *hopfferi* Neustetter, 1907	Brazil
i) euphone C. Felder & R. Felder, 1862	Colombia
idalion Weymer, 1894	Colombia
tleson Riffarth, 1901, nom. nud.	
aganippe Riffarth, 1901, nom. nud.	
aristiona lepidus Riffarth, 1907	Ecuador
idalion f. *confluens* Neustetter, 1912, preocc. (not Lathy, 1906)	Colombia
euphone ab. *nephele* Seitz, 1916	Colombia

euphone ab. *confluxus* Seitz, 1916, nom. nud.	
aristiona euphorbus Stichel, 1923	Colombia
mixta Apolinar, 1927	Colombia
j) euphrasius Weymer, 1890, hyb.	Colombia
gradatus Weymer, 1894, hyb.	Peru
k) geminatus Weymer, 1894	Brazil (AM)
l) holzingeri F. Fernández & K. S. Brown, 1976	Venezuela
m) ignotus Joicey & Kaye, 1917	Peru
n) illustris Weymer, 1894	Peru
silvana mirificus Stichel, 1906	Peru
novatus subnubilus Stichel, 1906	Peru
o) isabellinus H. W. Bates, 1862	Brazil (AM)
p) jiparanaensis Neustetter, 1931	Brazil (RO)
f. *mediatrix* Neustetter, 1931	Brazil (RO)
q) laura Neustetter, 1932	Colombia
r) lenaeus Weymer, 1891	'Colombia'
colepta Riffarth, 1901, nom. nud.	
s) lyrcaeus Weymer, 1891	Peru?
t) mavors Weymer, 1894	[Brazil (AM)]
numata superioris ab. *translata* Joicey & Kaye, 1917	Brazil (AM)
u) messene C. Felder & R. Felder, 1862	Colombia
sikinos Riffarth, 1901, nom. nud.	
aristiona messene f. *juncta* Neustetter, 1925	Colombia
aristiona colombiana Apolinar, 1927	Colombia
aristiona messene f. *euphrasinus* Neustetter, 1928	Colombia
v) mirus Weymer, 1894	Bolivia
leopardus Weymer, 1894	Bolivia
spadicarius A. G. Weeks, 1901	Bolivia
arethusa Riffarth, 1901, nom. nud.	
novatus obscurior Stichel, 1906	Bolivia
novatus artemis Riffarth, 1907	Bolivia
ainsolitus Avinoff, 1926	Bolivia
novatus f. *confluens* Neustetter, 1931, preocc. (not Lathy, 1906)	Bolivia
w) nubifer Butler, 1875	Brazil (AM)
x) peeblesi Joicey & Talbot, 1925	Venezuela
y) pratti Joicey & Kaye, 1917	Peru
z) robigus Weymer, 1875	'Venezuela'
numilia Herrich-Schäffer, 1865 (*Heliconia*), nom. nud.	?
aa) silvana (Stoll, 1781) (*Papilio*)	Surinam
diffusus Butler, 1873	Brazil (PA)
silvana var. *divisus* Kaye, 1906	Guyana

numata silvaniformis Joicey & Kaye, 1917	Brazil (PA)
silvana atakama Neustetter, 1931	Brazil (AM)
ab) sourensis K. S. Brown, 1976	Brazil (PA)
ac) superioris Butler, 1875	Brazil (AM/PA)
numata var. *maecenas* Weymer, 1894	Brazil (PA)
numatus sincerus Riffarth, 1907	Brazil?; Venezuela
ad) talboti Joicey & Kaye, 1917	Peru
ae) tarapotensis Riffarth, 1901	Peru
staudingeri Weymer, 1894, hyb.	Peru
staudingeri var. *pretiosus* Weymer, 1894, hyb.	Peru
aristiona lepidus f. *gracilis* Riffarth, 1907	Peru
aristiona timaeus f. *aristeus* Neustetter, 1931, hyb.	Peru
aristiona staudingeri f. *lutea* Neustetter, 1931, hyb.	Peru
af) timaeus Weymer, 1894	Peru
ag) zobrysi Fruhstorfer, 1910	Brazil (MT)
pachinus Salvin, 1871	Panama
pardalinus H. W. Bates, 1862	
a) pardalinus H. W. Bates, 1862	Brazil (AM)
b) ariadne Neukirchen, 1995	Bolivia
c) butleri K. S. Brown, 1976	Peru
d) dilatus Weymer, 1894	Peru
pardalinus radiosus f. *colorata* Stichel, 1919	Peru
e) julia Neukirchen, 2000	Ecuador
f) lucescens Weymer, 1894	Brazil (PA)
g) maeon Weymer, 1891	?
h) orteguaza K. S. Brown, 1976	Colombia
i) radiosus Butler, 1873	Brazil (PA)
fortunatus Weymer, 1883	[Brazil (AM)]
spurius Weymer, 1894	Brazil (AM)
j) sergestus Weymer, 1894	Peru
sergestus ab. *ninacura* Michael, 1926, hyb.	Peru
k) tithoreides Staudinger, 1900	Peru
tithoreides f. *minor* Riffarth, 1908, nom. nud.	
f. *garleppi* Neustetter, 1928	Peru
peruvianus C. Felder & R. Felder, 1859 (*Heliconia*)	Peru
aganice Riffarth, 1901, nom. nud.	
ricini (Linnaeus, 1758) (*Papilio*)	
a) ricini (Linnaeus, 1758) (*Papilio*)	America
myrti (Fabricius, 1775) (*Papilio*)	Surinam
polyhymnia (Shaw, 1806) (*Papilio*)	?
b) insulanus (Stichel, 1909) (*Eueides*)	Trinidad

sapho (Drury, 1782) (*Papilio*)

a) sapho (Drury, 1782) (*Papilio*)	'Jamaica'
b) candidus K. S. Brown, 1976	Ecuador
c) chocoensis K. S. Brown & Benson, 1975	Colombia
e) leuce Doubleday, 1847 (*Heliconia*)	'Brazil'
sappho [sic] Hübner [1831] (*Ajantis*) (not Drury, 1782)	
leuce H. W. Bates, 1864, preocc. (not Doubleday, 1847)	Guatemala

sara (Fabricius, 1793) (*Papilio*)

a) sara (Fabricius, 1793) (*Papilio*)	'Guyana'
rhea (Cramer, 1775) (*Papilio*), preocc. (not Poda, 1761)	Surinam
thamar (Hübner [1806]) (*Nerëis* [sic])	Brazil
rhea ab. *albinea* Riffarth, 1899	Surinam
sara praesignis Stichel, 1919	Brazil (PA)
sara thamar f. *nana* Stammeshaus, 1982	Surinam
b) apseudes (Hübner, 1818) (*Sicyonia*)	Brazil
apseudes (Hübner [1808]) (*Nerëis* [sic]), nom. nud.	
c) brevimaculata Riffarth, 1901 (ICZN, Art. 45.5.1)	Colombia
apseudes var. *magdalena* ab.? (var.?) *brevimaculata* Staudinger, 1897	Colombia
d) elektra Neukirchen, 1998	Colombia
e) fulgidus Stichel, 1906	Costa Rica
f) magdalena H. W. Bates, 1864	Colombia; Panama
apseudes var. *magdalena* ab. *albimaculata* Staudinger, 1897	Colombia
sara ab. *albula* Riffarth, 1900	Venezuela
sarae [sic] *lilianae* Emsley, 1965, nom. nud.	
g) sprucei H. W. Bates, 1864	Ecuador
h) theudela Hewitson, 1874 (*Heliconia*)	Panama
i) veraepacis H. W. Bates, 1864	Guatemala
f. *albifasciatus* C. C. Hoffmann, 1940	Mexico (CH)
j) williami Neukirchen, 1994	Trinidad

telesiphe Doubleday, 1847 (*Heliconia*)

a) telesiphe Doubleday, 1847 (*Heliconia*)	Bolivia
b) cretacea Neustetter, 1916	'French Guiana'
telesiphe ab. *nivea* Kaye, 1916	Peru
c) sotericus Salvin, 1871	Ecuador

timareta Hewitson, 1867 (*Heliconia*)

a) timareta Hewitson, 1867 (*Heliconia*)	Ecuador
contiguus Weymer, 1890	Ecuador
ab. *richardi* Riffarth, 1900	Ecuador
melpomene timareta ab. *virgata* Stichel, 1902	Ecuador
melpomene timareta f. *insolita* Riffarth, 1907	Ecuador
melpomene timareta f. *peregrina* Stichel, 1909	Ecuador

f. *strandi* Neustetter, 1928 — Ecuador
b) florencia Giraldo, C. Salazar, Jiggins, Bermingham & Linares, 2008 — Colombia
c) thelxinoe Lamas & Mérot, 2013 — Peru
d) timoratus Lamas, 1998 — Peru
e) tristero Brower, 1996 — Colombia

wallacei Reakirt, 1866
a) wallacei Reakirt, 1866 — [Brazil]
clytia ab. *parvimaculata* Riffarth, 1900, hyb. — Brazil (PA)
clytia f. *flavescens* ab. *wucherpfennigi* R. Krüger, 1933, hyb. — Brazil (PA)
ab. *erichi* R. Krüger, 1933, hyb. — Brazil (PA)
b) araguaia K. S. Brown, 1976 — Brazil (Go)
c) colon Weymer, 1891 — [Brazil (AM)]
clytia (Cramer, 1776) (*Papilio*), preocc. (not Linnaeus, 1758) — Surinam
clytia var. *elsa* Riffarth, 1899 — Surinam
wallacei brevimaculata ab. *halli* Kaye, 1919, hyb. — Brazil (AM)
d) flavescens Weymer, 1891 — Ecuador
sara. Hübner [1809] (*Nerëis* [sic]) (not Fabricius, 1793)
clytia var. *colon* Weymer, 1891, hyb. — [Brazil]
clytia var. *sulphurea* Staudinger, 1897, preocc. (not Weymer, 1894) — Peru
hagar W. F. Kirby [1899] — Guyana; 'Surinam'
f. *quadrimaculata* Neustetter, 1925 — Bolivia
clytia f. *flavescens* ab. *graphitica* R. Krüger, 1933 — Peru
f. *inez* Stammeshaus, 1982 — Surinam
e) kayei Neustetter, 1929, repl. name — Trinidad
wallacei latus Kaye, 1925, preocc. (not Riffarth, 1900) — Trinidad
f) mimulinus Butler, 1873 — Colombia

xanthocles H. W. Bates, 1862
a) xanthocles H. W. Bates, 1862 — Guyana; 'French Guiana'
b) buechei Neukirchen, 1992 — Venezuela
c) cleoxanthe H. Holzinger & R. Holzinger, 1972 — Venezuela
d) donatia Fruhstorfer, 1910 — Brazil (MT)
xanthocles melete f. *meridionalis* Neustetter, 1925 — Brazil (MT)
e) explicata K. S. Brown, 1976 (ICZN, Art. 45.5.1) — Colombia
xanthocles melete f. *explicata* Stichel, 1923 — Colombia
xanthocles melete f. *paranympha* Stichel, 1923 — Colombia
f) hippocrene H. Holzinger & K. S. Brown, 1982 — Bolivia
hippocrene Stichel, 1906, nom. nud.
g) melete C. Felder & R. Felder, 1865 — Colombia
melittus f. *fassli* Neustetter, 1912 — Colombia
cethosia Seitz, 1912 — Colombia

xanthocles flavosia Kaye, 1920	Colombia
cethosia completa Oberthür, 1920 (*Heliconia*)	Colombia
h) melior Staudinger, 1897	Peru
i) melittus Staudinger, 1897	Peru
j) napoensis H. Holzinger & K. S. Brown, 1982	Ecuador
xanthocles melete f. *latior* Neustetter, 1932	Colombia
k) paraplesius H. W. Bates, 1867	Brazil (MA)
olede Riffarth, 1901, nom. nud.	
l) quindecim Lamas, 1976	Peru
m) rindgei H. Holzinger & K. S. Brown, 1982	Colombia
n) similatus J. Zikán, 1937	Brazil (AM)
o) vala Staudinger, 1885	French Guiana
caternaulti Oberthür, 1902 (*Heliconia*)	French Guiana
p) zamora H. Holzinger & K. S. Brown, 1982	Ecuador

References

Abbott R, Albach D, Ansell S, et al. 2013. Hybridization and speciation. *Journal of Evolutionary Biology* **26**: 229–246.

Adams MD, Celniker SE, Holt RA, et al. 2000. The genome sequence of *Drosophila melanogaster*. *Science* **287**: 2185–2195.

Andersson J, Borg-Karlson AK & Wiklund C. 2004. Sexual conflict and anti-aphrodisiac titre in a polyandrous butterfly: male ejaculate tailoring and absence of female control. *Proceedings of the Royal Society of London B: Biological Sciences* **271**: 1765–1770.

Andersson S & Dobson HEM. 2003a. Behavioral foraging responses by the butterfly *Heliconius melpomene* to *Lantana camara* floral scent. *Journal of Chemical Ecology* **29**: 2303–2318.

Andersson S & Dobson HEM. 2003b. Antennal responses to floral scents in the butterfly *Heliconius melpomene*. *Journal of Chemical Ecology* **29**: 2319–2330.

Arias CF, Muñoz AG, Jiggins CD, et al. 2008. A hybrid zone provides evidence for incipient ecological speciation in *Heliconius* butterflies. *Molecular Ecology* **17**: 4699–4712.

Arias M, Mappes J, Théry M, et al. 2015. Inter-species variation in unpalatability does not explain polymorphism in a mimetic species. *Evolutionary Ecology* **30**: 419–433.

Arnold ML. 1997. *Natural Hybridization and Evolution*. New York: Oxford University Press.

Aymone ACB, Valente VLS & de Araújo AM. 2013. Ultrastructure and morphogenesis of the wing scales in *Heliconius erato phyllis* (Lepidoptera: Nymphalidae): what silvery/brownish surfaces can tell us about the development of color patterning? *Arthropod Structure & Development* **42**: 349–359.

Bacquet PMB, Brattström O, Wang HL, et al. 2015. Selection on male sex pheromone composition contributes to butterfly reproductive isolation. *Proceedings. Biological Sciences/The Royal Society* **282**: 20142734.

Baker HG & Baker I. 1973. Amino-acids in nectar and their evolutionary significance. *Nature* **241**: 543–545.

Barão KR, Gonçalves GL, Mielke OHH, et al. 2014. Species boundaries in *Philaethria* butterflies: an integrative taxonomic analysis based on genitalia ultrastructure, wing geometric morphometrics, DNA sequences, and amplified fragment length polymorphisms. *Zoological Journal of the Linnean Society* **170**: 690–709.

Barluenga M, Stölting KN, Salzburger W, et al. 2006. Sympatric speciation in Nicaraguan crater lake cichlid fish. *Nature* **439**: 719–723.

Barrett RDH & Schluter D. 2008. Adaptation from standing genetic variation. *Trends in Ecology & Evolution* **23**: 38–44.

Barton NH & Gale KS. 1993. Genetic analysis of hybrid zones. In: Harrison RG, ed. *Hybrid Zones and the Evolutionary Process*. New York: Oxford University Press, 13–45.

Barton NH & Hewitt GM. 1989. Adaptation, speciation and hybrid zones. *Nature* **341**: 497–503.

Bastock M. 1956. A gene mutation which changes a behavior pattern. *Evolution* **10**: 421–439.

Bateman AJ. 1948. Intra-sexual selection in *Drosophila*. *Heredity* **2**: 349–368.

Bates HW. 1862. Contributions to an insect fauna of the Amazon valley. Lepidoptera: Heliconidae. *Transactions of the Linnean Society of London* **23**: 495–566.

Bates HW. 1864. *The Naturalist on the River Amazons*. London: J. Murray.

Baxter SW, Johnston SE & Jiggins CD. 2009. Butterfly speciation and the distribution of gene effect sizes fixed during adaptation. *Heredity* **102**: 57–65.

Baxter SW, Nadeau NJ, Maroja LS, et al. 2010. Genomic hotspots for adaptation: the population genetics of Müllerian mimicry in the *Heliconius melpomene* clade. *PLoS Genetics* **6**: e1000794.

Baxter SW, Papa R, Chamberlain N, et al. 2008. Convergent evolution in the genetic basis of Müllerian mimicry in *Heliconius* butterflies. *Genetics* **180**: 1567–1577.

Bazin E, Dawson KJ & Beaumont MA. 2010. Likelihood-free inference of population structure and local adaptation in a bayesian hierarchical model. *Genetics* **185**: 587–602.

Beaumont MA & Balding DJ. 2004. Identifying adaptive genetic divergence among populations from genome scans. *Molecular Ecology* **13**: 969–980.

Beccaloni G. 1997. Vertical stratification of ithomiine butterfly (Nymphalidae: Ithomiinae) mimicry complexes: the relationship between adult flight height and larval host-plant height. *Biological Journal of the Linnean Society* **62**: 313–341.

Beebe W. 1950. Migration of Danaidae, Ithomiidae, Acraeidae and Heliconiidae (butterflies) at Rancho Grande, north-central Venezuela. *Zoologica, New York* **35**: 57–68.

Beebe W. 1955. Polymorphism in reared broods of *Heliconius* butterflies from Suriname and Trinidad. *Zoologica, New York* **40**: 139–143.

Beebe W, Crane J & Fleming H. 1960. A comparison of eggs, larvae and pupae in fourteen species of Heliconiine butterflies from Trinidad, W.I. *Zoologica, New York* **45**: 111–153.

Beldade P, Koops K & Brakefield PM. 2002. Developmental constraints versus flexibility in morphological evolution. *Nature* **416**: 844–847.

Belt T. 1874. *The Naturalist in Nicaragua*. New York: E. P. Dutton.

Beltran M, Jiggins CD, Brower AVZ, et al. 2007. Do pollen feeding, pupal-mating and larval gregariousness have a single origin in *Heliconius* butterflies? Inferences from multilocus DNA sequence data. *Biological Journal of the Linnean Society* **92**: 221–239.

Benson WW. 1971. Evidence for the evolution of unpalatability through kin selection in the Heliconiinae (Lepidoptera). *American Naturalist* **105**: 213–226.

Benson WW. 1972. Natural selection for Müllerian mimicry in *Heliconius erato* in Costa Rica. *Science* **176**: 936–939.

Benson WW. 1977. On the supposed spectrum between Batesian and Müllerian mimicry. *Evolution* **31**: 454–455.

Benson WW. 1978. Resource partitioning in passion vine butterflies. *Evolution* **32**: 493–518.

Benson WW. 1982. Alternative models for infrageneric diversification in the humid tropics: tests with passion vine butterflies. In: Prance GT, ed. *Biological Diversification in the Tropics*. New York, NY: Columbia University Press, 608–640.

Benson WW, Brown KS & Gilbert LE. 1975. Coevolution of plants and herbivores: passion flower butterflies. *Evolution* **29**: 659–680.

Benson WW, Haddad CFB & Cardoso MZ. 1989. Territorial behavior and dominance in some Heliconiine butterflies (Nymphalidae). *Journal of the Lepidopterists' Society* **43**: 33–49.

Benson WW & Hernandez MIM. 1991. Small-male territoriality in the tropical butterfly *Heliconius sara*. *American Zoologist* **31**: A110–A110.

Bernard GD & Remington CL. 1991. Color vision in *Lycaena* butterflies: spectral tuning of receptor arrays in relation to behavioral ecology. *Proceedings of the National Academy of Sciences* **88**: 2783–2787.

Bierne N, Welch J, Loire E, et al. 2011. The coupling hypothesis: why genome scans may fail to map local adaptation genes. *Molecular Ecology* **20**: 2044–2072.

Blum MJ. 2002. Rapid movement of a *Heliconius* hybrid zone: evidence for phase III of Wright's shifting balance theory? *Evolution* **56**: 1992–1998.

Blum MJ. 2008. Ecological and genetic associations across a *Heliconius* hybrid zone. *Journal of Evolutionary Biology* **21**: 330–341.

Boggs C & Gilbert LE. 1979. Male contribution to egg production in butterflies: evidence for transfer of nutrients at mating. *Science* **206**: 83–84.

Boggs CF. 1990. A general-model of the role of male-donated nutrients in female insects reproduction. *American Naturalist* **136**: 598–617.

Boggs CL. 1981a. Nutritional and life-history determinants of resource allocation in holometabolous insects. *American Naturalist* **117**: 692–709.

Boggs CL. 1981b. Selection pressures affecting male nutrient investment at mating in heliconiine butterflies. *Evolution* **35**: 931–940.

Boggs CL. 1997. Reproductive allocation from reserves and income in butterfly species with differing adult diets. *Ecology* **78**: 181–191.

Boggs CL, Smiley JT & Gilbert LE. 1981. Patterns of pollen exploitation by *Heliconius* butterflies. *Oecologia* **48**: 284–289.

Boppré M. 1984. Chemically mediated interactions between butterflies. In: Vane-Wright RI, Ackery PR, eds. *The Biology of Butterflies*. London: Academic Press.

Brévignon C. 1996. Description d'un nouvel *Heliconius* provenant de Guyane Française (Lepidoptera, Nymphalidae). *Lambillionea* **96**: 467–470.

Briscoe AD, Bybee SM, Bernard GD, et al. 2010. Positive selection of a duplicated UV-sensitive visual pigment coincides with wing pigment evolution in *Heliconius* butterflies. *Proceedings of the National Academy of Sciences* **107**: 3628–3633.

Briscoe AD & Chittka L. 2001. The evolution of color vision in insects. *Annual Review of Entomology* **46**: 471–510.

Briscoe AD, Macias-Muñoz A, Kozak KM, et al. 2013. Female behaviour drives expression and evolution of gustatory receptors in butterflies. *PLoS Genetics* **9**: *e* 1003620.

Brodie E, Thurman T, Evans E, et al. 2015, in preparation. Pupal mating is not obligate in *Heliconius erato*.

Brower AV. 1994b. Rapid morphological radiation and convergence among races of the butterfly *Heliconius erato* inferred from patterns of mitochondrial DNA evolution. *Proceedings of the National Academy of Sciences of the United States of America* **91**: 6491–6495.

Brower AVZ. 1994a. Phylogeny of *Heliconius* butterflies inferred from mitochondrial DNA sequences (Lepidoptera: Nymphalidae). *Molecular Phylogenetics and Evolution* **3**: 159–174.

Brower AVZ. 1996a. Parallel race formation and the evolution of mimicry in *Heliconius* butterflies: A phylogenetic hypothesis from mitochondrial DNA sequences. *Evolution* **50**: 195–221.

Brower AVZ. 1996b. A new mimetic species of *Heliconius* (Lepidoptera:Nymphalidae), from southeastern Colombia, revealed by cladistic analysis of mitochondrial DNA sequences. *Zoological Journal of the Linnean Society* **116**: 317–332.

Brower AVZ. 2010. Hybrid speciation in *Heliconius* butterflies? A review and critique of the evidence. *Genetica* **139**: 589–609.

Brower AVZ. 2013. Introgression of wing pattern alleles and speciation via homoploid hybridization in *Heliconius* butterflies: a review of evidence from the genome. *Proceedings of the Royal Society B: Biological Sciences* **280**: 20122302.

Brower AVZ & Egan MG. 1997. Cladistic analysis of *Heliconius* butterflies and relatives (Nymphalidae: Heliconiiti): a revised phylogenetic position for *Eueides* based on sequences from mtDNA and a nuclear gene. *Proceedings of the Royal Society B: Biological Sciences* **264**: 969–977.

Brower LP, Brower JVZ & Collins CT. 1963. Experimental studies of mimicry. 7. Relative palatability and Müllerian mimicry among neotropical butterflies of the subfamily Heliconiinae. *Zoologica, New York* **48**: 65–84.

Brower LP, Pough FH & Meck HR. 1970. Theoretical investigations of automimicry, I. Single trial learning. *Proceedings of the National Academy of Sciences* **66**: 1059–1066.

Brower LP, Ryerson WN, Coppinger LI, et al. 1968. Ecological chemistry and the palatability spectrum. *Science* **161**: 1342–1381.

Brown KS. 1967. Chemotaxonomy and chemomimicry: the case of 3-hydroxykynurenine. *Systematic Zoology* **16**: 213–216.

Brown KS. 1970. Rediscovery of *Heliconius nattereri* in eastern Brazil. *Entomological News* **81**: 129–140.

Brown KS. 1972a. Heliconians of Brazil (Lepidoptera-Nymphalidae). 3. Ecology and biology of *Heliconius nattereri*, a key primitive species near extinction, and comments on evolutionary development of *Heliconius* and *Eueides*. *Zoologica* **57**: 41–69.

Brown KS. 1972b. Maximizing daily butterfly counts. *Journal of the Lepidopterists' Society* **26**: 183–196.

Brown KS. 1976. An illustrated key to the silvaniform Heliconius (Lepidoptera: Nymphalidae) with descriptions of new subspecies. *Transactions of the American Entomological Society* **102**: 373–484.

Brown KS. 1979. *Ecologia Geográfica e Evoluçâo nas Florestas Neotropicais*. Campinas, Brazil: Universidade Estadual de Campinas.

Brown KS. 1981. The biology of *Heliconius* and related genera. *Annual Review of Entomology* **26**: 427–456.

Brown KS & Benson WW. 1974. Adaptive polymorphism associated with multiple Müllerian mimicry in *Heliconius numata* (Lepid.: Nymph.). *Biotropica* **6**: 205–228.

Brown KS & Benson WW. 1975. West Colombian biogeography. Notes on *Heliconius hecalesia* and *H. sapho* (Nymphalidae). *Journal of the Lepidopterists' Society* **29**: 199–212.

Brown KS, Emmel TC, Eliazar PJ, et al. 1992. Evolutionary patterns in chromosome-numbers in neotropical Lepidoptera. 1. Chromosomes of the Heliconiini (Family Nymphalidae, Subfamily Nymphalinae). *Hereditas* **117**: 109–125.

Brown KS & Holzinger H. 1973. The Heliconians of Brazil (Lepidoptera: Nymphalidae). Part IV. Systematics and biology of *Eueides tales* Cramer, with description of a new subspecies from Venezuela. *Zeitschrift der Arbeitsgemeinschaft Österreichischen Entomologen* **24**: 44–65.

Brown KS & Mielke OHH. 1972. The Heliconians of Brazil (Lepidoptera: Nymphalidae). Part II. Introduction and general comments, with a supplementary revision of the tribe. *Zoologica, New York* **57**: 1–40.

Brown KS, Sheppard PM & Turner JRG. 1974. Quaternary refugia in tropical America—evidence from race formation in *Heliconius* butterflies. *Proceedings of the Royal Society of London Series B-Biological Sciences* **187**: 369–378.

Brown KS, Trigo JR, Francini RB, et al. 1991. Aposematic insects on toxic host plants: coevolution, colonization, and chemical emancipation. In: Price PW, Lewinsohn TM, Fernandes GW, et al., eds. *Plant-Animal Interactions: Evolutionary Ecology in Tropical and Temperate Regions*. New York: John Wiley, 375–402.

Brucker RM & Bordenstein SR. 2012. Speciation by symbiosis. *Trends in Ecology & Evolution* **27**: 443–451.

Brunetti CR, Selegue JE, Monteiro A, et al. 2001. The generation and diversification of butterfly eyespot color patterns. *Current Biology* **11**: 1578–1585.

Buerkle CA & Lexer C. 2008. Admixture as the basis for genetic mapping. *Trends in Ecology & Evolution* **23**: 686–694.

Bull V, Beltran M, Jiggins CD, et al. 2006. Polyphyly and gene flow between non-sibling *Heliconius* species. *BMC Biology* **4**: 11.

Bush MB, Gosling WD & Colinvaux PA. 2011. Climate and vegetation change in the lowlands of the Amazon Basin. In: *Tropical Rainforest Responses to Climatic Change*. Springer Praxis Books, 61–84.

Bybee SM, Yuan F, Ramstetter MD, et al. 2012. UV photoreceptors and UV-yellow wing pigments in *Heliconius* butterflies allow a color signal to serve both mimicry and intraspecific communication. *The American Naturalist* **179**: 38–51.

Cain AJ & Sheppard PM. 1954. Natural selection in *Cepaea*. *Genetics* **39**: 89–116.

Caley MJ, Fisher R & Mengersen K. 2014. Global species richness estimates have not converged. *Trends in Ecology & Evolution* **29**: 187–188.

Campbell P, Good JM & Nachman MW. 2013. Meiotic sex chromosome inactivation is disrupted in sterile hybrid male house mice. *Genetics* **193**: 819–828.

Cardoso MZ & Gilbert LE. 2007. A male gift to its partner? Cyanogenic glycosides in the spermatophore of longwing butterflies (*Heliconius*). *Naturwissenschaften* **94**: 39–42.

Cardoso NZ. 2001. Patterns of pollen collection and flower visitation by *Heliconius* butterflies in southeastern Mexico. *Journal of Tropical Ecology* **17**: 763–768.

Carroll S, Gates J, Keys D, et al. 1994. Pattern formation and eyespot determination in butterfly wings. *Science* **265**: 109–114.

Carroll SB, Grenier JK & Weatherbee SD. 2004. *From DNA to Diversity: Molecular Genetics and the Evolution of Animal Design*. Malden, MA: Wiley-Blackwell.

Chai P. 1986. Field observations and feeding experiments on the responses of rufous-tailed jacamars, *Galbula ruficauda*, to free-flying butterflies in a tropical rainforest. *Biological Journal of the Linnean Society* **29**: 166–189.

Chai P. 1988. Wing coloration of free-flying neotropical butterflies as a signal learned by a specialized avian predator. *Biotropica* **20**: 20–30.

Chai P. 1996. Butterfly visual characteristics and ontogeny of responses to butterflies by a specialized tropical bird. *Biological Journal of the Linnean Society* **59**: 37–67.

Chai P & Srygley RB. 1990. Predation and the flight, morphology, and temperature of neotropical rain-forest butterflies. *American Naturalist* **135**: 748–765.

Chamberlain NL, Hill RI, Baxter SW, et al. 2011. Comparative population genetics of a mimicry locus among hybridizing *Heliconius* butterfly species. *Heredity* **107**: 200–204.

Chamberlain NL, Hill RI, Kapan DD, et al. 2009. Polymorphic butterfly reveals the missing link in ecological speciation. *Science* **326**: 847–850.

Chan YF, Marks ME, Jones FC, et al. 2010. Adaptive evolution of pelvic reduction in sticklebacks by recurrent deletion of a *Pitx1* enhancer. *Science* **327**: 302–305.

Chapman T, Liddle LF, Kalb JM, et al. 1995. Cost of mating in *Drosophila melanogaster* females is mediated by male accessory gland products. *Nature* **373**: 241–244.

Charlesworth B, Coyne JA & Barton NH. 1987. The relative rates of evolution of sex chromosomes and autosomes. *The American Naturalist* **130**: 113–146.

Charlesworth D & Charlesworth B. 1976. Theoretical genetics of Batesian mimicry. II. Evolution of supergenes. *Journal of Theoretical Biology* **55**: 305–324.

Chouteau M, Arias M & Joron M. 2016. Warning signals are under positive frequency-dependent selection in nature. *Proceedings of the National Academy of Sciences* **113**: 2164–2169.

Coley PD. 1980. Effects of leaf age and plant life history patterns on herbivory. *Nature* 284: 545–546.

Colinvaux PA, Oliveira PED, Moreno JE, et al. 1996. A long pollen record from lowland amazonia: forest and cooling in glacial times. *Science* **274**: 85–88.

Colosimo PF, Hosemann KE, Balabhadra S, et al. 2005. Widespread parallel evolution in sticklebacks by repeated fixation of ectodysplasin alleles. *Science* **307**: 1928–1933.

Condon MA & Gilbert LE. 1990. Reproductive biology and natural history of the neotropical vines *Gurania* and *Psiguria*. In: Bates D, Robinson RW, Jeffrey C, eds. *Biology and Utilization of the Cucurbitaceae*. Ithaca, NY: Cornell University Press, 150–166.

Cook LM, Thomason EW & Young AM. 1976. Population structure, dynamics and dispersal of tropical butterfly *Heliconius charitonius*. *Journal of Animal Ecology* **45**: 851–863.

Copp NH & Davenport D. 1978. *Agraulis* and *Passiflora*. II. Behavior and sensory modalities. *Biological Bulletin* **155**: 113–124.

Cornwallis CK & Uller T. 2010. Towards an evolutionary ecology of sexual traits. *Trends in Ecology & Evolution* **25**: 145–152.

Counterman BA, Araújo-Perez F, Hines HM, et al. 2010. Genomic hotspots for adaptation: the population genetics of Müllerian mimicry in *Heliconius erato*. *PLoS Genetics* **6**: e1000796.

Coyne JA, Barton NH & Turelli M. 1997. Perspective: a critique of Sewall Wright's shifting balance theory of evolution. *Evolution* **51**: 643–671.

Coyne JA, Barton NH & Turelli M. 2000. Is Wright's shifting balance process important in evolution? *Evolution* **54**: 306–317.

Coyne JA & Orr HA. 2004. *Speciation*. Sunderland, MA: Sinauer Associates, Inc.

Cramer P. 1775. *De uitlandsche Kapelleia voorkomende in de drie Waereld-Deelen Asia, Africa en America. Papillons exotiques de trois parties du Monde l'Asie, l'Afrique et l'Amerique*. Amsterdam: Deelen 1–4.

Crane J. 1954. Spectral reflectance characteristics of butterflies (Lepidoptera) from Trinidad, B.W.I. *Zoologica, New York* **39**: 85–115.

Crane J. 1955. Imaginal behaviour of a Trinidad butterfly, *Heliconius erato hydara* Hewitson, with special reference to the social use of color. *Zoologica, New York* **40**: 167–196.

Crane J. 1957a. Keeping house for tropical butterflies. *National Geographic Magazine* **112**: 193–217.

Crane J. 1957b. Imaginal behaviour in butterflies of the family Heliconiidae: changing social patterns and irrelevant actions. *Zoologica, New York* **42**: 135–145.

Crane J & Fleming H. 1953. Construction and operation of butterfly insectaries in the tropics. *Zoologica, New York* **38**: 161–172.

Cronin TW, Järvilehto M, Weckström M, et al. 2000. Tuning of photoreceptor spectral sensitivity in fireflies (Coleoptera: Lampyridae). *Journal of Comparative Physiology A***186**: 1–12.

Cruickshank TE & Hahn MW. 2014. Reanalysis suggests that genomic islands of speciation are due to reduced diversity, not reduced gene flow. *Molecular Ecology* **23**: 3133–3157.

Daborn PJ, Yen JL, Bogwitz MR, et al. 2002. A single P450 allele associated with insecticide resistance in *Drosophila*. *Science* **297**: 2253–2256.

Daly M & Wilson MI. 1982. Whom are newborn babies said to resemble? *Ethology and Sociobiology* **3**: 69–78.

Darwin C. 1863. [Review of] Contributions to an insect fauna of the Amazon Valley. By Henry Walter Bates, Esq. Transact. Linnean Soc. Vol. XXIII. 1862, p. 495. *Natural History Review* **3**: 219–224.

Darwin C. 1871. *The Descent of Man, and Selection in Relation to Sex*. London, UK: John Murray.

Dasmahapatra KK, Lamas G, Simpson F, et al. 2010. The anatomy of a 'suture zone' in Amazonian butterflies: a coalescent-based test for vicariant geographic divergence and speciation. *Molecular Ecology* **19**: 4283–4301.

Dasmahapatra KK, Silva-Vasquez A, Chung JW, et al. 2007. Genetic analysis of a wild-caught hybrid between non-sister *Heliconius* butterfly species. *Biology Letters* **3**: 660–663.

Davey JW, Chouteau M, Barker SL, et al. 2016. Major Improvements to the *Heliconius melpomene* genome assembly used to confirm 10 chromosome fusion events in 6 million years of butterfly evolution. *G3: Genes | Genomes | Genetics*: g3.115.023655.

Davidson EH & Erwin DH. 2006. Gene Regulatory Networks and the Evolution of Animal Body Plans. *Science* **311**: 796–800.

Davies N & Bermingham E. 2002. The historical biogeography of two Caribbean butterflies (Lepidoptera : Heliconiidae) as inferred from genetic variation at multiple loci. *Evolution* **56**: 573–589.

Davis RH & Nahrstedt A. 1987. Biosynthesis of cyanogenic glucosides in butterflies and moths—effective incorporation of 2-methylpropanenitrile and 2-methylbutanenitrile into linamarin and lotaustralin by *Zygaena* and *Heliconius* species (Lepidoptera). *Insect Biochemistry* **17**: 689–693.

Deinert EI. 2003. Sexual selection in *Heliconius hewitsoni*, a pupal mating butterfly. In: Boggs CL, Ehrlich PR, Watt WB, eds. *Ecology and Evolution Taking Flight: Butterflies as Model Study Systems*. Chicago: University of Chicago Press.

Deinert EI, Longino JT & Gilbert LE. 1994. Mate competition in butterflies. *Nature* **370**: 23–24.

Dhawan K, Dhawan S & Sharma A. 2004. *Passiflora*: a review update. *Journal of Ethnopharmacology* **94**: 1–23.

Dieckmann U & Doebeli M. 1999. On the origin of species by sympatric speciation. *Nature* **400**: 354–357.

Dinwiddie A, Null R, Pizzano M, et al. 2014. Dynamics of F-actin prefigure the structure of butterfly wing scales. *Developmental Biology* **392**: 404–418.

Dobzhansky T. 1936. Studies on hybrid sterility. II. Localization of sterility factors in *Drosophila pseudoobscura* hybrids. *Genetics* **21**: 113–135.

Dobzhansky T. 1937. *Genetics and the Origin of Species*. New York: Columbia University Press.

Douglas JM, Cronin TW, Chiou TH, et al. 2007. Light habitats and the role of polarized iridescence in the sensory ecology of neotropical nymphalid butterflies (Lepidoptera: Nymphalidae). *Journal of Experimental Biology* **210**: 788–799.

Drès M & Mallet J. 2002. Host races in plant-feeding insects and their importance in sympatric speciation. *Philosophical Transactions of the Royal Society of London B: Biological Sciences* **357**: 471–492.

Duenez-Guzman EA, Mavárez J, Vose MD, et al. 2009. Case studies and mathematical models of ecological speciation. 4. Hybrid speciation in butterflies in a jungle. *Evolution* **63**: 2611–2626.

Dukas R. 2008. Evolutionary biology of insect learning. *Annual Review of Entomology* **53**: 145–160.

Dunlap-Pianka HL, Boggs CL & Gilbert LE. 1977. Ovarian dynamics in heliconiine butterflies: programmed senescence versus eternal youth. *Science* **197**: 487–490.

Eberhard SH, Hrassnigg N, Crailsheim K, et al. 2007. Evidence of protease in the saliva of the butterfly *Heliconius melpomene* (L.) (Nymphalidae, Lepidoptera). *Journal of Insect Physiology* **53**: 126–131.

Eberhard SH & Krenn HW. 2003. Salivary glands and salivary pumps in adult Nymphalidae (Lepidoptera). *Zoomorphology* **122**: 161–167.

Eberhard SH & Krenn HW. 2005. Anatomy of the oral valve in nymphalid butterflies and a functional model for fluid uptake in Lepidoptera. *Zoologischer Anzeiger* **243**: 305–312.

Edwards WH. 1881. On certain habits of *Heliconia charitonia*, Linn., a species of butterfly found in Florida. *Papilio* **1**: 209–215.

Ehrlich PR & Gilbert LE. 1973. Population structure and dynamics of the tropical butterfly *Heliconius ethilla*. *Biotropica* **5**: 69–82.

Ehrlich PR & Raven PH. 1964. Butterflies and plants: a study in coevolution. *Evolution* **18**: 586–608.

Elena SF & Lenski RE. 2003. Evolution experiments with microorganisms: the dynamics and genetic bases of adaptation. *Nature Reviews Genetics* **4**: 457–469.

Elias M, Gompert Z, Jiggins CD, et al. 2008. Mutualistic interactions drive ecological niche convergence in a diverse butterfly community. *PLoS Biology* **6**: 2642–2649.

Ellegren H & Parsch J. 2007. The evolution of sex-biased genes and sex-biased gene expression. *Nature Reviews Genetics* **8**: 689–698.

Eltringham H. 1916. IV. On specific and mimetic relationships in the genus *Heliconius*, l. *Transactions of the Royal Entomological Society of London* **64**: 101–148.

Eltringham H. 1925. On the abdominal glands of *Heliconius*. *Transactions of the Royal Entomological Society, London* **1925**: 269–275.

Emsley MG. 1964. The geographical distribution of the color-pattern components of *Heliconius erato* and *Heliconius melpomene* with genetical evidence for the systematic relationship between the two species. *Zoologica, New York* **49**: 245–286.

Emsley MG. 1965. Speciation in *Heliconius* (Lep., Nymphalidae): morphology and geographic distribution. *Zoologica, New York* **50**: 191–254.

Emsley MG. 1970. An observation on the use of colour for species-recognition in *Heliconius besckei* (Nymphalidae). *Journal of the Lepidopterists' Society* **24**: 25.

Endler JA. 1982. Pleistocene forest refuges: fact or fancy? In: Prance GT, ed. *Biological Diversification in the Tropics*. New York: Columbia University Press, 641–657.

Endler JA & Basolo AL. 1998. Sensory ecology, receiver biases and sexual selection. *Trends in Ecology & Evolution* **13**: 415–420.

Engler-Chaouat HS & Gilbert LE. 2007. *De novo* synthesis vs. sequestration: negatively correlated metabolic traits and the evolution of host plant specialization in cyanogenic butterflies. *Journal of Chemical Ecology* **33**: 25–42.

Engler HS, Spencer KC & Gilbert LE. 2000. Preventing cyanide release from leaves. *Nature* **406**: 144–145.

Estrada C & Gilbert LE. 2010. Host plants and immatures as mate-searching cues in *Heliconius* butterflies. *Animal Behaviour* **80**: 231–239.

Estrada C & Jiggins CD. 2002. Patterns of pollen feeding and habitat preference among *Heliconius* species. *Ecological Entomology* **27**: 448–456.

Estrada C & Jiggins CD. 2008. Interspecific sexual attraction because of convergence in warning colouration: is there a conflict between natural and sexual selection in mimetic species? *Journal of Evolutionary Biology* **21**: 749–760.

Estrada C, Schulz S, Yildizhan S, et al. 2011. Sexual selection drives the evolution of antiaphrodisiac pheromones in butterflies. *Evolution* **65**: 2843–2854.

Estrada C, Yildizhan S, Schulz S, et al. 2010. Sex-specific chemical cues from immatures facilitate the evolution of mate guarding in *Heliconius* butterflies. *Proceedings of The Royal Society. Biological Sciences* **277**: 407–413.

Fabian D & Flatt T. 2012. Life history evolution. *Nature Education Knowledge* **3**: 24.

Fahrbach SE, Moore D, Capaldi EA, et al. 1998. Experience-expectant plasticity in the mushroom bodies of the honeybee. *Learning & Memory* **5**: 115–123.

Fedorka KM & Mousseau TA. 2002. Material and genetic benefits of female multiple mating and polyandry. *Animal Behaviour* **64**: 361–367.

Felsenstein J. 1981. Skepticism towards Santa Rosalia, or why are there so few kinds of animals? *Evolution* **35**: 124–138.

Feng M, Fang Y, Han B, et al. 2013. Novel aspects of understanding molecular working mechanisms of salivary glands of worker honeybees (*Apis mellifera*) investigated by proteomics and phosphoproteomics. *Journal of Proteomics* **87**: 1–15.

Ferguson L, Green J, Surridge A, et al. 2010b. Evolution of the insect yellow gene family. *Molecular Biology and Evolution* **28**: 257–272.

Ferguson L & Jiggins CD. 2009. Shared and divergent expression domains on mimetic *Heliconius* wings. *Evolution and Development* **11**: 498–512.

Ferguson L, Lee SF, Chamberlain N, et al. 2010a. Characterization of a hotspot for mimicry: assembly of a butterfly wing transcriptome to genomic sequence at the HmYb/Sb locus. *Molecular Ecology* **19**: 240–254.

Ferguson LC, Maroja L & Jiggins CD. 2011. Convergent, modular expression of *ebony* and *tan* in the mimetic wing patterns of *Heliconius* butterflies. *Development Genes and Evolution* **221**: 297–308.

Ferreira AA, Garcia RN & de Araújo AM. 2006. Pupal melanization in *Heliconius erato phyllis* (Lepidoptera; Nymphalidae): genetic and environmental effects. *Genetica* **126**: 133–140.

Feuillet C & MacDougal JM. 2007. Passifloraceae. In: Kubitzki PDK, ed. *Flowering Plants · Eudicots*. Springer Berlin Heidelberg, 270–281.

Finkbeiner SD. 2014. Communal roosting in *Heliconius* butterflies (Nymphalidae): roost recruitment, establishment, fidelity, and resource use trends based on age and sex. *Journal of the Lepidopterists' Society* **68**: 10–16.

Finkbeiner SD, Briscoe AD & Reed RD. 2012. The benefit of being a social butterfly: communal roosting deters predation. *Proceedings of the Royal Society B: Biological Sciences* **279**: 2769–2776.

Fisher RA. 1930. *The Genetical Theory of Natural Selection*. Oxford University Press, USA.

Fitzpatrick BM & Turelli M. 2006. The geography of mammalian speciation: mixed signals from phylogenies and range maps. *Evolution* **60**: 601–615.

Flanagan NS, Tobler A, Davison A, et al. 2004. Historical demography of Müllerian mimicry in the neotropical *Heliconius* butterflies. *Proceedings of the National Academy of Sciences of the United States of America* **101**: 9704–9709.

Fontaine MC, Pease JB, Steele A, et al. 2015. Extensive introgression in a malaria vector species complex revealed by phylogenomics. *Science* **347**: 1258524.

Frankel N, Erezyilmaz DF, McGregor AP, et al. 2011. Morphological evolution caused by many subtle-effect substitutions in regulatory DNA. *Nature* **474**: 598–603.

Freitas AVL & Oliveira PS. 1996. Ants as selective agents on herbivore biology: Effects on the behaviour of a non-myrmecophilous butterfly. *Journal of Animal Ecology* **65**: 205–210.

French V & Brakefield PM. 1995. Eyespot development on butterfly wings: the focal signal. *Developmental Biology* **168**: 112–123.

Futahashi R, Kawahara-Miki R, Kinoshita M, et al. 2015. Extraordinary diversity of visual opsin genes in dragonflies. *Proceedings of the National Academy of Sciences* **112**: E1247–E1256.

Futuyma DJ & Moreno G. 1988. The evolution of ecological specialization. *Annual Review of Ecology and Systematics* **19**: 207–233.

Galant R, Skeath JB, Paddock S, et al. 1998. Expression pattern of a butterfly *achaete-scute* homolog reveals the homology of butterfly wing scales and insect sensory bristles. *Current Biology* **8**: 807–813.

Gallant JR, Imhoff VE, Martin A, et al. 2014. Ancient homology underlies adaptive mimetic diversity across butterflies. *Nature Communications* **5**: 4817.

Garnett ST & Christidis L. 2007. Implications of changing species definitions for conservation purposes. *Bird Conservation International* **17**: 187–195.

Garzón-Orduña IJ, Silva-Brandão KL, Willmott KR, et al. 2015. Incompatible ages for clearwing butterflies based on alternative secondary calibrations. *Systematic Biology* 64: 752–767.

Gavrilets S. 2004. *Fitness Landscapes and the Origin of Species*. Princeton, NJ: Princeton University Press.

Gazave É, Chevillon C, Lenormand T, et al. 2001. Dissecting the cost of insecticide resistance genes during the overwintering period of the mosquito *Culex pipiens*. *Heredity* **87**: 441–448.

Gilbert LE. 1971. Butterfly-plant coevolution: has *Passiflora adenopoda* won the selectional race with heliconiine butterflies? *Science* **172**: 585–586.

Gilbert LE. 1972. Pollen feeding and reproductive biology of *Heliconius* butterflies. *Proceedings of the National Academy of Sciences of the United States of America* **69**: 1403–1407.

Gilbert LE. 1975. Ecological consequences of a coevolved mutualism between butterflies and plants. In: Gilbert LE, Raven PR, eds. *Coevolution of Animals and Plants*. Austin, TX: University of Texas Press, 210–240.

Gilbert LE. 1976. Postmating female odor in *Heliconius* butterflies—male-contributed anti-aphrodisiac. *Science* **193**: 419–420.

Gilbert LE. 1982. The coevolution of a butterfly and a vine. *Scientific American* **247**: 110–121.

Gilbert LE. 1984. The biology of butterfly communities. In: Vane-Wright RI, Ackery PR, eds. *The Biology of Butterflies*. London: Academic Press, 41–54.

Gilbert LE. 1991. Biodiversity of a Central American *Heliconius* community: pattern, process, and problems. In: Price PW, Lewinsohn TM, Fernandes TW, et al., eds. *Plant-Animal Interactions: Evolutionary Ecology in Tropical and Temperate Regions*. New York: John Wiley & Sons, 403–427.

Gilbert LE. 2003. Adaptive novelty through introgression in *Heliconius* wing patterns: evidence for shared genetic 'tool box' from synthetic hybrid zones and a theory of diversification. In: Boggs CL, Watt WB, Ehrlich PR, eds. *Ecology and Evolution Taking Flight: Butterflies as Model Systems*. Chicago: University of Chicago Press.

Gilbert LE, Forrest HS, Schultz TD, et al. 1988. Correlations of ultrastructure and pigmentation suggest how genes control development of wing scales of *Heliconius* butterflies. *Journal of Research on the Lepidoptera* **26**: 141–160.

Gilbert LE & Smiley JT. 1978. Determinants of local diversity in phytophagous insects: host specialists in tropical environments. In: Mound LA, Waloff N, eds. *Diversity of Insect Faunas*. Oxford: Blackwell Scientific, 89–104.

Gilbert SF. 2013. *Developmental Biology*. Sunderland, MA: Sinauer Associates.

Giraldo N, Salazar C, Jiggins CD, et al. 2008. Two sisters in the same dress: *Heliconius* cryptic species. *BMC Evolutionary Biology* **8**: 324.

Gleadow RM & Woodrow IE. 2002. Constraints on effectiveness of cyanogenic glycosides in herbivore defense. *Journal of Chemical Ecology* **28**: 1301–1313.

Glor RE. 2010. Phylogenetic insights on adaptive radiation. *Annual Review of Ecology, Evolution, and Systematics* **41**: 251–270.

Goldschmidt RB. 1945. Mimetic polymorphism, a controversial chapter of Darwinism (concluded). *The Quarterly Review of Biology* **20**: 205–230.

Gompel N & Prud'homme B. 2009. The causes of repeated genetic evolution. *Developmental Biology* **332**: 36–47.

Gompel N, Prud'homme B, Wittkopp PJ, et al. 2005. Chance caught on the wing: *cis*-regulatory evolution and the origin of pigment patterns in *Drosophila*. *Nature* **433**: 481–487.

Grant PR. 1999. *Ecology and Evolution of Darwin's Finches*. Princeton, NJ: Princeton University Press. (Original work published 1986)

Grant PR & Grant BR. 1992. Hybridization of bird species. *Science* **256**: 193–197.

Grant PR & Grant BR. 1995. Predicting microevolutionary responses to directional selection on heritable variation. *Evolution* **49**: 241–251.

Green RE, Krause J, Briggs AW, et al. 2010. A draft sequence of the Neandertal genome. *Science (New York, NY)* **328**: 710–722.

Gregory TR, Nicol JA, Tamm H, et al. 2007. Eukaryotic genome size databases. *Nucleic Acids Research* **35**: D332–D338.

Gronenberg W, Heeren S & Hölldobler B. 1996. Age-dependent and task-related morphological changes in the brain and the mushroom bodies of the ant *Camponotus floridanus*. *The Journal of Experimental Biology* **199**: 2011–2019.

Haffer J. 1969. Speciation in Amazonian forest birds. *Science* **165**: 131–137.

Haldane JBS. 1924. A mathematical theory of natural and artificial selection. *Transactions of the Cambridge Philosophical Society* **23**: 19–41.

Halpin CG, Skelhorn J & Rowe C. 2014. Increased predation of nutrient-enriched aposematic prey. *Proceedings of the Royal Society B: Biological Sciences* **281**: 20133255.

Hansen AK, Gilbert LE, Simpson BB, et al. 2006. Phylogenetic relationships and chromosome number evolution in *Passiflora*. *Systematic Botany* **31**: 138–150.

Hanski I, Saastamoinen M & Ovaskainen O. 2006. Dispersal-related life-history trade-offs in a butterfly metapopulation. *Journal of Animal Ecology* **75**: 91–100.

Hare EE, Peterson BK, Iyer VN, et al. 2008. Sepsid *even-skipped* enhancers are functionally conserved in *Drosophila* despite lack of sequence conservation. *PLoS Genetics* **4**: e1000106.

Harpel D, Cullen DA, Ott SR, et al. 2015. Pollen feeding proteomics: salivary proteins of the passion flower butterfly, *Heliconius melpomene*. *Insect Biochemistry and Molecular Biology* **63**: 7–13.

Hay-Roe MM & Mankin RW. 2004. Wing-click sounds of *Heliconius cydno alithea* (Nymphalidae : Heliconiinae) butterflies. *Journal of Insect Behavior* **17**: 329–335.

Hay-Roe MM & Nation J. 2007. Spectrum of cyanide toxicity and allocation in *Heliconius erato* and *Passiflora* host plants. *Journal of Chemical Ecology* **33**: 319–329.

Hayward P, Kalmar T & Arias AM. 2008. *Wnt/Notch* signalling and information processing during. *Development* **135**: 411–424.

He M, Sebaihia M, Lawley TD, et al. 2010. Evolutionary dynamics of *Clostridium difficile* over short and long time scales. *Proceedings of the National Academy of Sciences* **107**: 7527–7532.

Heard SB & Hauser DL. 1995. Key evolutionary innovations and their ecological mechanisms. *Historical Biology* **10**: 151–173.

Heinrich B. 1993. *The Hot-Blooded Insects*. Cambridge, MA: Harvard University Press.

Heinze S & Reppert SM. 2012. Anatomical basis of sun compass navigation I: The general layout of the monarch butterfly brain. *The Journal of Comparative Neurology* **520**: 1599–1628.

Hernandez MIM & Benson WW. 1998. Small-male advantage in the territorial tropical butterfly *Heliconius sara* (Nymphalidae): a paradoxical strategy? *Animal Behaviour* **56**: 533–540.

Hewitt G. 2000. The genetic legacy of the Quaternary ice ages. *Nature* **405**: 907–913.

Hewitt GM. 1996. Some genetic consequences of ice ages, and their role in divergence and speciation. *Biological Journal of the Linnean Society* **58**: 247–276.

Hey J & Machado CA. 2003. The study of structured populations—new hope for a difficult and divided science. *Nature Reviews Genetics* **4**: 535–543.

Higgie M, Chenoweth S & Blows MW. 2000. Natural selection and the reinforcement of mate recognition. *Science* **290**: 519–521.

Higham T, Douka K, Wood R, et al. 2014. The timing and spatiotemporal patterning of Neanderthal disappearance. *Nature* **512**: 306–309.

Hill RI, Gilbert LE & Kronforst MR. 2013. Cryptic genetic and wing pattern diversity in a mimetic *Heliconius* butterfly. *Molecular Ecology* **22**: 2760–2770.

Hines HM, Counterman BA, Papa R, et al. 2011. A wing patterning gene redefines the mimetic history of *Heliconius* butterflies. *Proceedings of the National Academy of Sciences* **108**: 19666–19671.

Hines HM, Papa R, Ruiz M, et al. 2012. Transcriptome analysis reveals novel patterning and pigmentation genes underlying *Heliconius* butterfly wing pattern variation. *BMC Genomics* **13**: 288.

Hoekstra HE, Hirschmann RJ, Bundey RA, et al. 2006. A single amino acid mutation contributes to adaptive beach mouse color pattern. *Science (New York, NY)* **313**: 101–104.

Hornett EA, Moran B, Reynolds LA, et al. 2014. The evolution of sex ratio distorter suppression affects a 25 cm genomic region in the butterfly *Hypolimnas bolina*. *PLoS Genetics* **10**: e1004822.

Horton BM, Moore IT & Maney DL. 2014. New insights into the hormonal and behavioural correlates of polymorphism in white-throated sparrows, *Zonotrichia albicollis*. *Animal Behaviour* **93**: 207–219.

Hoyal Cuthill J & Charleston M. 2012. Phylogenetic codivergence supports coevolution of mimetic *Heliconius* butterflies. *PLoS One* **7**: e36464.

Hsu R, Briscoe AD, Chang BSW, et al. 2001. Molecular evolution of a long wavelength-sensitive opsin in mimetic *Heliconius* butterflies (Lepidoptera : Nymphalidae). *Biological Journal of the Linnean Society* **72**: 435–449.

Huber B, Whibley A, Le Poul Y, et al. 2015. Conservatism and novelty in the genetic architecture of adaptation in *Heliconius* butterflies. *Heredity* **114**: 515–524.

Huheey JE. 1976. Studies in warning coloration and mimicry. VII. Evolutionary consequences of a Batesian-Müllerian spectrum: A model for Müllerian mimicry. *Evolution* **30**: 86–93.

Huheey JE. 1988. Mathematical models of mimicry. *The American Naturalist* **131**: S22–S41.

Hurst LD & Pomiankowski A. 1991. Causes of sex ratio bias may account for unisexual sterility in hybrids: a new explanation of Haldane's rule and related phenomena. *Genetics* **128**: 841–858.

Hutchinson GE. 1959. Homage to Santa Rosalia or why are there so many kinds of animals? *The American Naturalist* **93**: 145–159.

Huynh LY, Maney DL & Thomas JW. 2011. Chromosome-wide linkage disequilibrium caused by an inversion polymorphism in the white-throated sparrow (*Zonotrichia albicollis*). *Heredity* **106**: 537–546.

Ihalainen E, Lindström L, Mappes J, et al. 2008. Butterfly effects in mimicry? Combining signal and taste can twist the relationship of Müllerian co-mimics. *Behavioral Ecology and Sociobiology* **62**: 1267–1276.

Irwin DE. 2002. Phylogeographic breaks without geographic barriers to gene flow. *Evolution* **56**: 2383–2394.

Irwin DE. 2012. Local adaptation along smooth ecological gradients causes phylogeographic breaks and phenotypic clustering. *The American Naturalist* **180**: 35–49.

Jaenike J. 1990. Host specialization in phytophagous insects. *Annual Review of Ecology and Systematics* **21**: 243–273.

Janssen JM, Monteiro A & Brakefield PM. 2001. Correlations between scale structure and pigmentation in butterfly wings. *Evolution & Development* **3**: 415–423.

Janz N, Nyblom K & Nylin S. 2001. Evolutionary dynamics of host-plant specialization: a case study of the tribe nymphalini. *Evolution* **55**: 783–796.

Janz N, Nylin S & Wahlberg N. 2006. Diversity begets diversity: host expansions and the diversification of plant-feeding insects. *BMC Evolutionary Biology* **6**: 4.

Janzen DH. 1983. *Erblichia odorata* Seem. (Turneraceae) is a larval hostplant of *Eueides procula vulgiformis* (Nymphalidae: Heliconiini) in Santa Rosa National Park, Costa Rica. *Journal of the Lepidopterists' Society* **37**: 70–77.

Jarvis ED, Mirarab S, Aberer AJ, et al. 2014. Whole-genome analyses resolve early branches in the tree of life of modern birds. *Science* **346**: 1320–1331.

Jeffery WR. 2009. Regressive evolution in Astyanax cavefish. *Annual Review of Genetics* **43**: 25–47.

Jensen NB, Zagrobelny M, Hjernø K, et al. 2011. Convergent evolution in biosynthesis of cyanogenic defence compounds in plants and insects. *Nature Communications* **2**: 273.

Jeong C, Alkorta-Aranburu G, Basnyat B, et al. 2014. Admixture facilitates genetic adaptations to high altitude in Tibet. *Nature Communications* **5**: 3281.

Jiggins CD. 2008. Ecological speciation in mimetic butterflies. *BioScience* **58**: 541–548.

Jiggins CD & Davies N. 1998. Genetic evidence for a sibling species of *Heliconius charithonia* (Lepidoptera; Nymphalidae). *Biological Journal of the Linnean Society* **64**: 57–67.

Jiggins CD, Emelianov I & Mallet J. 2006. Assortative mating and speciation as pleiotropic effects of ecological adaptation: Examples in moths and butterflies. In: Fellowes MDE, Holloway GJ, Rolff J, eds. *Insect Evolutionary Ecology: Proceedings of the Royal Entomological Society's 22nd Symposium*. London, UK: CABI, 451–473.

Jiggins CD, Estrada C & Rodrigues A. 2004. Mimicry and the evolution of premating isolation in *Heliconius melpomene* Linnaeus. *Journal of Evolutionary Biology* **17**: 680–691.

Jiggins CD, Linares M, Naisbit RE, et al. 2001b. Sex-linked hybrid sterility in a butterfly. *Evolution* **55**: 1631–1638.

Jiggins CD & Mallet J. 2000. Bimodal hybrid zones and speciation. *Trends in Ecology and Evolution* **15**: 250–255.

Jiggins CD, Mavárez J, Beltran M, et al. 2005. A genetic linkage map of the mimetic butterfly *Heliconius melpomene*. *Genetics* **171**: 557–570.

Jiggins CD & McMillan WO. 1997. The genetic basis of an adaptive radiation: warning colour in two *Heliconius* species. *Proceedings of The Royal Society. Biological Sciences* **264**: 1167–1175.

Jiggins CD, McMillan WO, King P, et al. 1997. The maintenance of species differences across a *Heliconius* hybrid zone. *Heredity* **79**: 495–505.

Jiggins CD, McMillan WO & Mallet J. 1997. Host plant adaptation has not played a role in the recent speciation of *Heliconius himera* and *Heliconius erato*. *Ecological Entomology* **22**: 361–365.

Jiggins CD, McMillan WO, Neukirchen W, et al. 1996. What can hybrid zones tell us about speciation? The case of *Heliconius erato* and *H. himera* (Lepidoptera: Nymphalidae). *Biological Journal of the Linnean Society* **59**: 221–242.

Jiggins CD, Naisbit RE, Coe RL, et al. 2001a. Reproductive isolation caused by colour pattern mimicry. *Nature* **411**: 302–305.

Jiggins CD, Salazar C, Linares M, et al. 2008. Review. Hybrid trait speciation and *Heliconius* butterflies. *Philosophical Transactions of the Royal Society of London B: Biological Sciences* **363**: 3047–3054.

Johanson U, West J, Lister C, et al. 2000. Molecular analysis of FRIGIDA, a major determinant of natural variation in *Arabidopsis* flowering time. *Science* **290**: 344–347.

Johnson MS & Turner JRG. 1979. Absence of dosage compensation for a sex-linked enzyme in butterflies (*Heliconius*). *Heredity* **43**: 71–77.

Jones BM, Leonard AS, Papaj DR, et al. 2013. Plasticity of the worker bumblebee brain in relation to age and rearing environment. *Brain, Behavior and Evolution* **82**: 250–261.

Jones FM. 1930. The sleeping heliconians of Florida. *Natural History* **30**: 635–644.

Jones RT, Salazar PA, ffrench-Constant RH, et al. 2012. Evolution of a mimicry supergene from a multilocus architecture. *Proceedings of the Royal Society of London B: Biological Sciences* **279**: 316–325.

Joron M. 2005. Polymorphic mimicry, microhabitat use, and sex-specific behaviour. *Journal of Evolutionary Biology* **18**: 547–556.

Joron M, Frezal L, Jones RT, et al. 2011. Chromosomal rearrangements maintain a polymorphic supergene controlling butterfly mimicry. *Nature* **477**: 203–206.

Joron M & Mallet JLB. 1998. Diversity in mimicry: paradox or paradigm? *Trends in Ecology & Evolution* **13**: 461–466.

Joron M, Papa R, Beltran M, et al. 2006. A conserved supergene locus controls colour pattern diversity in *Heliconius* butterflies. *PLoS Biology* **4**: 1831–1840.

Joron M, Wynne IR, Lamas G, et al. 1999. Variable selection and the coexistence of multiple mimetic forms of the butterfly *Heliconius numata*. *Evolutionary Ecology* **13**: 721–754.

Kapan DD. 2001. Three-butterfly system provides a field test of Müllerian mimicry. *Nature* **409**: 338–340.

Kapan DD, Flanagan NS, Tobler A, et al. 2006. Localization of Müllerian mimicry genes on a dense linkage map of *Heliconius erato*. *Genetics* **173**: 735–757.

Kaye WJ. 1907. Notes on the dominant Müllerian group of butterflies from the Potaro district of British Guiana. *Transactions of the Entomological Society of London* **1906**: 411–439.

Kerpel SM & Moreira GRP. 2005. Absence of learning and local specialization on host plant selection by *Heliconius erato*. *Journal of Insect Behavior* **18**: 433–452.

Keys DN, Lewis DL, Selegue JE, et al. 1999. Recruitment of a hedgehog regulatory circuit in butterfly eyespot evolution. *Science* **283**: 532–534.

Khalifa A. 1949. Spermatophore production in Trichoptera and some other insects. *Transactions of the Royal Entomological Society of London* **100**: 449–471.

Killip EP. 1938. The American species of Passifloraceae. *Publications of the Field Museum of Natural History, Botanical Series* **19**: 1–163.

Kirkpatrick M & Barton N. 2006. Chromosome inversions, local adaptation and speciation. *Genetics* **173**: 419–434.

Kirkpatrick M & Ravigné V. 2002. Speciation by natural and sexual selection: models and experiments. *The American Naturalist* **159** Suppl 3: S22–35.

Klein AL & Araújo AM de. 2010. Courtship behavior of *Heliconius erato phyllis* (Lepidoptera, Nymphalidae) towards virgin and mated females: conflict between attraction and repulsion signals? *Journal of Ethology* **28**: 409–420.

Knapp S & Mallet J. 1998. A new species of *Passiflora* (Passifloraceae) from Ecuador with notes on the natural history of its herbivore, *Heliconius* (Lepidoptera : Nymphalidae : Heliconiiti). *Novon* **8**: 162–166.

Koch PB & Nijhout HF. 2002. The role of wing veins in colour pattern development in the butterfly *Papilio xuthus* (Lepidoptera: Papilionidae). *European Journal of Entomology* **99**: 67–72.

Kopp A. 2009. Metamodels and phylogenetic replication: a systematic approach to the evolution of developmental pathways. *Evolution* **63**: 2771–2789.

Koshikawa S, Giorgianni MW, Vaccaro K, et al. 2015. Gain of *cis*-regulatory activities underlies novel domains of wingless gene expression in *Drosophila*. *Proceedings of the National Academy of Sciences* **112**: 7524–7529.

Kozak KM. 2015. Macroevolution and phylogenomics in the adaptive radiation of Heliconiini butterflies. PhD thesis, University of Cambridge, UK.

Kozak KM, Wahlberg N, Neild A, et al. 2015. Multilocus species trees show the recent adaptive radiation of the mimetic *Heliconius* butterflies. *Systematic Biology* **64**: 505–524.

Krenn HW & Penz CM. 1998. Mouthparts of *Heliconius* butterflies (Lepidoptera : Nymphalidae): a search for anatomical adaptations to pollen-feeding behavior. *International Journal of Insect Morphology & Embryology* **27**: 301–309.

Kronforst M, Young LG & Gilbert LE. 2007. Reinforcement of mate preference among hybridizing *Heliconius* butterflies. *Journal of Evolutionary Biology* **20**: 278–285.

Kronforst MR, Hansen ME, Crawford NG, et al. 2013. Hybridization reveals the evolving genomic architecture of speciation. *Cell Reports* **5**: 666–677.

Kronforst MR, Kapan DD & Gilbert LE. 2006b. Parallel genetic architecture of parallel adaptive radiations in mimetic *Heliconius* butterflies. *Genetics* **174**: 535–539.

Kronforst MR, Young LG, Kapan DD, et al. 2006a. Linkage of butterfly mate preference and wing color preference cue at the genomic location of wingless. *Proceedings of the National Academy of Sciences of the United States of America* **103**: 6575–6580.

Krosnick SE, Porter-Utley KE, MacDougal JM, et al. 2013. New insights into the evolution of *Passiflora* subgenus Decaloba (Passifloraceae): phylogenetic relationships and morphological synapomorphies. *Systematic Botany* **38**: 692–713.

Kryvokhyzha, D. 2015. *Whole genome resequencing of Heliconius butterflies revolutionizes our view of the level of admixture between species*. Masters Thesis, University of Uppsala, Uppsala, Sweden.

Kunte K, Zhang W, Tenger-Trolander A, et al. 2014. *Doublesex* is a mimicry supergene. *Nature* **507**: 229–232.

Küpper C, Stocks M, Risse JE, et al. 2015. A supergene determines highly divergent male reproductive morphs in the ruff. *Nature Genetics* 48: 79–83.

Kuussaari M, van Nouhuys S, Hellmann JJ, et al. 2004. Larval biology of checkerspots. In: *On the Wings of Checkerspots: A Model System for Population Biology*. New York: Oxford University Press, 138–160.

Lamas G. 1997. Comentarios taxonómicos y nomenclaturales sobre Heliconiini neotropicales, con designación de lectotipos y descripción de cuatro subespecies nuevas. *Revista Peruana de Entomología* **40**: 111–125.

Lamas, G. 2004. *Checklist: Part 4A. Hesperioidea – Papilionoidea*. In J. B. Heppner [ed.], Atlas of Neotropical Lepidoptera. Volume 5A. Association for Tropical Lepidoptera/Scientific Publishers, Gainesville.

Lamas, G. 2016. *Annotated Bibliography of the Neotropical Butterflies and Skippers*, available at http://www.butterfliesofamerica.com/L/Biblio.htm

Lamichhaney S, Berglund J, Almén MS, et al. 2015a. Evolution of Darwin's finches and their beaks revealed by genome sequencing. *Nature* **518**: 371–375.

Lamichhaney S, Fan G, Widemo F, et al. 2015b. Structural genomic changes underlie alternative reproductive strategies in the ruff (*Philomachus pugnax*). *Nature Genetics* **48**: 84–88.

Langham GM. 2004. Specialized avian predators repeatedly attack novel color morphs of *Heliconius* butterflies. *Evolution* **58**: 2783–2787.

Langham GM. 2006. Rufous-tailed jacamars and aposematic butterflies: do older birds attack novel prey? *Behavioral Ecology* **17**: 285–290.

Leal IR, Fischer E, Kost C, et al. 2006. Ant protection against herbivores and nectar thieves in *Passiflora coccinea* flowers. *Ecoscience* **13**: 431–438.

Lenormand T, Bourguet D, Guillemaud T, et al. 1999. Tracking the evolution of insecticide resistance in the mosquito *Culex pipiens*. *Nature* **400**: 861–864.

Le Poul Y, Whibley A, Chouteau M, et al. 2014. Evolution of dominance mechanisms at a butterfly mimicry supergene. *Nature Communications* **5**: 5644.

Lihoreau M, Raine NE, Reynolds AM, et al. 2013. Unravelling the mechanisms of trapline foraging in bees. *Communicative & Integrative Biology* **6**: e22701.

Linares M. 1996. The genetics of the mimetic coloration in the butterfly *Heliconius cydno weymeri*. *Journal of Heredity* **87**: 142–149.

Linares M. 1997. The ghost of mimicry past: Laboratory reconstitution of an extinct butterfly 'race'. *Heredity* **78**: 628–635.

Lindström L, Alatalo RV, Lyytinen A, et al. 2004. The effect of alternative prey on the dynamics of imperfect Batesian and Müllerian mimicries. *Evolution* **58**: 1294–1302.

Lindström L, Lyytinen A, Mappes J, et al. 2006. Relative importance of taste and visual appearance for predator education in Müllerian mimicry. *Animal Behaviour* **72**: 323–333.

Lindström L, Rowe C & Guilford T. 2001. Pyrazine odour makes visually conspicuous prey aversive. *Proceedings of the Royal Society B: Biological Sciences* **268**: 159–162.

Linnen CR, Poh YP, Peterson BK, et al. 2013. Adaptive evolution of multiple traits through multiple mutations at a single gene. *Science* **339**: 1312–1316.

Llaurens V, Joron M & Billiard S. 2015. Molecular mechanisms of dominance evolution in Müllerian mimicry. *Evolution* **69**: 3097–3108.

Llaurens V, Joron M & Théry M. 2014. Cryptic differences in colour among Müllerian mimics: how can the visual capacities of predators and prey shape the evolution of wing colours? *Journal of Evolutionary Biology* **27**: 531–540.

Loh YHE, Bezault E, Muenzel FM, et al. 2013. Origins of shared genetic variation in African Cichlids. *Molecular Biology and Evolution* **30**: 906–917.

Longino JT. 1984. Shoots, parasitiods and ants as forces in the population dynamics of *Heliconius hewitsoni* in Costa Rica. PhD thesis, Austin, Texas, USA.

Losos J. 2011. *Lizards in an Evolutionary Tree: Ecology and Adaptive Radiation of Anoles*. Berkeley: University of California Press.

Ludwig MZ, Palsson A, Alekseeva E, et al. 2005. Functional evolution of a *cis*-regulatory module. *PLoS Biology* **3**: e93.

Macdonald WP, Martin A & Reed RD. 2010. Butterfly wings shaped by a molecular cookie cutter: evolutionary radiation of lepidopteran wing shapes associated with a derived *Cut/wingless* wing margin boundary system. *Evolution & Development* **12**: 296–304.

MacDougal JM & Hansen AK. 2003. A new section of *Passiflora*, subgenus Decaloba (Passifloraceae), from Central America, with two new species. *Novon* **13**: 459–466.

Mallet J. 1984. Population structure and evolution in *Heliconius* butterflies. PhD thesis, Austin, Texas, USA.

Mallet J. 1986a. Gregarious roosting and home range in *Heliconius* butterflies. *National Geographic Research* **2**: 198–215.

Mallet J. 1986b. Dispersal and gene flow in a butterfly with home range behavior: *Heliconius erato* (Lepidoptera: Nymphalidae). *Oecologia* **68**: 210–217.

Mallet J. 1986c. Hybrid zones of *Heliconius* butterflies in Panama and the stability and movement of warning color clines. *Heredity* **56**: 191–202.

Mallet J. 1989. The genetics of warning color in Peruvian hybrid zones of *Heliconius erato* and *Heliconius melpomene*. *Proceedings of the Royal Society of London Series B: Biological Sciences* **236**: 163–185.

Mallet J. 1991. Variations on a theme? *Nature* **354**: 368.

Mallet J. 1993. Speciation, raciation, and color pattern evolution in *Heliconius* butterflies: evidence from hybrid zones.

In: Harrison RG, ed. *Hybrid Zones and the Evolutionary Process*. New York: Oxford University Press, 226–260.

Mallet J. 1995. A species definition for the modern synthesis. *Trends in Ecology & Evolution* **10**: 294–299.

Mallet J. 1999. Causes and consequences of a lack of coevolution in Müllerian mimicry. *Evolutionary Ecology* **13**: 777–806.

Mallet J. **2001**. Gene flow. In: Woiwod IP, Reynolds DR, Thomas CD, eds. *Insect Movement: Mechanisms and Consequences*. Wallingford, UK: CABI, 337–360.

Mallet J. 2005. Hybridization as an invasion of the genome. *Trends in Ecology & Evolution* **20**: 229–237.

Mallet J. 2010. Shift happens! Shifting balance and the evolution of diversity in warning colour and mimicry. *Ecological Entomology* **35**: 90–104.

Mallet J & Barton NH. 1989a. Strong natural selection in a warning color hybrid zone. *Evolution* **43**: 421–431.

Mallet J & Barton N. 1989b. Inference from clines stabilized by frequency-dependent selection. *Genetics* **122**: 967–976.

Mallet J, Barton N, Lamas G, et al. 1990. Estimates of selection and gene flow from measures of cline width and linkage disequilibrium in *Heliconius* hybrid zones. *Genetics* **124**: 921–936.

Mallet J, Beltran M, Neukirchen W, et al. 2007. Natural hybridization in heliconiine butterflies: the species boundary as a continuum. *BMC Evolutionary Biology* **7**: 28.

Mallet J & Gilbert LE. 1995. Why are there so many mimicry rings—correlations between habitat, behavior and mimicry in *Heliconius* butterflies. *Biological Journal of the Linnean Society* **55**: 159–180.

Mallet J & Joron M. 1999. Evolution of diversity in warning color and mimicry: Polymorphisms, shifting balance, and speciation. *Annual Review of Ecology and Systematics* **30**: 201–233.

Mallet J & Longino JT. 1982. Hostplant records and descriptions of juvenile stages for two rare species of *Eueides* (Nymphalidae). *Journal of the Lepidopterists' Society* **36**: 136–144.

Mallet J, Longino JT, Murawski D, et al. 1987. Handling effects in *Heliconius*—where do all the butterflies go? *Journal of Animal Ecology* **56**: 377–386.

Mallet J & Singer MC. 1987. Individual selection, kin selection, and the shifting balance in the evolution of warning colours: the evidence from butterflies. *Biological Journal of the Linnean Society* **32**: 337–350.

Mallet JLB & Jackson DA. 1980. The ecology and social behavior of the neotropical butterfly *Heliconius xanthocles* Bates in Colombia. *Zoological Journal of the Linnean Society* **70**: 1–13.

Martin A, McCulloch KJ, Patel NH, et al. 2014. Multiple recent co-options of Optix associated with novel traits in adaptive butterfly wing radiations. *EvoDevo* **5**: 1–14.

Martin A & Orgogozo V. 2013. The loci of repeated evolution: a catalog of genetic hotspots of phenotypic variation. *Evolution* **67**: 1235–1250.

Martin A, Papa R, Nadeau NJ, et al. 2012. Diversification of complex butterfly wing patterns by repeated regulatory evolution of a Wnt ligand. *Proceedings of the National Academy of Sciences* **109**: 12632–12637.

Martin A & Reed RD. 2014. Wnt signaling underlies evolution and development of the butterfly wing pattern symmetry systems. *Developmental Biology* **395**: 367–378.

Martin CH, Cutler JS, Friel JP, et al. 2015. Complex histories of repeated gene flow in Cameroon crater lake cichlids cast doubt on one of the clearest examples of sympatric speciation. *Evolution* **69**: 1406–1422.

Martin SH, Dasmahapatra KK, Nadeau NJ, et al. 2013. Genome-wide evidence for speciation with gene flow in *Heliconius* butterflies. *Genome Research* **23**: 1817–1828.

Martin SH, Möst M, Palmer WJ, et al. 2016. Natural selection and genetic diversity in the butterfly *Heliconius melpomene*. *Genetics* **203**: 525–41.

Matute DR. 2010. Reinforcement of gametic isolation in *Drosophila*. *PLoS Biology* **8**: e1000341.

Mavárez J & Linares M. 2008. Homoploid hybrid speciation in animals. *Molecular Ecology* **17**: 4181–4185.

Mavárez J, Salazar CA, Bermingham E, et al. 2006. Speciation by hybridization in *Heliconius* butterflies. *Nature* **441**: 868–871.

Maynard-Smith J. 1966. Sympatric Speciation. *American Naturalist* **100**: 637–650.

Mayr E. 1963. *Animal Species and Evolution*. Harvard University, Cambridge MA: Belknap Press.

McCulloch KJ, Osorio D & Briscoe AD. 2016. Sexual dimorphism in the compound eye of *Heliconius erato*: a nymphalid with five spectral classes of photoreceptor. *Journal of Experimental Biology*. doi: 10.1242/jeb.136523.

McMillan WO, Jiggins CD & Mallet J. 1997. What initiates speciation in passion-vine butterflies? *Proceedings of the National Academy of Sciences of the United States of America* **94**: 8628–8633.

Mega NO & de Araújo AM. 2008. Do caterpillars of *Dryas iulia alcionea* (Lepidoptera, Nymphalidae) show evidence of adaptive behaviour to avoid predation by ants? *Journal of Natural History* **42**: 129–137.

Meise W. 1928. Die Verbreitung der Aaskraehe (Formenkreis *Corvus corone* L.). *Journal of Ornithology* **76**: 1–203.

Melo MC, Salazar C, Jiggins CD, et al. 2009. Assortative mating preferences among hybrids offers a route to hybrid speciation. *Evolution* **63**: 1660–1665.

Mendoza-Cuenca L & Macías-Ordóñez R. 2005. Foraging polymorphism in *Heliconius charitonia* (Lepidoptera : Nymphalidae): morphological constraints and behavioural compensation. *Journal of Tropical Ecology* **21**: 407–415.

Mendoza-Cuenca L & Macías-Ordóñez R. 2010. Female asynchrony may drive disruptive sexual selection on male mating phenotypes in a *Heliconius* butterfly. *Behavioral Ecology* **21**: 144–152.

Menna-Barreto Y & Araújo AM. 1985. Evidence for host plant preferences in *Heliconius erato phyllis* from Southern Brazil (Nymphalidae). *Journal of Research on the Lepidoptera* **24**: 41–46.

Menzel R. 1993. Associative learning in honey bees. *Aphidologie* **24**: 157–168.

Merian MS. 1705. *Metamorphosis insectorum Surinurnensium. In quae erucae ac vermes Surimmenses, cum omnibus suis Transformationibus, ad uiuum delineantur et describuntur, singulis eorum Plantas, flores et fructus collocatis, in quibus reperta sunt; tunc etiam Generatio Ranarum, Bufonurn rariorum, Lacertarum, Serpentum, Araneorum et Formicarum exhibetur*. Amstelodamum: G. Valk.

Mérot C. 2015. *La spéciation chez les papillons Heliconius : importance relative du mimétisme et de la divergence écologique*. Museum National d'Histoire Naturelle, Paris, France.

Mérot C, Mavárez J, Evin A, et al. 2013. Genetic differentiation without mimicry shift in a pair of hybridizing *Heliconius* species (Lepidoptera: Nymphalidae). *Biological Journal of the Linnean Society* **109**: 830–847.

Merrill RM, Dasmahapatra KK, Davey JW, et al. 2015. The diversification of *Heliconius* butterflies: what have we learned in 150 years? *Journal of Evolutionary Biology* **28**: 1417–1438.

Mérot C, Frérot B, Leppik E, et al. 2015. Beyond magic traits: multimodal mating cues in *Heliconius* butterflies. *Evolution* **69**: 2891–2904.

Merrill RM, Gompert Z, Dembeck LM, et al. 2011a. Mate preference across the speciation continuum in a clade of mimetic butterflies. *Evolution* **65**: 1489–1500.

Merrill RM, Naisbit RE, Mallet J, et al. 2013. Ecological and genetic factors influencing the transition between host-use strategies in sympatric *Heliconius* butterflies. *Journal of Evolutionary Biology* **26**: 1959–1967.

Merrill RM, Van Schooten B, Scott JA, et al. 2011b. Pervasive genetic associations between traits causing reproductive isolation in *Heliconius* butterflies. *Proceedings of The Royal Society. Biological Sciences* **278**: 511–518.

Merrill RM, Wallbank RWR, Bull V, et al. 2012. Disruptive ecological selection on a mating cue. *Proceedings of the Royal Society B: Biological Sciences* **279**: 4907–4913.

Michener CD. 1942. A generic revision of Heliconiinae (Lepidoptera, Nymphalidae). *American Museum Novitates* **1197**: 1–8.

Moczek AP & Emlen DJ. 2000. Male horn dimorphism in the scarab beetle, *Onthophagus taurus*: do alternative reproductive tactics favour alternative phenotypes? *Animal Behaviour* **59**: 459–466.

Moczek AP & Rose DJ. 2009. Differential recruitment of limb patterning genes during development and diversification of beetle horns. *Proceedings of the National Academy of Sciences* **106**: 8992–8997.

Moczek AP, Sultan S, Foster S, et al. 2011. The role of developmental plasticity in evolutionary innovation. *Proceedings of the Royal Society of London B: Biological Sciences* **278**: 2705–2713.

Moen D & Morlon H. 2014. From dinosaurs to modern bird diversity: extending the time scale of adaptive radiation. *PLoS Biology* **12**: e1001854.

Monteiro **A. 2012**. Gene regulatory networks reused to build novel traits. *BioEssays* **34**: 181–186.

Monteiro A. 2015. Origin, development, and evolution of butterfly eyespots. *Annual Review of Entomology* **60**: 253–271.

Montgomery SH, Merrill RM & Ott SR. 2016. Brain composition in *Heliconius* butterflies, post-eclosion growth and experience dependent neuropil plasticity. *Journal of Comparative Neurology* **9**: 1747–1769.

Moore JL & Tabashnik BE. 1989. Leg autotomy of adult diamondback moth (Lepidoptera: Plutellidae) in response to tarsal contact with insecticide residues. *Journal of Economic Entomology* **82**: 381–384.

Moran NA & Jarvik T. 2010. Lateral transfer of genes from fungi underlies carotenoid production in aphids. *Science* **328**: 624–627.

Moreira GRP & Mielke CGC. 2010. A new species of *Neruda* Turner, 1976 from northeast Brazil (Lepidoptera: Nymphalidae, Heliconiinae, Heliconiini). *Nachrichten des Entomol. Vereins Apollo* **31**: 85–91.

Moss AM. 1933. The gregarious sleeping habits of certain ithomiine and heliconiine butterflies in Brazil. *Proceedings of the Royal Entomological Society, London* **7**: 66–67.

Mugrabi-Oliveira E & Moreira GRP. 1996. Conspecific mimics and low host plant availability reduce egg laying by *Heliconius erato phyllis* (Fabricius) (Lepidoptera, Nymphalidae). *Revista Brasileira de Zoologia* **13**: 929–937.

Müller F. 1879. *Ituna* and *Thyridia*; a remarkable case of mimicry in butterflies. *Transactions of the Entomological Society of London*: xx–xxix.

Müller F. 1912. X. The scent-scales of the Male 'Maracujá butterflies'. In: Longstaff GB, ed. *Butterfly Hunting in Many Lands*. New York: Longmans, Green & Co., 655–659.

Muller HJ. 1942. Isolating mechanisms, evolution and temperature. *Biological Symposium* 6: 71–125.

Mundy NI. 2005. A window on the genetics of evolution: MC1R and plumage colouration in birds. *Proceedings of the Royal Society B: Biological Sciences* **272**: 1633–1640.

Muñoz AG, Baxter SW, Linares M, et al. 2011. Deep mitochondrial divergence within a *Heliconius* butterfly species is not explained by cryptic speciation or endosymbiotic bacteria. *BMC Evolutionary Biology* **11**: 358.

Muñoz AG, Salazar C, Castaño J, et al. 2010. Multiple sources of reproductive isolation in a bimodal butterfly hybrid zone. *Journal of Evolutionary Biology* **23**: 1312–1320.

Murawski DA. 1986. Pollination ecology of a Costa Rican population of *Psiguria warscewiczii* in relation to the foraging behaviour of *Heliconius* butterflies. PhD dissertation, University of Texas, Austin.

Murawski DA & Gilbert LE. 1986. Pollen flow in *Psiguria warscewiczii*—a comparison of *Heliconius* butterflies and hummingbirds. *Oecologia* **68**: 161–167.

Muyshondt A & Young AM. 1973. The biology of the butterfly *Dione juno huascama* (Nymphalidae: Heliconiinae) in El Salvador. *Journal of the New York Entomological Society* **81**: 137–151.

Nadeau NJ & Jiggins CD. 2010. A golden age for evolutionary genetics? Genomic studies of adaptation in natural populations. *Trends in Genetics* **26**: 484–492.

Nadeau NJ, Martin SH, Kozak KM, et al. 2013. Genome-wide patterns of divergence and gene flow across a butterfly radiation. *Molecular Ecology* **22**: 814–826.

Nadeau NJ, Pardo-Diaz C, Whibley A, et al. 2016. The gene *cortex* controls mimicry and crypsis in butterflies and moths. *Nature* **534**: 106–110.

Nadeau NJ, Ruiz M, Salazar P, et al. 2014. Population genomics of parallel hybrid zones in the mimetic butterflies, *H. melpomene* and *H. erato*. *Genome Research* **24**: 1316–1333.

Nadeau NJ, Whibley A, Jones RT, et al. 2012. Genomic islands of divergence in hybridizing *Heliconius* butterflies identified by large-scale targeted sequencing. *Philosophical Transactions of the Royal Society B: Biological Sciences* **367**: 343–353.

Nahrstedt A & Davis RH. 1983. Occurrence, variation and biosynthesis of the cyanogenic glucosides linamarin and lotaustralin in species of the Heliconiini (Insecta: Lepidoptera). *Comparative Biochemistry and Physiology B: Biochemistry & Molecular Biology* **75**: 65–73.

Naisbit R. 2001. Ecological divergence and speciation in *Heliconius cydno* and *H. melpomene*. PhD thesis; The University of London, London, UK.

Naisbit RE, Jiggins CD, Linares M, et al. 2002. Hybrid sterility, Haldane's rule and speciation in *Heliconius cydno* and *H. melpomene*. *Genetics* **161**: 1517–1526.

Naisbit RE, Jiggins CD & Mallet J. 2001. Disruptive sexual selection against hybrids contributes to speciation between *Heliconius cydno* and *Heliconius melpomene*. *Proceedings of the Royal Society of London Series B-Biological Sciences* **268**: 1849–1854.

Naisbit RE, Jiggins CD & Mallet J. 2003. Mimicry: developmental genes that contribute to speciation. *Evolution & Development* **5**: 269–280.

Narum SR & Hess JE. 2011. Comparison of FST outlier tests for SNP loci under selection. *Molecular Ecology Resources* **11**: 184–194.

Nason JD, Herre AE & Hamrick JL. 1998. The breeding structure of a tropical keystone plant resource. *Nature* **391**: 685–687.

Neafsey DE, Waterhouse RM, Abai MR, et al. 2015. Highly evolvable malaria vectors: The genomes of 16 Anopheles mosquitoes. *Science* **347**: 1258522.

Nel A, Roques P, Nel P, et al. 2013. The earliest known holometabolous insects. *Nature* **503**: 257–261.

Neu HC. 1992. The crisis in antibiotic resistance. *Science* **257**: 1064–1073.

Nicholson AJ. 1927. A new theory of mimicry in insects. *Australian Zoologist* **5**: 10–104.

Nieberding CM, de Vos H, Schneider MV, et al. 2008. The male sex pheromone of the butterfly *Bicyclus anynana*: towards an evolutionary analysis. *PLoS One* **3**: e2751.

Nieberding CM, Fischer K, Saastamoinen M, et al. 2012. Cracking the olfactory code of a butterfly: the scent of ageing. *Ecology Letters* **15**: 415–424.

Nielsen R & Beaumont MA. 2009. Statistical inferences in phylogeography. *Molecular Ecology* **18**: 1034–1047.

Nijhout HF. 1990. A comprehensive model for colour pattern formation in butterflies. *Proceedings of the Royal Society of London, B* **239**: 81–113.

Nijhout HF. 1991. *The Development and Evolution of Butterfly Wing Patterns*. Washington DC: Smithsonian Institution Scholarly Press.

Nijhout HF & Wray GA. 1988. Homologies in the colour patterns of the genus *Heliconius* (Lepidoptera: Nymphalidae). *Biological Journal of the Linnean Society* **33**: 345–365.

Nijhout HF, Wray GA & Gilbert LE. 1990. An analysis of the phenotypic effects of certain colour pattern genes in *Heliconius* (Lepidoptera: Nymphalidae). *Biological Journal of the Linnean Society* **40**: 357–372.

Nilsson DE. 2009. The evolution of eyes and visually guided behaviour. *Philosophical Transactions of the Royal Society of London B: Biological Sciences* **364**: 2833–2847.

Nishikawa H, Iijima T, Kajitani R, et al. 2015. A genetic mechanism for female-limited Batesian mimicry in *Papilio* butterfly. *Nature Genetics* **47**: 405–409.

Noda K ichi, Glover BJ, Linstead P, et al. 1994. Flower colour intensity depends on specialized cell shape controlled by a Myb-related transcription factor. *Nature* **369**: 661–664.

Norris LC, Main BJ, Lee Y, et al. 2015. Adaptive introgression in an African malaria mosquito coincident with the increased usage of insecticide-treated bed nets. *Proceedings of the National Academy of Sciences* **112**: 815–820.

Nosil P, Harmon LJ & Seehausen O. 2009. Ecological explanations for (incomplete) speciation. *Trends in Ecology & Evolution* **24**: 145–156.

Nüsslein-Volhard C & Wieschaus E. 1980. Mutations affecting segment number and polarity in *Drosophila*. *Nature* **287**: 795–801.

O'Brien DM, Boggs CL & Fogel ML. 2003. Pollen feeding in the butterfly *Heliconius charitonia*: isotopic evidence for essential amino acid transfer from pollen to eggs. *Proceedings of the Royal Society of London Series B: Biological Sciences* **270**: 2631–2636.

O'Brien DM, Boggs CL & Fogel ML. 2004. Making eggs from nectar: the role of life history and dietary carbon turnover in butterfly reproductive resource allocation. *Oikos* **105**: 279–291.

O'Brien DM, Boggs CL & Fogel ML. 2005. The amino acids used in reproduction by butterflies: A comparative study of dietary sources using compound-specific stable isotope analysis. *Physiological and Biochemical Zoology* **78**: 819–827.

Ochman H, Lawrence JG & Groisman EA. 2000. Lateral gene transfer and the nature of bacterial innovation. *Nature* **405**: 299–304.

Ohno S. 1970. *Evolution by Gene Duplication*. London, New York: Springer-Verlag.

Ohsaki N. 1995. Preferential predation of female butterflies and the evolution of Batesian mimicry. *Nature* **378**: 173–175.

Orr HA. 1996. Dobzhansky, Bateson, and the genetics of speciation. *Genetics* **144**: 1331–1335.

Orr HA. 1998. The population genetics of adaptation: the distribution of factors fixed during adaptive evolution. *Evolution* **52**: 935–949.

Orr HA. 2005. The genetic theory of adaptation: a brief history. *Nature Reviews Genetics* **6**: 119–127.

Palaisa K, Morgante M, Tingey S, et al. 2004. Long-range patterns of diversity and linkage disequilibrium surrounding the maize Y1 gene are indicative of an asymmetric selective sweep. *Proceedings of the National Academy of Sciences of the United States of America* **101**: 9885–9890.

Panchal M & Beaumont MA. 2007. The automation and evaluation of nested clade phylogeographic analysis. *Evolution* **61**: 1466–1480.

Papa R, Kapan DD, Counterman BA, et al. 2013. Multi-allelic major effect genes interact with minor effect QTLs to control adaptive color pattern variation in *Heliconius erato*. *PLoS One* **8**: e57033.

Papa R, Martin A & Reed RD. 2008b. Genomic hotspots of adaptation in butterfly wing pattern evolution. *Current Opinion in Genetics & Development* **18**: 559–564.

Papa R, Morrison CM, Walters JR, et al. 2008a. Highly conserved gene order and numerous novel repetitive elements in genomic regions linked to wing pattern variation in *Heliconius* butterflies. *BMC Genomics* **9**: 345.

Papageorgis C. 1975. Mimicry in neotropical butterflies. *American Scientist* **63**: 522–532.

Pardo-Diaz C & Jiggins CD. 2014. Neighboring genes shaping a single adaptive mimetic trait. *Evolution & Development* **16**: 3–12.

Pardo-Diaz C, Salazar C, Baxter SW, et al. 2012. Adaptive introgression across species boundaries in *Heliconius* butterflies. *PLoS Genetics* **8**: e1002752.

Parker GA & Birkhead TR. 2013. Polyandry: the history of a revolution. *Philosophical Transactions of the Royal Society of London B: Biological Sciences* **368**: 20120335.

Pemberton RW. 1989. Insects attacking *Passiflora mollissima* and other *Passiflora* species; Field Survey in the Andes. *Proceedings of the Hawaiian Entomological Society* **29**: 71–84.

Penz CM. 1999. Higher level phylogeny for the passion-vine butterflies (Nymphalidae; Heliconiinae) based on early stage and adult morphology. *Zoological Journal of the Linnean Society* **127**: 277–344.

Penz CM & Krenn HW. 2000. Behavioral adaptations to pollen-feeding in *Heliconius* butterflies (Nymphalidae, Heliconiinae): An experiment using *Lantana* flowers. *Journal of Insect Behavior* **13**: 865–880.

Petit RJ. 2008. The coup de grâce for the nested clade phylogeographic analysis? *Molecular Ecology* **17**: 516–518.

Pfennig DW, Wund MA, Snell-Rood EC, et al. 2010. Phenotypic plasticity's impacts on diversification and speciation. *Trends in Ecology & Evolution* **25**: 459–467.

Phillimore AB, Orme CDL, Thomas GH, et al. 2008. Sympatric speciation in birds is rare: insights from range data and simulations. *The American Naturalist* **171**: 646–657.

Pinheiro CE. 2009. Following the leader: how *Heliconius ethilla* butterflies exchange information on resource locations. *Journal of the Lepidopterists' Society* **63**: 179–181.

Pinheiro CEG. 1996. Palatability and escaping ability in neotropical butterflies: tests with wild kingbirds (*Tyrannus melancholicus*, Tyrannidae). *Biological Journal of the Linnean Society* **59**: 351–365.

Podos J. 2010. Acoustic discrimination of sympatric morphs in Darwin's finches: a behavioural mechanism for assortative mating? *Philosophical Transactions of the Royal Society of London B: Biological Sciences* **365**: 1031–1039.

Poelstra JW, Vijay N, Bossu CM, et al. 2014. The genomic landscape underlying phenotypic integrity in the face of gene flow in crows. *Science* **344**: 1410–1414.

Poulton EB. 1904. What is a species. *Proceedings of the Entomological Society of London* **1903**: lxxvii–cxvi.

Poulton EB. 1931a. The gregarious sleeping habits of *H. charitonia* L. *Proceedings of the Entomological Society of London* **6**: 4–10.

Poulton EB. 1931b. The gregarious resting habits of danaine butterflies in Australia; also of heliconiine and ithomiine butterflies in tropical America. *Proceedings of the Royal Entomological Society, London* **7**: 64–67.

Prance GT. 1974. Phytogeographic support for the theory of Pleistocene forest refuges in the Amazon basin, based on evidence from distribution patterns in Caryocaraceae, Chrysobalanaceae, Dichapetalaceae and Lecythidaceae. *Acta Amazonica* **3**: 5–28.

Protas M, Conrad M, Gross JB, et al. 2007. Regressive evolution in the Mexican cave Tetra, *Astyanax mexicanus*. *Current Biology* **17**: 452–454.

Provine WB. 2001. *The Origins of Theoretical Population Genetics*. Chicago: University of Chicago Press.

Prowell DP. 1998. Sex linkage and speciation in Lepidoptera. *Endless Forms: Species and Speciation* **90**: 309.

Punnet RC. 1915. *Mimicry in Butterflies*. Cambridge: Cambridge University Press.

Quek SP, Counterman BA, Albuquerque de Moura P, et al. 2010. Dissecting comimetic radiations in *Heliconius* reveals divergent histories of convergent butterflies. *Proceedings of the National Academy of Sciences of the United States of America* **107**: 7365–7370.

Qvarnstrom A & Bailey RI. 2008. Speciation through evolution of sex-linked genes. *Heredity* **102**: 4–15.

Racimo F, Sankararaman S, Nielsen R, et al. 2015. Evidence for archaic adaptive introgression in humans. *Nature Reviews Genetics* **16**: 359–371.

Rausher MD. 1978. Search image for leaf shape in a butterfly. *Science* **200**: 1071–1073.

Rausher MD & Delph LF. 2015. When does understanding phenotypic evolution require identification of the underlying genes? *Evolution* **69**: 1655–1664.

Ray TB & Andrews CC. 1980. Ant butterflies: butterflies that follow army ants to feed on antbird droppings. *Science* **210**: 1147–1148.

Read AF & Huijben S. 2009. Evolutionary biology and the avoidance of antimicrobial resistance. *Evolutionary Applications* **2**: 40–51.

Rebeiz M, Pool JE, Kassner VA, et al. 2009. Stepwise modification of a modular enhancer underlies adaptation in a *Drosophila* population. *Science (New York, NY)* **326**: 1663–1667.

Reed RD. 2003. Gregarious oviposition and clutch size adjustment by a *Heliconius* butterfly. *Biotropica* **35**: 555–559.

Reed RD. 2004. Evidence for Notch-mediated lateral inhibition in organizing butterfly wing scales. *Development Genes and Evolution* **214**: 43–46.

Reed RD & Gilbert LE. 2004. Wing venation and Distal-less expression in *Heliconius* butterfly wing pattern development. *Development Genes and Evolution* **214**: 628–634.

Reed RD, McMillan WO & Nagy LM. 2008. Gene expression underlying adaptive variation in *Heliconius* wing patterns: non-modular regulation of overlapping *cinnabar* and *vermilion* prepatterns. *Proceedings of the Royal Society B: Biological Sciences* **275**: 37–45.

Reed RD, Papa R, Martin A, et al. 2011. *optix* drives the repeated convergent evolution of butterfly wing pattern mimicry. *Science* **333**: 113–1141.

Renou M. 1983. Chemosensors of anterior tarsus of the female *Heliconius charitonis* (lep heliconiidae). *Annales De La Societe Entomologique De France* **19**: 101–106.

Reznick DN, Shaw FH, Rodd FH, et al. 1997. Evaluation of the rate of evolution in natural populations of guppies (*Poecilia reticulata*). *Science* **275**: 1934–1937.

Rieseberg LH & Carney SE. 1998. Plant hybridization. *New Phytologist* **140**: 599–624.

Rieseberg LH, Raymond O, Rosenthal DM, et al. 2003. Major ecological transitions in wild sunflowers facilitated by hybridization. *Science* **301**: 1211–1216.

Robinson GS, Ackery PR, Kitching IJ, et al. 2010. *HOSTS—A Database of the World's Lepidopteran Hostplants*. London: Natural History Museum.

Rockman MV. 2012. The QTN program and the alleles that matter for evolution: all that's gold does not glitter. *Evolution* **66**: 1–17.

Romanowsky HP, Gus R & Araújo AM. 1985. Studies on the genetics and ecology of *Heliconius erato* (Lepid.; Nymph.). III. Population size, preadult mortality, adult resources, and polymorphism in natural populations. *Revista Brasileira de Biologia* **45**: 563.

Ross GN, Fales HM, Lloyd HA, et al. 2001. Novel chemistry of abdominal defensive glands of nymphalid butterfly *Agraulis vanillae*. *Journal of Chemical Ecology* **27**: 1219–1228.

Rosser N, Dasmahapatra KK & Mallet J. 2014. Stable *Heliconius* butterfly hybrid zones are correlated with a local rainfall peak at the edge of the Amazon basin. *Evolution* **68**: 3470–3484.

Rosser N, Kozak KM, Phillimore AB, et al. 2015. Extensive range overlap between heliconiine sister species: evidence for sympatric speciation in butterflies? *BMC Evolutionary Biology* **15**: 125.

Rosser N, Phillimore AB, Huertas B, et al. 2012. Testing historical explanations for gradients in species richness in heliconiine butterflies of tropical America. *Biological Journal of the Linnean Society* **105**: 479–497.

Roubik DW & Moreno JE. 1991. *Pollen and spores of Barro Colorado Island*. St Louis, MO: Missouri Botanical Gardens.

Rowe C & Guilford T. 1996. Hidden colour aversions in domestic chicks triggered by pyrazine odours of insect warning displays. *Nature* **383**: 520–522.

Rowland HM, Ihalainen E, Lindström L, et al. 2007. Comimics have a mutualistic relationship despite unequal defences. *Nature* **448**: 64–67.

Rowland HM, Mappes J, Ruxton GD, et al. 2010. Mimicry between unequally defended prey can be parasitic: evidence for quasi-Batesian mimicry. *Ecology Letters* **13**: 1494–1502.

Rull V. 2011. Neotropical biodiversity: timing and potential drivers. *Trends in Ecology & Evolution* **26**: 508–513.

Rundle HD & Nosil P. 2005. Ecological speciation. *Ecology Letters* **8**: 336–352.

Ruxton GD, Sherratt TN & Speed M. 2004. *Avoiding Attack: The Evolutionary Ecology of Crypsis, Warning Signals and Mimicry*. Oxford: Oxford University Press.

Saenz JA. 1972. Toxic effect of fruit of *Passiflora adenopoda* DC. on humans: phytochemical determination. *Revista de Biologia Tropical* **20**: 137–140.

Salazar C, Baxter SW, Pardo-Diaz C, et al. 2010. Genetic evidence for hybrid trait speciation in *Heliconius* butterflies. *PLoS Genetics* **6**: e1000930.

Salazar PCA. 2012. Hybridization and the genetics of wing colour-pattern diversity in *Heliconius* butterflies. PhD thesis; University of Cambridge, UK.

Salcedo C. 2010a. Evidence of pollen digestion at nocturnal aggregations of *Heliconius sara* in Costa Rica (Lepidoptera: Nymphalidae). *Tropical Lepidoptera* **20**: 35–37.

Salcedo C. 2010b. Environmental elements involved in communal roosting in *Heliconius* butterflies (Lepidoptera: Nymphalidae). *Environmental Entomology* **39**: 907–911.

Salcedo C. 2011a. Behavioural traits expressed during *Heliconius* butterflies roost-assembly. *Tropical Lepidoptera* **21**: 80–83.

Salcedo C. 2011b. Evidence of predation and disturbance events at *Heliconius* (Insecta: Lepidoptera: Nymphalidae) nocturnal aggregations in Panama and Costa Rica. *Journal of Natural History* **45**: 1715–1721.

Sánchez-Gracia A, Vieira FG, Almeida FC, et al. 2001. Comparative Genomics of the Major Chemosensory Gene Families in Arthropods. In: *eLS*. John Wiley & Sons, Ltd.

Sankararaman S, Mallick S, Dannemann M, et al. 2014. The genomic landscape of Neanderthal ancestry in present-day humans. *Nature* **507**: 354–357.

Sargent TD. 1995. On the relative acceptabilities of local butterflies and moths to local birds. *Journal of the Lepidopterists' Society* **49**: 148–162.

Schliewen UK, Tautz D & Pääbo S. 1994. Sympatric speciation suggested by monophyly of crater lake cichlids. *Nature* **368**: 629–632.

Schulz S, Estrada C, Yildizhan S, et al. 2008. An antiaphrodisiac in *Heliconius melpomene* butterflies. *Journal of Chemical Ecology* **34**: 82–93.

Schulz S, Yildizhan S, Stritzke K, et al. 2007. Macrolides from the scent glands of the tropical butterflies *Heliconius cydno* and *Heliconius pachinus*. *Organic & Biomolecular Chemistry* **5**: 3434–3441.

Schumer M, Rosenthal GG & Andolfatto P. 2014. How common is homoploid hybrid speciation? *Evolution* **68**: 1553–1560.

Schwanwitsch BN. 1924. On the groundplan of wing-pattern in nymphalids and certain other families of rhopalocerous Lepidoptera. *Proceedings of the Zoological Society of London, series B* **34**: 509–528.

Seehausen O. 2004. Hybridization and adaptive radiation. *Trends in Ecology & Evolution* **19**: 198–207.

Seehausen O. 2006. Conservation: losing biodiversity by reverse speciation. *Current Biology* **16**: R334–R337.

Seehausen O, Terai Y, Magalhaes IS, et al. 2008. Speciation through sensory drive in cichlid fish. *Nature* **455**: 620–626.

Seigler DS, Spencer KC, Statler WS, et al. 1982. Tetraphyllin B and epitetraphyillin B sulphates: novel cyanogenic glucosides from *Passiflora caerulea* and *P. alato-caerulea*. *Phytochemistry* **21**: 2277–2285.

Seitz A. 1913. Heliconidae. In: Seitz A, ed. *Macrolepidoptera of the World*. Stuttgart: Kernen, 1139.

Seo HC, Curtiss J, Mlodzik M, et al. 1999. Six class homeobox genes in *Drosophila* belong to three distinct families and are involved in head development. *Mechanisms of Development* **83**: 127–139.

Servedio MR. 2001. Beyond reinforcement: the evolution of premating isolation by direct selection on preferences and postmating, prezygotic incompatibilities. *Evolution* **55**: 1909–1920.

Servedio MR & Noor MAF. 2003. The role of reinforcement in speciation: theory and data. *Annual Review of Ecology, Evolution, and Systematics* **34**: 339–364.

Servedio MR, Van Doorn GS, Kopp M, et al. 2011. Magic traits in speciation: 'magic' but not rare? *Trends in Ecology & Evolution* **26**: 389–397.

Shapiro MD, Marks ME, Peichel CL, et al. 2004. Genetic and developmental basis of evolutionary pelvic reduction in threespine sticklebacks. *Nature* **428**: 717–723.

Shaw KL & Lesnick SC. 2009. Genomic linkage of male song and female acoustic preference QTL underlying a rapid species radiation. *Proceedings of the National Academy of Sciences of the United States of America* **106**: 9737–9742.

Sheppard PM, Turner JRG, Brown KS, et al. 1985. Genetics and the evolution of Muellerian mimicry in *Heliconius* butterflies. *Philosophical Transactions of the Royal Society of London. B: Biological Sciences* **308**: 433–610.

Sherratt TN. 2011. The optimal sampling strategy for unfamiliar prey. *Evolution* **65**: 2014–2025.

Shimizu A, Dohzono I, Nakaji M, et al. 2014. Fine-tuned bee-flower coevolutionary state hidden within multiple pollination interactions. *Scientific Reports* **4**: 3988.

Shindo C, Aranzana MJ, Lister C, et al. 2005. Role of FRIGIDA and FLOWERING LOCUS C in determining variation in flowering time of *Arabidopsis*. *Plant Physiology* **138**: 1163–1173.

Silva DS da, Kaminski LA, Dell'Erba R, et al. 2008. Morfologia externa dos estágios imaturos de heliconíneos neotropicais : VII. *Dryadula phaetusa* (Linnaeus) (Lepidoptera, Nymphalidae, Heliconiinae). *Revista Brasileira de Entomologia* **52**: 500–509.

Singer MC, Thomas CD & Parmesan C. 1993. Rapid human-induced evolution of insect–host associations. *Nature* **366**: 681–683.

Sison-Mangus MP, Bernard GD, Lampel J, et al. 2006. Beauty in the eye of the beholder: the two blue opsins of lycaenid butterflies and the opsin gene-driven evolution of sexually dimorphic eyes. *Journal of Experimental Biology* **209**: 3079–3090.

Sivinski J. 1989. Mushroom body development in nymphalid butterflies: a correlate of learning. *Journal of Insect Behavior* **2**: 277.

Skelhorn J & Rowe C. 2007. Predators' toxin burdens influence their strategic decisions to eat toxic prey. *Current Biology* **17**: 1479–1483.

Smadja CM & Butlin RK. 2011. A framework for comparing processes of speciation in the presence of gene flow. *Molecular Ecology* **20**: 5123–5140.

Smiley J. 1978b. Plant chemistry and the evolution of host specificity: new evidence from *Heliconius* and *Passiflora*. *Science (New York, NY)* **201**: 745–747.

Smiley J. 1986. Ant constancy at *Passifora* extrafloral nectaries: effects on caterpillar survival. *Ecology* **67**: 516–521.

Smiley JT. 1978a. The host plant ecology of *Heliconius* butterflies in Northeastern Costa Rica. PhD thesis; University of Texas, Austin.

Smiley JT. 1985a. *Heliconius* caterpillar mortality during establishment on plants with and without attending ants. *Ecology* **66**: 845–849.

Smiley JT. 1985b. Are chemical barriers necessary for evolution of butterfly-plant associations? *Oecologia (Berlin)* **65**: 580–583.

Smith BT, McCormack JE, Cuervo AM, et al. 2014. The drivers of tropical speciation. *Nature* **515**: 406–409.

Snell-Rood EC, Papaj DR & Gronenberg W. 2009. Brain size: a global or induced cost of learning? *Brain, Behavior and Evolution* **73**: 111–128.

Sobel JM & Chen GF. 2014. Unification of methods for estimating the strength of reproductive isolation. *Evolution* **68**: 1511–1522.

Speed MP. 1993. Müllerian mimicry and the psychology of predation. *Animal Behaviour* **45**: 571–580.

Speed MP. 1999. Batesian, quasi-Batesian or Müllerian mimicry? Theory and data in mimicry Research. *Evolutionary Ecology* **13**: 755–776.

Speed MP, Alderson NJ, Hardman C, et al. 2000. Testing Müllerian mimicry: an experiment with wild birds. *Proceedings of the Royal Society of London B: Biological Sciences* **267**: 725–731.

Speed MP, Brockhurst MA & Ruxton GD. 2010. The dual benefits of aposematism: predator avoidance and enhanced resource collection. *Evolution* **64**: 1622–1633.

Speed MP & Turner JRH. 1999. Learning and memory in mimicry: II. Do we understand the mimicry spectrum? *Biological Journal of the Linnean Society* **67**: 281–312.

Spencer KC. 1987. Specificity of action of allelochemicals: diversification of glycosides. *American Chemistry Society, Symposium Series* **330**: 275–288.

Spencer KC. 1988. Chemical mediation of coevolution in the *Passiflora–Heliconius* interaction. In: Spencer KC, ed. *Chemical Mediation of Coevolution*. San Diego: Academic Press, 167–240.

Spencer KC & Seigler DS. 1984. Gynocardin from *Passiflora*. *Planta Medica* **50**: 356–357.

Spencer K & Seigler DS. 1985b. Passicoccin: a sulphated cyanogenic glycoside from *Passiflora coccinea*. *Phytochemistry* **24**: 2615–2617.

Spencer KC & Seigler DS. 1985a. Co-occurrence of valine/isoleucine-derived and cyclopentenoid cyanogens in a *Passiflora* species. *Biochemical Systematics and Ecology* **13**: 303–304.

Spencer KC & Seigler DS. 1985c. Passibiflorin, epipassibiflorin and passitrifasciatin: cyclopentenoid cyanogenic glycosides from *Passiflora*. *Phytochemistry* **24**: 981–986.

Spencer KC, Seigler DS & Nahrstedt A. 1986. Linamarin, lotaustralin, linustatin and neolinustatin from *Passiflora* species. *Phytochemistry* **25**: 645–647.

Sperling FAH. 1994. Sex-linked genes and species differences in Lepidoptera. *The Canadian Entomologist* **126**: 807–818.

Srygley RB. 1999. Locomotor mimicry in *Heliconius* butterflies: contrast analyses of flight morphology and kinematics. *Philosophical Transactions of the Royal Society of London B: Biological Sciences* **354**: 203–214.

Srygley RB. 2004. The aerodynamic costs of warning signals in palatable mimetic butterflies and their distasteful models. *Proceedings of the Royal Society of London Series B-Biological Sciences* **271**: 589–594.

Srygley RB & Chai P. 1990. Flight morphology of neotropical butterflies: palatability and the distribution of mass to the thorax and abdomen. *Oecologia (Berlin)* **84**: 491–499.

Stam LF & Laurie CC. 1996. Molecular dissection of a major gene effect on a quantitative trait: the level of alcohol dehydrogenase expression in *Drosophila melanogaster*. *Genetics* **144**: 1559–1564.

Stearns SC, Ackermann M, Doebeli M, et al. 2000. Experimental evolution of aging, growth, and reproduction in fruitflies. *Proceedings of the National Academy of Sciences* **97**: 3309–3313.

Stern DL. 2013. The genetic causes of convergent evolution. *Nature Reviews Genetics* **14**: 751–764.

Stern DL & Frankel N. 2013. The structure and evolution of *cis*-regulatory regions: the shavenbaby story. *Philo-*

sophical Transactions of the Royal Society B: Biological Sciences **368**: 20130028.

Stern DL & Orgogozo V. 2008. The loci of evolution: how predictable is genetic evolution? *Evolution* **62**: 2155–2177.

Stern DL & Orgogozo V. 2009. Is genetic evolution predictable? *Science (New York, NY)* **323**: 746–751.

Stichel H & Riffarth H. 1905. *Heliconiidae*. Berlin: R. Friedländer und Sohn.

Stone G & French V. 2003. Evolution: have wings come, gone and come again? *Current Biology* **13**: R436–R438.

Storz JF, Runck AM, Sabatino SJ, et al. 2009. Evolutionary and functional insights into the mechanism underlying high-altitude adaptation of deer mouse hemoglobin. *Proceedings of the National Academy of Sciences* **106**: 14450–14455.

Süffert F. 1927. Zur vergleichende Analyse der Schmetterlingszeichnung. *Biologisches Zentralblatt* **47**: 385–413.

Suomalainen E & Brown KS. 1984. Chromosome number variation within *Philaethria* butterflies (Lepidoptera: Nymphalidae, Heliconiini). *Chromosoma* **90**: 170–176.

Supple MA, Hines HM, Dasmahapatra KK, et al. 2013. Genomic architecture of adaptive color pattern divergence and convergence in *Heliconius* butterflies. *Genome Research* doi:10.1101/gr.**150615**.112.

Svensson EI, Eroukhmanoff F, Karlsson K, et al. 2010. A role for learning in population divergence of mate preferences. *Evolution* **64**: 3101–3113.

Swanson WJ & Vacquier VD. 2002. The rapid evolution of reproductive proteins. *Nature Reviews Genetics* **3**: 137–144.

Sweeney A, Jiggins C & Johnsen S. 2003. Insect communication: polarized light as a butterfly mating signal. *Nature* **423**: 31–32.

Swihart CA. 1971. Colour discrimination by the butterfly, *Heliconius charitonius* Linn. *Animal Behaviour* **19**: 156–164.

Swihart CA & Swihart SL. 1970. Colour selection and learned feeding preferences in the butterfly, *Heliconius charitonius* Linn. *Animal Behaviour* **18**: 60–64.

Swihart SL. 1967a. Neural adaptations in the visual pathway of certain heliconiine butterflies and related forms to variations in wing coloration. *Zoologica, New York* **52**: 1–14.

Swihart SL. 1967b. Hearing in butterflies (Nymphalidae: *Heliconius, Ageronia*). *Journal of Insect Physiology* **13**: 469–476.

Swihart SL. 1972. Neural basis of color vision in the butterfly *Heliconius erato*. *Journal of Insect Physiology* **18**: 1015–1025.

Swihart SL & Gordon WC. 1971. Red photoreceptor in butterflies. *Nature* **231**: 126–127.

Talekar NS & Shelton AM. 1993. Biology, ecology and management of the diamondback moth. *Annual Review of Entomology* **38**: 275–301.

Tarutani Y, Shiba H, Iwano M, et al. 2010. Trans-acting small RNA determines dominance relationships in *Brassica* self-incompatibility. *Nature* **466**: 983–986.

The Heliconius Genome Consortium. 2012. Butterfly genome reveals promiscuous exchange of mimicry adaptations among species. *Nature* **487**: 94–98.

The Marie Curie Speciation Network. 2012. What do we need to know about speciation? *Trends in Ecology & Evolution* **27**: 27–39.

Thomas CD. 1990. Fewer species. *Nature* **347**: 237.

Thomas CD, Bodsworth EJ, Wilson RJ, et al. 2001. Ecological and evolutionary processes at expanding range margins. *Nature* **411**: 577–581.

Thompson JN. 1999. Specific hypotheses on the geographic mosaic of coevolution. *The American Naturalist* **153**: S1–S14.

Thompson JN. 2001. Coevolution. In: *eLS*. Wiley online library. John Wiley & Sons, Ltd.

Thompson MJ & Jiggins CD. 2014. Supergenes and their role in evolution. *Heredity* **113**: 1–8.

Thompson MJ, Timmermans MJ, Jiggins CD, et al. 2014. The evolutionary genetics of highly divergent alleles of the mimicry locus in *Papilio dardanus*. *BMC Evolutionary Biology* **14**: 140.

Thorpe WH. 1956. *Learning and Instinct in Animals*. London: Methuen and Co. Ltd.

Tishkoff SA, Reed FA, Ranciaro A, et al. 2007. Convergent adaptation of human lactase persistence in Africa and Europe. *Nature Genetics* **39**: 31–40.

Tobias J. 2002. Family Galbulidae. In: J. del Hoyo, A. Elliot, J. Sargatal, eds. *Handbook of the Birds of the World*. Barcelona, Spain and Cambridge, UK: Lynx Editions, **7**: 74–101.

Tobler A, Kapan D, Flanagan NS, et al. 2005. First-generation linkage map of the warningly colored butterfly *Heliconius erato*. *Heredity* **94**: 408–417.

Toews DPL & Brelsford A. 2012. The biogeography of mitochondrial and nuclear discordance in animals. *Molecular Ecology* **21**: 3907–3930.

Tregenza T & Wedell N. 2002. Polyandrous females avoid costs of inbreeding. *Nature* **415**: 71–73.

Turner JRG. 1966a. A rare mimetic *Heliconius* (Lepidoptera: Nymphalidae). *Proceedings of the Royal Entomological Society of London, B* **35**: 128–132.

Turner JRG. 1966b. A little-recognized species of *Heliconius* butterfly (Nymphalidae). *Journal of Research on the Lepidoptera* **5**: 97–112.

Turner JRG. 1967a. Some early works on heliconiine butterflies and their biology (Lepidoptera, Nymphalidae). *Journal of the Linnean Society of London, Zoology* **46**: 255–266.

Turner JRG. 1967b. On supergenes. I. The evolution of supergenes. *The American Naturalist* **101**: 195–221.

Turner JRG. 1968a. Some new *Heliconius* pupae: their taxonomic and evolutionary significance in relation to mimicry (Lepidoptera: Nymphalidae). *Journal of Zoology (London)* **155**: 311–325.

Turner JRG. 1968b. Natural selection for and against a polymorphism which interacts with sex. *Evolution* **22**: 481–495.

Turner JRG. 1971a. Experiments on the demography of tropical butterflies. II. Longevity and home-range behaviour in *Heliconius erato*. *Biotropica* **3**: 21–31.

Turner JRG. 1971b. 2000 generations of hybridisation in a *Heliconius* butterfly. *Evolution* **25**: 471–482.

Turner JRG. 1973. The genetics of race formation in mimetic butterflies: is the gene the unit of selection? *Genetics* **74**: s 281.

Turner JRG. 1974. Breeding *Heliconius* in a temperate climate. *Journal of the Lepidopterists' Society* **28**: 26–33.

Turner JRG. 1975a. Communal roosting in relation to warning colour in two Heliconiine butterflies (Nymphalidae). *Journal of the Lepidopterists' Society* **29**: 221–226.

Turner JRG. 1975b. A tale of two butterflies. *Natural History* **84**: 28–37.

Turner JRG. 1976. Adaptive radiation and convergence in subdivisions of butterfly genus *Heliconius* (Lepidoptera-Nymphalidae). *Zoological Journal of the Linnean Society* **58**: 297–308.

Turner JRG. 1977. Butterfly mimicry—genetical evolution of an adaptation. In: Hecht MK, Steere WC, Wallace B, eds. *Evolutionary Biology*. New York: Plenum Press, 163–206.

Turner JRG. 1978. Why male butterflies are non-mimetic: natural selection, sexual selection, group selection, modification and sieving. *Biological Journal of the Linnean Society* **10**: 385–432.

Turner JRG. 1981. Adaptation and evolution in *Heliconius*—a defense of neodarwinism. *Annual Review of Ecology and Systematics* **12**: 99–121.

Turner JRG. 1984. Mimicry: the palatability spectrum and its consequences. In: Vane-Wright RI, Ackery PR, eds. *The Biology of Butterflies*. London: Academic Press, 141–161.

Turner JRG. 1987. The evolutionary dynamics of Batesian and Muellerian mimicry: similarities and differences. *Ecological Entomology* **12**: 81–95.

Turner JRG, Johnson MS & Eanes WF. 1979. Contrasted modes of evolution in the same genome—allozymes and adaptive change in *Heliconius*. *Proceedings of the National Academy of Sciences of the United States of America* **76**: 1924–1928.

Turner JRG & Mallet JLB. 1996. Did forest islands drive the diversity of warningly coloured butterflies? Biotic drift and the shifting balance. *Philosophical Transactions of the Royal Society of London Series B: Biological Sciences* **351**: 835–845.

Turner JRG & Sheppard PM. 1975. Absence of crossing-over in female butterflies (*Heliconius*). *Heredity* **34**: 265–269.

Tuttle EM, Bergland AO, Korody ML, et al. 2016. Divergence and functional degradation of a sex chromosome-like supergene. *Current Biology* **26**: 344–350.

Ulmer T & MacDougal JM. 2004. *Passiflora: Passionflowers of the World*. Portland, OR: Timber Press.

Valverde P, Healy E, Jackson I, et al. 1995. Variants of the melanocyte-stimulating hormone receptor gene are associated with red hair and fair skin in humans. *Nature Genetics* **11**: 328–330.

Vanjari S, Mann F, Merrill RM, et al. 2015. Male sex pheromone components in the butterfly *Heliconius melpomene*. *bioRxiv*: 033506.

van't Hof AE, Edmonds N, Dalíková M, et al. 2011. Industrial melanism in British peppered moths has a singular and recent mutational origin. *Science (New York, NY)* **332**: 958–960.

van Schooten B, Jiggins CD, Briscoe AD, et al. 2016. Genome-wide analysis of ionotropic receptors provides insight into their evolution in *Heliconius* butterflies. *BMC Genomics* **17**: 254.

Vargas HA, Barão KR, Massardo D, et al. 2014. External morphology of the immature stages of Neotropical heliconians: IX. *Dione glycera* (C. Felder & R. Felder) (Lepidoptera, Nymphalidae, Heliconiinae). *Revista Brasileira de Entomologia* **58**: 129–141.

Verzijden MN, ten Cate C, Servedio MR, et al. 2012. The impact of learning on sexual selection and speciation. *Trends in Ecology & Evolution* **27**: 511–519.

Vorobyev M & Osorio D. 1998. Receptor noise as a determinant of colour thresholds. *Proceedings of the Royal Society of London B: Biological Sciences* **265**: 351–358.

Wade N. 2011. A supergene paints wings for surviving biological war. *The New York Times*.

Wahlberg N, Leneveu J, Kodandaramaiah U, et al. 2009. Nymphalid butterflies diversify following near demise at the Cretaceous/Tertiary boundary. *Proceedings of the Royal Society of London B: Biological Sciences* **276**: 4295–4302.

Wakeley J. 2003. Inferences about the structure and history of populations: coalescents and intraspecific phylogeography. In: Singh RS, Uyenoyama MK, eds. *The Evolution of Population Biology*. Cambridge, UK: Cambridge University Press, 193–215.

Wallace AR. 1867. Mimicry and other protective resemblances among animals. *Westminster and Foreign Quarterly Review* **32**: 1–43.

Wallace AR. 1871. *Contributions to the Theory of Natural Selection. A Series of Essays*. London: Macmillan.

Wallbank RWR, Baxter SW, Pardo-Diaz C, et al. 2016. Evolutionary novelty in a butterfly wing pattern through enhancer shuffling. *PLoS Biology* **14**: e1002353.

Waller DA & Gilbert LE. 1982. Roost recruitment and resource utilization: observations on a *Heliconius charitonia* L. roost in Mexico (Nymphalidae). *Journal of the Lepidopterists' Society* **36**: 178–184.

Walters JR & Harrison RG. 2010. Combined EST and proteomic analysis identifies rapidly evolving seminal fluid proteins in *Heliconius* butterflies. *Molecular Biology and Evolution* **27**: 2000–2013.

Walters JR & Harrison RG. 2011. Decoupling of rapid and adaptive evolution among seminal fluid proteins in *Heliconius* butterflies with divergent mating systems. *Evolution* **65**: 2855–2871.

Walters JR, Stafford C, Hardcastle TJ, et al. 2012. Evaluating female remating rates in light of spermatophore degradation in *Heliconius* butterflies: pupal-mating monandry versus adult-mating polyandry. *Ecological Entomology* **37**: 257–268.

Wang J, Wurm Y, Nipitwattanaphon M, et al. 2013. A Y-like social chromosome causes alternative colony organization in fire ants. *Nature* **493**: 664–668.

Ward M, Dick CW, Gribel R, et al. 2005. To self, or not to self . . . A review of outcrossing and pollen-mediated gene flow in neotropical trees. *Heredity* **95**: 246–254.

Ward P & Zahavi A. 1973. The importance of certain assemblages of birds as 'information-centres' for food-finding. *Ibis* **115**: 517–534.

Watt WB. 1968. Adaptive significance of pigment polymorphisms in *Colias* butterflies. I. Variation of melanin pigment in relation to thermoregulation. *Evolution* **22**: 437–458.

Weatherbee SD, Frederik Nijhout H, Grunert LW, et al. 1999. Ultrabithorax function in butterfly wings and the evolution of insect wing patterns. *Current Biology* **9**: 109–115.

Weiss MR. 1995. Associative colour learning in a nymphalid butterfly. *Ecological Entomology* **20**: 298–301.

Welch JJ & Jiggins CD. 2014. Standing and flowing: the complex origins of adaptive variation. *Molecular Ecology* **23**: 3935–3937.

Werner T, Koshikawa S, Williams TM, et al. 2010. Generation of a novel wing colour pattern by the Wingless morphogen. *Nature* **464**: 1143–1148.

Westerman EL, Hodgins-Davis A, Dinwiddie A, et al. 2012. Biased learning affects mate choice in a butterfly. *Proceedings of the National Academy of Sciences* **109**: 10948–10953.

Whinnett A, Zimmermann M, Willmott KR, et al. 2005. Strikingly variable divergence times inferred across an Amazonian butterfly 'suture zone'. *Proceedings of the Royal Society B: Biological Sciences* **272**: 2525–2533.

Whiting MF, Bradler S & Maxwell T. 2003. Loss and recovery of wings in stick insects. *Nature* **421**: 264–267.

Whitlock MC & McCauley DE. 1999. Indirect measures of gene flow and migration: FST | [ne] | 1/(4Nm+1). *Heredity* **82**: 117–125.

Whitney HM & Glover BJ. 2013. Coevolution: Plant–Insect. In: *eLS*. Wiley online library. John Wiley & Sons, Ltd.

Wigby S & Chapman T. 2005. Sex peptide causes mating costs in female *Drosophila melanogaster*. *Current Biology* **15**: 316–321.

Wiklund C, Kaitala A, Lindfors V, et al. 1993. Polyandry and its effect on female reproduction in the green-veined white butterfly (*Pieris napi* L). *Behavioral Ecology and Sociobiology* **33**: 25–33.

Williams KS & Gilbert LE. 1981. Insects as selective agents on plant vegetative morphology: egg mimicry reduces egg laying by butterflies. *Science* **212**: 467–469.

Wittkopp PJ & Kalay G. 2012. Cis-regulatory elements: molecular mechanisms and evolutionary processes underlying divergence. *Nature Reviews Genetics* **13**: 59–69.

Wray GA. 2007. The evolutionary significance of cis-regulatory mutations. *Nature Reviews Genetics* **8**: 206–216.

Wright S. 1931. Evolution in Mendelian populations. *Genetics* **16**: 97–159.

Wright S. 1982. The shifting balance theory and macroevolution. *Annual Review of Genetics* **16**: 1–20.

Wu CI, Johnson NA & Palopoli MF. 1996. Haldane's rule and its legacy: why are there so many sterile males? *Trends in Ecology & Evolution* **11**: 281–284.

Yang J, Benyamin B, McEvoy BP, et al. 2010. Common SNPs explain a large proportion of the heritability for human height. *Nature Genetics* **42**: 565–569.

Yockteng R & Nadot S. 2004. Phylogenetic relationships among *Passiflora* species based on the glutamine synthetase nuclear gene expressed in chloroplast (ncpGS). *Molecular Phylogenetics and Evolution* **31**: 379–396.

Young AM & Thomason JH. 1975. Notes on communal roosting of *Heliconius charitonius* (Nymphalidae) in Costa Rica. *Journal of the Lepidopterists' Society* **29**: 243–255.

Zaccardi G, Kelber A, Sison-Mangus MP, et al. 2006a. Opsin expression in the eyes of *Heliconius erato*. *Perception* **35**: 142–143.

Zaccardi G, Kelber A, Sison-Mangus MP, et al. 2006b. Color discrimination in the red range with only one long-wavelength sensitive opsin. *Journal of Experimental Biology* **209**: 1944–1955.

Zagrobelny M, Bak S, Rasmussen AV, et al. 2004. Cyanogenic glucosides and plant–insect interactions. *Phytochemistry* **65**: 293–306.

Zagrobelny M & Møller BL. 2011. Cyanogenic glucosides in the biological warfare between plants and insects: the Burnet moth-Birdsfoot trefoil model system. *Phytochemistry* **72**: 1585–1592.

Zhang W, Dasmahapatra KK, Mallet J, et al. 2016. Genome-wide introgression among distantly related *Heliconius* butterfly species. *Genome Biology* **17**: 25.

Zhang G, Li C, Li Q, et al. 2014. Comparative genomics reveals insights into avian genome evolution and adaptation. *Science* **346**: 1311–1320.

Zhen Y, Aardema ML, Medina EM, et al. 2012. Parallel molecular evolution in an herbivore community. *Science* **337**: 1634–1637.

Zhu H, Sauman I, Yuan Q, et al. 2008. Cryptochromes define a novel circadian clock mechanism in monarch butterflies that may underlie sun compass navigation. *PLoS Biology* **6**: e4.

Zimmer, C. 2006. Darwin, Meet Frankenstein. The Loom, Discover Blog. http://blogs.discovermagazine.com/loom/2006/06/14/darwin-meet-frankenstein/#.V447B1ftBl0

Zirin JD & Mann RS. 2004. Differing strategies for the establishment and maintenance of teashirt and homothorax repression in the *Drosophila* wing. *Development* **131**: 5683–5693.

Index

Notes
vs. indicates a comparison